Invasive Plants: Impacts and Control

Invasive Plants: Impacts and Control

Edited by Ned Kennard

SYRAWOOD
PUBLISHING HOUSE

New York

Published by Syrawood Publishing House,
750 Third Avenue, 9th Floor,
New York, NY 10017, USA
www.syrawoodpublishinghouse.com

Invasive Plants: Impacts and Control
Edited by Ned Kennard

International Standard Book Number: 978-1-68286-781-5 (Hardback)

Cataloging-in-Publication Data

Invasive plants : impacts and control / edited by Ned Kennard.
     p. cm.
Includes bibliographical references and index.
ISBN 978-1-68286-781-5
1. Invasive plants. 2. Invasive plants--Control. I. Kennard, Ned.
SB613.5 .I58 2019
581.62--dc23

TABLE OF CONTENTS

PREFACE

All plants, which are not native to a specific location and have a potential to cause damage to the human economy, environment and human health are known as invasive plants. Exotic pest plants and invasive exotics are two primary categories of invasive plants. Kudzu vine, yellow starthistle and Andean pampas grass are some examples of such plants. They possess specific traits that guarantee their survival as compared to native species. Some of the traits of invasive species are rapid reproduction, fast growth, high tolerance and dispersal ability, etc. Some of the adverse effects of invasive species are extinction of native species, biodiversity loss, reduction in agricultural yield and forage loss. The geomorphological effects of invasive plants can be positive in nature such as bioprotection and bioconstruction. Restoration is an effective tool in the reduction of the spread of invasive plant species. This involves removing such species, reducing the invasive seeds that are dispersed to surrounding areas, using native species that are similar to invasive species to compete with invasive species, etc. The topics included in this book on invasive plants and their impacts and control are of utmost significance and bound to provide incredible insights to readers. It is appropriate for students seeking detailed information in this area as well as for experts.

This book is the end result of constructive efforts and intensive research done by experts in this field. The aim of this book is to enlighten the readers with recent information in this area of research. The information provided in this profound book would serve as a valuable reference to students and researchers in this field.

At the end, I would like to thank all the authors for devoting their precious time and providing their valuable contribution to this book. I would also like to express my gratitude to my fellow colleagues who encouraged me throughout the process.

Editor

Amaranthus palmeri, a second record for Africa and notes on *A. sonoriensis* nom. nov.

Authors:
Duilio Iamonico[1]
Ridha El Mokni[2,3] 🄳

Affiliations:
[1]Laboratory of Phytogeography and Applied Geobotany, Department DPTA, Section Environment and Landscape, Sapienza University of Rome, Italy

[2]Faculty of Sciences of Bizerta, University of Carthage, Tunisia

[3]Laboratory of Plant Biology, Faculty of Pharmacy of Monastir, University of Monastir, Tunisia

Corresponding author:
Duilio Iamonico,
d.iamonico@yahoo.it

Background: *Amaranthus* is a critical genus from a taxonomic point of view because of its high phenotypic variability, which has led to nomenclatural disorder, misapplication of names, and erroneous species identification. As a whole, floristic and taxonomic studies on this genus are still incomplete.

Objectives: The main objective was to record the North American species *Amaranthus palmeri* in Tunisia for the first time and to point out a second occurrence for Africa. At the same time, we highlight some nomenclatural confusion concerning the name *A. palmeri* var. *glomeratus* which appears to be untypified and which should be treated at species rank.

Method: The work is based on field surveys, analysis of relevant literature and examination of specimens preserved in the herbaria GH, FI, HFLA, K, MICH, NEBC, NY, MO, P, RO, US, and the Herbarium of Bizerta University.

Results: A population of *Amaranthus palmeri* was discovered in Bizerta Province, representing the first record for the national flora. This record also represents the second record for Africa. Morphological characters, as well as ecological data are provided. Nomenclatural notes are provided for *A. palmeri* var. *glomeratus* (lectotype at GH, here designated; isolectotypes at MICH and US), and a new name (*A. sonoriensis*) is proposed.

Conclusion: *Amaranthus palmeri* is an alien species in Tunisia, growing along roadsides and in public gardens. Future monitoring of the populations found will be necessary to verify possible naturalisation and spreading of *A. palmeri* in Tunisia. If this happens, actions for eradication of the plants will be necessary.

Introduction

Amaranthus L. (Amaranthaceae Juss.) is a genus of about 70 mostly annual, monoecious and dioecious species with worldwide distribution. Approximately 40 species are native to the Americas and the remaining species are native to the other continents (see e.g. Costea, Sanders & Waines 2001; Iamonico 2015a). Several American species are used as ornamentals, and others are cultivated for grains or as leafy vegetables. In many areas *Amaranthus* species have escaped cultivation and have become problematic for agricultural systems, causing economic loss and for biodiversity because the species compete with indigenous species for resources such as light and nutrients.

Amaranthus is a challenging genus from a taxonomic perspective because of its high phenotypic variability, which has led to nomenclatural disorder and misapplication of names (see e.g. Costea et al. 2001; Iamonico 2015a; Mosyakin & Robertson 1996; Nestor 2015). No comprehensive molecular study has been done yet, and the more recent infrageneric classification was proposed by Mosyakin and Robertson (1996) who recognised three subgenera: subgenus *Acnida* (L.) Aellen ex K.R. Robertson with three sections, subgenus *Albersia* (Kunth) Gren. & Godr. with four sections, and subgenus *Amaranthus*, with three sections and two subsections.

According to Mosyakin and Robertson (1996), the subgenus *Acnida* (≡ *Acnida* L.) would include the dioecious species, which were classified into three sections, sect. *Acnida* (L.) Mosyakin & K.R. Robertson (pistillate flowers with 0–2 tepals and mostly indehiscent fruits), sect. *Saueranthus* Mosyakin & K.R. Robertson (pistillate flowers with five tepals and dehiscent fruits) and sect. *Acanthochiton* (Torrey) Mosyakin & K.R. Robertson (pistillate flowers with broad, deltate and foliaceous bracts). All the taxa belonging to the subgenus *Acnida* are native to North America while they occur in other continents as aliens, sometimes as invasive (see e.g. Iamonico 2015b).

The Flora of Africa (see SANBI 2012 and literature therein; Iamonico 2015b) currently includes only one species belonging to the subgenus *Acnida*, *Amaranthus palmeri* S. Watson, which was recorded only in Egypt (Boulos 2009:57).

As part of an ongoing study on the genus *Amaranthus* (see e.g. Iamonico 2014a, 2014b, 2015a, 2015b, 2016a, 2016b, 2016c; Iamonico & Das 2014) and the Tunisian Amaranthaceae (Sukhorukov et al. 2016), field surveys were carried out and resulted in the identification of a population of *A. palmeri* and this represents the first record of the species for Tunisia and only the second for Africa. Morphological notes, data on the habitat, as a nomenclatural study of the name *Amaranthus palmeri* var. *glomeratus* Uline & W.L. Bray (which appears still untypified) and notes on the name *A. palmeri* are presented here.

Material and methods

The work is based on field surveys (July 2014–October 2015), analysis of relevant literature and examination of specimens preserved in the following herbaria: GH, FI, HFLA, K, MICH, NEBC, NY, MO, P, RO, US (acronyms according to Thiers 2016+) and the Herbarium of Bizerta University (not listed in Index Herbariorum).

The articles cited through the text refer to the Melbourne Code, ICN (McNeill et al. 2012).

Results and discussion

Typification of *Amaranthus palmeri* var. *glomeratus*

Uline and Bray (1894:272) described the var. *glomeratus* on the basis of specimens collected by E. Palmer in the Sonoran Desert in 1889, as reported in the protologue. According to the protologue, differential features are the habit (ascending), the plant height (shorter than the typical form) and the synflorescence structure [cymes aggregated in dense glomerules (3–5 cm in diameter) at the base of the plant].

Two specimens (No. 953 and 958) were cited in the protologue, and they are syntypes according to the Art. 9.5 of the ICN. We found these exsiccata at GH (barcodes 00037017 and 00037028, images, respectively, available at https://s3.amazonaws.com/huhwebimages/C29365768E19428/type/full/37027.jpg and https://s3.amazonaws.com/huhwebimages/20824DD7B74F449/type/full/37028.jpg), MICH (barcode 1115701) and US (barcodes 00106254 and 00106255). We here designate the GH-00037028 as the lectotype of the name *Amaranthus palmeri* var. *glomeratus* since it perfectly matches the diagnosis by Uline and Bray (1894:272). The specimens at MICH and US are isolectotypes.

Concerning the identity of the var. *glomeratus*, it cannot be associated with *A. palmeri* in our opinion. Unfortunately citations of this taxon are rare in the literature. No citation was found for Australia, Asia, Europe and Africa (see e.g. Bojian, Clemants & Borsch 2003; Iamonico 2015b; Palmer 2009;

SANBI 2012). Even the *Flora of North America* (Mosyakin & Robertson 2003) does not report this taxon. Shreve and Wiggins (1964:457–458) and Abrams (1994:99) listed the var. *glomeratus* as heterotypic synonym of *A. palmeri*. However, only the description reported in the *Vegetation and Flora of the Sonoran Desert* (Shreve & Wiggins 1964) corresponds to the original concept by Uline and Bray (1894), while that by Abrams (1994) completely refers to the typical form. Furthermore, the taxon *glomeratus* appears to occur only in the Sonoran Desert, a unique area of the world which includes part of the United States (Arizona, California) and Mexico (Baja California, Sonora, Sinaloa), and which is characterised by high levels of species richness and endemism (see e.g. Shreve & Wiggins 1964).

Based on the evidence presented here, we think that var. *glomeratus* should actually be at the rank of species. Since an *Amaranthus glomeratus* was already published by Pospichal (1897:375), we cannot propose a new combination of the var. *glmeratus* by Uline and Bray (1894:272), because this would make it a homonym and illegitimate name according to the Art. 53.1 of ICN. Moreover, Pospichal's concept of *A. glomeratus* cannot be related to that of the var. *glomeratus* by Uline and Bray (1894) since the Pospichal's species is monoecious and belongs to the *A. hybridus* group *sensu* Iamonico (2015b) (*Amaranthus* sect. *Amaranthus*). As a consequence, a new name has to be given (see the paragraph 'Taxonomic treatment').

Note of the typification of *Amaranthus palmeri*

Sauer (1955:31) correctly typified the name *Amaranthus palmeri* with a specimen collected by J.L. Berlandier on the banks of the Rio Grande, California, in July 1834. This specimen is preserved at GH (barcode 00037007) and represents one of the syntypes listed by Watson (1877:274) in the protologue ['on the banks of the Rio Grande, by Berlandier (n. 2407) in 1834']. We found a further two specimens, which are part of the Berlandier's collection: the first one at MO (barcode 247471, image available at http://www.tropicos.org/Image/66322), the other one at NY (barcode 1043131, image available at http://sweetgum.nybg.org/vh/specimen.php?irn=1187636). Both these specimens are numbered as no. 2407, and so they represent isolectotypes for the name *Amaranthus palmeri*.

Watson (1877:274) also listed another specimen, 'At Larkin's Station: San Diego County, California, by Dr. E. Palmer (n. 323 of his collection)'. We found three specimens at GH (barcode 00037006, image available at http://ids.lib.harvard.edu/ids/view/2520948), K (barcode 000814911, image available at http://apps.kew.org/herbcat/getImage.do?imageBarcode=K000814911) and MO (barcode 247470 image available at http://www.tropicos.org/Image/66321), which correspond (label data) to Watson's citation. These specimens are syntypes.

Taxonomic treatment

Amaranthus palmeri S. Watson, Proc. Amer. Acad. Arts 12:274 (1877).

Type: USA, California: banks of Rio Grande, July 1834, *Berlandier 2407* (GH-00037007! lectotype, designated by Sauer 1955:31, image available at http://ids.lib.harvard.edu/ids/view/2520952; isolectotypes at MO-247471, image available at http://www.tropicos.org/Image/66322, and NY-1043131, image available at http://sweetgum.nybg.org/vh/specimen.php?irn=1187636). Syntypes: USA, California: San Diego, Larkin station, 1875, *Palmer 323* (GH-00037006, MO-247470 and, K-000814911, images, respectively, available at http://ids.lib.harvard.edu/ids/view/2520948, http://www.tropicos.org/Image/66321, and http://apps.kew.org/herbcat/getImage.do?imageBarcode=K000814911).

Description: Herbs 5–20 cm tall, dioecious, annual (therophyte). Stems erect, glabrous or nearly so, green or reddish, branched. Leaves green; the lower ones ovate, rhomboidal (1.5–7.0 × 1.0–3.5 cm); the upper leaves lanceolate, glabrous, margins entire, apex obtuse to acute, mucronulate, base broadly cuneate, petioled (petiole 0.5–3.5 cm long). Synflorescences terminal, spike- or panicle-like type, drooping or erect, often interrupted in the proximal part, green, the main florescence up to 25 cm long. Floral bracts 1, usually light green, lanceolate [(0.5–)1.0–2.5(–3.0) × 4.0–6.0 mm], 2.0–2.5 times longer than the perianth, sometimes carinate, apex acuminate, margin entire, glabrous. Staminate flowers with 5 unequal tepals, lanceolate (2.0–4.0 × 0.5–1.5 mm), apex acute, awned (especially the inner tepals); stamens 5. Pistillate flowers with 5 tepals, obovate-spathulate [(1.5–)1.7–3.8(–4.0) × 0.4–0.6 mm]; style branches spreading, stigmas 2(–3). Fruit usually brown, subglobose or ellipsoidal (1.5–2.0 × 1.0–2.0 mm), shorter than the perianth, usually smooth, dehiscent. Seed lenticular [(1.0–1.2) mm in diameter], dark-reddish to brown.

Phenology: Flowering time July–August, fruiting time September–October.

Habitat: Ruderal on roadsides and public gardens.

Elevation: 3–8 m a.s.l.

Chromosome number: 2n = 34 (Reveal & Spellenberg 1976).

Distribution: 18–20 individuals were found all referring to a single scattered population, which occupies an area of about 3–4 ha.

Alien status: Neophyte species native to North America and which is considered casual in Tunisia according to Pyšek et al. (2002). Despite this, monitoring is necessary to verify a possible naturalisation and spreading of *A. palmeri* in Tunisia and, if this is found to be the case, actions for eradication of the plants will be required.

Occurrence in Tunisia: Bizerta (Nadhour, North of Tunisia).

Specimina visa: ITALY. **Emilia-Romagna:** Cervia, via Romea Nord (SS16), 2 m a.s.l, road embankment with ruderal vegetation, 08 Oct 2014, *Faggi, Iamonico & Ardenghi s.n.* (HFLA! five sheets). TUNISIA. **Bizerta:** North-Bizerta, Nadhour, 03–08 m a.s.l., 29 July 2015, *El Mokni & Iamonico s.n.* (HFLA!); *ibidem 00137* (Herb. Bizerta University). USA. **California:** banks of Rio Grande, July 1834, *Berlandier 2407* (GH-00037007!, lectotype); *ibidem* (MO-247471!, NY-1043131!); San Diego, Larkin station, 1875, *Palmer 323* (GH-00037006!, MO-247470! and K-000814911!, syntypes); **Massachusetts:** Malden, 04 Sep 1886, *Collins s.n.* (NEBC-00735067!); South Lawrence, 23 Sep 1902, *Pease 510* (NEBC-00735066!). **Los Angeles:** N La Verne, light sunny soil in Citrus grove, 01 Oct 1932, *Wheeler 1418* (P-05159602!).

Amaranthus sonoriensis Iamonico & El Mokni, *nom. nov. pro Amaranthus palmeri* var. *glomeratus*, Bot. Gaz. 19(7):272 (1894) non *Amaranthus glomeratus* Posp., Fl. Oesterr. Küstenl. i. 375 (1897).

Type: Mexico, Sonora, Lerdo, 24–26 Apr 1889, *Palmer 953* (GH-00037028! lectotype, here designated, image available at https://s3.amazonaws.com/huhwebimages/20824DD7B74F449/type/full/37028.jpg; isolectotypes at MICH-1115701, image available at http://plants.jstor.org/stable/history/10.5555/al.ap.specimen.mich1115701, and US-00106254, and US-00106255, image available at http://collections.nmnh.si.edu/search/botany/#new-search).

Description: Herbs 15–50 cm tall, dioecious, annual (therophyte). Stems ascending, glabrous or nearly so, green or brownish, branched. Leaves green, ovate to ovate-lanceolate, rhomboidal (0.5–2.5 × 0.3–1.0 cm), glabrous, margins entire, apex obtuse to acute, mucronulate, base cuneate, petioled (petiole 0.5–3.0 cm long). Synflorescences terminal, spike-like type, erect, the main florescence up to 15 cm long; cymes at the base of plants arranged in dense glomerules with diameters up to 5 cm. Floral bracts 1, green, lanceolate (0.5–2.0 × 4.0–6.0 mm), 2.0–2.5 longer than the perianth, apex acuminate, margin entire, glabrous. Staminate flowers with 5 tepals, ovate-lanceolate (1.5–3.0 × 0.5–1.5 mm), apex acute, awned; stamens 5. Pistillate flowers with 5 tepals, obovate-spathulate (1.0–4.0 × 0.4–0.5 mm); style branches spreading, stigmas 2(–3). Fruit brown, subglobose or ellipsoidal (1.0–1.5 × 0.5–1.5 mm), shorter than the perianth, usually smooth, dehiscent. Seed lenticular (about 1.0 mm in diameter), brown.

Etymology: The specific epithet refers to the Sonoran Desert.

Phenology: Flowering and fruiting times February–April.

Habitat: Desert washes.

Chromosome number: Unknown.

Distribution area: Lower Sonoran area[1].

Specimina visa: MEXICO. Baja California: Lerdo, Sonora Desert, 24–26 Apr 1889, *Palmer 953* (GH-00037028!, lectotype); *ibidem, Palmer 958* (GH-00037027!); Guaymas: Sonora Desert, 01 Mar 1905, *Palmer 321* (NY-324458!).

1. Further investigations need to define the exact distribution area.

Acknowledgements

We are grateful to directors and curators of all quoted official herbaria and colleagues of the cited personal herbaria for their support during our visits, loan of specimens or photographs, or for providing interesting information.

Competing interests

The authors declare that they have no financial or personal relationships that may have inappropriately influenced them in writing this article.

Authors' contributions

D.I. examined and confirmed the identification of the specimens as *A. palmeri*, as well as carried out the nomenclatural study on *Amaranthus palmeri* var. *glomeratus*. He also prepared and structured the original text of the manuscript. R.E.M. collected the plants in the field, provided data about distribution, ecology and status of naturalisation of *A. palmeri* in Tunisia, and amended the original text by the addition of this information.

References

Abrams, L., 1994, *Illustrated flora of the Pacific States Washington, Oregon, and California*, vol. 2 (Polygonaceae to Krameriaceae), Stanford University Press, Stanford, CA.

Bojian, B., Clemants, S.E. & Borsch, T., 2003, 'Amaranthus L.', In Z.Y. Wu, P.H. Raven & D.Y. Hong (eds.), *Flora of China*, vol. 5, pp. 415–429, Science Press, Beijing and Missouri Botanical Garden Press, St. Louis, MO.

Boulos, L., 2009, *Flora of Egypt checklist. Revised annotated edition*, Al Harara, Cairo.

Costea, M., Sanders, A. & Waines, G., 2001, 'Preliminary results towards a revision of the Amaranthus hybridus complex (Amaranthaceae)', *Sida* 19, 931–974.

Iamonico, D., 2014a, 'Amaranthus gangeticus (Amaranthaceae), a name *Incertae sedis*', *Phytotaxa* 162(5), 299–300. http://dx.doi.org/10.11646/phytotaxa.162.5.2

Iamonico, D., 2014b, 'Lectotypification of Linnaean names in the genus Amaranthus L. (Amaranthaceae)', *Taxon* 63(1), 146–150. http://dx.doi.org/10.12705/631.34

Iamonico, D., 2015a, 'Taxonomic revision of the genus Amaranthus (Amaranthaceae) in Italy', *Phytotaxa* 199(1), 1–84. http://dx.doi.org/10.11646/phytotaxa.199.1.1

Iamonico, D., 2015b, 'Amaranthaceae Juss.', In *Euro+Med Plantbase – the information resource for Euro-Mediterranean plant diversity*, viewed 21 March 2016, from http://ww2.bgbm.org/EuroPlusMed/PTaxonDetail.asp?NameCache=Amaranthus&PTRefFk=7300000

Iamonico, D., 2016a, 'Nomenclature survey of the genus Amaranthus (Amaranthaceae). 3. Names linked to the Italian flora', *Plant Biosystems* 150(3), 519–531. http://dx.doi.org/10.1080/11263504.2014.987188

Iamonico, D., 2016b, 'Nomenclature survey of the genus Amaranthus (Amaranthaceae). 4. The intricate questions around the name Amaranthus gracilis', *Botanica Serbica* 40(1), 61–68.

Iamonico, D., 2016c 'Nomenclature survey of the genus Amaranthus (Amaranthaceae). 5. Moquin-Tandon's names', *Phytotaxa* 273(2): 81–114. http://dx.doi.org/10.11646/phytotaxa.273.2.1

Iamonico, D. & Das, S., 2014, 'Amaranthus bengalense (Amaranthaceae) a new species from India, with taxonomical notes on A. blitum aggregate', *Phytotaxa* 181(5), 293–300. http://dx.doi.org/10.11646/phytotaxa.181.5.4

McNeill, J., Barrie, F.R., Buck, W.R., Demoulin, V., Greuter, D.L., Hawksworth, D.L., et al., (eds.), 2012, 'International Code of Nomenclature for algae, fungi and plants (Melbourne Code): Adopted by the Eighteenth International Botanical Congress, Melbourne, Australia, July 2011', *Regnum Vegetabile* 154, 1–274.

Mosyakin, S.L. & Robertson, K.R., 1996, 'New infrageneric taxa and combinations in Amaranthus (Amaranthaceae)', *Annales Botanici Fennici* 33, 275–281.

Mosyakin, S.L. & Robertson, K.R., 2003, 'Amaranthus L.', In Flora of North America Editorial Committee (eds.), *Flora of North America North of Mexico (Magnoliophyta: Caryophyllidae, part 1)*, vol. 4, pp. 410–435, Oxford University Press, Oxford.

Nestor, D.B., 2015, 'Revision taxonomica de las especies monoicas de Amaranthus (Amaranthaceae): Amaranthus subg. Amaranthus y Amaranthus subg. Albersia', *Annals of the Missouri Botanical Garden* 101(2), 261–383. http://dx.doi.org/10.3417/2010080

Palmer, J., 2009, 'A conspectus of the genus Amaranthus L. (Amaranthaceae) in Australia', *Nuytsia* 19, 107–128.

Pospichal, E., 1897, *Flora des oesterreichischen Küstenlandes*, vol. 1, F. Deuticke, Leipzig und Wien.

Pyšek, P., Richardson, D.M., Rejmanek, M., Webster, G.L., Williamson, M. & Kirschner, J., 2002, 'Alien plants in checklists and floras: Towards better communication between taxonomists and ecologists', *Taxon* 53, 131–143. http://dx.doi.org/10.2307/4135498

Reveal, J.L. & Spellenberg, R., 1976, 'Miscellaneous chromosome counts of Western American plants – III', *Rhodora* 78, 37–52.

Sanbi, 2012, *Biodiversity of life*, viewed 21 March 2016, from http://www.ville-ge.ch/musinfo/bd/cjb/africa/details.php?langue=an&id=187986

Sauer, J.D., 1955, 'Revision of the dioecious amaranths', *Madroño* 13, 5–46.

Shreve, F. & Wiggins, L., 1964, *Vegetation and flora of the Sonoran Desert*, vol. 1, Standford University Press, Standford, CA.

Sukhorukov, A.P., Martín-Bravo, S., Verloove, F., Maroyi, A., Iamonico, D., Catarino, L., et al., 2016, Chorological and taxonomic notes on African plants. *Acta Botanica Gallica* 163, 417–428. http://dx.doi.org/10.1080/23818107.2016.1224731

Thiers, B., 2016+, *Index herbariorum, a global directory of public herbaria and associated staff. New York Botanical Garden's Virtual Herbarium*, viewed 21 March 2016, from http://sweetgum.nybg.org/ih/

Uline, E.B. & Bray, W.L., 1894, 'A preliminary synopsis of the North American species of Amaranthus', *Botanical Gazzette* 19, 313–320. http://dx.doi.org/10.1086/327076

Watson, S., 1877, 'Description of nine species of plants, with revision of certain genera', *Proceedings of the American Academy of Arts and Sciences* 12, 246–278. http://dx.doi.org/10.2307/25138455

Terrestrial invasions on sub-Antarctic Marion and Prince Edward Islands

Authors:
Michelle Greve[1] ⓘ
Rabia Mathakutha[1] ⓘ
Christien Steyn[1] ⓘ
Steven L. Chown[2] ⓘ

Affiliations:
[1]Department of Plant and Soil Sciences, University of Pretoria, South Africa

[2]School of Biological Sciences, Monash University, Australia

Corresponding author:
Michelle Greve,
michelle_greve@yahoo.com

Background: The sub-Antarctic Prince Edward Islands (PEIs), South Africa's southernmost territories have high conservation value. Despite their isolation, several alien species have established and become invasive on the PEIs.

Objectives: Here we review the invasion ecology of the PEIs.

Methods: We summarise what is known about the introduction of alien species, what influences their ability to establish and spread, and review their impacts.

Results: Approximately 48 alien species are currently established on the PEIs, of which 26 are known to be invasive. Introduction pathways for the PEIs are fairly well understood – species have mainly been introduced with ship cargo and building material. Less is known about establishment, spread and impact of aliens. It has been estimated that less than 5% of the PEIs is covered by invasive plants, but invasive plants have attained circuminsular distributions on both PEIs. Studies on impact have primarily focussed on the effects of vertebrate invaders, of which the house mouse, which is restricted to Marion Island, probably has the greatest impact on the biodiversity of the islands. Because of the risk of alien introductions, strict biosecurity regulations govern activities at the PEIs. These are particularly aimed at stemming the introduction of alien species, and are likely to have reduced the rates of new introductions. In addition, some effort is currently being made to eradicate selected range-restricted species. However, only one species that had established and spread on the PEIs, the cat, has been successfully eradicated from the islands.

Conclusion: Given the ongoing threat of introductions, and the impacts of invaders, it is essential that future invasions to the PEIs are minimised, that the islands' management policies deal with all stages of the invasion process and that a better understanding of the risks and impacts of invasions is obtained.

Introduction

Islands are fragile ecosystems. Because of their isolation, islands are often taxonomically and functionally depauperate, that is, they house only a subset of the taxonomic or functional taxa compared to mainland ecosystem (Reaser et al. 2007). It is thought that this reduces biotic resistance to invasion and makes islands especially prone to invasions by alien species (Vitousek 1988; see also Gimeno, Vilà & Hulme 2006). Therefore, island species and ecosystems are thought to be particularly sensitive to the impacts of invasives (Donlan & Wilcox 2008; Reaser et al. 2007; see also Vilà et al. 2011). Despite better awareness of the risks of invasions and improved biosecurity regulations globally (García-de-Lomas & Vilà 2015; McGeoch et al. 2010), invasions continue unabated (Hulme 2009), with island invasions showing no sign of approaching a point of saturation (Sax & Gaines 2008).

Sub-Antarctic Islands represent some of the most isolated and least impacted habitats on the Earth; yet, they are also prone to invasions by nonindigenous species (Convey et al. 2006; Frenot et al. 2005; Shaw 2013). Indeed, alien invasions are thought to be the largest threat to the biodiversity of the sub-Antarctic Islands (Frenot et al. 2005), and a particularly high proportion of aliens that arrive on sub-Antarctic Islands appears to become invasive, of which many have negative impacts on native biodiversity (McGeoch et al. 2015).

The Prince Edward Islands (PEIs) constitute South Africa's southernmost, and only sub-Antarctic, territories (Figure 1). They consist of the larger Marion Island (46° 54′ S, 37° 45′ E; area: app. 270 km²) and the smaller Prince Edward Island (46° 38′ S, 37° 57′ E area: app. 45 km²), which are

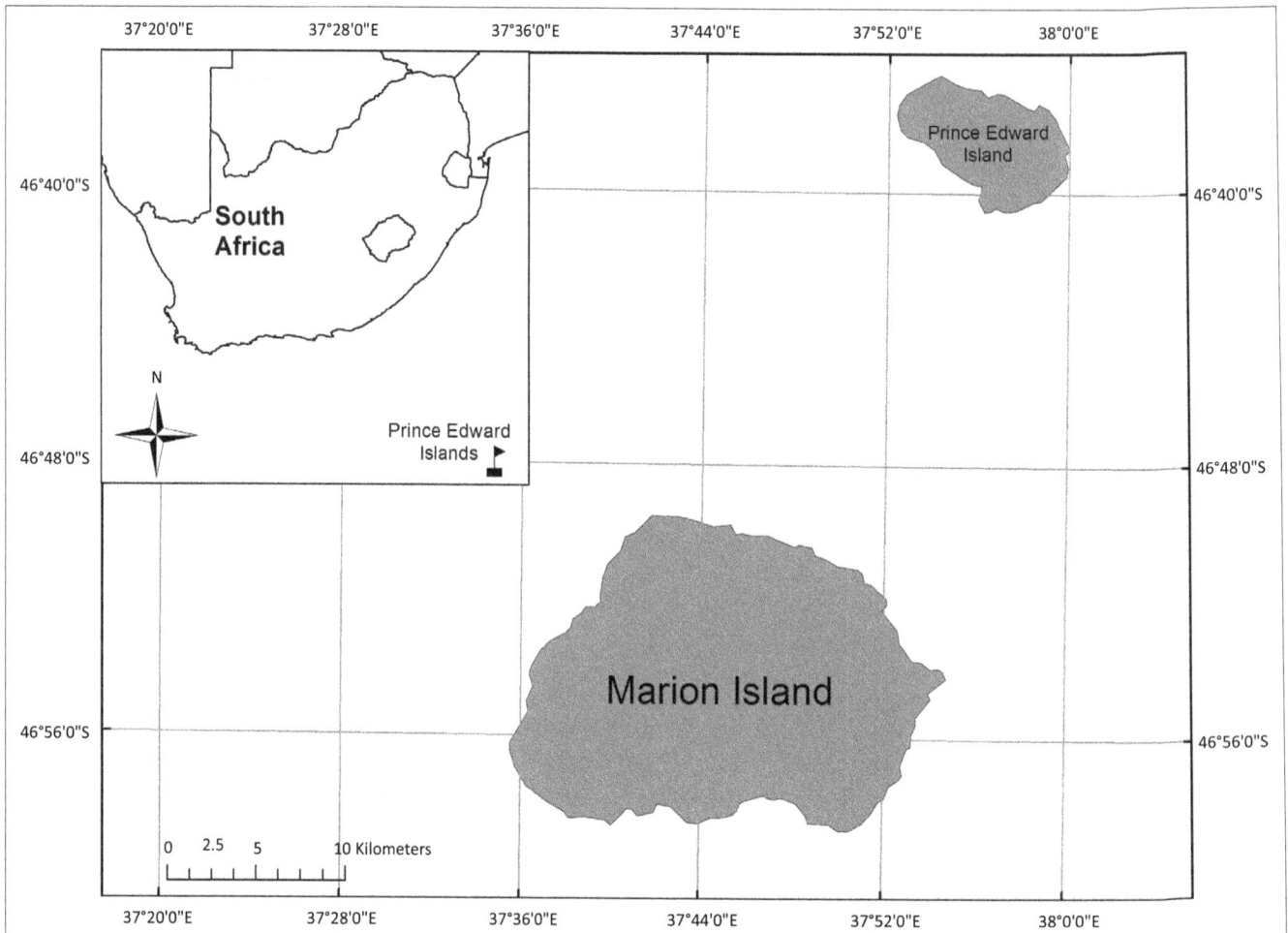

FIGURE 1: The location and relative positions of Marion Island and Prince Edward Island, which together constitute the Prince Edward Islands.

separated by approximately 19 km of ocean (Lutjeharms & Ansorge 2008). The islands are of volcanic origin, with an estimated age of 450 000 years (Boelhouwers et al. 2008). Marion Island reaches an altitude of 1230 m and Prince Edward Island an altitude of 672 m.

The PEIs have been a conservation area and a centre for scientific research since their occupation by South Africa in 1947. Along with other islands of the sub-Antarctic, the PEIs are of high conservation importance, where few small land masses support all terrestrial diversity in the region, including many of the top oceanic predators which use the islands as breeding grounds (de Villiers & Cooper 2008). In 1995, the PEIs were declared a Special Nature Reserve according to the Environment Conservation Act (No. 73 of 1989), which has now been superseded by the National Environmental Management: Protected Areas Act (No. 57 of 2003 NEMPA). Given the rich history of scientific activities on the PEIs (summarised in Chown & Froneman 2008b), their setting in a climatically dynamic area of the Southern Ocean (Fraser et al. 2009) and the fairly low anthropogenic pressures on the islands, the PEIs offer a valuable opportunity for studying and understanding ecological processes and management and research challenges, and how these challenges can be managed and forecast.

A sizeable number of alien species have been introduced to the PEIs (Table 1) (Chown & Froneman 2008b). Currently, approximately 18 (3) terrestrial plants, one (0) terrestrial vertebrate, 28 (5) terrestrial invertebrates and one (0) terrestrial fungus are thought to be established on Marion (and Prince Edward) Islands (Table 1). Some of the introduced species have a restricted distribution and have little effect on the native biodiversity on the islands (Gremmen 1997; Huntley 1971), but several (6 plants, 1 vertebrate and 19 invertebrates) have become invasive (invasion status D2 or E, Table 1) on the islands. (These are conservative estimates as the invasion status of other alien species is unknown, Table 1). Many of the invasives severely impact the native biodiversity and the ecosystem functioning of the islands. Indeed, along with climate change, invasive species probably pose the greatest conservation risk to the PEIs.

While a number of reviews about invasions to the greater sub-Antarctic region have been written (e.g. Convey et al. 2006; Frenot et al. 2005; Hughes & Convey 2010; Shaw 2013), we aim to review the state of knowledge of invasions for the PEIs. We specifically focus on invasions of terrestrial habitats, although the islands are also at risk of marine invasions (e.g. Lee & Chown 2007). The main aims of this paper are (1) to review what is known about alien and invasive organisms on the islands at different stages of the invasion process

TABLE 1: Alien species recorded on the Prince Edward Islands.

Order	Family	Species	Invasion status	Impact	Confidence level for impact	Islands	Introduction notes	Additional info	References
Plants: CLADE – Angiosperma: Monocotyledonae									
Alismatales	Potamogetonaceae	*Potamogeton nodosus* Poir.	status uncertain	Impact unknown	-	MI	First recorded in 1965	-	6, 24
Poales	Poaceae (Gramineae)	*Elymus repens* (L.) Gould	C3	Major	High	MI	First recorded in 1965, likely to have been introduced by sealers	-	6, 24
		Agrostis castellana Boiss. & Reut.	status uncertain	Impact unknown	-	MI	First recorded in 1975	-	6, 24
		Agrostis gigantea Roth	C3	Minimal	Medium	MI	First recorded in 1994, likely introduced with building equipment	-	6, 24
		Agrostis stolonifera L.	E	Major	High	MI	First recorded in 1965	-	6, 24
		Alopecurus geniculatus L.	C2–C3	Impact unknown	-	MI	First recorded in 1965, likely to have been introduced by sealers	Eradicated	14, 15, 24
		Avena sativa L.	C0	Impact unknown	-	MI	First recorded in 1965/1966	Transient	6, 14
		Festuca rubra L.	C3	Major	High	MI	First recorded in 1965, likely to have been introduced by sealers	-	6, 24
		Holcus lanatus L.	C0	Impact unknown	-	MI	First recorded in 1953	Eradicated/Transient	14
		Poa annua L.	E	Moderate	Medium	MI & PEI	First recorded in 1948, likely to have been introduced by sealers	-	6, 24
		Poa pratensis L.	D2	Moderate	Low	MI	First recorded in 1965, likely introduced with building equipment	-	6, 24
	Juncaceae	*Juncus effusus* L.	C1 or D1 (uncertain)	Minimal	High	MI	First recorded in 1965	-	6, 24
		Luzula cf. multiflora (Ehrh) Lej.	C3	Minimal	Medium	MI	First recorded in 1999	-	6, 24
Plants: CLADE – Angiosperma: Dicotyledonae									
Asterales	Asteraceae	*Hypochaeris radicata* L.	status uncertain	Impact unknown	-	MI	First recorded in 1953	Eradicated/Transient	14
		Senecio sp.	status uncertain	Impact unknown	-	MI	First recorded in 1988, likely introduced with building equipment	Eradicated	14, 15
		Sonchus sp.	status uncertain	Impact unknown	-	MI	First recorded in 1983, likely introduced with building equipment	Eradicated	14, 15
Caryophyllales	Caryophyllaceae	*Cerastium fontanum* Baumg.	E	Minor	High	MI & PEI	First recorded in 1873, likely to have been introduced by sealers	-	6, 24
		Sagina procumbens L.	E	Major	High	MI & PEI	First recorded in 1965, likely introduced with building equipment	-	6, 24
		Stellaria media (L.) Vill	D2	Minor	High	MI	First recorded in 1873, likely to have been introduced by sealers	-	6, 15, 24
	Polygonaceae	*Rumex acetosella* L.	D1	Impact unknown	-	MI	First recorded in 1953, likely to have been introduced by sealers	-	6, 24
Lamiales	Plantaginaceae	*Plantago lanceolata* L.	status uncertain	Impact unknown	-	MI	First recorded in 1983, likely introduced with building equipment	Eradicated	14
Rosales	Rhamnaceae	*Ochetophila trinervis* (Gillies ex Hook.) Poepp. ex Endl.	C2	Minimal	Medium	MI	First recorded in 2004	-	6, 24
Unknown	Unknown	Unidentified plant.	status uncertain	Impact unknown	-	MI	First recorded in 2016	-	
DIVISION – Bryophyta									
Hypnales	Thuidiaceae	*Thuidium delicatulum* (Hedw.) Schimp.	C3	Minimal	Low	MI	-	-	6
Pottiales	Pottiaceae	*Leptodontium gemmascens* (Mitt.) Braithw.	status uncertain	Impact unknown	-	MI	-	-	6
Vertebrates: CLASS: Mammalia									
Rodentia	Muridae	*Mus musculus domesticus* Schwarz & Schwarz	E – fully invasive	Massive	High	MI	First recorded in 1818, likely introduced by sealers	-	7
Artiodactyla	Bovidae	*Ovis aries* L.	C1	Impact unknown	-	MI	Deliberately introduced in 1950	Eradicated	13

Table 1 continues on the next page →

TABLE 1 (Continues...): Alien species recorded on the Prince Edward Islands.

Order	Family	Species	Invasion status	Impact	Confidence level for impact	Islands	Introduction notes	Additional info	References
Carnivora	Felidae	Felis catus L.	E – fully invasive	Massive	High	MI	Deliberately introduced in 1949	Eradicated	7
Artiodactyla	Suidae	Sus scrofa L.	C3–E	Impact unknown	Low	MI	Deliberately introduced in 1804	Absent – hunted to extinction	7
Invertebrates: PHYLUM – Arthropoda (Subphylum: Crustacea; Class: Malacostraca)									
Isopoda	Porcellionidae	Porcellio scaber Latreille	C3	Impact unknown	Low	MI	Probably introduced in 2001 in building material	-	6, 18, 29
Invertebrates: PHYLUM – Arthropoda (Class: Arachnida)									
Mesostigmata	Digamasellidae	Dendrolaelaps sp.nov.	Status uncertain	Impact unknown	-	MI & PEI	First recorded in 1997/1998	-	6, 18, 29
Prostigmata	Scutacaridae	Disparipes antarcticus Richters	Status uncertain	Impact unknown	-	MI	-	-	6
	Pygmephoridae	Neopygmephorus sp.	Status uncertain	Impact unknown	-	MI (status for PEI unclear)	-	-	6
Astigmata	Glycyphagidae	Glycyphagus domesticus (de Geer)	Status uncertain	Impact unlikely	-	MI	-	-	6
Araneae	Miturgidae	Cheiracanthium furculatum Karsch	Status uncertain	Impact unknown	-	MI	First recorded in 2004	-	37
	Linyphiidae	Prinerigone vagans (Audouin)	D2–E	Impact unknown	-	MI	First recorded in 1965	-	35, 36
Invertebrates: PHYLUM – Arthropoda (Class: Entognatha; Subclass: Collembola)									
Poduromorpha	Hypogastruridae	Ceratophysella denticulata (Bagnall)	E	Impact unknown	-	MI & PEI	Probably introduced after annexation	Under climate change (drier conditions), invasive species may be favoured over indigenous species	12, 21, 33
Entomobryomorpha	Isotomidae	Isotomurus maculatus Müller	D2–E	Impact unknown	-	MI	-	Under climate change (drier conditions), invasive species may be favoured over indigenous species	12, 21, 33
		Parisotoma notabilis (Schäffer)	D2–E	Impact unknown	-	MI	-	Under climate change (drier conditions), invasive species may be favoured over indigenous species	12, 21
Neelipleona	Tomoceridae	Pogonognathellus flavescens (Tullberg)	D2	Impact unknown	-	MI	First recorded in 1983	Under climate change (drier conditions), invasive species may be favoured over indigenous species	6, 21, 32, 33
	Neelidae	Megalothorax minimus Willem	E	Impact unknown	-	MI	-	Under climate change (drier conditions), invasive species may be favoured over indigenous species	12, 21
Invertebrates: PHYLUM – Arthropoda (Class: Insecta)									
Hemiptera	Aphididae	Macrosiphum euphorbiae (Thomas, C.)	E	Minor	Low	MI	-	Uses indigenous plants as host, although effects on host unknown	1, 5, 17
		Myzus ascalonicus Doncaster	E	Minor	Low	MI	-	Uses indigenous plants as host, although effects on host unknown	1, 5, 17
		Rhopalosiphum padi (L.)	E	Minor	Low	MI	Probably introduced with station supplies in 1940s	Uses indigenous plants as host, although effects on host unknown	1, 5, 6, 10, 17
Thysanoptera	Thripidae	Apterothrips apteris (Daniel)	E	Impact unknown	-	MI	-	-	5, 17
Coleoptera	Anobiidae sp.		B3	Impact unlikely	-	MI	Introduced in food stores	Not established	17
	Dermestidae sp.		B3	Impact unlikely	-	MI	-	Not established	6
	Chrysomelidae sp.		B3	Impact unlikely	-	MI	First recorded in 1996, introduced in food stores	Not established	6

Table 1 continues on the next page →

TABLE 1 (Continues...): Alien species recorded on the Prince Edward Islands.

Order	Family	Species	Invasion status	Impact	Confidence level for impact	Islands	Introduction notes	Additional info	References
Lepidoptera	Noctuidae	Agrotis ipsilon Hufnagel	D1–D2	Impact unknown	-	MI	First recorded in 1996, introduced in food stores	-	3, 5, 6
		Agrotis segetum Schiffermüller	D2–E	Impact unknown	-	MI	-		5, 6
		Cosmophila sabulifera (Geunee)	B3–C2	Impact unlikely	-	MI	-	Not established	5, 6
		Helicoverpa armigera Hübner	B3	Impact unlikely	-	MI	-	Not established	5, 6
	Nymphalidae	Vanessa cardui (L.)	D2	Impact unknown	-	MI	First recorded in 1970s		3, 5, 17
	Plutellidae	Plutella xylostella (L.)	D2–E	Moderate	Low	MI	-	Larval feeding causes considerable damage to the native Pringlea antiscorbutica	4, 5, 6, 9, 17
Diptera	Pyralidae	Nomophila sp.	B3	Impact unlikely	-	MI	First recorded in 1996	-	5, 6
	Calliphoridae	Calliphora vicina Robineau-Desvoidy	C3	Impact unlikely	-	MI			5, 17
	Chironomidae	Limnophyes minimus (Meigen)	E	Impact unknown	-	MI & PEI	First recorded in 1939		5, 6, 8, 16
	Drosophilidae	Scaptomyza oxyphallus Tsacas	C3	Impact unlikely	-	MI			5, 6, 17
	Faniidae	Fannia canicularis (L.)	D2–E	Impact unknown	-	MI & PEI			5, 6
	Lonchaeidae	Lamprolonchaea smaragdi (Walker)	B3	Impact unknown	-	MI	First recorded in 1996, introduced in food stores	Not established	6
	Psychodidae	Psychoda parthenogenetica Tonnoir	E	Impact unknown	-	MI & PEI			5, 6, 17
Hymenoptera	Braconidae	Aphidius matricariae Haliday	E	Minor	Low	MI	Probably introduced in 2001 in food stores	Controls populations of invasive aphid Rhopalosiphum padi	5, 25, 26
	Formicidae	Lepisiota capensis (Mayr)	B3	Impact unlikely	-	MI	First recorded in 1996	Not established	6
Invertebrates: PHYLUM – Mollusca	Limacidae	Deroceras panormitanum (Lessona & Pollonera)	E	Minor	-	MI	First recorded in 1972	Decomposer, but mineralises N at a considerable slower rate than a comparative indigenous decomposer	6, 30, 31
Microbes: KINGDOM: Fungi									
Helotiales	Sclerotiniaceae	Botryotinia fuckeliana (de Bary) Whetzel status uncertain		Moderate	Low	MI	First recorded in 1988	-	38

MI, Marion Island; PEI, Prince Edward Island.

The invasion status and impact of each species, the confidence for the impact, the islands that each species has been recorded on, any available information about dates and pathways of the introduction of each species to the PEIs and the references (Table 1-A1) that were consulted to fill out the table for every species are indicated. The invasion status was classified according to Blackburn et al. (2011); a key to the invasion status column is provided in Table 2-A1. Impact was classified according to Blackburn et al. (2014); a key to the impacts is provided in Table 3-A1. The date of introduction is the first record for the Prince Edward Island archipelago.

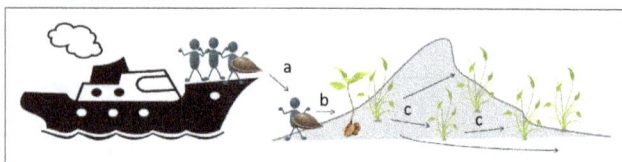

FIGURE 2: The different stages of the invasion process to the Prince Edward Islands. A ship arrives at an island. A propagule is carried by one of the passengers. a) The passenger disembarks from the ship and carries the propagule with them. b) The propagule establishes and c) spreads to other parts of the island.

(Figure 2) (Blackburn et al. 2011). The first stage, introduction, deals with the arrival of organisms at the islands. In the second stage, establishment, we discuss how the alien species survive the environmental conditions on the island, grow and reproduce. Finally, the third stage comprises the spread of alien species, when they extend their ranges from the initial point of introduction to spread and occupy a larger range. (2) To summarise the known impacts of alien organisms. (3) To provide lists of the terrestrial alien species on the islands and what is known about their introduction, invasion status and impact. This information is especially relevant to South Africa's National Status Report on Biological Invasions, which will be released in 2017. (4) In light of what is known about invasions to the islands, we comment on management policies and current and future challenges for the PEIs.

Human activities on the PEIs

Introductions of alien species to the PEIs are closely linked to the human activities on the islands. It is thought that the PEIs were first discovered in March 1663, but it was to be more than 100 years before the first person set foot on the islands. By 1802 the first sealers had made temporary camps on the islands (Gremmen 1975). Sealing activities continued on the island for approximately 50 years, during which time sealers were left to hunt seals on the islands for months at a time (Cooper 2008). After the seal populations had been decimated, human visits were relatively infrequent until 1947 when the islands were annexed by the South African government (Cooper 2008). Since annexation, Marion Island has been continuously inhabited by a small, though variable-sized (most recently in the old base approximately 15, since the building of the new base around 20), team of scientists and support staff, with no permanent residents living on the island (the maximum amount of time usually spent on the island is 13 months, although in exceptional circumstances individuals have stayed for 2 years). Prince Edward Island is unoccupied and only rarely visited. Due to the absence of a resident human population on the PEIs, the impact of humans has been limited compared to similar islands in the sub-Antarctic region (Chown, Gremmen & Gaston 1998). Nevertheless, human activities on Marion Island have been increasing steadily since the establishment of a permanent meteorological station on Marion Island in 1948. These anthropogenic activities have negatively impacted ecosystems and species on the island, with the introduction of alien or nonindigenous species being one particularly detrimental consequence (Chown & Froneman 2008a).

Human activities have been responsible for the introduction to the PEIs of a range of species from a variety of taxonomic groups (Chown & Froneman 2008b). The first alien introductions to the islands were associated with sealing vessels in the early 1800s. Subsequent military and scientific expeditions resulted in further introductions (Cooper 2008). Of all animal introductions in the PEIs, none have had a more substantial impact on the island's flora and fauna than the house mouse *Mus musculus* L. and the feral cat *Felis catus* L. The house mouse was accidentally introduced to Marion Island by sealers (Cooper 2008). Because of its tolerance of the sub-Antarctic climate and the absence of natural enemies on Marion Island, the house mouse was able to thrive on an omnivorous diet of plants and animals (Avenant & Smith 2003; Smith, Avenant & Chown 2002). In 1949, domestic cats were introduced to Marion Island in an attempt to deal with the mouse problem at the research station. These cats soon became feral, consuming large numbers of burrowing seabirds (Bester et al. 2002). They were also responsible for the extirpation of the common diving petrel (*Pelecanoides urinatrix* Gmelin) in 1965 (Bester et al. 2002). The cats were successfully removed in 1991, following a protracted eradication programme (Bester et al. 2002).

As human activities on the PEIs have increased over the years, so have the number of alien species and their impacts, and we are still coming to understand the invasion process and the impacts of invasives on the PEIs.

The stages of invasion
Introduction

The first stage of the invasion pathway is the introduction stage, when an organism is transported to an area to which it is non-native (Blackburn et al. 2011). Although many species that are introduced to areas outside their native range do not become invasive, some will do. Usually the greater the propagule pressure of non-native species to an area, the greater the chance that one of these species will become invasive (Simberloff et al. 2013). Therefore, the most effective way of curbing new alien introductions to an area is by gaining a thorough understanding of introduction pathways and vectors and using this information to inform the implementation of appropriate management actions and regulations to prevent future invasions (Faulkner et al. 2016; Hulme 2009; Keller & Kumschick 2017; Simberloff et al. 2013).

Islands are discreet landmasses; therefore, introduction pathways and the movement of vectors of alien species (i.e. people and goods) can often be effectively quantified (e.g. Christy, Savidge & Rodda 2007; Frenot, Gloaguen & Massé 2001). Likewise, the transport of goods and people to the PEIs is limited and well documented (DEA: Natural Resources Management Programme et al. 2012). In addition, there has been long-term and ongoing scientific research on the islands, with a recent focus on introduction pathways for the islands (e.g. Lee & Chown 2009a, 2009b, 2011). As a consequence, a fair amount is known about the timing, determinants and pathways of invasions to the PEIs.

An important determinant of alien introductions to an area is the nature of human activities the area is exposed to. Across the sub-Antarctic Islands, invasions are strongly influenced by the intensity of human activities: islands with more human visitors tend to have more introduced species (Chown et al. 2009; Frenot et al. 2005). The PEIs have no permanent human settlements; activities on the islands are restricted to scientific research and supporting activities, the protection or management of the islands and to reporting and educational activities, all of which are strictly regulated (Department of Environmental Affairs Directorate: Antarctica and Islands 2010). In addition, Marion Island is usually visited only once annually by one vessel. The current vessel transports approximately 80 scientists and support staff to the island, while Prince Edward Island is visited infrequently – at the most once every 4 years, and then only by a team of up to 10 scientists for a maximum duration of 8 days (Cooper et al. 2009; Department of Environmental Affairs Directorate: Antarctica and Islands 2010). Because visits to the PEIs are relatively infrequent, fewer alien species have been introduced to the islands compared to similar sub-Antarctic Islands with permanent

settlements or more visitors (after controlling for climate, Chown et al. 1998). In addition, fewer alien species occur on the less visited Prince Edward Island than on Marion Island (de Villiers & Cooper 2008).

Aliens on the PEIs arrived at the islands via ships, their cargo and their occupants. A number of introductions occurred before the annexation of the PEIs, probably at the hands of sealers (Table 1, le Roux et al. 2013) although most introductions happened shortly after annexation in 1947/1948 (de Villiers & Cooper 2008). The building of the first research station in the 1960s coincided with a wave of new species being introduced to Marion Island (Huntley 1971; le Roux et al. 2013). Although existing biosecurity measures (see below) have been effective at reducing rates of introduction (see le Roux et al. 2013 for dates of discovery of alien plants on the PEIs), they have not prevented new species from arriving to the region.

Introductions to the islands have been intentional (e.g. cats) and unintentional (e.g. the birdeye pearlwort *Sagina procumbens* L., Figure 3a). Intentional introductions

Source: (a–b) Michelle Greve; (c) Rabia Mathakutha/Christien Steyn; (d) Christien Steyn; (e–f) Michelle Greve; (g) Michelle Greve; (h) Michelle Greve.

FIGURE 3: a) *Sagina procumbens* is one of the most widespread alien plants on Marion Island. b) It occupies a wide variety of habitats, including the recently unvegetated Kaalkoppie on the west of the island. c) Areas of high disturbance and nutrient content, such as the Biotic Complex which experience high seal activity, have higher abundance and cover of alien plants. d) *Agrostis stolonifera* growing around a field hut. e) Mouse paths running through *Blechnum penna-marina* slopes, connecting the entrances to mouse burrows. f) The remains of an *Azorella selago* cushion that has been destroyed by mice. g) Dense stands of *Agrostis stolonifera*. h) The biosecurity inspection of footwear or 'bootwashing ceremony' on the research vessel *SA Agulhas*. The soles of all shoes are scrubbed with bleach liquid and bags and clothing are checked for propagules.

Source: (a–b) Michelle Greve; (c) Rabia Mathakutha/Christien Steyn; (d) Christien Steyn; (e–f) Michelle Greve; (g) Michelle Greve; (h) Michelle Greve.

FIGURE 3 (Continues...): a) *Sagina procumbens* is one of the most widespread alien plants on Marion Island. b) It occupies a wide variety of habitats, including the recently unvegetated Kaalkoppie on the west of the island. c) Areas of high disturbance and nutrient content, such as the Biotic Complex which experience high seal activity, have higher abundance and cover of alien plants. d) *Agrostis stolonifera* growing around a field hut. e) Mouse paths running through *Blechnum penna-marina* slopes, connecting the entrances to mouse burrows. f) The remains of an *Azorella selago* cushion that has been destroyed by mice. g) Dense stands of *Agrostis stolonifera*. h) The biosecurity inspection of footwear or 'bootwashing ceremony' on the research vessel *SA Agulhas*. The soles of all shoes are scrubbed with bleach liquid and bags and clothing are checked for propagules.

took place in the earlier years of the islands' occupation and included several vertebrate species. For example, the domestic pig (*Sus scrofa* L.) was deliberately introduced to Marion Island in 1804 where it bred successfully until it was eradicated some years later (Cooper 2008), while sheep, chickens and trout were introduced after annexation to provide fresh food to the overwintering teams (de Villiers & Cooper 2008; Gremmen 1975). None of these species have persisted. The only intentionally introduced species to the island that significantly impacted the native biodiversity was the cat (Cooper 2008; van Aarde 1980).

Some plant species were also intentionally introduced to the PEIs. After annexure, an attempt at planting several tree species, mostly of Northern hemisphere origin, was made; however, the conditions on the island (possibly the soils) proved unsuitable for their establishment (Gremmen 1975; La Grange 1954).

Most aliens were introduced accidentally. Introduction pathways are likely to have varied from rodents probably

escaping off ships (Cooper 2008) to propagules being brought in clothing or cargo (Lee & Chown 2009a, 2009b, 2011), with building material (Bergstrom & Smith 1990; Slabber & Chown 2002), or in fresh produce (Smith 1992). During recent searches of cargo containers, passenger luggage and passenger clothing on voyages from Cape Town to Marion Island, live insects and plant seeds were retrieved (Lee & Chown 2009a, 2009b, 2011). The container packing location seemed to be a major source of propagules found in containers, and field clothing and field luggage contained a significant number of alien seeds (Lee & Chown 2009a).

Alien species are typically weedy Holarctic species which are also established in South Africa, from where the vessels visiting the PEIs most often depart (Lee & Chown 2011). However, because several scientists working on Marion Island also travel more widely in the greater Antarctic region, intra-regional transport of propagules to Marion Island is also likely (Lee & Chown 2011). Other studies have shown that people travelling to the region often visit and work in other cold climatic regions (Chown et al. 2012; Lee & Chown 2011).

Despite increasingly stringent policies to reduce introductions of alien species to the PEIs (DEA: Natural Resources Management Programme et al. 2012), new introductions are continuing (Table 1). Unsuccessful establishments of alien species are difficult to detect, but several new establishments have been recorded since stringent biosecurity regulations were put in place in the mid-1990s (e.g. Lee & Chown 2016; Slabber & Chown 2002), suggesting that new species continued being introduced. As a consequence, the main pathway of introductions, the annual relief voyages, is now subject to even greater biosecurity protocol than previously (Department of Environmental Affairs Directorate: Antarctica and Islands 2010). No new arrivals have been recorded since 2010, but systematic surveys for aliens on Marion Island have fallen away in the last few years – therefore, it is not possible to ascertain whether the absence of new records is because of increased biosecurity efforts or because of lack of surveys.

Introduction of aliens to Marion Island is mainly because of direct introductions by humans (whether intentional or unintentional). However, it is likely that some introductions of alien species to Prince Edward Island have been the result of secondary transport of alien propagules by natural vectors, such as wind or birds, from Marion to Prince Edward Island (Gremmen & Smith 1999; Ryan, Smith & Gremmen 2003). For example, the first record of *S. procumbens* L. on Prince Edward Island was in an area of high seal and bird activity, some distance from where scientists usually land on their expeditions to the island, making it unlikely that humans were responsible for the introduction (Gremmen & Smith 1999).

Which species reach the sub-Antarctic Islands such as the PEIs may be determined by dispersal ability. Vagility influences the nestedness of the different alien taxa of the sub-Antarctic Islands (Greve et al. 2005); in other words, assemblages of groups that are most mobile are more nested than those of less mobile groups. This suggests that organisms with better vagility are more likely to reach more sub-Antarctic Islands, whereas the arrival of species with lower vagility on the islands may be more haphazard (Greve et al. 2005; Shaw et al. 2010). Nevertheless, some overlap in alien assemblage composition across taxa is present across different sub-Antarctic Islands despite the large distances between them, with many of the alien species composing European weedy species (Frenot et al. 2005; Shaw et al. 2010).

Establishment

After introduction, the next stage of the invasion sequence is establishment; an organism must be able to survive and be viable in its new environment (Blackburn et al. 2011). This means that the environmental conditions must be favourable for the organism and that the organism is not driven to extinction by predators or pathogens in its new habitat.

The establishment of alien species on the PEIs is less well understood than the introduction vectors and pathways.

The alien vascular plant species that have established on the PEIs tend to originate from plant families with weedy tendencies (Shaw 2013), indicating that species from specific taxa are predisposed to being successful establishment.

The environment also plays an important role in establishment. Species richness of aliens on sub-Antarctic Islands is strongly influenced by climate: more alien species occur on warmer islands (Chown et al. 1998; Chown, Hull & Gaston 2005). This suggests that temperature is a major limiting factor to the establishment of alien species. Indeed, most alien species on the PEIs originate from the similarly cold Northern hemisphere (Shaw et al. 2010), rather than from milder South Africa, where most ship voyages originate from. (It is, however, important to note that many of the weedy plants established on the PEIs occur in South Africa and are common at the wharf precinct in Cape Town – the departure point of the supply ship; Lee & Chown 2009a). In addition, on Marion Island, many of the alien taxa are found in the milder lowlands, rather than in the more climatically harsh high-altitude areas (Chown et al. 2013; le Roux et al. 2013). Moreover, Lee et al. (2009) showed that the abundance of the invasive slug, *Deroceras panormitanum* (Lessona and Pollonera), was most strongly influenced by microclimatic conditions, with slugs being most abundant in conditions that corresponded to their physiological tolerances (Lee et al. 2009).

In many instances, plant invasions are promoted by disturbances and nutrient enrichment (Pyšek & Richardson 2008; Richardson et al. 2000). On the PEIs, the main agents of disturbance and nutrient enrichment are seals and seabirds (Smith 1978), which disturb the environment by trampling and burrowing (some seabirds) and enrich the soils through manuring. Accordingly, the establishment of alien plants on Marion Island is greater in areas that are prone to disturbances and nutrient enrichment by seals (Figure 3b and c) (Haussmann et al. 2013; Huntley 1971), and it is common to find alien species such as *Poa annua* L. growing at the entrance to bird burrows in regions where the incidence of alien species is otherwise low or non-existent.

Species that establish on the PEIs must possess adaptations to survive the conditions on the islands. These include low temperatures and strong winds. One of the most aggressive invasive plants on Marion Island is the grass *P. annua* (le Roux et al. 2013). It has spread extensively across the greater sub-Antarctic region and is the only vascular plant that has established and spread on the Antarctic continent (Chwedorzewska et al. 2015; Shaw et al. 2010). Its success in the region has been attributed to its high levels of phenotypic plasticity, which allow it to survive a range of different conditions. For example, it possesses the ability to switch between a perennial and annual life cycle depending on environmental conditions (reviewed in Chwedorzewska et al. 2015). In another example, the cold tolerance and the rapid generation time of the moth *Plutella xylostella* L. are considered pre-adaptations of the species to successfully establish on Marion Island (Crafford & Chown 1990).

Alien species that become invasive often possess a different set of traits to non-invasive aliens and native species (Ordonez, Wright & Olff 2010; van Kleunen, Weber & Fischer 2010), suggesting that some traits promote invasiveness. Identifying what traits confer higher invasiveness assists in identifying which species may be pre-adapted to becoming established and invasive to an area. Little is known about the role that functional traits play in the establishment of alien species on the PEIs. This has only been explored qualitatively for individual species of the PEIs (Crafford 1986; van Aarde 1986). For example, van Aarde (1986) suggested that cats could successfully establish on Marion Island because of a selective advantage of the founder population. The founder group showed a high incidence of the dark-coloured phenotype, the proportion of which increased over time, suggesting that this phenotype is associated with a selective advantage (discussed in van Aarde 1986). Therefore, not only the presence of the dark-coloured phenotype and its associated advantages in the founder population but also the selection for this phenotype as the population went through successive breeding cycles on Marion Island is thought to have aided the establishment of cats (van Aarde 1986).

A comparison of the traits of invasive species and indigenous species could provide insights into whether the establishment success of invasives is because they possess novel traits (i.e. they occupy niches not occupied by indigenous species) or because of their pre-adaptations to conditions on the island (under which circumstances one would expect these species to have traits similar to those of indigenous species) (Tecco et al. 2010). Only one published study has undertaken a semblance of such a study (Pammenter, Drennan & Smith 1986), although it entailed a comparison of only one pair of alien and indigenous con-generics (*Agrostis stolonifera* L. and *Agrostis magellanica* Lam., respectively). Therefore, its findings that the invasive species assimilated CO_2 more effectively at low photon flux densities but showed less investment in sclerophyllous leaf tissue than the indigenous species cannot be generalised. Preliminary findings of a more recent study indicate that invasive plant species on Marion Island occupy functional niches not occupied by native species (Mathakutha et al., unpublished).

Dispersal and spread

After establishment, a non-native species must spread and extend its range in the new environment before it is classified an invasive (Blackburn et al. 2011). After an alien organism has started spreading, it usually becomes difficult and expensive to effectively control (Simberloff et al. 2013). It is thus recommended that efforts to eradicate alien species should be initiated before they reach the spread stage (Hughes & Convey 2012). In addition, factors that promote spread need to be thoroughly understood in order to prioritise eradication efforts.

Not all alien species introduced to the PEIs have spread on the islands – a number of species remain range-restricted (Table 1). Not much is known about the mode of spread of

alien organisms on the PEIs once they have established. The rate and geography of spread are best understood for plants. Of the alien plant species that have successfully established on Marion Island, six have spread extensively since first detection (Table 1) (le Roux et al. 2013). Four of these are grasses. *Sagina procumbens* is thought to have spread at a rate of more than 1.5 km and 2 km per year on Marion and Prince Edward Island, respectively.

The rate of spread of alien plants on the uninhabited and infrequently visited Prince Edward Island suggests that natural (i.e. non-human) vectors play a significant role in the spread of invasive species (Ryan et al. 2003). Most alien plants on the PEIs possess a wind dispersal mechanism, and there is evidence that wind is a major determinant for the direction in which species disperse on the island (Born et al. 2012; Chown & Avenant 1992; Hedding, Nel & Anderson 2015). However, birds may also act as dispersal vectors (Hughes & Convey 2012; Ryan et al. 2003). For example, *P. annua* is frequently found growing at the entrances to bird burrows, also at higher altitudes, where the plant most likely arrives on the feathers or legs of seabirds (Bergstrom & Smith 1990). In addition to seabirds, seals are thought to aid alien spread by providing suitable high nutrient habitats which are favoured by many of the alien plant species (Haussmann et al. 2013; Ryan et al. 2003). Finally, on Marion Island, the spread of alien plants is most certainly aided by humans. This is obvious from the distribution of some alien plants growing along frequently used paths and their occurrence around island huts frequented by researchers (Figure 3d) (Gremmen & Smith 1999; le Roux et al. 2013).

The role of dispersal ability in driving the spread of alien species is well illustrated by two invasive invertebrates on Marion Island. The winged wasp *Aphidius matricariae* Haliday spread around Marion Island at a rapid rate of between 3 km and 5 km year[-1] (Lee & Chown 2016). In contrast, the invasive springtail *Pogonognathellus flavescens* (L.) spread at a slow rate and has not yet occupied most niches suitable to it because of dispersal limitation (Treasure & Chown 2013). This can be explained by the fact that, in the absence of obvious dispersal vectors, the species is limited by low rates of movement between suitable habitats (Treasure & Chown 2013).

Some alien species on the PEIs may currently still be in a lag phase; that is, their populations are fairly restricted and stationery in extent, but they may spread more extensively in the future (Ryan et al. 2003). Other species are still filling suitable niches that have been uninhabited (Treasure & Chown 2013). For example, in 2013, the range of *Poa pratensis* L. on Marion Island was known to extend between approximately Long Ridge in the north and Bullard Beach in the south. By 2016, the species had also been recorded at the huts at Kildalkey in the south and Mixed Pickle and Swartkops in the west of the island (Steyn, Mathakutha & Greve, pers. obs., 2015&2016), suggesting that the species is still spreading.

It is difficult to ascertain what percentage of the island is invaded. For animal species, this has been poorly quantified (with the exception of individual species, e.g. Treasure & Chown 2013). For plants, it was estimated in 2008 that the overall cover of alien plants on the island was below 5% (Gremmen & Smith 2008). Another study which recorded localities of alien vascular plants in half-minute (app. 926 × 635 m) grid cells showed that by 2011 alien plant species had been recorded in approximately 42% and 53% of grid cells on Marion and Prince Edward Islands, respectively (le Roux et al. 2013). The lower percentage recorded on Marion Island is attributed to the higher altitudes of the island, at which few to no vascular plants grow. However, on Marion Island, almost all high nutrient lowland areas are invaded by one or several alien plant species that have circuminsular distributions (Gremmen & Smith 2008; le Roux et al. 2013).

Impacts

Invasive species are often considered to be a threat to native diversity and ecosystems because of the impacts they have on said diversity. Therefore, it is essential to understand what underlies these impacts (Levine et al. 2003). For the sub-Antarctic region, the impacts of introduced vertebrates on the biodiversity of islands have been most extensively studied (Frenot et al. 2005; Shaw 2013). Nevertheless, impacts of invasives on most indigenous taxa, with the exception perhaps of seabirds, are not well understood (Table 1) (McGeoch et al. 2015). On the PEIs, the impacts of only a handful of aliens have been investigated (Table 1), with the greatest focus being given to the study of mouse invasions (also see review in Chown & Froneman 2008a). The impacts of mouse invasions are fairly well understood for not only seabird populations but also some invertebrates, plants and ecosystem functioning and structure. In contrast, the impact of most other invasive species has received little attention.

Impacts of vertebrate predators on native biota are particularly severe because these species fulfil functional roles otherwise absent from the islands (Convey 2011; Courchamp, Chapuis & Pascal 2003). Organisms indigenous to the islands evolved in the absences of vertebrate predators and are thus poorly suited to defend themselves against these predators (Convey 2011; Frenot et al. 2005). Although the diet of cats on Marion Island included some mice, it mainly comprised seabirds, especially burrowing petrels (van Aarde 1980; van Rensburg 1983). Indeed, before cats were eradicated from Marion Island, their impact on seabird populations, especially burrowing species, was particularly severe, leading to decreased breeding success and reductions in their populations (van Rensburg 1983), and the extinction of at least one species on Marion Island (Bester et al. 2002). Furthermore, although seabird populations have seen some recovery after the reduction and eventual extirpation of cat populations on Marion Island (Cooper & Fourie 1991; Cooper et al. 1995), cat impacts have been long-lasting. Populations of burrowing petrels, which were particularly impacted by cats on Marion Island, have not recovered two decades after the eradication of the cats (Cerfonteyn & Ryan 2016; Smith 2008).

The house mouse, which occurs only on Marion Island, is probably the best-researched invader of the PEIs. One of its most obvious effects on the Marion Island system is through the extensive burrows that the animals construct (Figure 3e) (Avenant & Smith 2003; Eriksson & Eldridge 2014; Phiri, McGeoch & Chown 2009). Mouse activity causes direct and indirect damage to plants. The most notable direct damage to plants is to the keystone cushion plant, *Azorella selago* Hook. f. (Figure 3f) (Phiri et al. 2009). Mice burrow into the cushions, with the damage to cushions ranging from partial damage through entrance burrows to complete excavation, and thus disintegration of cushions (Phiri et al. 2009). The microclimate experienced by the cushion plants is also impacted by burrowing activities (Eriksson & Eldridge 2014). In addition, mice consume significant quantities of seeds, especially of the sedge *Uncinia compacta* R. Br., which has suffered significant population declines (Chown & Smith 1993; Smith & Steenkamp 1990). The impact of burrows on plants may also be indirect. Estimates of the amount of sediment removed through mouse burrows range between 2.4 t ha^{-1} and 8.4 t ha^{-1}, with the latter probably being the more accurate estimate (Eriksson & Eldridge 2014). The large amount of sediment movement affects soil microclimate, with excavated soils being drier and warmer, and surface stability, resulting in higher levels of erosion (Eriksson & Eldridge 2014). Such changes to the soils are likely to impact the establishment of seedlings and the stability of plant communities (Eriksson & Eldridge 2014; Gremmen 1981), with likely knock-on effects on invertebrate assemblages (Hugo et al. 2004).

Invertebrates make up the bulk of the diet of the house mouse on Marion Island (Smith et al. 2002). The impacts of mice on invertebrates have mainly been assessed by comparing invertebrate assemblages on Marion Island and Prince Edward Island (free from mice) and by using mouse exclosures. It has been estimated that mice on Marion Island consume up to 2% of the invertebrate standing stock daily, and that mice predation has driven changes in population densities and mating strategies of some invertebrate species and a decline in the body size of weevil prey species (Chown & Smith 1993; Crafford & Scholtz 1987; Treasure & Chown 2014). Such predation is considered to have far-reaching effects on nutrient cycling (Crafford 1990b; Smith & Steenkamp 1990). A short-term (4-year) mouse exclosure experiment failed to reveal a significant effect of predation on invertebrate densities, although this could be because of poor statistical power of the analyses (van Aarde, Ferreira & Wassenaar 2004). By contrast, analyses of mouse and invertebrate densities over 40 years revealed significant and substantial impacts of mouse predation on key invertebrate populations (up to two orders of magnitude decline) (McClelland et al. in review).

More recently, it has been shown that mice are predating on albatross chicks on Marion Island (Jones & Ryan 2010). This behaviour was first recorded in 2003, but appears to be increasing in extent (it has been observed in colonies across the island) and frequency, and is resulting in chick mortalities on Marion Island (Dilley et al. 2016). This may partially be a consequence of the eradication of cats from the island, resulting in increased mouse populations which then give rise to mouse predatory behaviour (Dilley et al. 2016; Jones & Ryan 2010; Wanless et al. 2007). Predation by mouse on neighbouring Gough Island, first documented not long before the first record of mouse predation on Marion Island, has resulted in extensive losses of albatross and petrel chicks, causing populations of some seabirds to become unsustainable (Wanless et al. 2007). Seabirds on sub-Antarctic Islands tend to show some degree of naivety, probably as a result of a lack of natural predators in their native breeding habitat, possibly making them especially prone to mouse predation.

On islands, invasive plants cause fewer extinctions than invasive vertebrates (Sax & Gaines 2008). Nonetheless, plant invaders often alter whole habitats, thereby reducing populations of native species and affecting ecosystem processes and function (Reaser et al. 2007). The effects of invasive plants on the native diversity of the PEIs are poorly studied. Invasives occur across most vegetation complexes on the island (Gremmen 1997), with their effects most pronounced at low altitudes and in areas with high nutrient input (Gremmen & Smith 2008). Most of the invasive plants reach dominance in one or more vegetation complexes, in effect smothering the native vegetation and negatively impacting the diversity and abundance of native plants (Gremmen 1997; Gremmen et al. 1998). In some vegetation types, this may be because the leaves of the invasive grasses decompose more slowly than those of the indigenous species; thus the dry leaves accumulate and, in effect, block out the light (Gremmen 1997). One of the only studies on the impacts of an invasive plant concluded that the invasive grass, *A. stolonifera*, does not pose any threats of extinction to native species of plants or insects, although it forms dense stands which negatively affects the abundance of many native species (Figure 3g) (Gremmen et al. 1998). However, *A. stolonifera* is leading to the loss of native plant communities and changes in abiotic conditions (most especially light availability) for the bryophytes that grow in these communities, thereby fundamentally changing community structure (Gremmen et al. 1998). Although the impacts of invasive plants on Prince Edward Island biodiversity are only partially understood, many of the Prince Edward Island plant invaders are known to be invasive elsewhere (McGeoch et al. 2015) where they negatively impact native biodiversity.

Most studies on the impacts of alien species have focussed on impacts on individual native taxa. However, the knock-on effect of some impacts is evident in taxa that are indirectly affected by alien species. The insectivorous common sheathbill, *Chionis minor* Hartlaub, has seen significant declines, apparently because of invertebrate predation by mice (Huyser, Ryan & Cooper 2000). Indirect impacts of alien species can also be because of changes in the function and structure of ecosystems (Raymond et al. 2011). Of all alien taxa currently present on Marion Island, mice have the most obvious impact on ecosystem structure and functioning. Ecosystem functioning is most heavily impacted through changes in nutrient cycling because of the predation of macroinvertebrates that are instrumental in the decomposition process of the island (Smith 2008; Smith & Steenkamp 1990). This results in reduced rates of mineralisation and concomitant changes in the formation of peats and soils, which, in turn, affects vegetation succession and thus vegetation structure (Smith 2008).

Although some alien decomposers are present on the island, they do not necessarily fulfil identical functions to the indigenous macroinvertebrates that are heavily preyed upon by mice. For example, the introduced slug *D. panormitanum* is a decomposer; however, it mineralises N at a considerably slower rate than a comparative indigenous decomposer, the moth *Pringleophaga marioni* Viette (Smith 2007). In addition, mice preferentially feed on the indigenous *P. marioni* over the introduced *D. panormitanum*. This means that the build-up and the quality of peat on Marion Island are fundamentally altered by the higher abundance of the slug than *P. marioni* (Smith 2007).

While the known impacts of alien species are predominantly negative, alien species can also perform important ecosystem functions. Work by Hänel and Chown (1998) indicates that on Marion Island the introduced chironomid midge *Lymnophyes minimus* Meigen, also a litter feeder, is partly fulfilling the role of the heavily predated *P. marioni*, a keystone litter decomposer on the island (Crafford 1990a).

Invasions and climate change

It has been suggested that the synergy between two or more extinction drivers can exacerbate the risks to biodiversity that one of these drivers would otherwise pose alone (Brook, Sodhi & Bradshaw 2008). On the PEIs, such a synergy, in this case between alien invasions and climate change, is likely. The PEIs have experienced significant decreases in rainfall and increases in temperatures since climate monitoring on Marion Island started in 1951 (le Roux & McGeoch 2008; Smith & Steenkamp 1990). Climate change, especially the amelioration of temperatures, is expected to increase the vulnerability of sub-Antarctic systems to alien invasions (Convey 2011; Frenot et al. 2005). Indeed, the range expansions of several alien plant species on Marion Island (Chown et al. 2013; le Roux et al. 2013) and increases in mouse populations and their impacts (Chown & Smith 1993) are thought to have been driven by a warmer climate. Furthermore, physiological experiments on several invertebrate taxa indicate that alien species often display higher fitness or physiological tolerances than indigenous species under the warmer and drier conditions that may be expected under climate change (e.g. Chown et al. 2007; Janion et al. 2010; Slabber et al. 2007). Thus, there is strong evidence that climate change will exacerbate invasions and their impacts on Marion Island: alien species benefit from a milder climate, while native species decline, either because there are fewer areas where they can survive or because they are outcompeted by alien species (Chown & Froneman 2008a).

Taxonomic biases and gaps in knowledge

Globally, a strong taxonomic bias in invasion research exists (Pyšek et al. 2008). This is also true for the PEIs (Table 1), where the impacts of vertebrates are best understood, followed by plants and insects (Table 1). For these taxa, the listed species (Table 1) present a fairly comprehensive account of aliens on the islands. However, other taxa have generally been poorly studied, and therefore little is known about species in these taxa. For example, while there is a high diversity of lichens on the islands (Øvstedal & Gremmen 2011), records of new species are still being added for the island (N.J. Gremmen, pers. comm., 2015). None of the lichens that have been identified on the PEIs are known to have been introduced by humans. Similarly, no alien algae have been described (Van de Vijver, Gremmen & Smith 2008; van Staden 2011), although there is evidence that these groups are being transported to the region (Huiskes et al. 2014).

Microbes are not visible, yet they are ubiquitous and easily transported organisms (Cowan et al. 2011). The microbiology of the PEIs is poorly known (Sanyika, Stafford & Cowan et al. 2012), but it is highly probable that a number of microbes have been introduced to the PEIs by humans (see discussion in Cowan et al. 2011; Hughes, Cowan & Wilmotte 2015). The only alien microbe known to have been introduced to the PEIs is the fungal pathogen *Botryotina fuckeliana* (de Bary) Whetzel on Marion Island (Kloppers & Smith 1998). It is thought to have been introduced in fresh produce and attacks the Kerguelen cabbage, *Pringlea antiscorbutica* R. Br. (Kloppers & Smith 1998). Other possible pathways by which microbes may have been introduced to the PEIs are diverse (Cowan et al. 2011) and could include footwear, scientific equipment, foods with yeasts, human excrement or the soils that were brought from South Africa to Marion Island for greenhouse experiments shortly after annexation (La Grange 1954). It is thus essential to understand what microbes are being introduced to the islands by which pathways, how readily they spread on the islands and what their impacts are.

Invasion policies and management

Globally, the territories of the Antarctic region, including the PEIs, have some of the strictest regulations governing introduced species (de Villiers et al. 2006; Hughes & Pertierra 2016). The South African Department of Environmental Affairs manages alien species on the PEIs, with most efforts directed at aliens on Marion Island which has more introduced species and a permanent scientific contingent.

The best way to minimise invasions and their impacts is to prevent them (Simberloff et al. 2013). Accordingly, the largest efforts in alien control in the Antarctic region are concentrated on reducing introductions. In contrast, monitoring and managing existing invasive plants receive less attention (McGeoch et al. 2015). This is also the case for the PEIs. The Prince Edward Islands Management Plan (Department of Environmental Affairs Directorate: Antarctica and Islands 2010) deals most stringently with the introduction stage of the invasion process. The management plan currently sets detailed regulations with the aim of preventing introductions of alien species to the islands. A stringent vessel inspection is required prior to departure to the islands, and the vessel has many trapping techniques deployed for animal (rodent and invertebrate) propagules. Likewise, cargo, personal equipment and clothing are all inspected prior to departure, en route to the island (Figure 3h), and on landing at the scientific station, where all cargo is delivered prior to field deployment to other parts of the island (Department of Environmental Affairs Directorate: Antarctica and Islands 2010). In addition, there is a ban on the import of organic materials, soils or rocks to the islands, and the release of ballast water or food waste from the supply vessel is prohibited in the proximity of the islands (Department of Environmental Affairs Directorate: Antarctica and Islands 2010). While these biosecurity regulations probably contribute to lower introduction rates to the PEIs, they are not always strictly enforced and do not entirely preclude the introduction of new species.

Once an alien species has arrived on the PEIs, it must be detected to be controlled. Detection of many species is dependent on scientists with knowledge of the taxa on the island spotting new species. Currently, no specific surveys for the purpose of detection take place; instead, detection is dependent on organisms being recorded during other scientific or monitoring activities. Identification can be aided by genetic methods: in cases where species identification is difficult, DNA barcoding or similar methods can be employed to aid identification of new organisms (Chown, Sinclair & van Vuuren 2008).

Some alien species, especially those first recorded at the base, have been eradicated from Marion Island upon first detection before they had the opportunity to establish (Hänel, Chown & Davies 1998). Other species have probably disappeared from the islands because they were not able to establish (e.g. Gremmen & Smith 1999), or through a combination of eradication efforts and climatic conditions unfavourable to the species (e.g. Cooper et al. 1992). Currently, there are ongoing efforts to eradicate some alien species on Marion Island (Table 2). These efforts are especially focussed on alien plants with restricted distributions, that is, species that have established but not spread extensively, with the preferred method of control being targeted herbicide use (DEA: Natural Resources Management Programme et al. 2012). Because they have restricted ranges and relatively low rates of dispersal, plants are easier to eradicate than most animal taxa. At present eradication efforts of restricted-range species are effective in preventing these species from spreading, and some species have been eradicated (Tables 1 and 2).

The eradication of species that have established and become widespread is difficult, and accordingly little management of

TABLE 2: List of alien species that are listed in *National Environmental Management: Biodiversity Act* (NEM:BA) or those subject to eradication or control efforts on Marion Island.

Species name	NEM:BA: Category/Area	Species intervention	Effectiveness of intervention	Notes	Source
Vascular plants					
Luzula cf. *multiflora* (Juncaceae)	1a Prince Edward and Marion Islands	Chemical control	5	(60% successful after first application)	(DEA: Natural Resources Management Programme et al., 2012)
Agrostis gigantea (Gramineae)	1a Prince Edward and Marion Islands	Chemical control including spraying	5	(90% successful – still monitoring effects of second application)	(DEA: Natural Resources Management Programme et al., 2012)
Agropyron repens (Gramineae)	1a Prince Edward and Marion Islands	Chemical control including spraying	5	(100% successful – no regrowth observed)	(DEA: Natural Resources Management Programme et al., 2012)
Rumex acetosella (Polygonaceae)	1a Prince Edward and Marion Islands	Manual removal followed by chemical control to reduce the standing biomass	5	(40% successful – still monitoring effects of second application)	(DEA: Natural Resources Management Programme et al., 2012)
Festuca rubra (Gramineae)	1a Prince Edward and Marion Islands	Chemical control including spraying	5	(80% successful – still monitoring after effects)	(DEA: Natural Resources Management Programme et al., 2012)
Alopecurus geniculatus (Gramineae)	1a Prince Edward and Marion Islands	Manual removal	5	(No longer present)	(DEA: Natural Resources Management Programme et al., 2012)
Stellaria media (Caryophyllaceae)	1a Prince Edward Island. 1b Marion Island	Chemical control including spraying	5		(DEA: Natural Resources Management Programme et al., 2012)
Holcus lanatus (Gramineae)	Not listed in NEM:BA	Manual removal	5	(No longer present since 2012)	(DEA: Natural Resources Management Programme et al., 2012)
Cerastium fontanum (Caryophyllaceae)	1b Prince Edward and Marion Islands	None	1		
Poa pratensis (Gramineae)	1a Prince Edward Island. 1b Marion Island	None	1		
Agrostis stolonifera (Gramineae)	1a Prince Edward Island. 1b Marion Island	None	1	(Considered unfeasible)	(Gremmen and Smith 2008)
Sagina procumbens (Caryophyllaceae)	1b Prince Edward and Marion Islands	None	1	(Considered unfeasible)	(Ryan et al. 2003)
Agrostis castellana (Gramineae)	1a Prince Edward Island. 1b Marion Island	None	1		
Invertebrates					
Porcellio scaber (Isopoda, Porcellionidae)	Not listed in NEM:BA	Pesticide use	5		(DEA: Natural Resources Management Programme et al., 2012)
Mammals					
Mus musculus (Muridae)	1b for off-shore islands (Marion Island)	Rodent traps (e.g. snap traps), poison bait stations. Mainly around research station and huts	2		(DEA: Natural Resources Management Programme et al., 2012)

NEM:BA categories, current interventions and the effectiveness of interventions for control or eradication of each species are indicated.

NEM:BA category: 1a, Listed species which must be combatted or eradicated; 1b, Listed species which must be controlled. Effectiveness 1: No species management programmes or assessment of the species; 2: Interventions are present but having no discernible impact; 5: For species targeted for eradication, progress towards eventual eradication.

these species takes place. Indeed, it has been argued that the removal of many plant species and most invertebrates from sub-Antarctic Islands such as the PEIs is impractical (Convey 2011). The eradication of cats from Marion Island constitutes the most ambitious attempt at eradicating a species on the island, as well as the only eradication of a widespread and established species. The biological control agent feline panleukopenia virus was introduced to Marion Island in 1977 to reduce cat populations. The remaining, but diminished, cat population was then exterminated through hunting (Bester et al. 2002). Interestingly, one other biological control agent was introduced to Marion Island accidentally, which has led to the inadvertent control of an invasive aphid. The wasp *A. matricariae*, a parasitoid of aphids, was accidentally introduced to Marion Island in cargo between 2001 and 2003 (Lee et al. 2007) and has rapidly spread across the island where it appears to be effectively controlling populations of the invasive aphid *Rhopalosiphum padi* (L.) (Lee & Chown 2016).

More recently, investigations into the feasibility of eradicating mice from Marion Island by chemical means have been conducted. Rodent eradications have been successful on other islands (Howald et al. 2007). Such an operation on Marion Island would be expensive and logistically complex given the difficult terrain and weather, but not impossible (e.g. Springer 2016).

Finally, while humans are the main agents of species introductions to the PEIs, it must be kept in mind that new species may arrive to the islands from their native range by natural means and that these species should thus not be the subject of control measures as they are colonisations (Lee, Terauds & Chown 2014). Examples of what are thought to be natural dispersal events to Marion Island include the globally invasive diamondback moth *P. xylostella* (Chown & Avenant 1992) and the shrub *Ochetophila trinervis* (Gillies ex Hook.) Poepp. ex Endl. (Gremmen & Smith 2008; Kalwij et al. unpublished data). To determine the mode of arrival of an alien species at the PEIs may be difficult. However, if a new species is first detected at a considerable distance from sites of high human activities on the islands, it may indicate that the species arrived by natural means (Chown &

Avenant 1992; Gremmen & Smith 1999; Lee et al. 2014). In some cases, genetic tools may be used to narrow down the region of origin of the species, providing further clues as to whether the species arrived by natural or by human-mediated means (Chown & Avenant 1992; Lee et al. 2014). However, where uncertainty exists as to whether a species is a natural colonist, a precautionary approach has been recommended, where a species is eradicated before it spreads (Hughes & Convey 2012).

Implications for the national status report on biological invasions

The current pathways of human-assisted introductions of alien species to the PEIs are straightforward because all species arrive by ship. Thus, pathways can be narrowed down to unintentional introductions, mainly through cargo, building material and clothes. For those organisms that have established on the islands, the status of invasion is well known for plants and the one vertebrate on the island, but only for some insect species. Not all introduced species establish and spread, but those that do have spread extensively across both islands. The mouse has the largest impact on the biodiversity of the PEIs – impacts span trophic levels and affects island-scale ecosystem functioning. The impacts of invasive plants are thought to be high in some vegetation types, while they are negligible in others. In contrast, impacts of invertebrates are poorly understood.

Current control interventions of some range-restricted alien species on Marion Island are effective at preventing their populations from spreading, although interventions for widely distributed species are absent. No interventions are being implemented on Prince Edward Island, which has received a number of alien species by natural vectors from Marion Island.

In summary, Marion Island is fairly heavily impacted by alien species, with extensive effects recorded for native species and ecosystem functioning. Prince Edward Island is currently invaded by a comparatively small number of species. While no works on the impacts of invasive species have been conducted on Prince Edward Island, the impacts are likely to be significantly less than those on Marion Island on account of the number and the identity of the invasive species found on Prince Edward Island.

Future prospects and conclusion

Along with climate change, the introduction of alien species probably entails the greatest threat to the biodiversity of the PEIs. Despite existing biosecurity regulations, transport of nonindigenous propagules to the region continues (Houghton et al. 2016). Therefore, to ensure the continued conservation of the PEIs, it is of utmost importance that alien introductions and invasions are kept to a minimum. Prevention remains the easiest, and cheapest, way to keep alien species off the PEIs (Hughes & Pertierra 2016). It must

thus be ensured that strict biosecurity regulations aimed at reducing propagule and live individual entrainment (Department of Environmental Affairs Directorate: Antarctica and Islands 2010) are implemented by the Department of Environmental Affairs. Furthermore, it is critical that these regulated biosecurity protocols are enforced and that participants and managers of voyages to the islands are aware of the requirements. In addition, while the focus of this article has been on terrestrial organisms, the threats of marine and, perhaps to a lesser extent, freshwater invasions must not be overlooked (Table 4-A1) (Dartnall & Smith 2012; Lee & Chown 2007).

While efforts at controlling range-limited species have managed to contain the spread of a number of alien species on Marion Island, regular field surveys are important for the detection and recording of new occurrences or recent range expansions of alien species. Such activities would be most effective around sites of high human activities to where alien species are most likely to be introduced and spread from (Hughes & Pertierra 2016; le Roux et al. 2013). This allows for early detection and the implementation of rapid response measures to contain spread (Hughes & Pertierra 2016; Huntley 1971). On Marion Island, regular surveys aimed at detecting new arrivals to the island were conducted between the early 1980s and 2011 (Lee & Chown 2016). These surveys no longer take place. If such surveys by field experts should be regularly undertaken and be a standard component of the formal management plan of the islands, it would provide better continuity.

A current challenge to undertaking such surveys is the recent loss of capacity of entomological and other invertebrate knowledge, and possibly the imminent loss of bryophyte knowledge, from the scientific community within the South African National Antarctic Programme (SANAP) community. A lack of capacity to identify species from these taxonomic groups can limit the ability to detect new arrivals to the islands. Thus, retention of taxonomic knowledge within the SANAP programme is essential for continued alien species management. Alternatively, efforts to build capacity should be supported, such as inviting international experts to train new cohorts of researchers.

In addition, monitoring of widespread alien species should be a priority to ensure that management policies incorporate control or mitigation measures where possible (Hughes & Pertierra 2016). Currently, monitoring activities of widespread aliens is conducted on an *ad hoc* basis every few years as part of the research programmes of researchers in SANAP (e.g. Gremmen 1975; le Roux et al. 2013; Ryan et al. 2003). A more formalised approach may be desirable.

Finally, it is important that scientific studies on the extent and impacts of introduced species on the PEIs should continue. To effectively channel efforts for invasion management, it is essential to understand which invasive species pose the greatest threat to diversity, what their precise impacts are, and to identify in which systems such impacts are most pronounced (Gurevitch & Padilla 2004). Research on biological invasions on the PEIs has enjoyed steady attention in recent decades. A survey of a number of papers about alien and invasive species on the PEIs shows a steady increase from the early 1990s to the late 2000s, with a recent decline since 2009 (Figure 4). Given the ongoing threat of invasions to the PEIs, and the need to better understand and manage their impacts, it is imperative that the invasion biology of the PEIs remains an important focus of research on the islands. For example, a risk assessment of which species might become invasive would assist in preventing future invasions. Such risk assessments can be based on climatic suitability or on species traits known to be predictors of invasive capability. While the introduction stage of invasions is reasonably well understood for the PEIs, less attention has been given to the process of alien establishment and spread. In addition, the impacts of many of the alien species need to be better understood. Finally, basic science on the biology of native species and ecosystem processes on the islands needs to continue. Without such an understanding, the impacts of invasions cannot be fully appreciated.

Acknowledgements

Two anonymous reviewers provided helpful comments on earlier versions of this review. The authors thank Peter le Roux for discussions.

The South African NRF-funded National Antarctic Programme is thanked for providing financial support to M.G. (Grant No. 93065).

Competing interests

The authors declare that they have no financial or personal relationship(s) that may have inappropriately influenced them in writing this article.

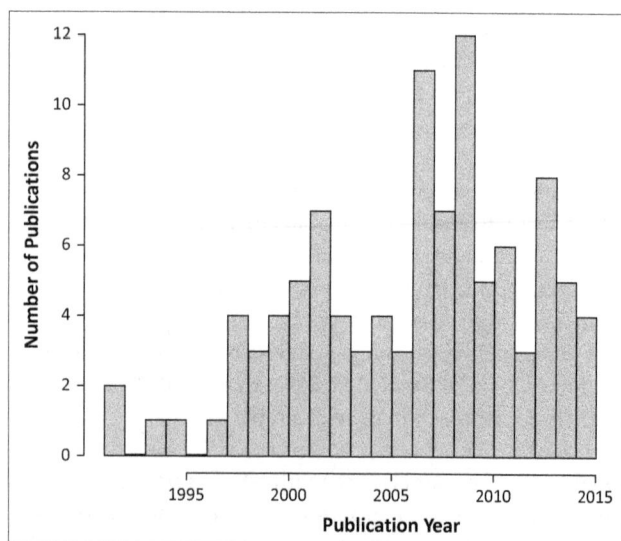

For methods, see Appendix 1, Box 1.

FIGURE 4: Annual count of ISI-indexed publications dealing with alien and invasive species on the Prince Edward Islands between 1990 and 2015.

Authors' contributions

M.G. led the writing, with contributions from R.M., C.S. and S.L.C. M.G., R.M. and C.S. collated information on current invasions with input from S.L.C.

References

Avenant, N.L. & Smith, V.R., 2003, 'The microenvironment of house mice on Marion Island (sub-Antarctic)', *Polar Biology* 26(2), 129–141.

Bergstrom, D.M. & Smith, V.R., 1990, 'Alien vascular flora of Marion and Prince Edward Islands: New species, present distribution and status', *Antarctic Science* 2(4), 301–308. https://doi.org/10.1017/S0954102090000426

Bester, M.N., Bloomer, J.P., van Aarde, R.J., Erasmus, B.H., van Rensburg, P.J.J., Skinner, J.D. et al., 2002, 'A review of the successful eradication of feral cats from sub-Antarctic Marion Island, Southern Indian Ocean', *South African Journal of Wildlife Research* 32(1), 65–73.

Blackburn, T.M., Essl, F., Evans, T., Hulme, P.E., Jeschke, J.M., Kühn, I. et al., 2014, 'A unified classification of alien species based on the magnitude of their environmental impacts', *PLoS Biology* 12(5), e1001850. https://doi.org/10.1371/journal.pbio.1001850

Blackburn, T.M., Pyšek, P., Bacher, S., Carlton, J.T., Duncan, R.P., Jarošík, V. et al., 2011, 'A proposed unified framework for biological invasions', *Trends in Ecology & Evolution* 26(7), 333–339. https://doi.org/10.1016/j.tree.2011.03.023

Boelhouwers, J.C., Meiklejohn, K.I., Holness, S.D. & Hedding, D.W., 2008, 'Geology, geomorphology and climate change', in S.L. Chown & P.W. Froneman (eds.), *The Prince Edward Islands: Land-sea interactions in a changing ecosystem*, pp. 65–96, SUN PReSS, Stellenbosch.

Born, C., le Roux, P.C., Spohr, C., McGeoch, M.A. & Jansen van Vuuren, B., 2012, 'Plant dispersal in the sub-Antarctic inferred from anisotropic genetic structure', *Molecular Ecology* 21(1), 184–194. https://doi.org/10.1111/j.1365-294X.2011.05372.x

Brook, B.W., Sodhi, N.S. & Bradshaw, C.J.A., 2008, 'Synergies among extinction drivers under global change', *Trends in Ecology & Evolution* 23(8), 453–460. https://doi.org/10.1016/j.tree.2008.03.011

Cerfonteyn, M. & Ryan, P.G., 2016, 'Have burrowing petrels recovered on Marion Island two decades after cats were eradicated? Evidence from sub-Antarctic skua prey remains', *Antarctic Science* 28(1), 51–57. https://doi.org/10.1017/S0954102015000474

Chown, S.L. & Avenant, N., 1992, 'Status of *Plutella xylostella* at Marion Island six years after its colonisation', *South African Journal of Antarctic Research* 22(1&2), 37–40.

Chown, S.L. & Froneman, P.W., 2008a, 'Conclusion: Change in terrestrial and marine systems', in S.L. Chown & P.W. Froneman (eds.), *The Prince Edward Islands: Land-sea interactions in a changing ecosystem*, pp. 351–372, African SunMedia, Stellenbosch.

Chown, S.L. & Froneman, P.W. (eds.), 2008b, *The Prince Edward Islands: Land-Sea Interactions in a Changing Ecosystem*, African SunMedia, Stellenbosch.

Chown, S.L., Gremmen, N.J.M. & Gaston, K.J., 1998, 'Ecological biogeography of southern ocean islands: Species-area relationships, human impacts, and conservation', *American Naturalist* 152(4), 562–575. https://doi.org/10.1086/286190

Chown, S.L., Huiskes, A.H.L., Gremmen, N.J.M., Lee, J.E., Terauds, A., Crosbie, K. et al., 2012, 'Continent-wide risk assessment for the establishment of nonindigenous species in Antarctica', *Proceedings of the National Academy of Sciences of the United States of America* 109(13), 4938–4943. https://doi.org/10.1073/pnas.1119787109

Chown, S.L., Hull, B. & Gaston, K.J., 2005, 'Human impacts, energy availability and invasion across Southern Ocean Islands', *Global Ecology and Biogeography* 14(6), 521–528. https://doi.org/10.1111/j.1466-822x.2005.00173.x

Chown, S.L., le Roux, P.C., Ramaswiela, T., Kalwij, J.M., Shaw, J.D. & McGeoch, M.A., 2013, 'Climate change and elevational diversity capacity: Do weedy species take up the slack?', *Biology Letters* 9(1), 20120806. https://doi.org/10.1098/rsbl.2012.0806

Chown, S.L., Sinclair, B.J. & Vuuren, B.J., 2008, 'DNA barcoding and the documentation of alien species establishment on sub-Antarctic Marion Island', *Polar Biology* 31(5), 651–655. https://doi.org/10.1007/s00300-007-0402-z

Chown, S.L., Slabber, S., McGeoch, M.A., Janion, C. & Leinaas, H.P., 2007, 'Phenotypic plasticity mediates climate change responses among invasive and indigenous arthropods', *Proceedings of the Royal Society B: Biological Sciences* 274(1625), 2531–2537. https://doi.org/10.1098/rspb.2007.0772

Chown, S.L. & Smith, V.R., 1993, 'Climate change and the short-term impact of feral house mice at the sub-Antarctic Prince Edward Islands', *Oecologia* 96(4), 508–516. https://doi.org/10.1007/BF00320508

Chown, S.L., Spear, D., Lee, J.E. & Shaw, J.D., 2009, 'Animal introductions to southern systems: Lessons for ecology and for policy', *African Zoology* 44(2), 248–262. https://doi.org/10.3377/004.044.0213

Christy, M.T., Savidge, J.A. & Rodda, G.H., 2007, 'Multiple pathways for invasion of anurans on a Pacific island', *Diversity and Distributions* 13(5), 598–607. https://doi.org/10.1111/j.1472-4642.2007.00378.x

Chwedorzewska, K.J., Giełwanowska, I., Olech, M., Molina-Montenegro, M.A., Wódkiewicz, M. & Galera, H., 2015, 'Poa annua L. in the maritime Antarctic: An overview', Polar Record 51(6), 637–643. https://doi.org/10.1017/S0032247414000916

Convey, P., 2011, 'Antarctic terrestrial biodiversity in a changing world', Polar Biology 34(11), 1629–1641. https://doi.org/10.1007/s00300-011-1068-0

Convey, P., Frenot, Y., Gremmen, N.J.M. & Bergstrom, D.M., 2006, 'Biological invasions', in D.M. Bergstrom, P. Convey & A.H.L. Huiskes (eds.), Trends in Antarctic terrestrial and limnetic ecosystems: Antarctica as a global indicator, pp. 193–220, Springer, Dordrecht.

Cooper, J., 2008, 'Human history', in S.L. Chown & P.W. Froneman (eds.), The Prince Edward Islands: Land-sea interactions in a changing ecosystem, pp. 331–350, SUN PReSS, Stellenbosch.

Cooper, J., Bester, M.N., Chown, S.L., Crawford, R.J.M., Daly, R., Heyns, E. et al., 2009, 'Biological survey of the Prince Edward Islands, December 2008', South African Journal of Science 105(7&8), 317–320.

Cooper, J., Crafford, J.E. & Hecht, T., 1992, 'Introduction and extinction of brown trout (Salmo trutta L.) in an impoverished subantarctic stream', Antarctic Science 4(1), 9–14. https://doi.org/10.1017/S095410209200004X

Cooper, J. & Fourie, A., 1991, 'Improved breeding success of great-winged petrels Pterodroma macroptera following control of feral cats Felis catus at subantarctic Marion Island', Bird Conservation International 1(2), 171–175. https://doi.org/10.1017/S0959270900002033

Cooper, J., Marais, A.V.N., Bloomer, J.P. & Bester, M.N., 1995, 'A success story: Breeding of burrowing petrels (Procellariidae) before and after the eradication of feral cats Felis catus at subantarctic Marion Island', Marine Ornithology 23(1), 33–37.

Courchamp, F., Chapuis, J.L. & Pascal, M., 2003, 'Mammal invaders on islands: Impact, control and control impact', Biological Reviews 78(3), 347–383. https://doi.org/10.1017/S1464793102006061

Cowan, D.A., Chown, S.L., Convey, P., Tuffin, M., Hughes, K., Pointing, S. et al., 2011, 'Non-indigenous microorganisms in the Antarctic: Assessing the risks', Trends in Microbiology 19(11), 540–548. https://doi.org/10.1016/j.tim.2011.07.008

Crafford, J.E., 1986, 'A case study of an alien invertebrate (Limnophyes pusillus, Diptera, Chironomidae) introduced on Marion Island: Selective advantages', South African Journal of Antarctic Research 16(3), 115–117. https://doi.org/10.1016/0006-3207(87)90119-4

Crafford, J.E., 1990a, 'Patterns of energy flow in populations of the dominant insect consumers on Marion Island', DPhil thesis, Department of Entomology, University of Pretoria.

Crafford, J.E., 1990b, 'The role of feral house mice in ecosystem functioning on Marion Island', in K.R. Kerry & G. Hempel (eds.), Antarctic ecosystems. Ecological change and conservation, pp. 359–364, Springer-Verlag, Berlin.

Crafford, J.E. & Chown, S.L., 1990, 'The introduction and establishment of the diamondback moth (Plutella xylostella L., Plutellidae) on Marion Island', in K.R. Kerry & G. Hempel (eds.), Antarctic ecosystems. Ecological change and conservation, pp. 354–358, Springer-Verlag, Berlin.

Crafford, J.E. & Scholtz, C.H., 1987, 'Quantitative differences between the insect faunas of Sub-Antarctic Marion and Prince Edward Islands: A result of human intervention?', Biological Conservation 40(4), 255–262.

Dartnall, H.J.G. & Smith, V.R., 2012, 'Freshwater invertebrates of sub-Antarctic Marion Island', African Zoology 47(2), 203–215. https://doi.org/10.3377/004.047.0207

de Villiers, M.S. & Cooper, J., 2008, 'Conservation and management', in S.L. Chown & P.W. Froneman (eds.), The Prince Edward Islands: Land-sea interactions in a changing ecosystem, pp. 301–330, SUN PReSS, Stellenbosch.

de Villiers, M.S., Cooper, J., Carmichael, N., Glass, J.P., Liddle, G.M., McIvor., E. et al., 2006, 'Conservation management at Southern Ocean Islands: Towards the development of best-practice guidelines', Polarforschung 75(2/3), 113–131.

DEA: Natural Resources Management Programme, DEA: Southern Oceans and Antarctic Support, DEA: Environmental Impact Evaluation & DST-NRF Centre of Excellence for Invasion Biology: Stellenbosch University, 2012, Eradication, monitoring and control of alien and invasive alien species on Marion Island, Unpublished.

Department of Environmental Affairs Directorate: Antarctica and Islands, 2010, Prince Edward Islands Management Plan Version 0.2, Department of Environmental Affairs, Directorate: Antarctica and the Islands.

Dilley, B.J., Schoombie, S., Schoombie, J. & Ryan, P.G., 2016, '"Scalping" of albatross fledglings by introduced mice spreads rapidly at Marion Island', Antarctic Science 28(2), 73–80. https://doi.org/10.1017/S0954102015000486

Donlan, C.J. & Wilcox, C., 2008, 'Diversity, invasive species and extinctions in insular ecosystems', Journal of Applied Ecology 45(4), 1114–1123. https://doi.org/10.1111/j.1365-2664.2008.01482.x

Eriksson, B. & Eldridge, D.J., 2014, 'Surface destabilisation by the invasive burrowing engineer Mus musculus on a sub-Antarctic island', Geomorphology 223, 61–66. https://doi.org/10.1016/j.geomorph.2014.06.026

Faulkner, K.T., Robertson, M.P., Rouget, M. & Wilson, J.R.U., 2016, 'Understanding and managing the introduction pathways of alien taxa: South Africa as a case study', Biological Invasions 18(1), 73–87. https://doi.org/10.1007/s10530-015-0990-4

Fraser, C.I., Nikula, R., Spencer, H.G. & Waters, J.M., 2009, 'Kelp genes reveal effects of subantarctic sea ice during the Last Glacial Maximum', Proceedings of the National Academy of Sciences 106(9), 3249–3253. https://doi.org/10.1073/pnas.0810635106

Frenot, Y., Chown, S.L., Whinam, J., Selkirk, P.M., Convey, P., Skotnicki, M. et al., 2005, 'Biological invasions in the Antarctic: Extent, impacts and implications', Biological Reviews 80(1), 45–72. https://doi.org/10.1017/S1464793104006542

Frenot, Y., Gloaguen, J.-C. & Massé, L., 2001, 'Human activities, ecosystem disturbances and plant invasions in the French islands of the southern Indian Ocean (Crozet, Kerguelen and Amsterdam Islands)', Biological Conservation 101(1), 33–50. https://doi.org/10.1016/S0006-3207(01)00052-0

García-de-Lomas, J. & Vilà, M., 2015, 'Lists of harmful alien organisms: Are the national regulations adapted to the global world?', Biological Invasions 17(11), 3081–3091. https://doi.org/10.1007/s10530-015-0939-7

Gimeno, I., Vilà, M. & Hulme, P.E., 2006, 'Are islands more susceptible to plant invasion than continents? A test using Oxalis pes-caprae L. in the western Mediterranean', Journal of Biogeography 33(9), 1559–1565. https://doi.org/10.1111/j.1365-2699.2006.01525.x

Gremmen, N.J.M., 1975, 'The distribution of alien vascular plant on Marion and Prince Edward Islands', South African Journal of Antarctic Research 5, 25–30.

Gremmen, N.J.M., 1981, The vegetation of the Subantarctic islands Marion and Prince Edward, Dr. W. Junk, The Hague.

Gremmen, N.J.M., 1997, 'Changes in the vegetation of sub-Antarctic Marion Island resulting from introduced vascular plants', in B. Battaglia, J. Valencia & D.W.H. Walton (eds.), Antarctic communities: Species, structure and survival, pp. 417–423, Cambridge University Press, Cambridge.

Gremmen, N.J.M., Chown, S.L. & Marshall, D.J., 1998, 'Impact of the introduced grass Agrostis stolonifera on vegetation and soil fauna communities at Marion Island, sub-Antarctic', Biological Conservation 85(3), 223–231. https://doi.org/10.1016/S0006-3207(97)00178-X

Gremmen, N.J.M. & Smith, V.R., 1999, 'New records of alien vascular plants from Marion and Prince Edward Islands, Sub-Antarctic', Polar Biology 21(6), 401–409. https://doi.org/10.1007/s003000050380

Gremmen, N.J.M. & Smith, V.R., 2008, 'Terrestrial vegetation and dynamics', in S.L. Chown & P.W. Froneman (eds.), The Prince Edward Islands: Land-sea interactions in a changing ecosystem, pp. 215–244, African SunMedia, Stellenbosch.

Greve, M., Gremmen, N.J.M., Gaston, K.J. & Chown, S.L., 2005, 'Nestedness of Southern Ocean island biotas: Ecological perspectives on a biogeographical conundrum', Journal of Biogeography 32(1), 155–168. https://doi.org/10.1111/j.1365-2699.2004.01169.x

Gurevitch, J. & Padilla, D.K., 2004, 'Are invasive species a major cause of extinctions?', Trends in Ecology & Evolution 19(9), 470–474. https://doi.org/10.1016/j.tree.2004.07.005

Hänel, C. & Chown, S.L., 1998, 'The impact of a small, alien invertebrate on a sub-Antarctic terrestrial ecosystem: Limnophyes minimus (Diptera, Chironomidae) at Marion Island', Polar Biology 20(2), 99–106. https://doi.org/10.1007/s003000050282

Hänel, C., Chown, S.L. & Davies, L., 1998, 'Records of alien insect species from sub-Antarctic Marion and South Georgia Islands', African Entomology 6(2), 366–369.

Haussmann, N.S., Rudolph, E.M., Kalwij, J.M. & McIntyre, T., 2013, 'Fur seal populations facilitate establishment of exotic vascular plants', Biological Conservation 162, 33–40. https://doi.org/10.1016/j.biocon.2013.03.024

Hedding, D.W., Nel, W. & Anderson, R.L., 2015, 'Aeolian processes and landforms in the sub-Antarctic: Preliminary observations from Marion Island', Polar Research 34, 26365. https://doi.org/10.3402/polar.v34.26365

Houghton, M., McQuillan, P., Bergstrom, D., Frost, L., van den Hoff, J. & Shaw, J., 2016, 'Pathways of alien invertebrate transfer to the Antarctic region', Polar Biology 39(1), 23–33. https://doi.org/10.1007/s00300-014-1599-2

Howald, G., Donlan, C.J., Galván, J.P., Russell, J.C., Parkes, J., Samaniego, A. et al., 2007, 'Invasive rodent eradication on islands', Conservation Biology 21(5), 1258–1268. https://doi.org/10.1111/j.1523-1739.2007.00755.x

Hughes, K.A. & Convey, P., 2010, 'The protection of Antarctic terrestrial ecosystems from inter- and intra-continental transfer of non-indigenous species by human activities: A review of current systems and practices', Global Environmental Change 20(1), 96–112. https://doi.org/10.1016/j.gloenvcha.2009.09.005

Hughes, K.A. & Convey, P., 2012, 'Determining the native/non-native status of newly discovered terrestrial and freshwater species in Antarctica – Current knowledge, methodology and management action', Journal of Environmental Management 93(1), 52–66. https://doi.org/10.1016/j.jenvman.2011.08.017

Hughes, K.A., Cowan, D.A. & Wilmotte, A., 2015, 'Protection of Antarctic microbial communities – "Out of sight, out of mind"', Frontiers in Microbiology 6, 151. https://doi.org/10.3389/fmicb.2015.00151

Hughes, K.A. & Pertierra, L.R., 2016, 'Evaluation of non-native species policy development and implementation within the Antarctic Treaty area', Biological Conservation 200, 149–159. https://doi.org/10.1016/j.biocon.2016.03.011

Hugo, E.A., McGeoch, M.A., Marshall, D.J. & Chown, S.L., 2004, 'Fine scale variation in microarthropod communities inhabiting the keystone species Azorella selago on Marion Island', Polar Biology 27(8), 466–473. https://doi.org/10.1007/s00300-004-0614-4

Huiskes, A.H.L., Gremmen, N.J.M., Bergstrom, D.M., Frenot, Y., Hughes, K.A., Imura, S. et al., 2014, 'Aliens in Antarctica: Assessing transfer of plant propagules by human visitors to reduce invasion risk', Biological Conservation 171, 278–284. https://doi.org/10.1016/j.biocon.2014.01.038

Hulme, P.E., 2009, 'Trade, transport and trouble: Managing invasive species pathways in an era of globalization', Journal of Applied Ecology 46(1), 10–18. https://doi.org/10.1111/j.1365-2664.2008.01600.x

Huntley, B.J., 1971, 'Vegetation', in E.M. Van Zinderen Bakker, J.M. Winterbottom & R.A. Dyer (eds.), Marion and Prince Edward Islands. Report on the South African biological & geological expedition 1965–1966, pp. 98–160, A.A. Balkema, Cape Town.

Huyser, O., Ryan, P.G. & Cooper, J., 2000, 'Changes in population size, habitat use and breeding biology of lesser sheathbills *Chionis minor* at Marion Island: Impacts of cats, mice and climate change?', *Biological Conservation* 92(3), 299–310. https://doi.org/10.1016/S0006-3207(99)00096-8

Janion, C., Leinaas, H.P., Terblanche, J.S. & Chown, S.L., 2010, 'Trait means and reaction norms: The consequences of climate change/invasion interactions at the organism level', *Evolutionary Ecology* 24(6), 1365–1380. https://doi.org/10.1007/s10682-010-9405-2

Jones, M.G.W. & Ryan, P.G., 2010, 'Evidence of mouse attacks on albatross chicks on sub-Antarctic Marion Island', *Antarctic Science* 22(1), 39–42. https://doi.org/10.1017/S0954102009990459

Keller, R.P. & Kumschick, S., 2017, Promise and challenges of risk assessment as an approach for preventing the arrival of harmful alien species. *Bothalia* 47(2), a2136. https://doi.org/10.4102/abc.v47i2.2136

Kloppers, F.J. & Smith, V.R., 1998, 'First Report of *Botryotinia fuckeliana* on Kerguelen Cabbage on the Sub-Antarctic Marion Island', *Plant Disease* 82(6), 710. https://doi.org/10.1094/PDIS.1998.82.6.710A

La Grange, J.J., 1954, 'The South African station on Marion Island', *Polar Record* 7(48), 155–158. https://doi.org/10.1017/S0032247400043606

le Roux, P.C. & McGeoch, M.A., 2008, 'Changes in climate extremes, variability and signature on sub-Antarctic Marion Island', *Climatic Change* 86(3–4), 309–329. https://doi.org/10.1007/s10584-007-9259-y

le Roux, P.C., Ramaswiela, T., Kalwij, J.M., Shaw, J.D., Ryan, P.G., Treasure, A.M. et al., 2013, 'Human activities, propagule pressure and alien plants in the sub-Antarctic: Tests of generalities and evidence in support of management', *Biological Conservation* 161, 18–27. https://doi.org/10.1016/j.biocon.2013.02.005

Lee, J.E. & Chown, S.L., 2007, '*Mytilus* on the move: Transport of an invasive bivalve to the Antarctic', *Marine Ecology Progress Series* 339, 307–310. https://doi.org/10.3354/meps339307

Lee, J.E. & Chown, S.L., 2009a, 'Breaching the dispersal barrier to invasion: Quantification and management', *Ecological Applications* 19(7), 1944–1959. https://doi.org/10.1890/08-2157.1

Lee, J.E. & Chown, S.L., 2009b, 'Quantifying the propagule load associated with the construction of an Antarctic research station', *Antarctic Science* 21(05), 471–475. https://doi.org/10.1017/S0954102009990162

Lee, J.E. & Chown, S.L., 2011, 'Quantification of intra-regional propagule movements in the Antarctic', *Antarctic Science* 23(04), 337–342. https://doi.org/10.1017/S0954102011000198

Lee, J.E. & Chown, S.L., 2016, 'Range expansion and increasing impact of the introduced wasp *Aphidius matricariae* Haliday on sub-Antarctic Marion Island', *Biological Invasions* 18(5), 1235–1246. https://doi.org/10.1007/s10530-015-0967-3

Lee, J.E., Janion, C., Marais, E., Jansen van Vuuren, B. & Chown, S.L., 2009, 'Physiological tolerances account for range limits and abundance structure in an invasive slug', *Proceedings of the Royal Society B-Biological Sciences* 276(1661), 1459–1468. https://doi.org/10.1098/rspb.2008.1240

Lee, J.E., Slabber, S., Jansen van Vuuren, B., van Noort, S. & Chown, S.L., 2007, 'Colonisation of sub-Antarctic Marion Island by a non-indigenous aphid parasitoid *Aphidius matricariae* (Hymenoptera, Braconidae)', *Polar Biology* 30(9), 1195–1201. https://doi.org/10.1007/s00300-007-0277-z

Lee, J.E., Terauds, A. & Chown, S.L., 2014, 'Natural dispersal to sub-Antarctic Marion Island of two arthropod species', *Polar Biology* 37(6), 781–787. https://doi.org/10.1007/s00300-014-1479-9

Levine, J.M., Vilà, M., Antonio, C.M.D., Dukes, J.S., Grigulis, K. & Lavorel, S., 2003, 'Mechanisms underlying the impacts of exotic plant invasions', *Proceedings of the Royal Society of London. Series B: Biological Sciences* 270(1517), 775–781. https://doi.org/10.1098/rspb.2003.2327

Lutjeharms, J.R.E. & Ansorge, I.J., 2008, 'Oceanographic setting of the Prince Edward Islands', in S.L. Chown & P.W. Froneman (eds.), *The Prince Edward Islands*, pp. 17–38, SUN PReSS, Stellenbosch.

McClelland, G.T.W., Altwegg, R., van Aarde, R.J., Ferreira, S., Burger, A.E. & Chown, S.L., in review, 'Climate change leads to increasing population density and impacts of a key island invader'.

McGeoch, M.A., Butchart, S.H.M., Spear, D., Marais, E., Kleynhans, E.J., Symes, A. et al., 2010, 'Global indicators of biological invasion: Species numbers, biodiversity impact and policy responses', *Diversity and Distributions* 16(1), 95–108. https://doi.org/10.1111/j.1472-4642.2009.00633.x

McGeoch, M.A., Shaw, J.D., Terauds, A., Lee, J.E. & Chown, S.L., 2015, 'Monitoring biological invasion across the broader Antarctic: A baseline and indicator framework', *Global Environmental Change* 32, 108–125. https://doi.org/10.1016/j.gloenvcha.2014.12.012

Ordonez, A., Wright, I.J. & Olff, H., 2010, 'Functional differences between native and alien species: A global-scale comparison', *Functional Ecology* 24(6), 1353–1361. https://doi.org/10.1111/j.1365-2435.2010.01739.x

Øvstedal, D.O. & Gremmen, N.J.M., 2011, 'The lichens of Marion and Prince Edward islands', *South African Journal of Botany* 67(4), 552–572. https://doi.org/10.1016/S0254-6299(15)31187-X

Pammenter, N.W., Drennan, P.M. & Smith, V.R., 1986, 'Physiological and anatomical aspects of photosynthesis of two *Agrostis* species at a sub-Antarctic island', *New Phytologist* 102(1), 143–160. https://doi.org/10.1111/j.1469-8137.1986.tb00806.x

Phiri, E.E., McGeoch, M.A. & Chown, S.L., 2009, 'Spatial variation in structural damage to a keystone plant species in the sub-Antarctic: Interactions between *Azorella selago* and invasive house mice', *Antarctic Science* 21(3), 189–196. https://doi.org/10.1017/S0954102008001569

Pyšek, P. & Richardson, D.M., 2008, 'Invasive plants', in S.E. Jørgensen & B.D. Fath (eds.), *Encyclopedia of ecology*, Elsevier, Oxford.

Pyšek, P., Richardson, D.M., Pergl, J., Jarošík, V., Sixtová, Z. & Weber, E., 2008, 'Geographical and taxonomic biases in invasion ecology', *Trends in Ecology & Evolution* 23(5), 237–244. https://doi.org/10.1016/j.tree.2008.02.002

Raymond, B., McInnes, J., Dambacher, J.M., Way, S. & Bergstrom, D.M., 2011, 'Qualitative modelling of invasive species eradication on subantarctic Macquarie Island', *Journal of Applied Ecology* 48(1), 181–191. https://doi.org/10.1111/j.1365-2664.2010.01916.x

Reaser, J.K., Meyerson, L.A., Cronk, Q., De Poorter, M., Eldrege, L.G., Green, E. et al., 2007, 'Ecological and socioeconomic impacts of invasive alien species in island ecosystems', *Environmental Conservation* 34(2), 98–111. https://doi.org/10.1017/S0376892907003815

Richardson, D.M., Pyšek, P., Rejmánek, M., Barbour, M.G., Panetta, F.D. & West, C.J., 2000, 'Naturalization and invasion of alien plants: Concepts and definitions', *Diversity and Distributions* 6(2), 93–107. https://doi.org/10.1046/j.1472-4642.2000.00083.x

Ryan, P.G., Smith, V.R. & Gremmen, N.J.M., 2003, 'The distribution and spread of alien vascular plants on Prince Edward Island', *African Journal of Marine Science* 25(1), 555–562. https://doi.org/10.2989/18142320309504045

Sanyika, T.W., Stafford, W. & Cowan, D.A., 2012, 'The soil and plant determinants of community structures of the dominant actinobacteria in Marion Island terrestrial habitats, Sub-Antarctica', *Polar Biology* 35(8), 1129–1141. https://doi.org/10.1007/s00300-012-1160-0

Sax, D.F. & Gaines, S.D., 2008, 'Species invasions and extinction: The future of native biodiversity on islands', *Proceedings of the National Academy of Sciences of the United States of America* 105(Supplement 1), 11490–11497. https://doi.org/10.1073/pnas.0802290105

Shaw, J.D., 2013, 'Southern Ocean Islands invaded: Conserving biodiversity in the world's last wilderness', in L.C. Foxcroft, P. Pyšek, D.M. Richardson & P. Genovesi (eds.), *Plant invasions in protected areas: Patterns, problems and challenges*, pp. 449–470, Springer Science+Business Media, Dordrecht.

Shaw, J.D., Spear, D., Greve, M. & Chown, S.L., 2010, 'Taxonomic homogenization and differentiation across Southern Ocean Islands differ among insects and vascular plants', *Journal of Biogeography* 37(2), 217–228. https://doi.org/10.1111/j.1365-2699.2009.02204.x

Simberloff, D., Martin, J.-L., Genovesi, P., Maris, V., Wardle, D.A., Aronson, J. et al., 2013, 'Impacts of biological invasions: What's what and the way forward', *Trends in Ecology & Evolution* 28(1), 58–66. https://doi.org/10.1016/j.tree.2012.07.013

Slabber, S. & Chown, S.L., 2002, 'The first record of a terrestrial crustacean, *Porcellio scaber* (Isopoda, Porcellionidae), from sub-Antarctic Marion Island', *Polar Biology* 25(11), 855–858. https://doi.org/10.1016/j.jinsphys.2006.10.010

Slabber, S., Worland, M.R., Leinaas, H.P. & Chown, S.L., 2007, 'Acclimation effects on thermal tolerances of springtails from sub-Antarctic Marion Island: Indigenous and invasive species', *Journal of Insect Physiology* 53(2), 113–125. https://doi.org/10.1016/j.jinsphys.2006.10.010

Smith, V., Avenant, N. & Chown, S., 2002, 'The diet and impact of house mice on a sub-Antarctic island', *Polar Biology* 25(9), 703–715.

Smith, V.R., 1978, 'Animal-plant-soil nutrient relationships on Marion Island (Subantarctic)', *Oecologia* 32(2), 239–253. https://doi.org/10.1007/BF00366075

Smith, V.R., 1992, 'Terrestrial slug recorded from sub-Antarctic Marion Island', *Journal of Molluscan Studies* 58(1), 80–81. https://doi.org/10.1093/mollus/58.1.80

Smith, V.R., 2007, 'Introduced slugs and indigenous caterpillars as facilitators of carbon and nutrient mineralisation on a sub-Antarctic island', *Soil Biology & Biochemistry* 39(2), 709–713. https://doi.org/10.1016/j.soilbio.2006.09.026

Smith, V.R., 2008, 'Energy flow and nutrient cycling in the Marion Island terrestrial ecosystem: 30 years on', *Polar Record* 44(3), 211–226. https://doi.org/10.1017/S0032247407007218

Smith, V.R. & Steenkamp, M., 1990, 'Climatic change and its ecological implications at a subantarctic island', *Oecologia* 85(1), 14–24. https://doi.org/10.1007/BF00317338

Springer, K., 2016, 'Methodology and challenges of a complex multi-species eradication in the sub-Antarctic and immediate effects of invasive species removal', *New Zealand Journal of Ecology* 40(2), 273–278. https://doi.org/10.20417/nzjecol.40.30

Tecco, P.A., Díaz, S., Cabido, M. & Urcelay, C., 2010, 'Functional traits of alien plants across contrasting climatic and land-use regimes: Do aliens join the locals or try harder than them?', *Journal of Ecology* 98(1), 17–27. https://doi.org/10.1111/j.1365-2745.2009.01592.x

Treasure, A.M. & Chown, S.L., 2013, 'Contingent absences account for range limits but not the local abundance structure of an invasive springtail', *Ecography* 36(2), 146–156. https://doi.org/10.1111/j.1600-0587.2012.07458.x

Treasure, A.M. & Chown, S.L., 2014, 'Antagonistic effects of biological invasion and temperature change on body size of island ectotherms', *Diversity and Distributions* 20(2), 202–213. https://doi.org/10.1111/ddi.12153

van Aarde, R.J., 1980, 'The diet and feeding-behavior of feral cats, *Felis catus* at Marion Island', *South African Journal of Wildlife Research* 10(3&4), 123–128.

van Aarde, R.J., 1986, 'A case study of an alien predator (*Felis catus*) introduced on Marion Island: Selective advantages', *South African Journal of Antarctic Research* 16(3), 113–114.

van Aarde, R.J., Ferreira, S.M. & Wassenaar, T.D., 2004, 'Do feral house mice have an impact on invertebrate communities on sub-Antarctic Marion Island?', *Austral Ecology* 29(2), 215–224. https://doi.org/10.1111/j.1442-9993.2004.01341.x

Van de Vijver, B., Gremmen, N. & Smith, V., 2008, 'Diatom communities from the sub-Antarctic Prince Edward Islands: Diversity and distribution patterns', *Polar Biology* 31(7), 795–808. https://doi.org/10.1007/s00300-008-0418-z

van Kleunen, M., Weber, E. & Fischer, M., 2010, 'A meta-analysis of trait differences between invasive and non-invasive plant species', *Ecology Letters* 13(2), 235–245. https://doi.org/10.1111/j.1461-0248.2009.01418.x

van Rensburg, P.J.J., 1983, 'The feeding ecology of a decreasing feral house cat, *Felis catus*, population at Marion Island', in W.R. Siegfried, P.R. Condy & R.M. Laws (eds.), *Antarctic nutrient cycles and food webs*, pp. 620–624, Springer Verlag, Berlin.

van Staden, W., 2011, 'Limnoecology of the freshwater algal genera (excluding diatoms) on Marion Island (sub-Antarctic)', MSc thesis, North-West University.

Vilà, M., Espinar, J.L., Hejda, M., Hulme, P.E., Jarošík, V., Maron, J.L. et al., 2011, 'Ecological impacts of invasive alien plants: A meta-analysis of their effects on species, communities and ecosystems', *Ecology Letters* 14(7), 702–708. https://doi.org/10.1111/j.1461-0248.2011.01628.x

Vitousek, P.M., 1988, 'Diversity and biological invasions of oceanic islands', in E.O. Wilson & F.M. Peters (eds.), *Biodiversity*, pp. 181–190, National Academies Press, Washington, DC.

Wanless, R.M., Angel, A., Cuthbert, R.J., Hilton, G.M. & Ryan, P.G., 2007, 'Can predation by invasive mice drive seabird extinctions?', *Biology Letters* 3(3), 241–244. https://doi.org/10.1098/rsbl.2007.0120

Appendix 1

Terrestrial invasions on Sub-Antarctic Marion and Prince Edward Islands M. Greve, R. Mathakutha, C. Steyn & S.L. Chown

TABLE 1-A1: List of references cited in Table 1.

Reference	Citation
1	Abraham, S., Somers, M.J. & Chown, S.L. 2011. Seasonal, altitudinal and host plant-related variation in the abundance of aphids (Insecta, Hemiptera) on sub-Antarctic Marion Island. Polar Biology 34:513–520.
2	Bester, M.N., Bloomer, J.P. van Aarde, R.J. Erasmus, B.H. van Rensburg, P.J.J. Skinner, J.D. Howell, P.G. & Naude, T.W. 2002. A review of the successful eradication of feral cats from sub-Antarctic Marion Island, Southern Indian Ocean. South African Journal of Wildlife Research 32:65–73.
3	Chown, S.L., & Language, K. 1994. Recently established Diptera and Lepidoptera on sub-Antarctic Marion Island: short communication. African Entomology 2:57–60.
4	Chown, S.L., & Avenant, N. 1992. Status of *Plutella xylostella* at Marion Island six years after its colonisation. South African Journal of Antarctic Research 22:37–40.
5	Chown, S.L., & Convey, P. 2016. Antarctic entomology. Annual Review of Entomology 61:119–137.
6	Chown, S.L., & Froneman, P.W. editors. 2008. The Prince Edward Islands: Land-Sea Interactions in a Changing Ecosystem. African SunMedia, Stellenbosch.
7	Cooper, J. 2008. Human history. Pages 331–350 in Chown S.L. & Froneman, P.W. editors. The Prince Edward Islands. SUN PReSS, Stellenbosch.
8	Crafford, J.E. 1986. A case study of an alien invertebrate (*Limnophyes pusillus*, Diptera, Chironomidae) introduced on Marion Island: selective advantages. South African Journal of Antarctic Research 16:115–117.
9	Crafford, J.E., & Chown, S.L. 1987. *Plutella xylostella* L. (Lepidoptera: Plutellidae) on Marion Island. Journal of the Entomological Society of Southern Africa 50:259–260.
10	Crafford, J.E., Scholtz, C.H. & Chown, S.L. 1986. The insects of sub-Antarctic Marion and Prince Edward Islands; with a bibliography of entomology of the Kerguelen Biogeographical Province. South African Journal of Antarctic Research 16:42–84.
11	de Villiers, M.S., & Cooper, J. 2008. Conservation and management. Pages 301–330 in S.L. Chown and P.W.Froneman, editors. The Prince Edward Islands. SUN PReSS, Stellenbosch.
12	Gabriel, A.G.A., Chown, S.L. Barendse, J. Marshall, D.J. Mercer, R.D. Pugh, P.J.A. & Smith, V.R. 2001. Biological invasions of Southern Ocean islands: the Collembola of Marion Island as a test of generalities. Ecography 24:421–430.
13	Gremmen, N.J.M. 1975. The distribution of alien vascular plant on Marion and Prince Edward Islands. South African Journal of Antarctic Research 5:25–30.
14	Gremmen, N.J.M., & Smith, V.R. 1999. New records of alien vascular plants from Marion and Prince Edward Islands, sub-Antarctic. Polar Biology 21:401–409.
15	Gremmen, N.J.M., & Smith, V.R. 2008. Terrestrial vegetation and dynamics. Pages 215–244 in Chown, S.L. & Froneman, P.W. editors. The Prince Edward Islands: Land-Sea Interactions in a Changing Ecosystem. African SunMedia, Stellenbosch.
16	Hänel, C., & Chown, S.L. 1998. The impact of a small, alien invertebrate on a sub-Antarctic terrestrial ecosystem: Limnophyes minimus (Diptera, Chironomidae) at Marion Island. Polar Biology 20:99–106.
17	Hänel, C., Chown, S.L. & Davies, L. 1998. Records of alien insect species from sub-Antarctic Marion and South Georgia Islands. African Entomology 6:366–369.
18	Hugo, E.A., Chown, S.L. & McGeoch, M.A. 2006. The microarthropods of sub-Antarctic Prince Edward Island: a quantitative assessment. Polar Biology 30:109.
19	Khoza, T.T., Dippenaar, S.M. & Dippenaar-Schoeman, A.S. 2005. The biodiversity and species composition of the spider community of Marion Island, a recent survey (Arachnida: Araneae). 2005 48:103–107.
20	Kloppers, F.J., & Smith, V.R. 1998. First Report of *Botryotinia fuckeliana* on Kerguelen Cabbage on the Sub-Antarctic Marion Island. Plant Disease 82:710–710.
21	Janion-Scheepers, C., Deharveng, L. Bedos, A. & Chown, S.L. 2015. Updated list of Collembola species currently recorded from South Africa. ZooKeys 503:55–88.
22	Jumbam, K.R., Terblanche, J.S. Deere, J.A. Somers, M.J. & Chown, S.L. 2008. Critical thermal limits and their responses to acclimation in two sub-Antarctic spiders: *Myro kerguelenensis* and *Prinerigone vagans*. Polar Biology 31:215–220.
23	Lawrence, R.F. 1971. Araneida. Pages 301–313 in Van Zinderen Bakker, E.M. Winterbottom, J.M. & Dyer, R.A. editors. Marion and Prince Edward Islands. Report on the South African Biological & Geological Expedition 1965–1966. Cape Town, A.A. Balkema.
24	le Roux, P.C., Ramaswiela, T. Kalwij, J.M. Shaw, J.D. Ryan, P.G. Treasure, A.M. McClelland, G.T.W. McGeoch, M.A. & Chown, S.L. 2013. Human activities, propagule pressure and alien plants in the sub-Antarctic: tests of generalities and evidence in support of management. Biological Conservation 161:18–27.
25	Lee, J.E., & Chown, S.L. 2016. Range expansion and increasing impact of the introduced wasp *Aphidius matricariae* Haliday on sub-Antarctic Marion Island. Biological Invasions 18:1235–1246.
26	Lee, J.E., Slabber, S. Jansen van Vuuren, B. van Noort, S. & Chown, S.L. 2007. Colonisation of sub-Antarctic Marion Island by a non-indigenous aphid parasitoid *Aphidius matricariae* (Hymenoptera, Braconidae). Polar Biology 30:1195–1201.
27	Marshall, D.J., Gremmen, N.J.M. Coetzee, L. O'Connor, B.M. Pugh, P.J.A. Theron, P.D. & Ueckermann, E.A. 1999. New records of Acari from the sub-Antarctic Prince Edward Islands. Polar Biology 21:84–89.
28	Ramaswiela, T. 2010. The spatial distribution of alien and invasive vascular plant species on sub-Antarctic Marion Island. MSc Thesis, Stellenbosch University, Stellenbosch.
29	Slabber, S., & Chown, S.L. 2002. The first record of a terrestrial crustacean, *Porcellio scaber* (Isopoda, Porcellionidae), from sub-Antarctic Marion Island. Polar Biology 25:855–858.
30	Smith, V.R. 1992. Terrestrial slug recorded from sub-Antarctic Marion Island. Journal of Molluscan Studies 58:80–81.
31	Smith, V.R. 2007. Introduced slugs and indigenous caterpillars as facilitators of carbon and nutrient mineralisation on a sub-Antarctic island. Soil Biology & Biochemistry 39:709–713.
32	Treasure, A.M., & Chown, S.L. 2013. Contingent absences account for range limits but not the local abundance structure of an invasive springtail. Ecography 36:146–156.
33	www.collembola.co.za (accessed 27/11/2016).

TABLE 2-A1: Meaning of invasion status classification in Table 1, taken directly from Blackburn et al. (2011).

Status code	Meaning of invasion status from Blackburn et al. 2011, Table 1
A	'Not transported beyond limits of native range'
B1	'Individuals transported beyond limits of native range, and in captivity or quarantine (i.e. individuals provided with conditions suitable for them, but explicit measures of containment are in place)'
B2	'Individuals transported beyond limits of native range, and in cultivation (i.e. individuals provided with conditions suitable for them but explicit measures to prevent dispersal are limited at best)'
B3	'Individuals transported beyond limits of native range, and directly released into novel environment'
C0	'Individuals released into the wild (i.e. outside of captivity or cultivation) in location where introduced, but incapable of surviving for a significant period'
C1	'Individuals surviving in the wild (i.e. outside of captivity or cultivation) in location where introduced, no reproduction'
C2	'Individuals surviving in the wild in location where introduced, reproduction occurring, but population not self-sustaining'
C3	'Individuals surviving in the wild in location where introduced, reproduction occurring, and population self-sustaining'
D1	'Self-sustaining population in the wild, with individuals surviving a significant distance from the original point of introduction'
D2	'Self-sustaining population in the wild, with individuals surviving and reproducing a significant distance from the original point of introduction'
E	'Fully invasive species, with individuals dispersing, surviving and reproducing at multiple sites across a greater or lesser spectrum of habitats and extent of occurrence'

TABLE 3-A1: Impact criteria used to assign species to different impact categories. The scheme for this classification was taken from Blackburn et al. (2014).

Impact category	Categories adhere to the following general meaning
Massive (MA)	Causes at least local extinction of species and irreversible changes in community composition; even if the alien species is removed, the system does not recover its original state
Major (MR)	Causes changes in community composition, which are reversible if the alien species is removed
Moderate (MO)	Causes declines in population densities, but no changes in community composition
Minor (MI)	Causes reductions in individual fitness, but no declines in native population densities
Minimal (ML)	No effect on fitness of individuals of native species

TABLE 4-A1: Two alien freshwater species recorded from the Prince Edward Islands.

	Mite species	Brown Trout
Phylum (Class)	Arthropoda (Arachnida)	Chordata (Actinopterygii)
Order	Prostigmata	Salmoniformis
Family	Halacaridae	Salmonidae
Species	*Peregrinacarus reticulatus* Bartsch	*Salmo trutta* L.
Status	Status uncertain	C3
Impact	Impact unknown	Impact unknown
Island	Marion Island	Marion Island
Introduction Notes		Deliberately introduced 1964 (c)
Additional Information		Eradicated
References	Bartsch, I. 1999. *Peregrinacarus reticulatus* gen. nov. spec. nov., a freshwater halacarid mite (Acari, Halacaridae) from Marion Island. Hydrobiologia 392:225-232.	Cooper, J., Crafford, J.E. & Hecht, T. 1992. Introduction and extinction of brown trout (*Salmo trutta* L.) in an impoverished subantarctic stream. Antarctic Science 4:9-14.

A status of C3 means 'individuals surviving in the wild in location where introduced, reproduction occurring, and population self-sustaining' (Blackburn et al. 2011).

BOX 1: How papers on aliens and invasives were extracted from ISI Web of Knowledge.

To obtain a count of the number of publications dealing with invasive species, a search was conducted on ISI Web of Knowledge on 26 July 2016. The search terms used were *(alien OR invasi* OR introduc*) AND TOPIC:(prince edward)*, and *(alien OR invasi* OR introduce* OR exotic) AND TOPIC:('marion island')*. The references were downloaded to EndNote. Because there is a Prince Edward Island in Canada, all references with the word *Canada* in the title, abstract or keywords were removed. Finally, the titles of all references were checked to confirm that they dealt with the Prince Edward Islands. If it was obvious from the title that the study had been conducted in a location other than the Prince Edward Islands (e.g. Bouvet), the reference was removed. References that dealt with broad topics (e.g. aliens in the greater Antarctic region) were retained. The numbers of references per calendar year were then tallied.

3

Bertiera sinoensis Jongkind (Rubiaceae), a new forest liana from Liberia

Author:
Carel Jongkind[1]

Affiliation:
[1]Botanic Garden Meise, Belgium

Corresponding author:
Carel Jongkind,
carel.jongkind@kpnmail.nl

Background: Fieldwork in Liberia in recent years has improved our knowledge of the local endemic species.

Objectives: To describe a new species in *Bertiera* to accommodate material from the south-east of Liberia that cannot be included in any known species.

Methods: Existing herbarium collections were studied, the new species was studied in the field and the relevant published literature was consulted.

Results: The new species *Bertiera sinoensis* is described and illustrated here based on six specimens.

Conclusions: The new species adds one more species to the botanical hotspot in south-east Liberia. It is assigned a preliminary conservation status of 'Endangered' (IUCN).

Introduction

Bertiera Aubl. (Rubiaceae) is a genus of *ca.* 55 species from Africa, including Madagascar and the Mascarenes, and from America (World Checklist of Selected Plant Families [WCSP] 2016). Recently, several new African species have been published, as well as the taxonomic revisions for Cameroon and Madagascar (Nguembou et al. 2003, 2006, 2009; Sonké et al. 2005; Wittle & Davis 2010). Recent phylogenetic research suggests that the genus is monophyletic, although less than 10% of its species have been sampled (Tosh et al. 2009).

Until now, there were only two climbing *Bertiera* species known, *B. bracteolata* Hiern and *B. chevalieri* Hutch. & Dalziel, and a third climbing species, *B. sinoensis*, is described here. The new species was first discovered in the herbarium by William Hawthorne when we were working on a guide to the forest trees, shrubs and lianes from Senegal to Ghana (Hawthorne & Jongkind 2006). Recently, several more specimens have been studied and collected in the field by the author and his team.

The area in south-east Liberia, where the new species is found is a biodiversity hotspot known to be home to several other local endemic species, most of them discovered only recently, including *Napoleonaea sapoensis* Jongkind (Prance & Jongkind 2015), *Pauridiantha liberiensis* Ntore (Ntore 2008), *Pavetta sapoensis* Hawthorne (Hawthorne 2013), *Psychotria tetragonopus* O.Lachenaud & Jongkind (Lachenaud & Jongkind 2013) and *Soyauxia kwewonii* Breteler & Jongkind (Breteler, Bakker & Jongkind 2015). At the moment, an important part of the forest in this area being replaced by oil palm plantations.

Research method and design

All relevant herbarium collections at the BR, K, P and WAG herbaria were examined. *Bertiera sinoensis* has recently been studied in the field on several occasions. Preliminary assessment of the International Union for the Conservation of Nature (IUCN) Red List categories was performed and species according to the criteria through the RBG Kew website (http://geocat.kew.org).

Taxonomic treatment

Bertiera sinoensis Jongkind, *sp. nov.*

Bertiera 'nimbae', Hawthorne & Jongkind (2006:618, 619).

Type: Liberia. Sinoe County, east of Greenville-Zwedru road, 126 m, 5°30.50'N, 8°39.31'W; forest edge. fl., fr., 26 Sep. 2013, *Jongkind, de Wet & Sambolah 12157* (WAG, holo; BR, FHO, K, MO, P, iso.).

FIGURE 1: (a-c, e) *Bertiera sinoensis*. From *Jongkind 12157*. (a) Part of inflorescence, from several flowers, the corolla dropped and the green, glabrous disk became visible; (b) inflorescence; (c) infructescence with immature, green fruits; (d) *B. bracteolata*, infructescence. From *Jongkind 11845*; (e) leaves from *B. sinoensis*.

Source: Photos by Jongkind

Description

Slender winding liana; twigs densely pubescent. *Stipules* 9–13 mm long, acuminate, sheating for 3–4 mm, pubescent. *Leaves* opposite, petiole 3–4 mm long, pubescent, blade oblanceolate to narrowly elliptic, 6–10 cm × 2–3.3 cm, pubescent on both sides but hairs more densely and soft to the touch adaxially, with 5–7 pairs of lateral nerves, smaller nerves invisible on both sides, base cuneate, apex shortly acuminate. *Inflorescence* terminal, pedunculate, up to 160 mm long, with many short, cymose branches on a much longer, straight, bracteate axis, bracts 5–9 mm long, pubescent. *Flowers* sessile or shortly pedicellate, hermaphrodite, 5-merous; bracteoles 4–5 mm long, up to 1 mm wide, pubescent; calyx *ca.* 1 mm long, with short, triangular lobes, pubescent, green; corolla tube *ca.* 5 mm long, green to white, pubescent outside, lobes spreading, 1.5 × 1 mm, white, pubescent at both sides, hairs on corolla *ca.* 0.5 mm long, corolla in bud acutely pointed; anthers included, *ca.* 2 mm long, subsessile, linear, with an apical connective appendage; disk equalling the calyx, annular, glabrous, fleshy; ovary 2-celled with numerous ovules. *Fruit* subglobose, *ca.* 8 mm diameter, green to white to purplish, shiny, almost glabrous with many seeds. *Seeds* angular, *ca.* 1 mm, brown (Figure 1).

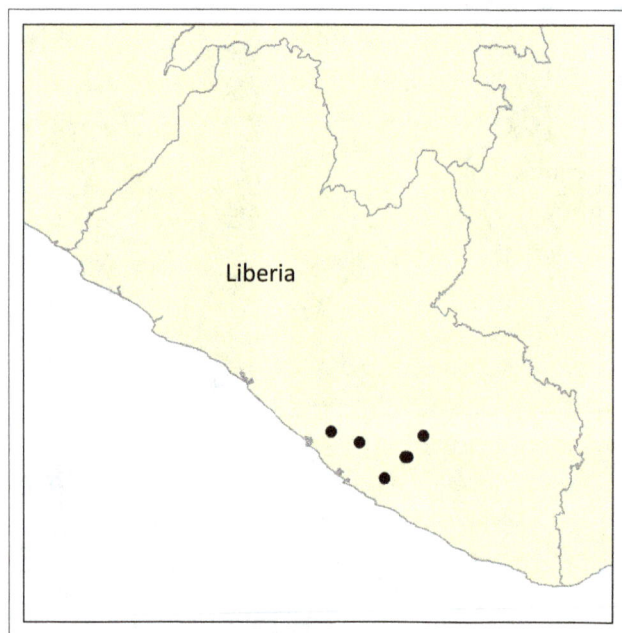

FIGURE 2: Map showing the geographic distribution of *B. sinoensis*.

Etymology

The species epithet refers to the Sinoe River that runs through the area where the new species is found.

Ecology and distribution

Bertiera sinoensis is only known from the evergreen forest in south-east Liberia (Figure 2).

Diagnosis

Bertiera sinoensis resembles *B. bracteolata* and *B. chevalieri*. All three are slender woody climbers with elongated inflorescences and are the only climbing species in the genus. The leaves of *B. sinoensis* are much more hairy and more slender than those of the other two species, the stipules and bracteoles are longer and more hairy, and the corolla is more conspicuously hairy. Both *B. bracteolata* and *B. chevalieri* are known from both Upper and Lower Guinea (sensu White 1979), but the new species is restricted to south-east Liberia. The important taxonomic differences between these three species are summarised in Table 1.

Apart from the inflorescence branching pattern, there is very little difference between *B. bracteolata* and *B. chevalieri*, and the relationship between the two taxa in Cameroon requires investigation. *Bertiera bracteolata* and *B. chevalieri* in Lower Guinea share the faint, closed pattern of very fine parallel veinlets on the lower side of the dried leaves, a pattern not seen on the leaves of the two taxa from Upper Guinea.

The fruits of *B. bracteolata* are often described as strongly ribbed (Hepper 1963:158), but this is only the case when they are dry and not when they are fresh (Figure 1c-d).

Additional specimens seen

Liberia: River Cess County – Cestos-Sanguin area, Logging Concession of the Cooper's. fr., 7 Dec. 2002, *Jongkind and staff & students of the University of Liberia 5696* (BR, G, WAG). Sinoe County – African Fruit Company plantation. fl., 29 Jul. 1977, *de Gier & Goll 52* (WAG); Sapo NP, buffer zone, on short distance of the Sinoe River. fl., 28 Nov. 2002, *Jongkind and staff & students of the University of Liberia 5490* (WAG); South of Sayon Town. fr., 1 Dec. 2010, *Jongkind, Bilivogui & Daniels 9969* (BR, MO, K, WAG); Sinoe River bank east of Jalay's Town. ster., 8 Feb. 2016, *Jongkind & Sambolah 13051* (BR).

TABLE 1: Comparison of important taxonomic differences between *B. sinoensis*, *B. bracteolata* and *B. chevalieri*.

Variable	B. sinoensis	B. bracteolata	B. chevalieri
Leaf blade, indumentum	Pubescent on both sides, on and in between the nerves	Pubescent on the midrib and main lateral nerves below, glabrous between them	Pubescent on the midrib and main lateral nerves below, glabrous between them
Leaf blade, size	60–100 mm × 20–33 mm	70–120 mm × 25–60 mm	60–100 mm × 25–55 mm
Stipule lengths	9–13 mm	5–8 mm	4–8 mm
Inflorescence: number of flowers on side branches	Two or more flowers	Two or more flowers	Only one flower
Inflorescence: bracteole length	4–5 mm	1.5–3.0 mm	1–2 mm
Corolla indumentum	*ca.* 0.5 mm long hairs on the outside	*ca.* 0.2 mm long hairs on the outside	*ca.* 0.2 mm long hairs on the outside
Distribution	South-east Liberia only	From Guinea to Gabon but not in Togo and Benin	From Guinea to Cameroon but not in Ghana, Togo and Benin

Conservation notes

Bertiera sinoensis is known from six locations. It is not known from protected areas but it is found close to Sapo National Park. Forest near some localities where it was previously collected has been transformed into oil palm plantations. With an extent of occurrence (EOO) of 1582 km2 and an area of occupancy (AOO) of 24 km2 (based on a cell width of 2 km), *B. sinoensis* is assigned a preliminary conservation status of 'Endangered' [EN B1b(i, ii, iii)+2b(i, ii, iii)] following IUCN Red List Categories and Criteria (IUCN 2015).

Acknowledgements

The author wishes to thank the staff of the Liberian office of Fauna and Flora International for their support for the field work. The most recent expedition on which I could study the new species was funded by the 'Hugo de Vries Fonds'.

Competing interests

The author declares that he has no financial or personal relationships that may have inappropriately influenced him in writing this article.

References

Breteler, F.J., Bakker, F.T. & Jongkind, C.C.H., 2015, 'A synopsis of *Soyauxia* (Peridiscaceae, formerly Medusandraceae) with a new species from Liberia', *Plant Ecology and Evolution* 148, 409–419. https://doi.org/10.5091/plecevo.2015.1040

Hawthorne, W.D., 2013, 'Six new *Pavetta* (Rubiaceae), including three "litter-bin" species from the evergreen forests of Western Africa', *Kew Bulletin* 68, 559–577. https://doi.org/10.1007/S12225-013-9484-7

Hawthorne, W.D. & Jongkind, C.C.H., 2006, *Woody plants of Western African forests: A guide to the forest trees, shrubs and lianes from Senegal to Ghana*, Kew Publishing, Richmond.

Hepper, F.N., 1963, 'Bertiera', in F.N. Hepper (ed.), *Flora of West Tropical Africa*, pp. 158–160, 2nd edn., vol. 2, Crown Agents for Overseas Governments, London.

IUCN, 2015, *IUCN Red List categories and criteria: Version 3.1*, IUCN, Gland, Switzerland.

Lachenaud, O. & Jongkind, C.C.H., 2013, 'New and little-known *Psychotria* (Rubiaceae) from West Africa, and notes on litter-gathering angiosperms', *Plant Ecology and Evolution* 146(2), 219–233. https://doi.org/10.5091/plecevo.2013.765

Nguembou, K.C., Esono, P., Onana, J.M. & Sonké, B., 2006, 'Un *Bertiera* (Rubiaceae) nouveau heterophylle du Cameroun et du Gabon', *Systematics and Geography of Plants* 76, 211–216.

Nguembou, K.C., Ewedje, E.-E.B.K., Droissart, V., Stévart, T. & Sonké, B., 2009, 'Une espèce nouvelle de *Bertiera* (sous-genre *Bertierella*, Rubiaceae) d'Afrique centrale atlantique', *Adansonia* 31(2), 397–406. https://doi.org/10.5252/a2009n2a9

Nguembou, K.C., Sonké, B., Zapfack, L. & Lejoly, J., 2003, 'Les especes camerounaises du genre *Bertiera* (Rubiaceae)', *Systematics and Geography of Plants* 73, 237–280.

Ntore, S., 2008, 'Révision du genre Afrotropical *Pauridiantha* (Rubiaceae)', *Opera Botanica Belgica* 15, 227 pp.

Prance, G.T. & Jongkind, C.C.H., 2015, 'A revision of African Lecythidaceae', *Kew Bulletin* 70, 6. https://doi.org/10.1007/s12225-014-9547-4

Sonké, B., Esono, P., Nguembou, K.C. & Stévart, T., 2005, 'Une nouvelle espece de *Bertiera* Aubl. (Rubiaceae) du sous-genre *Bertierella* decouverte en Guinee Equatoriale et au Cameroun', *Adansonia* 25(2), 309–315.

Tosh, J., Davis, A.P., Dessein, S., De Block, P., Huysmans, S., Fay, M.F. et al. 2009, 'Phylogeny of *Tricalysia* (Rubiaceae) and its relationships with allied genera based on plastid DNA data: Resurrection of the genus *Empogona*', *Annals of the Missouri Botanical Garden* 96, 194–213. https://doi.org/10.3417/2006202

White, F., 1979, 'The Guineo-Congolian Region and its relationships to other phytochoria', *Bulletin van den nationale plantentuin van België* 49, 11–55. https://doi.org/10.2307/3667815

Wittle, P.M. & Davis, A.P., 2010, 'A revision of Madagascan *Bertiera* (Rubiaceae)', *Blumea* 55, 105–110. https://doi.org/10.3767/000651910X525259

World Checklist of Selected Plant Families (WCSP), 2016, *Facilitated by the Royal Botanic Gardens, Kew*, viewed 27 July 2016, from http://apps.kew.org/wcsp/

4

A proposed national strategic framework for the management of Cactaceae in South Africa

Authors:
Haylee Kaplan[1]
John R.U. Wilson[1,2] ⊕
Hildegard Klein[3] ⊕
Lesley Henderson[3] ⊕
Helmuth G. Zimmermann[4]
Phetole Manyama[1]
Philip Ivey[1]
David M. Richardson[2]
Ana Novoa[1,2] ⊕

Affiliations:
[1]Invasive Species Programme, South African National Biodiversity Institute, Kirstenbosch Research Centre, South Africa

[2]Centre for Invasion Biology, Department of Botany and Zoology, Stellenbosch University, South Africa

[3]Agricultural Research Council – Plant Protection Research Institute, South Africa

[4]Helmuth Zimmermann & Associates, South Africa

Corresponding author:
Haylee Kaplan,
haylee.kaplan@gmail.com

Background: South Africa has a long history of managing biological invasions. The rapid increase in the scale and complexity of problems associated with invasions calls for new, more strategic management approaches. This paper explores strategic management approaches for cactus invasions in South Africa. Cacti (Cactaceae) have had a long history of socio-economic benefits, considerable negative environmental and socio-economic impacts, and a wide range of management interventions in South Africa.

Objectives: To guide the future management of cactus invasions, a national strategic framework was developed by the South African Cactus Working Group. The overarching aim of this framework is to reduce the negative impacts of cacti to a point where their benefits significantly outweigh the losses.

Method: Four strategic objectives were proposed: (1) all invasive and potentially invasive cactus species should be prevented from entering the country, (2) new incursions of cactus species must be rapidly detected and eradicated, (3) the impacts of invasive cacti must be reduced and contained and (4) socio-economically useful cacti (both invasive and non-invasive species) must be utilised sustainably to minimise the risk of further negative impacts.

Results: There are currently 35 listed invasive cactus species in the country; 10 species are targeted for eradication and 12 are under partial or complete biological control. We discuss approaches for the management of cactus species, their introduction and spread pathways and spatial prioritisation of control efforts.

Conclusion: A thorough understanding of context-specific invasion processes and stakeholder support is needed when implementing strategies for a group of invasive species.

Introduction

Biological invasions need to be appropriately managed to prevent and reduce negative environmental and socio-economic impacts (Simberloff et al. 2013). However, in most cases, the effort required to manage all invasions far exceeds the available resources. Moreover, such management options can create conflicts between stakeholder groups both directly (by taking away a desired resource) and indirectly (e.g. opposition to the release of chemicals) (Zengeya et al. 2017). Such conflicts can impede management interventions. Management must, therefore, be strategic such that (1) interventions are appropriate and sufficient to meet the goals of management and (2) management efforts are spatially and temporally consistent and coordinated.

A useful approach to strategic planning for biological invasions is to jointly consider groups of species with similar management requirements (van Wilgen et al. 2011). Grouping species for management identifies not only common goals but also common stakeholders. This allows for the simplification of decision-making processes. Such strategies require a good understanding of the target species (i.e. which species need to be managed and how), their pathways (i.e. the routes and vectors of introduction and spread) and the spatial distribution of impacts (i.e. areas containing resources that are susceptible to threats by invasions) (Visser et al. 2017). Effective management interventions should be planned in this context and should incorporate pathway-, species- and area-based approaches (Wilson et al. 2017).

This article explores strategic management planning using the family Cactaceae in South Africa as a case study. Cacti form a distinct taxonomic group that, with a few exceptions, share similar physiological traits, habitat preferences, spread pathways and negative impacts (Novoa et al. 2015b, 2016b) and, importantly, are also managed in similar ways (Walters et al. 2011). The aim of this article is to explore the process of developing a national strategic framework for a group of invasive species. This strategic planning process relies heavily on a good understanding of invasion processes specific to the target group and effective stakeholder engagement.

The history and status of Cactaceae in South Africa

Cacti are among the most widespread and dominant groups of invasive plants in South Africa (Nel et al. 2004; van Wilgen et al. 2012), with 35 species already listed as invaders under the *National Environmental Management: Biodiversity Act* no. 10 of 2004 (NEM:BA). This is a result of their long history of introduction and utilisation in South Africa for agriculture and ornamental horticulture (Walters et al. 2011). Benefits of cacti in South Africa are derived from a range of socio-economically important activities, such as the horticulture trade, and commercial and subsistence agriculture. Their impacts are mainly related to costs associated with losses of biodiversity, ecological functioning and agricultural productivity. The country's arid interior offers favourable conditions for the establishment of drought-adapted species, such as cacti. Consequently, South Africa is a global hotspot of cactus invasions, with one of the highest diversities of naturalised cactus species outside the family's native range (Novoa et al. 2015b).

There is also a long history of cactus management in South Africa (Zimmermann, Moran & Hoffmann 2004). The genus *Opuntia* was among the first invasive taxa to be regulated (under the *Agricultural Pests Act* no. 11 of 1911), and South Africa's first biological control programme was implemented against the invasive cactus *Opuntia monacantha* in 1913 (Zimmermann et al. 2004). Since then, 15 additional cactus species have been targeted for biological control (Klein 2011; Paterson et al. 2011), and several widespread cacti are controlled through physical and chemical clearing, as part of the Working for Water programme (van Wilgen et al. 2012). Ten cactus species are listed as category 1a invasive species under NEM:BA and have thus been targeted for eradication (Wilson et al. 2013). However, the coordination and prioritisation of these cactus control programmes at a national level has yet to be implemented.

Management decisions need to be based on a clear understanding of the underlying invasion processes involved. Here we discuss the distribution and abundance, the benefits and impacts, and the pathways of cacti in South Africa to provide suitable context for management planning.

Distribution and abundance

An estimated 400 cacti taxa have been introduced to South Africa (Walters et al. 2011), many of which are currently present in gardens and private collections. Thirty-five cactus taxa are invasive in South Africa (Figure 1). Distributions per taxon vary; some occur country-wide (e.g. *Opuntia ficus-indica* is widely cultivated for fodder and fruit and has naturalised at many of these sites), while other taxa are confined to relatively small ranges (e.g. *Opuntia pubescens* has naturalised at a single site in a botanical garden) or occur in low numbers at several isolated localities (e.g. *Opuntia microdasys*, a popular garden ornamental, has naturalised at over 50 sites). However, introductions have occurred continuously for several decades without formal risk assessment, and some invasive cacti have shown long lag phases (>50 years between introduction and the start of invasive spread; Walters et al. 2011). It is therefore likely that a large invasion debt (sensu Rouget et al. 2016) has accumulated, which must be considered when formulating long-term management plans.

Benefits and impacts

Cacti have many important socio-economic benefits in South Africa. Around 300 species of cacti are imported to South Africa annually for ornamental horticultural purposes (Novoa et al. 2017). Cacti are highly valued for their use as ornamentals and are widely popular in gardens and as curiosity plants for collectors. Several cactus species are also used for commercial and subsistence agriculture where they are farmed for food, fruit and livestock fodder. These drought-resistant crops enable significant increases in the productivity of marginal land (Brutsch & Zimmermann 1993). These industries (i.e. horticultural and agricultural) are noteworthy contributors to the economy and food security of a developing country such as South Africa.

Cacti also have substantial negative environmental and socio-economic impacts (Barbera, Inglese & Pimienta 1995; Novoa et al. 2016b). Their ability to spread vegetatively results in the formation of large, dense invasive stands (in some cases up to 100% canopy cover) that exclude other vegetation and animals. Spines and glochids which are present on most species are damaging to small wildlife and livestock that have not coevolved with cacti (Walters et al. 2011). This translates into costly negative impacts, particularly to agricultural systems. Invasions of rangelands result in reductions in productivity and capacity of commercial and subsistence grazing (Lloyd & Reeves 2014). Added to these impacts are the considerable costs involved in controlling cactus invasions. Although the full cost of the impacts of cactus invasions in South Africa has not yet been quantified, control efforts between 1995 and 2008 cost nearly ZAR100 million (in 2008 equivalent Rand; van Wilgen et al. 2012).

Pathways

There are several main pathways along which cacti are introduced and spread (intentionally or unintentionally) around South Africa (Table 1). Most introduction pathways are related to utilisation of cacti (e.g. for ornamental horticulture, food production and livestock fodder) and involve intentional introductions. Local scale spread is often unintentional, for

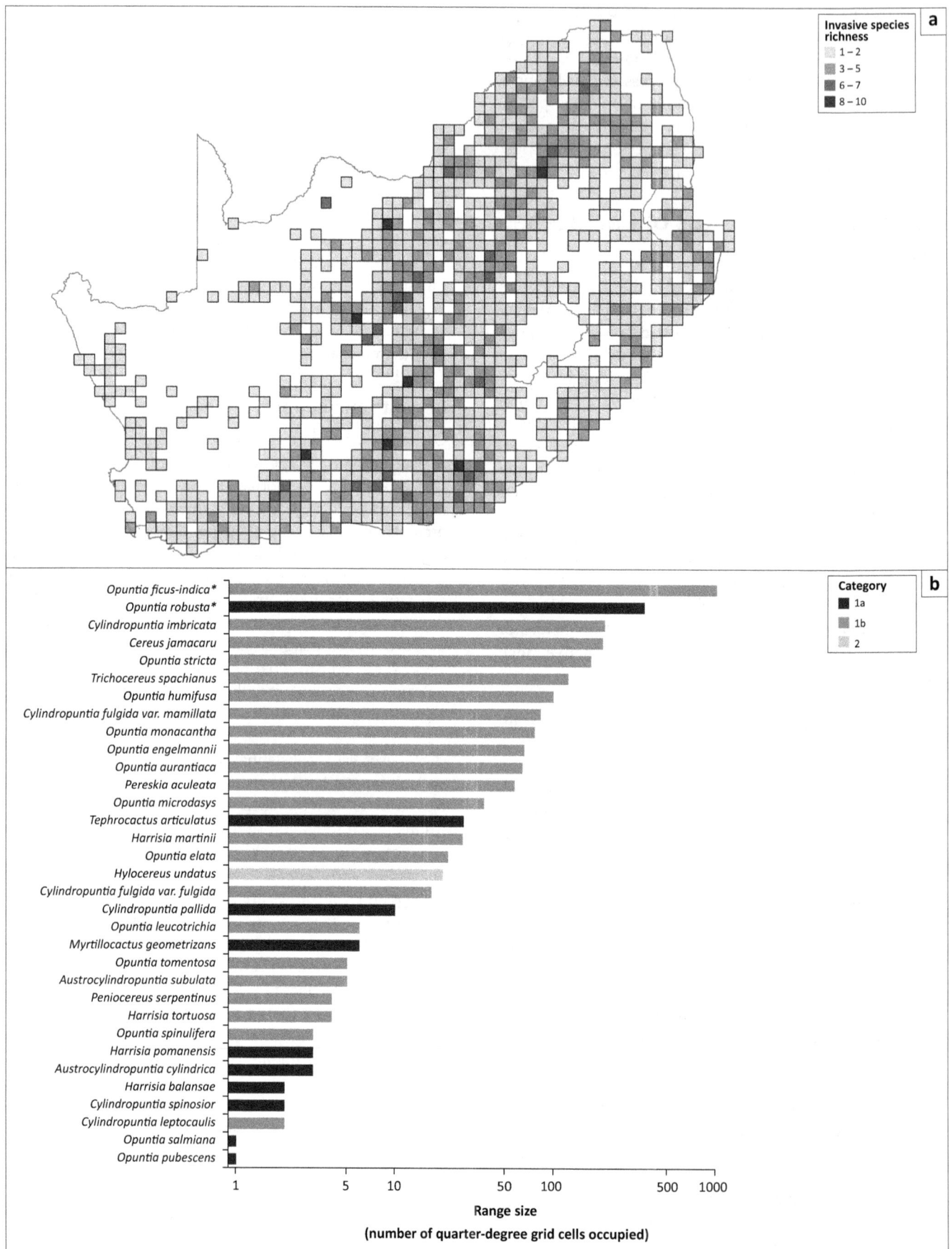

Source: Data sourced from Southern African Plant Invaders Atlas (SAPIA) database, accessed March 2016, for an updated list see Henderson and Wilson (2017), ARC-Plant Protection Research Institute, Pretoria
Shadings in (b) indicate categories in the 2016 Regulations of the *National Environmental Management: Biodiversity Act*.

*, Excludes spineless cultivars.

FIGURE 1: The current extent of cactus invasions in South Africa, showing (a) the distribution and species richness of listed invasive cacti per quarter degree grid cell (QDGC) and (b) range sizes of invasive cacti in South Africa.

TABLE 1: Invasion pathways for cacti at different scales and recommended management interventions to limit introduction and spread of invasive species. Pathways are listed in descending order of ease of management.

Pathway	International scale	Regional scale	Local scale	Management interventions
Horticulture (legal)	Permitted import of seeds by wholesalers	Distribution of plants to towns through nursery industry	Informal trade at garden clubs, markets	Correct listing of prohibited species
				Engagement with nursery stakeholders
				Enforcement of regulations
Food production	Import of spineless *Opuntia* cladodes	Distribution of *Opuntia* fruits for sale in supermarkets	Escape from cultivation and reversion to spiny forms	Biological and chemical control of escaped spiny plants
Livestock farming (fodder)	Not applicable	Farmers share plant material for fodder production	Movement of plant material by animals.	Increased public awareness
				Enforcement of regulations
				Livestock exclusion in invaded areas
Horticulture (illegal)	Import of seeds and live plants through online trade without permit	Illegal sale of plants in the nursery trade	Informal trade at garden clubs, markets	At-border screening of imports
			Exchange of plants among gardeners and succulent enthusiasts	Post-border inspections of nurseries and enforcement of regulations
				Improved public awareness of illegal cacti
Animal dispersal	Not applicable	Long-distance dispersal of seeds by birds	Movement of plant material and seeds by animals, for example, elephants and baboons	Contain spread by detecting and removing outlier populations
Abiotic dispersal	Not applicable	Not applicable	Dispersal of seeds by water and wind	Contain spread by detecting and removing outlier infestations

example, via animal-mediated dispersal or escape from cultivation. These pathways and their associated vectors need to be managed to limit the movement and subsequent invasions by invasive and potentially invasive species.

Almost all contemporary introductions of cacti are via the horticultural trade to meet a growing demand for ornamental cacti. The horticulture pathway has contributed the most invasive cacti to South Africa (Walters et al. 2011). Legal importation of cacti is conducted by a small number of nursery wholesalers in South Africa (Novoa et al. 2017). However, most of the approximately 300 ornamental cactus species imported by wholesalers are unlikely to become invasive (Novoa et al. 2015b). Those that are potentially invasive make up a small proportion of the horticulture trade, which means that prohibition of trade in selected species or genera is unlikely to have a large enough impact on the horticulture industry to cause resistance.

Illegal horticultural introductions of invasive species are more challenging to manage. Most cacti are imported as seed, which makes detection of illegal imports difficult. Screening seed imports by seed size has recently been proposed as an accurate method of discriminating invasive and potentially invasive from non-invasive cacti: larger seed size is correlated with invasiveness (Novoa et al. 2016c). However, we believe that illegal imports make up a small proportion of the total horticultural trade in cacti. A major threat is the local dissemination of invasive species already present in the country. While registered nurseries and growers generally avoid trade in listed invasive species, there is a large, unregulated trade in cacti which are sold and exchanged informally (personal observation).

A source of local scale spread of invasive cacti is the utilisation of certain opuntioid cacti for food production and livestock fodder. In the majority of cases, spineless cultivars are used although there is a risk of reversion to spiny populations

(Flepu et al. n.d.), which increases the likelihood of escape and invasion. Escape from cultivation is facilitated primarily by animal dispersal of seeds and plant material through fruit consumption and vegetative propagation (Dean & Milton 2000; Foxcroft & Rejmánek 2007).

The South African Cactus Working Group

In response to the need for strategic cactus management, a national working group (the South African Cactus Working Group; SACWG – see Appendix 1) was established in 2013 to develop and coordinate the implementation of a national management strategy for invasive cacti. The working group consists of representatives from all relevant organisations in South Africa involved in research, policy or management of cactus invasions. The primary benefit of the working group is its use as a forum to exchange ideas and current knowledge and to inform ongoing research and interventions. This enhances collaboration and cooperation among government departments and organisations and provides a more conducive environment for strategic decision-making. For instance, a recent proposal by the Department of Environmental Affairs to prohibit the entire Cactaceae family (with considerable socio-economic repercussions) was retracted in favour of a more nuanced and risk-appropriate listing proposed by the SACWG through a strategic decision-making process (Novoa et al. 2015a). This ability for cohesive, expert-driven decision-making based on current evidence makes the cactus working group a suitable entity for national scale strategic planning.

Stakeholder engagement

Although strategic decision-making is overseen by the SACWG, which represents organisations that are mandated to regulate and control cacti, the outcomes will clearly affect multiple external stakeholders (Figure 2). Proposing control

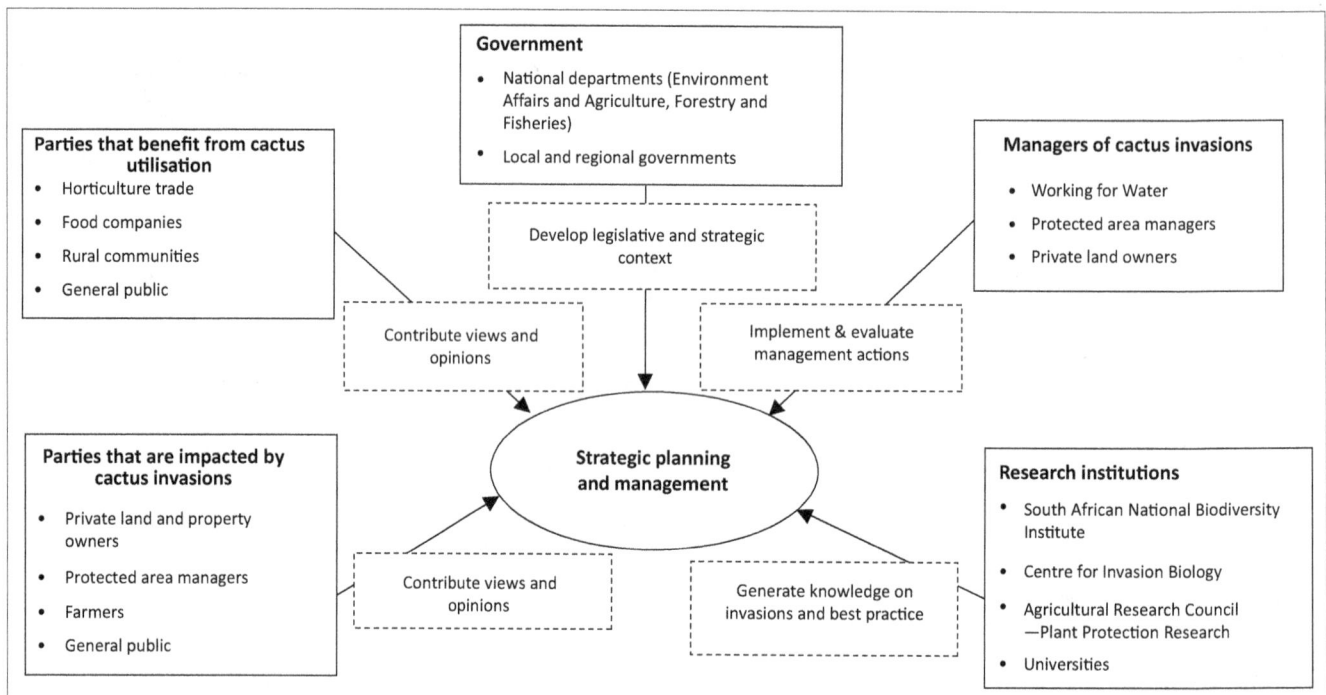

FIGURE 2: Stakeholder groups and their representative organisations involved in strategic planning and management of cactus invasions in South Africa.

or regulation of non-native species with benefits can create conflicts of interest that hinder management success (Estévez et al. 2015). Overcoming these conflicts can be difficult, especially where livelihoods are at stake, as described by Beinart (2003) for prickly pear in South Africa. The challenge is to garner wide stakeholder support for management interventions when values around certain invasive species may differ. Effective stakeholder engagement during the strategic planning process is therefore essential (García-Llorente et al. 2011).

To achieve such engagement in South Africa, all the stakeholders were invited to participate in a workshop with the aim of increasing awareness of different viewpoints and values associated with cactus impacts and benefits (Novoa et al. 2016a). Following this workshop, there was somewhat of a convergence of stakeholder perceptions, which facilitated a smoother decision-making process. This was encouraging, as altering behaviours to support strategic management of cacti (e.g. not propagating and selling invasive cacti) involves a change in perceptions and values associated with invasions (Selge, Fischer & van der Wal 2011).

Strategic framework

A national strategic framework (Figure 3) was constructed by the SACWG over four consecutive workshops during 2013–2015, taking all the stakeholders' opinions into account. The overarching vision of the strategy is to reduce the negative impacts of cacti to a point where the benefits of having them in the country would significantly outweigh the losses. To achieve this vision, we considered that four strategic objectives need to be met based on the approach by van Wilgen et al. (2011): (1) all invasive and potentially invasive cactus species are prevented from entering the country,

(2) new incursions are detected and eradicated, (3) the invasive impacts of species are reduced and contained and (4) socio-economically useful cacti are utilised sustainably.

A species-based approach

The next step was to determine what management action should be taken for each species. To do this, we developed a protocol with five endpoints (Table 2; Figure 4): (1) Do nothing (i.e. no regulation of species needed unless further evidence to the contrary), (2) Prevention (i.e. prohibition and preventing entry of potentially invasive species into the country), (3) Eradication (i.e. eradication of new incursions), (4) Containment (i.e. stopping or slowing the spread of invasive species) and (5) Impact reduction (i.e. maintaining invasive populations at densities with tolerable impacts). Before implementing the strategies outlined in the framework, species-based management goals must be assigned to the each taxon (i.e. all species in the Cactaceae family) using the decision protocol in Figure 4.

Unlike within other taxonomic groups (e.g. Australian acacias, van Wilgen et al. 2011), within the family Cactaceae, all species considered as useful by the legal trade are not invasive or harmful – that is, the species used in agriculture are the non-invasive spineless cultivars of O. ficus-indica and Opuntia robusta, and no invasive species are considered as useful by the international legal trade (Novoa et al. n.d.). Therefore, the decision protocol in Figure 4 will result in 'do nothing' for useful species. However, this might need to be amended in the future.

Preventing introductions of high risk species is an important and often highly cost-effective step in reducing the potential impacts of invasions. Risk assessment is needed to distinguish

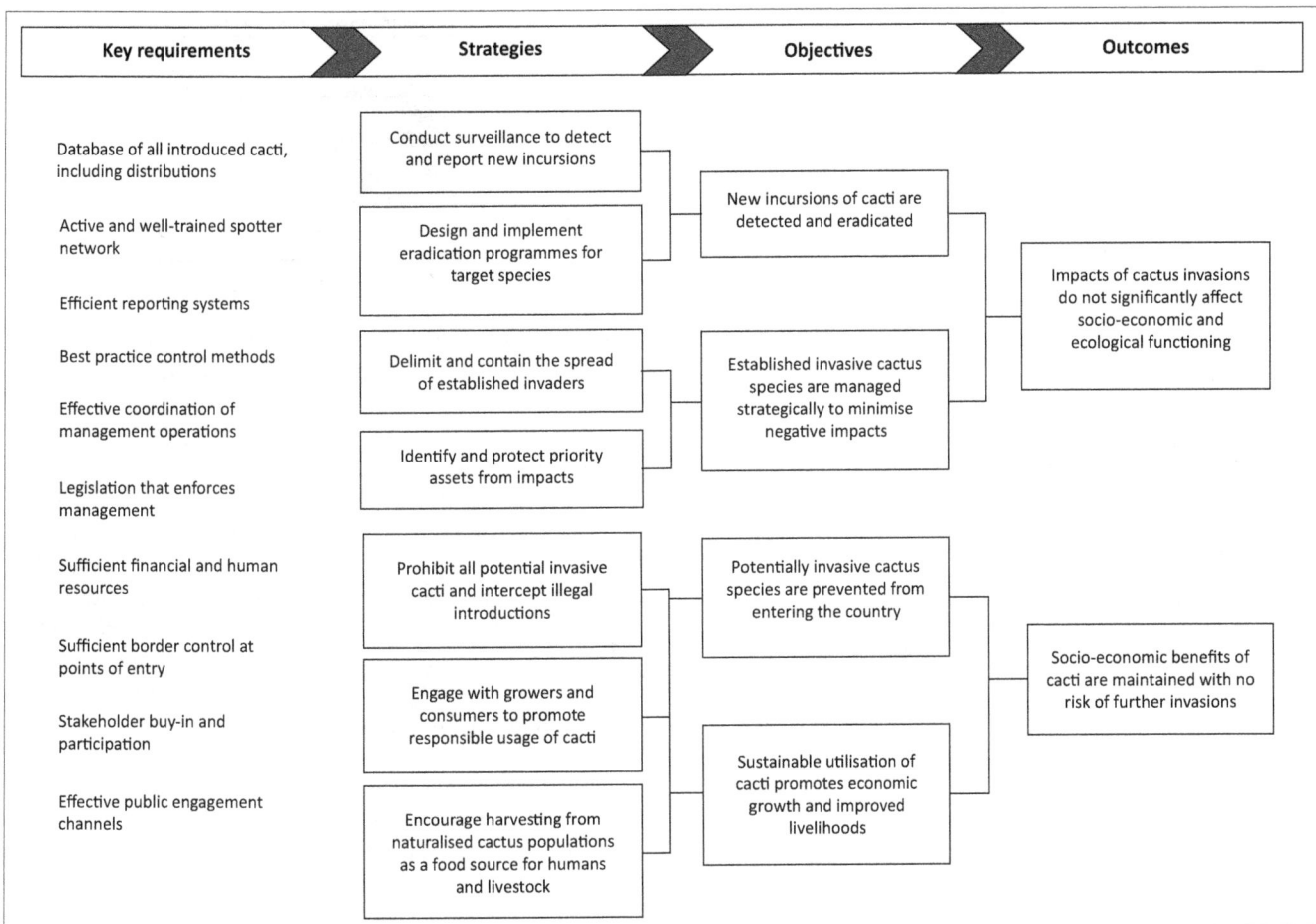

FIGURE 3: National strategic framework and key requirements for the management of cactus invasions in South Africa.

TABLE 2: Implementation of species-based management of cacti in South Africa.

Species-based goal/endpoint	Implementation	Actions to date in South Africa	Reference
Do nothing	Species with low risk of invasiveness are not regulated	Risk assessment of Cactaceae	Novoa et al. 2015b
Prevention	Species with high risk of invasiveness or invasive elsewhere are prohibited	Risk assessment of Cactaceae	Novoa et al. 2015b
		Four invasive genera prohibited under NEM:BA regulations	
Eradication	New incursions of cacti and naturalised species with limited distributions are eradicated from the country	New incursions detected and recorded in SAPIA	Henderson 2007
		Ten cactus species currently being assessed for eradication feasibility	Wilson et al. 2013
Containment	The spread of species with high risk of range expansion is stopped or slowed	No formal assessment of containment feasibility for cacti to date	
		Use of biological control to prevent seed set in some species	Paterson et al. 2011
Impact reduction	Populations of widespread invasive species are reduced to tolerable levels	Biological control of 16 cactus species	Paterson et al. 2011
		Strategies for integrated management of cacti in protected areas developed	Lotter and Hoffmann 1998

NEM:BA, *National Environmental Management: Biodiversity Act*; SAPIA, Southern African Plant Invaders Atlas.

species that pose significant invasive threats from those species that are safe to utilise. All invasive or potentially harmful species should be prohibited. A global assessment of the Cactaceae by Novoa and colleagues (2015b) has shown that invasiveness in cacti is correlated with growth form (i.e. morphological traits that increase the ability to propagate, e.g. segmented stems) and native range size (Novoa et al. 2015b), and most invasive species in the family belong to 13 genera (out of 130), particularly in the Opuntioideae subfamily. Species-specific preventative measures, such as prohibition, should thus be targeted at those species identified as potential invaders. Species from four genera (*Cylindropuntia, Harrisia, Opuntia* and *Pereskia*) are currently on the NEM:BA prohibited list (i.e. they may not be introduced to South Africa) because of the prevalence of globally invasive taxa in these groups.

Given the potentially large invasion debt of cacti in South Africa, new instances of naturalisation are likely to occur. New incursions of cacti should be eradicated where feasible. Feasibility of eradication is assessed on an individual species basis and broadly depends on reproductive and dispersal characteristics and the eradication effort required relative to available resources (Panetta 2015). Approaches for evaluating the feasibility of eradication of invasive taxa in South Africa have been developed (Jacobs, Richardson & Wilson 2014; Kaplan et al. 2012) and can be adapted for application to

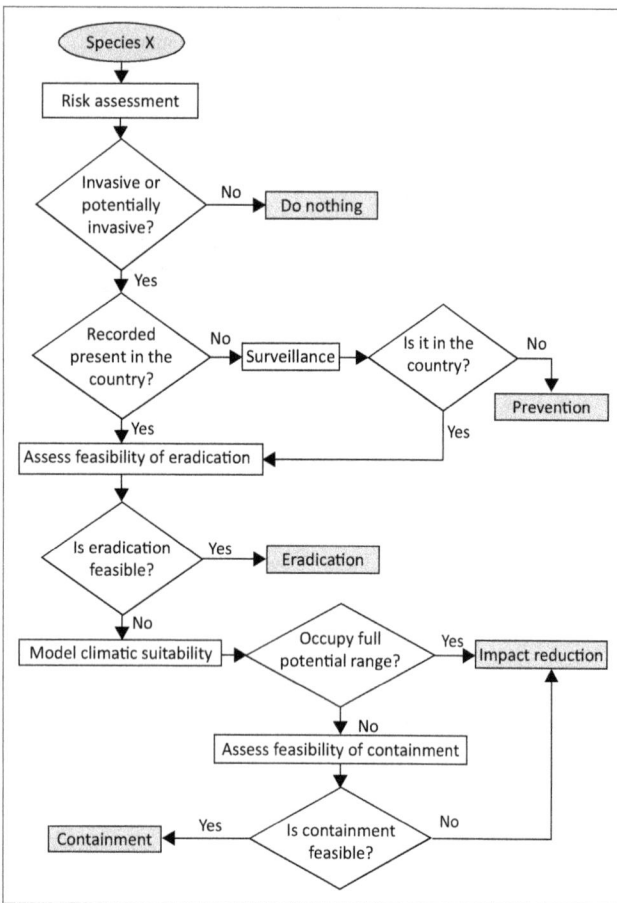

FIGURE 4: Decision framework for setting species-based management goals for invasive cactus species.

TABLE 3: Effectiveness of control of listed invasive cacti in South Africa.

Variables	NEM:BA category			Total
	1a	1b	2	
Taxa with registered herbicides	0	8	0	8
Taxa under complete biological control	0	7	0	7
Taxa under partial biological control	3	8	0	11
Taxa where the only current management option is physical removal	7	2	1	10
Total Taxa listed	10	25	1	36

NEM:BA, *National Environmental Management: Biodiversity Act.*

cactus taxa. Although cacti do not form long-lived seed banks, their propensity for vegetative propagation is likely to complicate eradication efforts; small pieces of plant material can break off, spread and root easily (Bobich & Nobel 2001), forming new and often inconspicuous plants. Species considered as feasible eradication targets should have few, localised populations (e.g. *O. pubescens*; Cindi & Jaca 2016) and highly effective control treatments available.

Species for which eradication is not feasible should be controlled by containment of spread and reduction of negative impacts. Feasibility of containment is also assessed on a species-by-species basis. Containment is considered only for those species that do not occupy their full potential invasive range in South Africa (based on bioclimatic models) and which have the ability to spread. Preventing further range expansion of cacti or slowing the spread would involve stopping seed production through biological control, and setting up barrier zones around existing infestations and regular monitoring of these zones to detect and remove extra-limital incursions (Sharov & Liebhold 1998). As with eradication, containment of a species relies on complete removal of populations to limit further spread and should be considered only for species for which there are effective control methods (Table 3). To date, there have been no attempts to contain cacti in South Africa, although there have

been instances where fruit production has been successfully hindered by biological control agents (Paterson et al. 2011).

Impact reduction is an appropriate goal for widespread, dominant cactus species. Thresholds for impact tolerance may vary by land use type because of relative susceptibility to impacts. For example, protected areas and rangelands will likely have a much lower tolerance to impacts than transformed or urban areas. Consequently, populations will require different management interventions to reduce densities to the required maintenance levels, although further research is needed to determine optimal maintenance levels under different land use scenarios.

A highly effective management tool for reducing the impacts of cacti is biological control (McFadyen 1998). Twelve cactus species are currently under complete or partial control in South Africa by three groups of biological control agents: cochineal insects (*Dactylopius* sp.), a mealybug (*Hypogeococcus festerianus*) and cactus moth (*Cactoblastis cactorum*) (Table 3). In some cases, novel associations between biocontrol agents and host species can result in almost complete extirpation of populations, such as with *Cylindropuntia fulgida* where augmented release of control agents resulted in a kill rate of up to 99% (T. Xivuri et al., unpublished data). At least eight cactus invaders are now at full maintenance levels (with no or very limited control measures required) because of successful biological control which necessitates a residue population of the weed for the survival of the biological control agents. Under such levels of biological control, beneficial species such as *O. ficus-indica* and *O. robusta*, which are used as fodder sources, could potentially be delisted from the regulations as they no longer pose significant threats. In cases where biological control is less effective, an integrated management approach combining biological, chemical and mechanical control should be implemented. For example, biological control of *Opuntia aurantiaca* is less effective in higher rainfall areas (Moran & Zimmermann 1991), necessitating integrated management interventions. Investment in biological control will be a key element of any strategy to deal with cactus invasions (Zachariades et al. 2017).

A pathway-based approach

Introduction and spread pathways must be identified and prioritised to prevent and contain the impacts of cacti (Table 1). The most difficult pathways to manage are those that involve abiotic and animal vectors of spread. Migrating animals, such as birds, or floodwaters can disperse cactus propagules over long distances (Walters et al. 2011). These are fairly stochastic processes and should therefore be managed on a case-by-case basis. This will most likely involve species-based containment or impact reduction approaches, although certain vectors, such as livestock animals and transport vehicles, should be inspected and controlled in high risk areas.

Human-mediated pathways (including both intentional and unintentional introductions) can be managed more strategically. The two broad human-mediated pathways of intentional cactus introductions and dissemination in South Africa are horticulture and agriculture (including legal and illegal trade). To ensure that no prohibited cacti are introduced intentionally by growers through the legal cactus trade, the existing permit-regulation process needs to be well managed. However, intentional illegal trade of cacti is facilitated mainly by international online trading (such as Ebay.com) where suppliers are not necessarily held to the same import regulations as 'legal' importers (Humair et al. 2015). Moreover, unintentional introductions of invasive cacti may still occur, for instance, because of misidentification or incorrect labelling of seed imports.

We believe that legal horticultural trade of cacti would be relatively easy to regulate through increased awareness of prohibited and regulated cacti among importers and their international suppliers (Novoa et al. 2017), and enforcement of compliance where necessary. Illegal trade of cacti can be best managed through at-border screening of seed imports, increased public awareness of the risks of cactus invasions and the promotion of safer, non-invasive cacti and succulents for gardening and landscaping. Managing escape from cultivation and subsequent spread of impacts of cacti requires responsible utilisation. Engaging with livestock farmers and growers of cacti to increase awareness of the threats posed by cactus invasions and ways to prevent them, such as animal exclusion from invaded areas, is recommended. The movement of cacti across the land borders between South Africa and other African countries is a potential threat, although cactus biological control agents can also move in this way (Faulkner et al. 2017).

An area-based approach

Because of lack of funding and capacity for managing invasions, often species cannot be managed across their entire invasive range, particularly those species that are very widespread. An alternative is to spatially prioritise management efforts to areas where the majority of impacts or potential impacts are likely to occur (Downey et al. 2010). These priority areas contain ecologically or socio-economically important assets that are at highest risk from impacts by cactus invasions. These include protected areas and livestock production regions (Figure 5). Criteria for prioritisation of certain areas would include level of conservation concern (e.g. a formally protected area vs. a

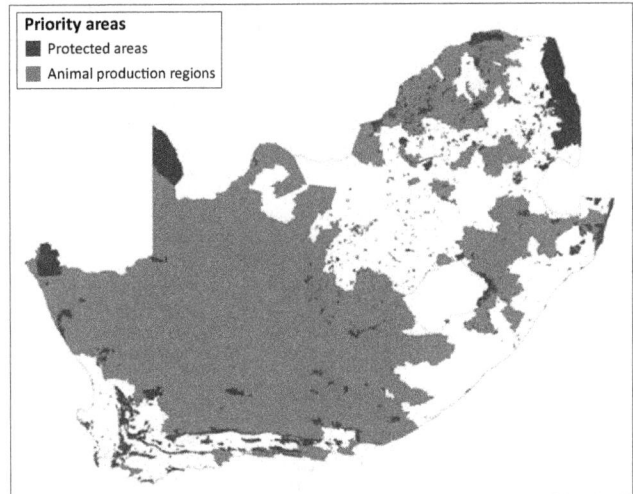

Source: Data sourced from Department of Agriculture, Forestry and Fisheries.

FIGURE 5: Priority areas for cactus management in South Africa based on assets most vulnerable to the negative impacts of cactus invasions.

critical biodiversity area), grazing capacity or agricultural productivity, and eco-tourism value. These criteria should be rated and ranked following a multi-criteria decision analysis framework (e.g. Forsyth et al. 2012).

For simplicity of management, prioritisation should occur at the scale of land ownership, for example, individual farms or nature reserves. Further research on fine-scale impacts and local invasion hotspots would assist in refining priority areas to enable better allocation of resources. If available resources are insufficient, further prioritisation and trade-offs will need to be made to ensure that strategic objectives are being met at a national scale (van Wilgen et al. 2016).

The way forward

Strategic management needs to incorporate a means of monitoring and assessing the efficacy of strategies towards achieving the desired outcomes. An effective monitoring programme should assess the accuracy of the problem definition, audit the achievement of goals and provide feedback to evaluate policy (Rogers & Biggs 1999). Although targets and endpoints are not explicitly defined here, we propose a set of indicators that can be used to evaluate progress towards achieving strategic objectives (Table 4).

For management to be adaptive, strategies must be based on clear evidence. To this end, the SACWG must ensure that management operations that are encompassed by this framework are well documented and aligned with current knowledge of cacti and best practice. This requires effective cooperation and collaboration among the partner organisations within the SACWG. Likewise, collaboration with relevant international experts is essential. Accordingly, several members of the SACWG are represented on an international cactus working group (http://academic.sun.ac.za/cib/projects/cactuswg/index.asp), which has recently been initiated to collate information on cactus invasions and their management globally.

TABLE 4: Indicators of progress towards achieving the national strategic objectives for cactus management in South Africa.

Management objective	Indicator	Action
Prevention of invasive cactus species	Number of new imports of potentially invasive species	Refine risk assessments based on current data
Sustainable utilisation of cacti	Number of instances where nursery industry does not comply with invasive species policy	Audit nursery compliance with NEM:BA Alien and Invasive Species regulations
Eradication of new incursions	New incursions of cacti detected before naturalisation or spread occurs	Active surveillance of cactus invasion hotspots
	Successful eradication of cactus species	Monitor progress of eradication programmes
Reduction of impacts of widespread invasive cacti	Stakeholder support for control of invasive cacti	Engage stakeholders in the management of invasive cacti
	Effective control methods are available for invasive species	Develop and test biological and chemical control for all listed invasive species
	Stable and decreasing population densities	Monitor changes in population densities over time at fixed points

NEM:BA, *National Environmental Management: Biodiversity Act.*

Importantly, this strategy requires ownership in order to ensure implementation and continuity over time. Ownership of the strategy should lie with the South African National Department of Environmental Affairs that represents NEM:BA and the Alien and Invasive Species Regulations.

General conclusions

Strategic management of invasions requires integration of pathway-, area- and species-based interventions. To explicitly incorporate these approaches into strategic planning and management objectives, a good understanding of invasion processes is necessary. We demonstrated this for cacti which have benefited from both a well documented history of management and a large body of research in South Africa. Unfortunately, this is not the case for many other groups of invasive species requiring management. In instances where data and knowledge are insufficient, the formation of taxon-specific working groups, such as the SACWG, is recommended to bring together stakeholders to build the expertise and knowledge necessary for strategic planning at a national level. Coordination and buy-in from stakeholders is essential for successful management of invasive species, especially at a national scale. We believe that future management of cacti in South Africa will be greatly enhanced through the adoption of this proposed strategic framework and with continual coordination and engagement between SACWG stakeholders. More work is needed to improve the framing of issues and problems relating to invasive cacti with ongoing consultation with all stakeholders to identify innovative solutions (Zengeya et al. 2017).

Acknowledgements

We thank members of the South African Cactus Working Group for their inputs into the development of the national strategic framework and for their useful insights and discussion. We also thank E. van Wyk, I. Nänni, T. Xivuri and J. Renteria for their useful comments on the manuscript.

Funding for this work was provided by the Department of Environmental Affairs, through the South African National Biodiversity Institute's Invasive Species Programme. H.K. and L.H. thank the Agricultural Research Council for facilitating their involvement. D.M.R. acknowledges funding from the DST-NRF Centre of Excellence for Invasion Biology and the National Research Foundation (grant no. 85417).

Competing interests

The authors declare that they have no financial or personal relationship(s) that may have inappropriately influenced them in writing this article.

Authors' contributions

H.Kaplan led the development of the strategic framework with A.N. and J.R.U.W. H.Kaplan and A.N. wrote the article. All other authors contributed to the strategic framework and the article during workshops held by the SACWG. H.Klein H.G.Z. and P.M. made inputs regarding biological, chemical and integrated control of cacti. P.I. and D.M.R. provided guidance on conceptualisation of the strategic framework. L.H. provided distribution and abundance data for cacti.

References

Barbera, G., Inglese, P. & Pimienta, E., 1995, *Agro-ecology, cultivation and uses of cactus pear*, Food and Agriculture Organization of the United Nations, Rome.

Beinart, W., 2003, *The rise of conservation in South Africa: Settlers, livestock, and the environment 1770–1950*, Oxford University Press, New York.

Bobich, E.G. & Nobel, P.S., 2001, 'Vegetative reproduction as related to biomechanics, morphology and anatomy of four cholla cactus species in the Sonoran Desert', *Annals of Botany* 87, 485–493. https://doi.org/10.1006/anbo.2000.1360

Brutsch, M.O. & Zimmermann, H.G., 1993, 'The prickly pear (*Opuntia ficus-indica* [Cactaceae]) in South Africa: Utilization of the naturalized weed, and of the cultivated plants', *Economic Botany* 47, 154–162.

Cindi, D.D. & Jaca, T.P., 2016, 'First record of *Opuntia pubescens* H. L. Wendland ex Pfeiffer, 1835 naturalised in South Africa', *BioInvasions Records* 5(4), 213–219. https://doi.org/10.3391/bir.2016.5.4.04

Dean, W.R.J. & Milton, S.J., 2000, 'Directed dispersal of *Opuntia* species in the Karoo, South Africa: Are crows the responsible agents?', *Journal of Arid Environments* 45, 305–314. https://doi.org/10.1006/jare.2000.0652

Downey, P.O., Williams, M.C., Whiffen, L.K., Auld, B.A., Hamilton, M.A., Burley, A.L. et al., 2010, 'Managing alien plants for biodiversity outcomes – The need for triage', *Invasive Plant Science and Management* 3, 1–11. https://doi.org/10.1614/IPSM-09-042.1

Estévez, R.A., Anderson, C.B., Pizarro, J.C. & Burgman, M.A., 2015, 'Clarifying values, risk perceptions, and attitudes to resolve or avoid social conflicts in invasive species management', *Conservation Biology* 29, 19–30. https://doi.org/10.1111/cobi.12359

Faulkner, K.T., Hurley, B.P., Robertson, M.P., Rouget, M. & Wilson, J.R.U., 2017, 'The balance of trade in alien species between South Africa and the rest of Africa', *Bothalia* 47(2), a2157.

Forsyth, G.G., Le Maitre, D.C., O'Farrell, P.J. & van Wilgen, B.W., 2012, 'The prioritisation of invasive alien plant control projects using a multi-criteria decision model informed by stakeholder input and spatial data', *Journal of Environmental Management* 103, 51–57. https://doi.org/10.1016/j.jenvman.2012.01.034

Foxcroft, L.C. & Rejmánek, M., 2007, 'What helps *Opuntia stricta* invade Kruger National Park, South Africa: Baboons or elephants?', *Applied Vegetation Science* 10, 265–270. https://doi.org/10.1111/j.1654-109X.2007.tb00525.x

García-Llorente, M., Martín-López, B., Nunes, P.A.L.D., González, J.A., Alcorlo, P. & Montes, C., 2011, 'Analyzing the social factors that influence willingness to pay for invasive alien species management under two different strategies: Eradication and prevention', *Environmental Management* 48, 418–435. https://doi.org/10.1007/s00267-011-9646-z

Henderson, L. 2007. Invasive, naturalized and casual alien plants in southern Africa: A summary based on the Southern African Plant Invaders Atlas (SAPIA). *Bothalia.* 37(2):215–248. https://doi.org/10.4102/abc.v37i2.322

Humair, F., Humair, L., Kuhn, F. & Kueffer, C., 2015, 'E-commerce trade in invasive plants', *Conservation Biology* 29(6), 1658–1665. https://doi.org/10.1111/cobi.12579

Jacobs, L.E.O., Richardson, D.M. & Wilson, J.R.U., 2014, '*Melaleuca parvistaminea* Byrnes (Myrtaceae) in South Africa: Invasion risk and feasibility of eradication', *South African Journal of Botany* 94, 24–32. https://doi.org/10.1016/j.sajb.2014.05.002

Kaplan, H., Van Zyl, H.W.F., Le Roux, J.J., Richardson, D.M. & Wilson, J.R.U., 2012, 'Distribution and management of *Acacia implexa* (Benth.) in South Africa: A suitable target for eradication?', *South African Journal of Botany* 83, 23–35. https://doi.org/10.1016/j.sajb.2012.07.016

Klein, H., 2011, 'A catalogue of the insects, mites and pathogens that have been used or rejected, or are under consideration, for the biological control of invasive alien plants in South Africa', *African Entomology* 19, 515–549. https://doi.org/10.4001/003.019.0214

Lloyd, S. & Reeves, A., 2014, *Situation statement on opuntioid cacti* (Austrocylindropuntia spp., Cylindropuntia spp. and Opuntia spp.) in Western Australia, Invasive Species Program, Department of Agriculture and Food, Western Australia.

Lotter, W. & Hoffmann, J.H., 1998, 'An integrated management plan for the control of *Opuntia stricta* (Cactaceae) in the Kruger National Park, South Africa', *Koedoe* 41, 63–68.

McFadyen, R.E.C., 1998, 'Biological control of weeds', *Annual Review of Entomology* 43, 369–393. https://doi.org/10.1146/annurev.ento.43.1.369

Moran, V.C. & Zimmermann, H.G., 1991, 'Biological control of jointed cactus, *Opuntia aurantiaca* (Cactaceae), in South Africa', *Agriculture, Ecosystems and Environment* 37, 5–27. https://doi.org/10.1016/0167-8809(91)90136-L

Nel, J.L., Richardson, D.M., Rouget, M., Mgidi, T.N., Mdzeke, N., Le Maitre, D.C. et al., 2004, 'A proposed classification of invasive alien plant species in South Africa: Towards prioritizing species and areas for management action', *South African Journal of Science* 100, 53–64

Novoa, A., Kaplan, H., Kumschick, S., Wilson, J.R.U. & Richardson, D.M., 2015a, 'Soft touch or heavy hand? Legislative approaches for preventing invasions: Insights from cacti in South Africa', *Invasive Plant Science and Management* 8, 307–316. https://doi.org/10.1614/IPSM-D-14-00073.1.

Novoa, A., Kaplan, H., Wilson, R.U. & Richardson, D.M., 2016a, 'Resolving a prickly situation: Involving stakeholders in invasive cactus management in South Africa', *Environmental Management* 57, 998–1008. https://doi.org/10.1007/s00267-015-0645-3

Novoa, A., Kumschick, S., Richardson, D.M., Rouget, M. & Wilson, J.R.U., 2016b, 'Native range size and growth form in Cactaceae predict invasiveness and impact', *NeoBiota* 30, 75–90. https://doi.org/10.3897/neobiota.30.7253

Novoa, A., Le Roux, J.J., Richardson, D.M. & Wilson, J.R.U. 2017 'Does commercial horticultural trade in Cactaceae pose a major environmental threat?', *Conservation Biology*. https://doi.org/10.1111/cobi.12892

Novoa, A., Le Roux, J.J., Robertson, M.P., Wilson, J.R.U. & Richardson, D.M., 2015b, 'Introduced and invasive cactus species: A global review', *AoB PLANTS* 7, plu078. https://doi.org/10.1093/aobpla/plu078

Novoa, A., Rodriguez, J., Lopez-Nogueira, A., Richardson, D.M. & Gonzalez, L., 2016c, 'Seed characteristics in Cactaceae: Useful diagnostic features for screening species for invasiveness?', *South African Journal of Botany* 105, 61–65. https://doi.org/10.1016/j.sajb.2016.01.003

Panetta, F.D., 2015, 'Weed eradication feasibility: Lessons of the 21st century', *Weed Research* 55, 226–238. https://doi.org/10.1111/wre.12136

Paterson, I.D., Hoffmann, J.H., Klein, H., Mathenge, C.W., Neser, S. & Zimmermann, H.G., 2011, 'Biological control of Cactaceae in South Africa', *African Entomology* 19, 230–246. https://doi.org/10.4001/003.019.0221

Rogers, K. & Biggs, H., 1999, 'Integrating indicators, endpoints and value systems in strategic management of the rivers of the Kruger National Park', *Freshwater Biology* 41, 439–451. https://doi.org/10.1046/j.1365-2427.1999.00441.x

Rouget, M., Robertson, M.P., Wilson, J.R.U., Cang Hui, Essl, F., Renteria, J.L. et al., 2016, 'Invasion debt – Quantifying future biological invasions', *Diversity and Distributions* 22, 445–456. https://doi.org/10.1111/ddi.12408

Selge, S., Fischer, A. & van der Wal, R., 2011, 'Public and professional views on invasive non-native species – A qualitative social scientific investigation', *Biological Conservation* 144, 3089–3097. https://doi.org/10.1016/j.biocon.2011.09.014

Sharov, A.A. & Liebhold, A.M., 1998, 'Bioeconomics of managing the spread of exotic pest species with barrier zones', *Ecological Applications* 8, 833–845. https://doi.org/10.1111/j.0272-4332.2004.00486.x

Simberloff, D., Martin, J.-L., Genovesi, P., Maris, V., Wardle, D.A., Aronson, J. et al., 2013, 'Impacts of biological invasions: What's what and the way forward', *Trends in Ecology & Evolution* 28, 58–66. https://doi.org/10.1016/j.tree.2012.07.013

Van Wilgen, B., Fill, J.M., Baard, J., Cheney, C., Forsyth, A.T. & Kraaij, T., 2016, 'Historical costs and projected future scenarios for the management of invasive alien plants in protected areas in the Cape Floristic Region', *Biological Conservation* 200, 168–177. https://doi.org/10.1016/j.biocon.2016.06.008

Van Wilgen, B.W., Dyer, C., Hoffmann, J.H., Ivey, P., Le Maitre, D.C., Moore, J.L. et al., 2011, 'National-scale strategic approaches for managing introduced plants: Insights from Australian acacias in South Africa', *Diversity and Distributions* 17, 1060–1075. https://doi.org/10.1111/j.1472-4642.2011.00785.x

Van Wilgen, B.W., Forsyth, G.G., Le, D.C., Wannenburgh, A., Kotzé, J.D.F., van den Berg, E. et al., 2012, 'An assessment of the effectiveness of a large, national-scale invasive alien plant control strategy in South Africa', *Biological Conservation* 148, 28–38. https://doi.org/10.1016/j.biocon.2011.12.035

Visser, V., Wilson, J.R.U., Canavan, K., Canavan, S., Fish, L., Le Maitre, D. et al., 2017, 'Grasses as invasive plants in South Africa revisited: Patterns, pathways and management', *Bothalia* 47(2), a2169.

Walters, M., Figueiredo, E., Crouch, N.R., Winter, P.J.D., Smith, G.F., Zimmermann, H.G. et al., 2011, 'Naturalised and invasive succulents of Southern Africa', *ABC Taxa* 11, 91–217.

Wilson, J.R.U., Gaertner, M., Richardson, D.M. & van Wilgen, B.W., 2017, 'Contributions to the National Status Report on Biological Invasions in South Africa', *Bothalia* 47(2), a2207.

Wilson, J.R.U., Ivey, P., Manyama, P. & Nänni, I., 2013, 'A new national unit for invasive species detection, assessment and eradication planning', *South African Journal of Science* 109, 1–13. https://doi.org/10.1590/sajs.2013/20120111

Zachariades, C., Paterson, I.D., Strathie, L.W., Hill, M.P. & van Wilgen, B.W., 2017, 'Assessing the status of biological control as a management tool for suppression of invasive alien plants in South Africa', *Bothalia* 47(2), a2142.

Zengeya, T., Ivey, P., Woodford, D.J., Weyl, O., Novoa, A., Shackleton, R. et al., 2017, 'Managing conflict-generating invasive species in South Africa: Challenges and trade-offs', *Bothalia* 47(2), a2160.

Zimmermann, H.G., Moran, V.C. & Hoffmann, J.H., 2004, 'Biological control in the management of invasive alien plants in South Africa, and the role of the Working for Water programme', *South African Journal of Science* 100, 34–40.

Appendix 1

Background, role, and composition of the South African Cactus Working Group (SACWG)

The *National Environmental Management: Biodiversity Act* (10 of 2004) requires the South African National Biodiversity Institute (SANBI) to regularly monitor and report on the status of listed invasive species in South Africa.

Towards this end, SANBI initiated a South African National Cactus Working Group (SACWG) to strategically monitor and coordinate management of cactus species in South Africa. The role of the SACWG is to:

1. Develop a national cactus management strategy;
2. Co-ordinate nationally work done on cactus;
3. Assess the risks and management feasibility of cacti;
4. Ensure best practice control methods are used against target cactus species;
5. Improve co-ordination and communication among research institutes, invasive species managers and relevant government departments; and
6. Engage with external stakeholders.

Representatives from all relevant organisations involved in the management or research of cactus and invasive species policy-makers are included in the working group (Table 1).

The SACWG convened in June 2012 to constitute itself. They will continue to meet biannually with SANBI serving as secretariat.

TABLE1-A1: Member organisations of the South African Cactus Working Group.

Organisation	Relevant expertise	Web link
Department of Environmental Affairs – Environmental Programmes	Design and implementation of policies on alien and invasive species in terms of the *National Environmental Management: Biodiversity Act* no. 10 of 2004	https://www.environment.gov.za/branches/environmental_programmes
Department of Agriculture, Forestry and Fisheries – Directorate of Land Use and Soil Management	Design and implementation of policies on alien and invasive species in terms of the *Conservation of Agricultural Resources Act* no. 43 of 1983	http://www.daff.gov.za/ http://www.daff.gov.za/daffweb3/Branches/Forestry-Natural-Resources-Management/LUSAM
South African National Biodiversity Institute: Invasive Species Programme	Detection and assessment of invasive species for eradication	http://www.sanbi.org/biodiversity-science/state-biodiversity/biodiversity-monitoring-assessment/invasive-aliens-early-det
DST-NRF Centre for Invasion Biology	Conduct research and development and training in biodiversity science especially as it applies to understanding the impacts of, and managing and preventing biological invasions	http://academic.sun.ac.za/cib/
Agricultural Research Council – Plant Protection Research Institute	Research on the ecology and control of invasive alien plants in South Africa with emphasis on non-native problem plants in conservation and pasture situations.	http://www.arc.agric.za/arc-ppri/Pages/ARC-PPRI-Homepage.aspx
South African National Parks	Management of invasive species in protected areas in South Africa	https://www.sanparks.org/

5

The biological control of aquatic weeds in South Africa: Current status and future challenges

Authors:
Martin P. Hill[1] ⓘ
Julie Coetzee[2] ⓘ

Affiliations:
[1]Department of Zoology and Entomology, Rhodes University, South Africa

[2]Department of Botany, Rhodes University, South Africa

Corresponding author:
Martin Hill,
M.hill@ru.ac.za

Background: Aquatic ecosystems in South Africa are prone to invasion by several invasive alien aquatic weeds, most notably, *Eichhornia crassipes* (Mart.) Solms-Laub. (Pontederiaceae) (water hyacinth); *Pistia stratiotes* L. (Araceae) (water lettuce); *Salvinia molesta* D.S. Mitch. (Salviniaceae) (salvinia); *Myriophyllum aquaticum* (Vell. Conc.) Verd. (parrot's feather); and *Azolla filiculoides* Lam. (Azollaceae) (red water fern).

Objective: We review the biological control programme on waterweeds in South Africa.

Results: Our review shows significant reductions in the extent of invasions, and a return on biodiversity and socio-economic benefits through the use of this method. These studies provide justification for the control of widespread and emerging freshwater invasive alien aquatic weeds in South Africa.

Conclusions: The long-term management of alien aquatic vegetation relies on the correct implementation of biological control for those species already in the country and the prevention of other species entering South Africa.

Introduction

Aquatic ecosystems in South Africa have been prone to invasion by introduced macrophytes since the late 1800s, when water hyacinth, *Eichhornia crassipes* (Mart.) Solms-Laub. (Pontederiaceae), was first recorded as naturalised in KwaZulu-Natal (Cilliers 1991). Several other species of freshwater aquatic plants, all notorious weeds in other parts of the world, have also become invasive in many of the rivers, man-made impoundments, lakes and wetlands of South Africa (Hill 2003). These are *Pistia stratiotes* L. (Araceae) (water lettuce); *Salvinia molesta* D.S. Mitch. (Salviniaceae) (salvinia); *Myriophyllum aquaticum* (Vell. Conc.) Verd. (parrot's feather); and *Azolla filiculoides* Lam. (Azollaceae) (red water fern) (Hill 2003), which along with water hyacinth comprise the 'Big Bad Five' (Henderson & Cilliers 2002). Recently, new invasive aquatic plant species have been recorded which are still at their early stages of invasion, including the submerged species, *Egeria densa* Planch. (Hydrocharitaceae) (Brazilian water weed) and *Hydrilla verticillata* (L.f.) Royle (Hydrocharitaceae); the emergent species, *Sagittaria platyphylla* (Engelm.) J.G.Sm. and *S. latifolia* Willd. (Alismataceae); *Lythrum salicaria* L. (Lythraceae) (purple loosestrife), *Nasturtium officinale* W.T. Aiton. (Brassicaceae) (watercress); *Iris pseudacorus* L. (Iridaceae) (yellow flag); and *Hydrocleys nymphoides* (Humb. & Bonpl. ex Willd.) Buchenau (Alismataceae) (water poppy); and the new floating weeds, *Salvinia minima* Baker (Salviniaceae) and *Azolla cristata* Kaulf. (Azollaceae) (Mexican azolla); and the rooted floating *Nymphaea mexicana* Zucc. (Nymphaeceae) (Mexican water lily) (Coetzee et al. 2011a; Coetzee, Bownes & Martin 2011b). The mode of introduction of these species is mainly through the horticultural and aquarium trade (Martin & Coetzee 2011), and two issues contribute to the invasiveness of these macrophytes following establishment: the lack of co-evolved natural enemies in their adventive range (McFadyen 1998); and disturbance, the presence of nitrate- and phosphate-enriched waters, associated with urban, agricultural and industrial pollution that promotes plant growth (Coetzee & Hill 2012).

Aquatic weeds in South Africa are found throughout the country including the winter rainfall areas of the western part of the country, the more subtropical eastern parts and the cool, temperate areas of the Highveld plateau (Henderson 2001). Although the alteration of hydrological flows in South African river systems through the construction of impoundment walls, gauging weirs, culverts and low-water bridges where constant slow-flowing waters have facilitated population build-up and thus problems caused by aquatic weeds (Hill & Olckers 2001), infestations are also found in unimpacted habitats, such as *A. filiculoides* infestations in wetlands in the southern Free State and *I. pseudocorus* in wetlands of the Cape Peninsula. Here, we review the current status of aquatic weeds in South Africa, their socio-economic and environmental impacts and the benefits of their control.

Drivers of invasive aquatic plant invasions

It is important to understand the invasion biology of an organism, if effective control measures are to be implemented. Several authors (e.g. Bauer 2012; MacDougall & Turkington 2005) have grouped invasive alien species into three broad categories *viz.* (1) passengers, which are solely dependent on a disturbance for establishment and proliferation, and if the disturbance is removed, the invasion and associated impacts cease; (2) drivers of biodiversity loss, which include species that do not need any disturbance to establish; and (3) back-seat drivers whereby an initial disturbance is required for an invasive alien plant species to establish, but once established, even if the disturbance is removed, the invasion continues. Aquatic weed invasions in South Africa are examples of back-seat drivers. These invasive species rely on the broad ecosystem disturbance of slow-flowing permanent waters caused by impoundments and eutrophication which facilitates establishment and, linked with enemy release, allows them to proliferate, thereby gaining a competitive advantage over indigenous aquatic plants (Coetzee & Hill 2012). The resulting large continuous mats significantly impact all aspects of aquatic biodiversity and ecosystem functioning (see below).

Impacts of aquatic invasive plant invasions

Aquatic weeds cause various environmental (or ecological) and socio-economic impacts (which are in their majority negative), affecting floral and faunal diversity and ecosystem functioning and services. The impact mechanisms and effects of aquatic weeds differ between species, which is largely based on differences in their growth form and the habitat that they have invaded. We applied the generic impact scoring system (GISS) presented by Nentwig et al. (2016) to assess the impacts of eight water weed species in South Africa before and after biological control. This is not the intended use of GISS, which was designed to prioritise invasive alien species for control, but does allow a comparison and ranking of the impacts of water weeds in South Africa and an assessment of

the success of the biological control programmes. GISS relies on published evidence and comprises 12 impact categories divided evenly between environmental and socio-economic impacts. Under the environmental impacts, the effect that the invasive alien species has on native fauna and flora, either directly or through competition, disease transmission, hybridisation and the ecosystem services, is rated. Under the socio-economic categories, the impact that the invasive alien species has on agriculture, forestry, infrastructure, human health and social well-being is scored. A six-level scoring system is applied (Nentwig et al. 2016):

0 – no data available, no impacts known, not detectable or not applicable
1 – minor impacts, only locally, only on common species and negligible economic loss
2 – minor impacts, more widespread, also on rarer species and minor economic loss
3 – medium impacts, large-scale, several species concerned, relevant decline, relevant ecosystem modifications and medium economic loss
4 – major impact with high damage, major changes in ecosystem functions, decrease of species and major economic loss
5 – major large-scale impact with high damage and complete destruction, threat to species including local extinctions and high economic costs

Table 1 presents an analysis of the socio-economic and environmental impacts caused by eight of the most invasive and well-studied species of aquatic weeds established in South Africa. The list of weed species chosen for this study was not exhaustive, but represents three of the main habits, free-floating species, emergent species and submerged species. Scoring systems have their flaws, in that they are constrained by time and locality, but provide valuable benchmarks. We, therefore, scored aquatic weeds in South Africa based on both the worst-case scenario and the current status (i.e. before and after biological control for the weeds that have biocontrol programmes), thereby presenting a measure of the value of biological control. The scores were based on published studies using South African data captured in review papers or chapters, except for *S. platyphylla*, *E. densa* and *H. verticillata*, where little data exist on their impacts in South Africa, and thus, we relied on data from elsewhere in the world (e.g. Adair et al. 2012; Langeland 1996; Yarrow et al. 2009). Where the

TABLE 1: The impact scores with level of confidence per impact categories of the GISS (Nentwig et al. 2016) for eight water weeds in South Africa, presenting the worst-case scenario in the absence of any biological control, and the current situation in South Africa, post biological control, where applicable.

Weed	Prior to biological control			Post biological control		
	Environmental impact (level of confidence)	Socio-economic impact (level of confidence)	Total	Environmental impact (level of confidence)	Socio-economic impact (level of confidence)	Total
Eichhornia crassipes	22 (2.67)	21 (3.00)	43 (2.83)	12 (2.50)	11 (2.67)	23 (2.58)
Pistia stratiotes	22 (2.67)	16 (2.83)	38 (2.75)	2 (3.00)	4 (2.83)	6 (2.92)
Salvinia molesta	22 (2.67)	16 (2.83)	38 (2.75)	2 (3.00)	4 (2.83)	6 (2.92)
Azolla filiculoides	20 (2.83)	20 (2.83)	40 (2.83)	0 (3.00)	0 (3.00)	0 (3.00)
Myriophyllum aquaticum	18 (2.83)	20 (2.83)	38 (2.83)	8 (2.83)	7 (2.83)	15 (2.83)
Sagittaria platyphylla	18 (1.30)	17 (1.30)	35 (1.30)	-	-	-
Egeria densa	21 (1.30)	21 (1.30)	42 (1.30)	-	-	-
Hydrilla verticillata	22 (1.30)	22 (1.30)	44 (1.30)	-	-	-

impact was unknown, largely because it was unstudied, the impact was assigned a neutral score of zero. Further, confidence limits were based on Nentwig et al. (2016) where 1 = low confidence (no empirical data or literature to support the impact score), 2 = medium confidence (no empirical data from South Africa, but literature from elsewhere to support the impact score) and 3 = high confidence (empirical and published data from South Africa support the impact score) (Appendix 1).

This analysis shows that of the floating macrophytes, *E. crassipes* had the biggest impact on aquatic ecosystems in South Africa, followed by *A. filiculoides*, *P. stratioes* and *S. molesta*. Although based on literature, this result is supported by annual field surveys throughout South Africa and is probably because of the fact that water hyacinth is the largest of the macrophytes that warrant control, it is the most widespread, we have studied its impacts (e.g. Coetzee, Jones & Hill 2014; Fraser, Martin & Hill 2016; Midgley, Hill & Villet 2006) and it has historically been the most difficult of the water weeds to control (e.g. Coetzee et al. 2011a; Hill 2003; Hill & Cilliers 1999). Although not considered to be under complete biological control (Klein 2011), the ecological and socio-economic impact of the weed has been significantly reduced through the introduction of eight biological control agents (Coetzee et al. 2011a; Paterson et al. 2016). On the contrary, the impacts of *A. filiculoides* on South African freshwater systems were quantified by Ashton and Walmsley (1984) and McConnachie et al. (2003). Based on this evidence, this weed achieved a score of 40 on the GISS, just below *E. crassipes*, and was considered more damaging than either *P. stratiotes* or *S. molesta* in the absence of biological control. Following the introduction of the highly successful agent, *Stenopelmus rufinasus* Gyllenhal (Coleoptera: Curculionidae), *A. filiculoides* no longer poses a threat to aquatic ecosystems of the country (Hill & McConnachie 2009; McConnachie et al. 2003; McConnachie, Hill & Byrne 2004); indeed, we could not find a single negative impact of this weed in this country and thus scores 0. Furthermore, biological control has significantly reduced the impact scores of *P. stratiotes*, *S. molesta* and *M. aquaticum*, highlighting the ecological and economic benefits of biological control.

Interestingly, the two submerged species analysed, *E. densa* and *H. verticillata*, recorded the two of the highest impact scores. This is largely because of their fairly recent invasion status in South Africa, and thus, we relied heavily on the published literature. Although *H. verticillata* is only confined to one site in South Africa (Coetzee et al. 2009b), and its impact at this site has not been quantified, its impact in the United States suggests that it should be given a very high priority in terms of impact and thus the need for control (Balciunas et al. 2002; Langeland 1996). The emergent species, *S. platyphylla*, scored the lowest in comparison with the other macrophytes possibly because it is a new invader still in the lag phase, not yet dominating the riparian zone, and is also not yet considered a major weed elsewhere in the world.

Impact of water weeds on biodiversity loss

Although the socio-economic impacts of water weeds have been fairly well reported (reviewed in Villamagna & Murphy 2010), there are very few specific examples that have documented their impacts on biodiversity. Below we present two case studies of the direct impact of water hyacinth on aquatic biodiversity in South Africa.

Case study 1: Midgley et al. (2006)

In this first case study, the benthic invertebrate community and algal biomass were sampled under water hyacinth mats and in water hyacinth-free water over a 13-month period, using artificial substrates in New Year's Dam, Eastern Cape Province, a cool temperate region of the country. The number of families and the number of individuals per substrate were significantly lower under the mats. Further, measures of biodiversity, including Shannon-Weiner diversity index, Margalef's richness index, Pielou's evenness index and chlorophyll *a*, were all significantly lower under water hyacinth mats than in water hyacinth-free zones, demonstrating the impact of water hyacinth on benthic biodiversity.

Case study 2: Coetzee et al. (2014)

Although similar to the previous study, this study aimed to determine whether the presence of water hyacinth altered the diversity and assemblage structure of benthic macroinvertebrates in a conservation area in a subtropical region of the country, the Nseleni Nature Reserve near Richard's Bay. The benthic macroinvertebrate assemblage was sampled over 1 year at five sites under water hyacinth mats and at five sites without water hyacinth in the Nseleni River. Once again, artificial substrates were placed beneath water hyacinth mats or in the open water to allow for colonisation by freshwater macroinvertebrates, and left for a period of 6 weeks, repeated on seven occasions over 10 months. Twenty-nine families comprising 18 797 individuals were collected, 817 (13 families) individuals were from under water hyacinth mat sites compared with 17 980 (27 families) individuals from open water sites. However, 98% of individuals collected were the invasive snail, *Tarebia granifera* L. (Thiaridae). This study again highlights that the presence of water hyacinth has a significantly negative impact on aquatic macroinvertebrate biodiversity, but in a conservation area.

Control of aquatic invasive plant invasions

In South Africa, water weeds have been controlled through the use of mechanical and manual removal, herbicide application and biological control. Although manual removal using rakes and pitchforks can be successful, it is labour intensive. Although one of the pillars of the Working for Water Programme of the Natural Resources Management

Programmes of the Department of Environmental Affairs is job creation through alien plant removal, this method is really ineffective for water weeds and this work force is better used on controlling terrestrial weeds in South Africa. Manual removal of submerged aquatic species such as *H. verticillata* and *E. densa* invariably leads to fragmentation of the weed mat and subsequent dispersal and increased infestation of the weed (Dayan & Netherland 2005).

Herbicidal control, using formulations containing the active ingredient glyphosate, is still used to control water hyacinth in some of the larger dams and river systems in South Africa. Herbicidal control of water hyacinth depends on skilled operators who maintain a long-term follow-up programme continually to control re-infestation from scattered plants and those germinating from seed. Therefore, any herbicide programme against the weed requires a commitment to an ongoing operation of unlimited duration. It is the lack of a follow-up regime that has often led to the failure of herbicidal control programmes (Hill & Olckers 2001). Although herbicide application is often used as part of an integrated management approach (Hill & Coetzee 2008), Hill, Coetzee and Ueckermann (2012) showed that a number of herbicide formulations used in South Africa were toxic to some of the biological control agents that have been released against this weed.

The biological control programme against water weeds in South Africa was initiated in 1973 and the weevil, *Neochetina eichhorniae* Warner, was released in 1974 (Cilliers 1991). Since that time, 13 agent species (11 insects, one mite and one pathogen) have been released against five weeds (Table 2). The biological control programme against water weeds in South Africa has been highly successful with four of the five weeds targeted (water lettuce, salvinia, parrot's feather and azolla) considered to be under complete control whereby no other control methods are required to keep the weed populations at a level where they no longer impact the aquatic biodiversity and water utilisation (see above) (Coetzee et al. 2011a). Although water hyacinth is not considered to be under complete biological control, in some areas biological control has controlled the weed, whereas in other areas it has reduced populations and impact such that alternative control methods such as herbicide applications are required far less frequently (Coetzee et al. 2011a; Hill & Cilliers 1999).

The biological control programme on water weeds in South Africa is co-ordinated through Rhodes University in collaboration with University of the Witwatersrand and the Plant Protection Research Institute of the Agricultural Research Council. This programme comprises about 7 research staff, 14 support and technical staff, and 12 postgraduate

TABLE 2: Biological control agents released for the control of freshwater alien aquatic weed species in South Africa (after Klein 2011).

Target weed	Natural enemy	Feeding guild	Agent status	Weed status	Key references
Alismataceae					
Sagittaria platyphylla	*Listronotus appendiculatus* (Coleoptera: Curculionidae	Flower feeder	Under investigation	-	-
	Listronotus frontalis (Coleoptera: Curculionidae)	Tuber feeder	Under investigation	-	-
	Listronotus lutulentus (Coleoptera: Curculionidae)	Leaf feeder	Under investigation	-	-
	Listronotus sordidus (Coleoptera: Curculionidae)	Root crown feeder	Under investigation	-	-
Araceae					
Pistia stratiotes	*Neohydronomus affinis* (Coleoptera: Curculionidae)	Leaf and stem borer	Released 1985, extensive	Complete	Coetzee et al. 2011a
Azollaceae					
Azolla filiculoides	*Stenopelmus rufinasus* (Coleoptera: Curculionidae)	Frond feeder	Released 1997, extensive	Complete	McConnachie et al. 2004
Azolla cristata	*Stenopelmus rufinasus* (Coleoptera: Curculionidae)	Frond feeder	First recorded on this species in 2004, extensive	Substantial	Madeira et al. 2016
Haloragaceae					
Myriophyllum aquaticum	*Lysathia* sp. (Coloptera: Chrysomelidae)	Leaf feeder	Released 1994, extensive	Complete	Coetzee et al. 2011a
Hydrocharitaceae					
Egeria densa	*Hydrellia egeriae* (Diptera: Ephydridae)	Leaf miner	Under investigation	-	Coetzee et al. 2011b
Hydrilla verticillata	*Hydrellia purcelli* (Diptera: Ephydridae)	Leaf miner	Release permit issued	-	Coetzee et al. 2011b
Pontederiaceae					
Eichhornia crassipes	*Cercospora piaropi* (Mycosphaerellales: Mycosphaerellaceae)	Leaf pathogen	Released 1992, considerable	Substantial	Morris, Wood & den Breeÿen 1999
	Cornops aquaticum (Orthoptera: Acrididae)	Lead feeder	Released 2011, establishment unconfirmed	-	Coetzee et al. 2011a
	Eccritotarsus catarinensis (Hemiptera: Miridae)	Leaf sucker	Released 1996, considerable	-	Coetzee et al. 2011a
	Eccritotarsus sp. nov. (Hemiptera: Miridae)	Leaf sucker	Released 2008, establishment unconfirmed	-	Paterson et al. 2016
	Megamelus scutellaris (Hemiptera: Delphacidae)	Leaf sucker	Released 2013, established		
	Neochetina bruchi (Coleoptera: Curculionidae)	Stem borer	Released 1990, considerable	-	Coetzee et al. 2011a; Hill and Cilliers 1999
	Neochetina eichhorniae (Coleoptera: Curculionidae)	Stem borer	Released 1974, considerable	-	Coetzee et al. 2011a; Hill and Cilliers 1999
	Niphograpta albiguttalis (Lepidoptera: Crambidae)	Petiole borer	Released 1990, considerable	-	Coetzee et al. 2011a; Hill and Cilliers 1999
	Orthogalumna terebrantis (Acari: Sarcoptiformes: Glamunidae)	Leaf miner	Released 1989, considerable	-	Coetzee et al. 2011a; Hill and Cilliers 1999
Salviniaceae					
Salvinia molesta	*Cyrtobagous salviniae* (Coleoptera: Curculionidae)	Stem borer	Released 1985, considerable	Complete	Cilliers et al. 2003; Coetzee et al. 2011a

students and postdoctoral fellows. The activities carried out by this research group include pre-release studies on new agents for several species, including *S. platyphylla*, *I. pseudocorus* and *E. densa*, and qualitative post-release evaluation studies on all of the weeds on which agents have been released. The most significant aspect of the post-release evaluation studies is an annual country-wide survey of all water weed sites (~450 infested water bodies) assessing weed and agent populations. These surveys that have been carried out since 2008 provide the guidance for the water weed biological control programme and a measure of success or failure.

Results of the surveys show that since 2008, there has been a substantial increase in the number of recorded invaded sites, but more importantly, the percentage of these sites where the respective biocontrol agents are present has increased significantly because of enhanced efforts to release agents from mass rearing centres (Figure 1). This has led to an increase in the control of the weeds and a reduction in their ecological and environmental impacts (Table 1). Another unintended benefit of these surveys is that they have served as an ideal early detection platform for additional freshwater invasive macrophytes. For example, since 2008, the number of locality records for *E. densa*, *S. platyphylla* and *I. pseudacorus* has increased (from 0 to 14, 16 and 14 sites, respectively) to the point that these species are no longer considered eradication targets (Wilson et al. 2013). All three species are now targets for biological control. On the contrary, *H. verticillata* remains confined to one system, Jozini Dam (KZN).

Some new developments arising from these field surveys since the 2011 review paper (Coetzee et al. 2011a) are presented below. Mass rearing and implementation also forms an important part of the research programme as in many areas of the country that are prone to cold winters and eutrophic waters, classical biological control is not as effective as an augmentive programme whereby high numbers of healthy agents are released at the onset of summer when field populations of the agents are low. Part of the mass rearing programme involves the employment of people living with disabilities (Weaver et al. 2016).

The biological control programme against *A. filiculoides* in South Africa using the Azolla specialist *S. rufinasus* has been highly successful (McConnachie et al. 2003, 2004). However, field surveys showed that the agent utilised another Azolla species, thought to be the native *Azolla pinnata* subsp. *africana* (Desv.) Baker, which contradicted the host specificity trials (Hill 1998). However, molecular analysis showed that what we thought was the native species, *A. pinnata subsp. africana*, was a new invasive species, *A. cristata* Kaulfuss, a close relative of *A. filiculoides* (Madeira et al. 2016). Field surveys have shown that *S. rufinasus* is capable of establishing populations on *A. cristata* in the warmer, eastern part of the country and will likely result in this plant never becoming highly invasive.

Most of the biological control research on water weeds is centred around water hyacinth, and the plant hopper, *Megamelus scutellaris* Berg (Hemiptera: Delphacidae), is the most recent agent to have been released in 2013. This agent has now established and is impacting the plant in the cooler areas of the country where the other agents have traditionally struggled to establish and have an effect (Coetzee, Byrne & Hill 2007). Recent molecular work has revealed that two separate populations of the mirid, *Eccritotarsus catarinensis* Carvahlo (Hemiptera: Miridae), collected from Brazil (collected in 1994) and Peru (collected in 1999), respectively, are in fact cryptic species (Paterson et al. 2016). Fortunately, these populations were kept separate and both subjected to impact and host specificity testing. However, this finding does show that the importation of multiple consignments of the same species for biological control should be conducted with caution.

Benefits of biological control of water hyacinth

Weed biological control has traditionally suffered from a lack of quantitative post-release evaluation studies that show economic or ecological benefit. Where the benefits of a biological control programme have been measured, it has focussed on economic benefits (e.g. Van Wilgen et al. 2004). For aquatic weeds, McConnachie et al. (2003) quantified the benefits of the biological control programme against red water fern using the weevil, *S. rufinasus* Gyllenhal, in South Africa and showed that the agent removed the impact of the weed on water supply, stock health and recreational activities (see above). Further, De Groote et al. (2003) demonstrated that the successful biological control of water hyacinth in southern Benin significantly increased the yearly income of the population of this region through

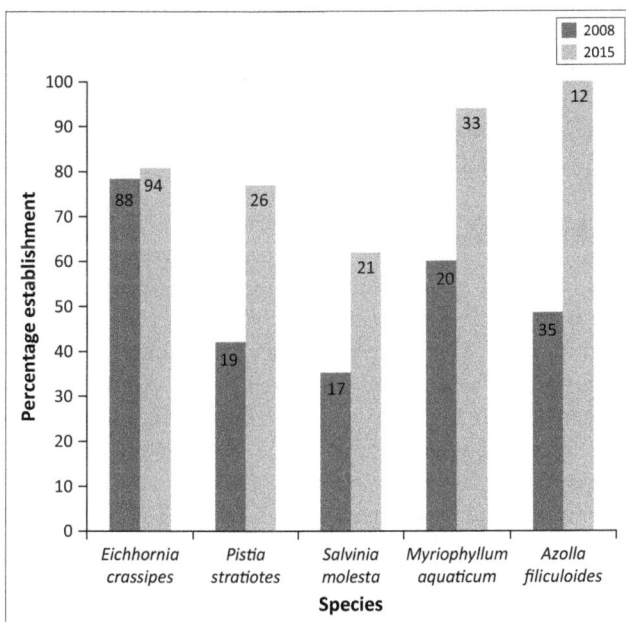

The bars represent the percentage of sites infested by the weed where at least one control agent species was present. The value embedded in each bar is the number of sites invaded by each species.

FIGURE 1: Results of the first (2008) and most recent (2015) nationwide surveys on aquatic weeds in South Africa.

increased crop and fish production. Also, Van Wyk and Van Wilgen (2002) compared the costs and benefits of three control interventions for *E. crassipes* and showed that biological control, along with integrated control, offered the best return on investment.

New threats to the aquatic environment

Coetzee et al. (2011b) highlighted the significance that the delays in promulgating appropriate legislation against a suite of new aquatic invaders could have in allowing their unmitigated establishment and spread in South African water bodies. In 2014, however, the promulgation of the *National Environmental Management: Biodiversity Act* (10/2004) (NEM:BA) and the publication of the Alien and Invasive Species List in 2014 resulted in the listing of 10 Category 1a aquatic plant species, including *H. verticillata*, *I. pseudacorus* and *S. platyphylla*; 16 Category 1b aquatic plant species, including *A. cristata* and *S. minima*; and one Category 2 aquatic plant species. This legislation will provide much needed impetus to curb the spread and impacts of this suite of invaders in South Africa. As these five species are no longer considered targets for eradication (Coetzee et al. 2011a, 2011b; Jaca & Mkhize 2015), biological control programmes have been initiated against these species and are currently at various stages of development: from surveying for potential natural enemies, in the case of *I. pseudacorus*; screening for host specificity in quarantine, in the case of *S. platyphylla* and *S. minima*; pending release of a suitable agent in the case of *H. verticillata*; to assessing the impact of an agent already released against *A. filiculoides* that has subsequently been found on *A. cristata* (Table 2).

Despite the NEM:BA legislation, there are a number of additional unlisted aquatic plant species whose introduction and establishment must be prevented at all costs. The role that pet traders, aquarists, boating enthusiasts and fishermen play in the spread of invasive aquatic species has been highlighted from around the world (Cohen et al. 2007; Maki & Galatowitsch 2004; Padilla & Williams 2004) and is a significant channel for the introduction and spread of aquatic plants throughout South Africa too (Martin & Coetzee 2011). Species such as *Cabomba caroliniana* Gray (Cabombaceae), *Alternanthera philoxeroides* Griseb. (Amaranthaceae) and *Stratiotes aloides* L. (Hydrocharitaceae) are widespread invaders elsewhere in the world (e.g. Julien et al. 2012; Schooler, Cabrera-Walsh & Julien 2009; Thiebaut 2007) and pose a threat to South African waterways, should they be introduced. Awareness and publicity programmes on potential new threats could go a long way in preventing their introduction and trade, as well as improved phytosanitary efforts and border control.

Discussion

Hill and Olckers (2001) critiqued the biological control programme on water hyacinth in South Africa. Although their emphasis was water hyacinth, the points made in that paper are pertinent to all invasive alien water weeds in South Africa. Hill and Olckers stated that there were four issues that mitigated against the sustainable biological control of water hyacinth; the injudicious use of herbicides that was antagonistic to the biological control agents; the cold winters in the temperate regions of the country that was deleterious to the build-up of agent populations; eutrophic waters that allowed the weeds to compensate for herbivory; and the fact that many of the systems infested by these weeds were small and lacked the necessary wind fetch to break up mats of agent infested weed. In the 15 years since the publication of Hill and Olckers, a considerable amount of research has been undertaken to better understand these four issues (summarised and reviewed in Byrne et al. 2010; Coetzee et al. 2011a, 2011b; Coetzee & Hill 2012). The implementation of this research has resulted in the release of additional agents that are better adapted to the diversity of habitats in South Africa (e.g. *M. scutellaris* which is able to establish on water hyacinth in cooler regions), and an emphasis on inundative releases of high numbers of agents at appropriate times of year (e.g. in spring and after herbicide application). This has been made possible through the construction of three mass-rearing facilities (City of Cape Town, Rhodes University and the South African Sugar Research Institute) that produce agents on demand.

The biological control programme against water weeds in South Africa has been highly successful, as measured by an increase in the number of sites under biological control, coupled with a significant reduction in the percentage cover of these weeds and a recovery of ecosystem services. However, unless the primary driver of disturbance (i.e. eutrophication by nitrates and phosphates) in aquatic ecosystems is addressed, we anticipate, rather than control, a succession of invasions by a suite of water weeds (Coetzee et al. 2011a, 2011b). Although we have shown that biological control has played a significant role in the recovery of aquatic biodiversity, these biodiversity benefits will be short-lived in impacted ecosystems unless an integrated catchment management approach is adopted which addresses eutrophication.

Acknowledgements

This research was funded through the Department of Environmental Affairs, Natural Resource Management Programme's Working for Water programme. Further funding for this work was provided by the South African Research Chairs Initiative of the Department of Science and Technology and the National Research Foundation of South Africa. Any opinion, finding, conclusion or recommendation expressed in this material is that of the authors, and the National Research Foundation does not accept any liability in this regard.

Competing interests

The authors declare that they stand to gain no financial benefit from the publication of this article.

Authors' contributions

J.C. and M.P.H. contributed equally to the conception, analysis and writing of the article.

References

Adair, R.A., Keener, B.R., Kwong R.M., Sagliocco, J.L. & Flower, G.E., 2012, 'The Biology of Australian weeds, 60. *Sagittaria platyphylla* (Engelmann) J.G. Smith and *Sagittaria calycina* Engelmann', *Plant Protection Quarterly* 27(2), 47–58.

Ashton, P.J. & Walmsley, R.D., 1984, 'The taxonomy and distribution of *Azolla* species in southern Africa', *Botanical Journal of the Linnean Society* 89, 239–247. https://doi.org/10.1111/j.1095-8339.1984.tb02198.x

Balciunas, J.K., Grodowitz, M.J., Cofrancesco, A.F. & Shearer, J.F., 2002, 'Hydrilla', in R. Van Driesch, B. Blossey, M. Hoddle, S. Lyon & R. Reardon (eds.), *Biological control of invasive plants in the Eastern United States*, pp. 91–114, USDA Forest Service, Morgantown, WV.

Bauer, J., 2012, 'Invasive species: "Back-seat drivers" of ecosystem change?', *Biological Invasions* 14, 1295–1304. https://doi.org/10.1007/s10530-011-0165-x

Byrne, M.J., Hill, M.P., Robertson, M., King, A., Jadhav, A., Katembo, N. et al., 2010, *Integrated Management of Water Hyacinth in South Africa: Development of an integrated management plan for water hyacinth control, combining biological control, herbicidal control and nutrient control, tailored to the climatic regions of South Africa*, Report to the Water Research Commission. WRC Report No. No TT 454/10, Water Research Commission, Pretoria, South Africa.

Cabrera Walsh, G., Magal Dalto, Y., Mattioli, F.M., Carruthers, R.I. & Anderson, L.W., 2013, 'Biology and ecology of Brazilian elodea (*Egeria densa*) and its specific herbivore, *Hydrellia* sp., in Argentina', *Biocontrol* 58, 133–147. https://doi.org/10.1007/s10526-012-9475-x

Center, T.D., Hill, M.P., Cordo, H. & Julien, M.H., 2002, 'Waterhyacinth', in R. G. van Driesche, S. Lyon, B. Blossey, M. S. Hoddle & R. Reardon (eds.), *Biological control of invasive plants in the Eastern United States*, pp. 41–64, USDA Forest Service, Morgantown, WV.

Chapman, M. & Dore, D., 2009, *Sagittaria strategic plan 2009*, Unpublished report to Goulburn Murray Water, Boulburn Broken Catchment Management Authority and the tri-state Sagittaria taskforce, RuralPlan Pty Ltd, Goomalibee, VIC.

Cilliers, C.J., 1991, 'Biological control of water hyacinth, *Eichhornia crassipes* in South Africa', *Agriculture, Ecosystems and Environment* 37, 207–217. https://doi.org/10.1016/0167-8809(91)90149-R

Cilliers, C.J., 1999, 'Biological control of parrot's feather, *Myriophyllum aquaticum* (Vell.) Verdc. (Haloragaceae), in South Africa', in T. Olckers & M.P. Hill (eds.), *Biological control of weeds in South Africa (1990–1998)*, *African Entomology Memoir* 1, pp. 113–118.

Cilliers, C.J., Hill, M.P., Ogwang, J.A. & Ajuonu, O., 2003, 'Aquatic weeds in Africa and their control', in P. Neuenschwander, C. Borgemeister & J. Langewald (eds.), *Biological control in IPM systems in Africa*, pp. 161–178, CAB International, Wallingford, CT.

Coetzee, J.A., Bownes, A. & Martin, G.D., 2011b, 'Prospects for the biological control of submerged macrophytes in South Africa', *African Entomology* 19(2), 469–487. https://doi.org/10.4001/003.019.0203

Coetzee, J.A., Byrne, M.J. & Hill, M.P., 2007, 'Predicting the distribution of *Eccritotarsus catarinensis*, a natural enemy released on water hyacinth in South Africa', *Entomologia Experimentalis et Applicata* 125, 237–247. https://doi.org/10.1111/j.1570-7458.2007.00622.x

Coetzee, J.A., Center, T.D., Byrne, M.J. & Hill, M.P., 2005, 'Impact of the biological control agent *Eccritotarsus catarinesis*, a sap-feeding mirid, on the competitive performance of water hyacinth, *Eichhornia crassipes*', *Biological Control* 32, 90–96. https://doi.org/10.1016/j.biocontrol.2004.08.001

Coetzee, J.A. & Hill, M.P., 2012, 'The role of eutrophication in the biological control of water hyacinth, *Eichhornia crassipes*, in South Africa', *BioControl* 57, 247–261. https://doi.org/10.1007/s10526-011-9426-y

Coetzee, J.A., Hill, M.P., Byrne, M.J. & Bownes, A., 2011a, 'A review of the biological control programmes on *Eichhornia crassipes* (C. Mart.) Solms (Pontederiaceae), *Salvinia molesta* D.S. Mitch. (Salviniaceae), *Pistia stratiotes* L. (Araceae), *Myriophyllum aquaticum* (Vell.) Verdc. (Haloragaceae) and *Azolla filiculoides* Lam. (Azollaceae) in South Africa', *African Entomology* 19(2), 451–468. https://doi.org/10.4001/003.019.0202

Coetzee, J.A., Hill, M.P., Julien, M.H., Center, T.D. & Cordo, H.A., 2009b, '*Eichhornia crassipes* (Mart.). Solms-Laub. (Pontederiaceae)', in R. Muniappan, G.V.P. Reddy & A. Raman (eds.), *Biological control of tropical weeds using arthropods*, pp.183–210, Cambridge University Press, Cambridge, UK.

Coetzee, J.A., Jones, R.W. & Hill, M.P., 2014, 'Water hyacinth, *Eichhornia crassipes* (Mart.) Solms-Laub. (Pontederiaceae), reduces benthic macroinvertebrate diversity in a protected subtropical lake in South Africa', *Biodiversity and Conservation* 23, 1319–1330. https://doi.org/10.1007/s10531-014-0667-9

Coetzee, J.A., Schlange, D. & Hill, M.P., 2009a, 'Potential spread of the invasive plant *Hydrilla verticillata* in South Africa based on anthropogenic spread and climate suitability', *Biological Invasions* 11, 801–812. https://doi.org/10.1007/s10530-008-9294-2

Cohen, J., Mirotchnick, N. & Leung, B., 2007, 'Thousands introduced annually: The aquarium pathway for non-indigenous plants to the St. Lawrence Seaway', *Frontiers in Ecology and the Environment* 5, 528–532. https://doi.org/10.1890/060137

Dayan, F.E. & Netherland, M.D., 2005, 'Hydrilla, the perfect aquatic weed, becomes more noxious than ever', *Outlooks on Pest Management* 16, 277–282. https://doi.org/10.1564/16dec11

De Groote, H., Ajuonu, O., Attignon, S., Djessou, R. & Neuenschwander, P., 2003, 'Economic impact of biological control of water hyacinth in Southern Benin', *Ecological Economics* 45, 105–117. https://doi.org/10.1016/S0921-8009(03)00006-5

Diop, O., Coetzee, J.A. & Hill, M.P., 2010, 'Impact of different densities of *Neohydronomus affinis* (Coleoptera: Cuculionidae) on *Pistia stratiotes* (Araceae) under laboratory conditions', *African Journal of Aquatic Science* 32, 267–271. https://doi.org/10.2989/16085914.2010.538505

Doeleman, J.A., 1989, *Biological control of* Salvinia molesta *in Sri Lanka: An assessment of costs and benefits*, Australian Centre for International Agricultural Research, Technical Report 12, 14 pp, Union Offset, Canberra, Australia.

Fraser, G., Martin, J. & Hill, M.P., 2016, 'Economic evaluation of water loss saving due to the biological control of water hyacinth at New Year's Dam, Eastern Cape Province, South Africa', *African Journal of Aquatic Science* 41(2), 227–234. https://doi.org/10.2989/16085914.2016.1151765

Henderson, L., 2001, *Alien weeds and invasive plants: A complete guide to declared weeds and invaders in South Africa*, Plant protection research institute handbook No. 12, Agricultural Research Council, Pretoria, South Africa.

Henderson, L. & Cilliers, C.J., 2002, *Invasive aquatic plants*, Plant protection research institute handbook no. 16, pp. 1–88, Agricultural Research Council, Pretoria, South Africa.

Hill, M.P., 1998, 'Life history and laboratory host range of *Stenopelmus rufinasus*, a natural enemy for *Azolla filiculoides* in South-Africa', *BioControl* 43, 215–224. https://doi.org/10.1023/A:1009903704275

Hill, M.P., 2003, 'The impact and control of alien aquatic vegetation in South African aquatic ecosystems', *African Journal of Aquatic Science* 28, 19–24. https://doi.org/10.2989/16085914.2003.9626595

Hill, M.P. & Cilliers, C.J., 1999, 'A review of the arthropod natural enemies, and factors that influence their efficacy, in the biological control of water hyacinth, *Eichhornia crassipes* (Mart.) Solms-Laubach (Pontederiaceae), in South Africa', in T. Olckers & M.P. Hill (eds.), *Biological control of weeds in South Africa (1990–1998)*, *African Entomology Memoir* 1, 103–112.

Hill, M.P. & Coetzee J.A., 2008, 'Integrated control of water hyacinth (*Eichhornia crassipes*) in Africa', *EPPO Bulletin/Bulletin OEPP* 38(3), 452–457. https://doi.org/10.1111/j.1365-2338.2008.01263.x

Hill, M.P., Coetzee, J.A. & Ueckermann, C., 2012, 'Toxic effect of herbicides used for water hyacinth control on two insects released for its biological control in South Africa', *Biocontrol Science and Technology* 22, 1321–1333. https://doi.org/10.1080/09583157.2012.725825

Hill, M.P. & McConnachie, A.J., 2009, '*Azolla filiculoides* Lamarck (Azollaceae)', in R. Muniappan, G.V.P. Reddy & A. Raman (eds.), *Biological control of tropical weeds using arthropods*, pp. 74–87, Cambridge University Press, Cambridge, UK.

Hill, M.P. & Olckers, T., 2001, 'Biological control initiatives against water hyacinth in South Africa: Constraining factors, success and new courses of action', in M.H. Julien, M.P. Hill, T.D. Center & D. Jianqing (eds.), *Biological and integrated control of water hyacinth, Eichhornia crassipes*, ACIAR Proceedings, pp. 33–38, Australian Centre for International Agricultural Research, Canberra.

Jaca, T. & Mkhize, V., 2015 'Distribution of *Iris pseudacorus* (Linnaeus, 1753) in South Africa', *BioInvasions Records* 4, 249–253. https://doi.org/10.3391/bir.2015.4.4.03

Julien, M., Sosa, A., Chan, R., Schooler, S. & Traversa, G., 2012, '*Alternanthera philoxeroides* (Martius) Grisebach – Alligator weed', in M. Julien, R. McFadyen & J. Cullen (eds.), *Biological control of weeds in Australia*, pp. 43–51, CSIRO Publishing, Melbourne.

Julien, M.H., Hill, M.P. & Tipping, P.W., 2009, '*Salvinia molesta* D. S. Mitchell (Salviniaceae)', in R. Muniappan, G.V.P. Reddy & A. Raman (eds.), *Biological control of tropical weeds using arthropods*, pp. 378–407, Cambridge University Press, Cambridge, UK.

Klein, H., 2011, 'A catalogue of the insects, mites and pathogens that have been used or rejected, or are under consideration, for the biological control of invasive alien plants in South Africa', *African Entomology* 19, 515–549. https://doi.org/10.4001/003.019.0214

Lambertini, C., Riis, T., Olesen, B., Clayton, J.S., Sorrell, B.K. & Brix, H., 2010, 'Genetic diversity in three invasive clonal aquatic species in New Zealand', *BMC Genetics* 11, 52–70. https://doi.org/10.1186/1471-2156-11-52

Langa, S.D.F., 2013, 'Impact and control of waterweeds in the Southern Mozambique Basin rivers', PhD thesis, Rhodes University, 283 pp.

Langeland, K.A., 1996, '*Hydrilla verticillata* (L.F.) Royle (Hydrocharitaceae), "the perfect aquatic weed"', *Castanea* 61, 293–304.

MacDougall, A.S. & Turkington, R., 2005, 'Are invasive species the drivers or passengers of change in degraded ecosystems?', *Ecology* 86, 42–55. https://doi.org/10.1890/04-0669

Madeira, P.T., Center, T.D., Coetzee, J.A., Pemberton, R.W., Purcell, M.A. & Hill, M.P., 2013, 'Identity and origins of introduced and native *Azolla* species in Florida', *Aquatic Botany* 111, 9–15. https://doi.org/10.1016/j.aquabot.2013.07.009

Madeira, P.T., Hill, M.P., Dray, F.A., Jr., Coetzee, J.A., Paterson, I.D. & Tipping, P.W., 2016, 'Molecular identification of *Azolla* invasions in Africa: The Azolla specialist, *Stenopelmus rufinasus* proves to be an excellent taxonomist', *South African Journal of Botany* 105, 299–305. https://doi.org/10.1016/j.sajb.2016.03.007

Mailu, A.M., 2001, 'Preliminary assessment of the social, economic and environmental impacts of water hyacinth in the Lake Victoria Basin and the status of control', in M.H. Julien, M.P. Hill, T.D. Center & D. Jianqing (eds.), *Biological and integrated control of water hyacinth, Eichhornia crassipes*, ACIAR Proceedings, pp. 130–139, Australian Centre for International Agricultural Research, Canberra.

Maki, K. & Galatowitsch, S., 2004, 'Movement of invasive aquatic plants into Minnesota through the horticultural trade', *Biological Conservation* 118, 389–396. https://doi.org/10.1016/j.biocon.2003.09.015

Martin, G.D. & Coetzee, J.A., 2011, 'Fresh water aquatic plant invasion risks posed by the aquarium trade, aquarists and the internet trade in South Africa', *Water SA* 37, 371–380.

McConnachie, A.J., de Wit, M.P., Hill, M.P. & Byrne, M.J., 2003, 'Economic evaluation of the successful biological control of *Azolla filiculoides* in South Africa', *Biological Control* 28, 25–32. https://doi.org/10.1016/S1049-9644(03)00056-2

McConnachie, A.J., Hill, M.P. & Byrne, M.J., 2004, 'Field assessment of a frond-feeding weevil, a successful biological control agent of red water fern, *Azolla filiculoides*, in southern Africa', *Biological Control* 29, 326–331. https://doi.org/10.1016/j.biocontrol.2003.08.010

McFadyen, R.E., 1998, 'Biological control of weeds', *Annual Review of Entomology* 43, 369–393. https://doi.org/10.1146/annurev.ento.43.1.369

Midgley, J.M., Hill, M.P. & Villet, M.H., 2006, 'The effect of water hyacinth, *Eichhornia crassipes* (Martius) Solms-Laubach (Pontederiaceae), on benthic biodiversity in two impoundments on the New Year's River, South Africa', *African Journal of Aquatic Science* 31(1), 25–30. https://doi.org/10.2989/16085910609503868

Moore, G.R. & Hill, M.P., 2012, 'A quantitative post-release evaluation of biological control of water lettuce, *Pistia stratiotes* L. (Araceae) by the weevil *Neohydronomus affinis* Hustache (Coleoptera: Curculionidae) at Cape Recife Nature Reserve, Eastern Cape Province, South Africa', *African Entomology* 20(2), 380–385. https://doi.org/10.4001/003.020.0217

Morris, M.J., Wood, A.R. & den Breeÿen, A., 1999, 'Plant pathogens and biological control of weeds in South Africa: A review of projects and progress during the last decade', in T. Olckers & M.P. Hill (eds.), *Biological control of weeds in South Africa (1990–1998)*, African Entomology Memoir 1, 129–137.

Nentwig, W., Bacher, S., Pyšek, P., Vilà, M. & Kumschick, S., 2016, 'The generic impact scoring system (GISS): A standardization tool to quantify the impacts of alien species', *Environmental Monitoring and Assessment* 188, 315. https://doi.org/10.1007/s10661-016-5321-4

Neuenschwander, P., Julien, M.H., Center, T.D. & Hill, M.P., 2009, '*Pistia stratiotes* L. (Araceae)', in R. Muniappan, G.V.P. Reddy & A. Raman (eds.), *Biological control of tropical weeds using arthropods*, pp. 332–352, Cambridge University Press, Cambridge, UK.

Padilla, D.K. & Williams, S.L., 2004, 'Beyond ballast water: Aquarium and ornamental trades as sources of invasive species in aquatic ecosystems', *Frontiers in Ecology and the Environment* 2, 131–138, https://doi.org/10.1890/1540-9295(2004)002[0131:BBWAAO]2.0.CO;2

Paterson, I., Mangan, R., Downie, D., Coetzee, J., Hill, M., Burke, A. et al., 2016, 'Two in one: Cryptic species discovered in biological control agent populations using molecular data and crossbreeding experiments', *Ecology and Evolution* 6, 6139–6150. https://doi.org/10.1002/ece3.2297

Pieterse, A.H., 1981, '*Hydrilla verticillata* – A review', Abstracts on Tropical Agriculture 7, 9–34.

Schooler, S., Cabrera-Walsh, W. & Julien, M., 2009, '*Cabomba caroliniana* Gray (Cabombaceae)', in R. Muniappan, G.V.P. Reddy & A. Raman (eds.), *Biological control of tropical weeds using arthropods*, pp. 88–107, Cambridge University Press, Cambridge, UK.

Souza, W.T.Z., 2011, '*Hydrilla verticillata* (Hydrocharitaceae), a recent invader threatening Brazil's freshwater environments: A review of the extent of the problem', *Hydrobiologia* 669, 1–20. https://doi.org/10.1007/s10750-011-0696-2

Thiebaut, G., 2007, 'Non-indigenous aquatic and semiaquatic plant species in France', in F. Gherardi (ed.), *Biological invaders in inland waters: Profiles, distribution, and threats*, vol. 2, pp. 209–229, Invading Nature – Springer Series in Invasion Ecology, Berlin.

Thomas, P.A. & Room, P.M., 1986, 'Taxonomy and control of *Salvinia molesta*', *Nature* 320(6063), 581–584.

Van Driesche, R.G., Carruthers, R.I., Center, T., Hoddle, M.S., Hough-Goldstein, J., Morin, L. et al., 2010, 'Classical biological control for the protection of natural ecosystems', *Biological Control* 54, S2–S33. https://doi.org/10.1016/j.biocontrol.2010.03.003

Van Wilgen, B.W., De Wit, M.P., Anderson, H.J., Le Maitre, D.C, Kotze, I.M., Ndala, S. et al., 2004, 'Costs and benefits of biological control of invasive alien plants: Case studies from South Africa: Working for Water', *South African Journal of Science* 100, 113–122.

Van Wyk, E. & Van Wilgen, B.W., 2002, 'The cost of water hyacinth control in South Africa: A case study of three options', *African Journal of Aquatic Science* 27, 141–149. https://doi.org/10.2989/16085914.2002.9626585

Villamagna, A.M. &. Murphy, B.R., 2010, 'Ecological and socio-economic impacts of invasive water hyacinth (*Eichhornia crassipes*): A review', *Freshwater Biology* 55, 282–298. https://doi.org/10.1111/j.1365-2427.2009.02294.x

Weaver, K., Hill, M.P., Hill, J., Coetzee, J.A., Paterson, I.D. & Martin, G.D., 2016, 'Project and practice narrative: Using insects to bridge the gap between science and the community', *Journal for New Generation Sciences (JNGS)*.

Wilson, J.R.U., Ivey, P., Manyama, P. & Nänni, I., 2013, 'A new national unit for invasive species detection, assessment and eradication planning', *South African Journal of Science* 109(5/6), 1–13. https://doi.org/10.1590/sajs.2013/20120111

Yarrow, M., Marın, V.H., Finlayson, M., Tironi, A., Delgado, L.E. & Fischer, F., 2009, 'The ecology of *Egeria densa* Planchon (Liliopsida: alismatales): A wetland ecosystem engineer?', *Revista Chilena de Historia Natural* 82, 299–313. https://doi.org/10.4067/S0716-078X2009000200010

Appendix 1

TABLE 1-A1: Summary of GISS detailed impact levels (Nentwig et al. 2016), for eight freshwater invasive alien aquatic weeds, in South Africa. Scoring was conducted in both the absence of biological control (BC) (worst-case scenario – no BC) and the presence of biological control (current status – BC), where applicable. References are included. Where there is a paucity on impact data and no implementation of biological control to date in South Africa, impacts realised elsewhere have been considered (i.e. for *Sagittaria platyphylla*, *Egeria densa* and *Hydrilla verticillata*).

Impact	Species and impact category	Description	Impact level, no BC	Confidence[a], no BC	Impact level, BC	Confidence[a], BC	References
Eichhornia crassipes							
1. Environmental impacts	1.1. On vegetation	Habitat alteration	5	3	3	2	Coetzee et al. 2009b; Villamagna and Murphy 2010
	1.2. On animals	Habitat alteration	5	3	2	3	Coetzee et al. 2009b, 2014; Midgley et al. 2006; Villamagna and Murphy 2010
	1.3. On competition	Space, light, nutrients	5	3	2	3	Coetzee et al. 2005, 2009b; Villamagna and Murphy 2010
	1.4. Transmission of disease	Provide habitat for vectors of disease	2	1	2	1	Mailu 2001; Villamagna and Murphy 2010
	1.5. Hybridisation	None	0	3	0	3	No indigenous *Eichhornia* spp
	1.6. Ecosystems	Ecosystem alteration	5	3	3	3	Coetzee et al. 2009b, 2014; Midgley et al. 2006; Van Driesche et al. 2010; Villamagna and Murphy 2010
Subtotal	-	-	22	2.67	12	2.5	-
2. Economic impacts	2.1. Agricultural production	Reduced irrigation	4	3	2	3	Fraser et al. 2016
	2.2. Animal production	Reduced irrigation	4	3	2	3	Fraser et al. 2016
	2.3. Forestry production	None	0	3	0	3	NA
	2.4. Human infrastructure	Bridges, weirs, hydropower, pumps	5	3	3	3	Hill 2003
	2.5. Human health	Disease vectors	3	3	2	1	Mailu 2001; Villamagna and Murphy 2010
	2.6. Human social life	Reduced water-based recreation	5	3	2	3	Center et al. 2002; Coetzee et al. 2009b
Subtotal	-	-	21	3	11	2.67	-
Total	-	-	43	2.83	23	2.58	-
Pistia stratiotes							
1. Environmental impacts	1.1. On vegetation	Habitat alteration	5	3	1	3	Diop, Coetzee and Hill 2010; Moore and Hill 2012; Neuenschwander et al. 2009
	1.2. On animals	Habitat alteration	5	3	0	3	Langa 2013; Neuenschwander et al. 2009
	1.3. On competition	Space, light, nutrients	5	3	0	3	Neuenschwander et al. 2009; Strange unpublished data
	1.4. Transmission of disease	Provide habitat for vectors of disease	2	1	0	3	Neuenschwander et al. 2009
	1.5. Hybridisation	None	0	3	0	3	Sole member of the Pistioidea
	1.6. Ecosystems	Ecosystem alteration	5	3	1	3	Coetzee et al. 2011a; Neuenschwander et al. 2009
Subtotal			22	2.67	2	3	
2. Economic impacts	2.1. Agricultural production	Reduced irrigation	4	3	1	3	Coetzee et al. 2011a; Neuenschwander et al. 2009
	2.2. Animal production	Reduced irrigation	3	3	0	3	Coetzee et al. 2011a; Neuenschwander et al. 2009
	2.3. Forestry production	None	0	3	0	3	NA
	2.4. Human infrastructure	Bridges, weirs, hydropower, pumps	3	3	1	3	Hill 2003
	2.5. Human health	Disease vectors	2	2	1	2	Neuenschwander et al. 2009
	2.6. Human social life	Reduced water-based recreation	4	3	1	3	Coetzee et al. 2011a; Neuenschwander et al. 2009
Subtotal	-	-	16	2.83	4	2.83	-
Total	-	-	38	2.75	6	2.92	-
Salvinia molesta							
1. Environmental impacts	1.1. On vegetation	Habitat alteration	5	3	1	3	Coetzee et al. 2011a; Doeleman 1989; Julien, Hill and Tipping 2009
	1.2. On animals	Habitat alteration	5	3	0	3	Coetzee et al. 2011a; Doeleman 1989; Julien et al. 2009
	1.3. On competition	Space, light, nutrients	5	3	0	3	Coetzee et al. 2011a; Doeleman 1989; Julien et al. 2009
	1.4. Transmission of disease	Provide habitat for vectors of disease	2	1	0	3	Coetzee et al. 2011a; Thomas and Room 1986

Table 1-A1 continues on next page →

TABLE 1-A1 (Continues...): Summary of GISS detailed impact levels (Nentwig et al. 2016), for eight freshwater invasive alien aquatic weeds, in South Africa. Scoring was conducted in both the absence of biological control (BC) (worst-case scenario – no BC) and the presence of biological control (current status – BC), where applicable. References are included. Where there is a paucity on impact data and no implementation of biological control to date in South Africa, impacts realised elsewhere have been considered (i.e. for *Sagittaria platyphylla*, *Egeria densa* and *Hydrilla verticillata*).

Impact	Species and impact category	Description	Impact level, no BC	Confidence[a], no BC	Impact level, BC	Confidence[a], BC	References
	1.5. Hybridisation	None	0	3	0	3	No indigenous Salviniaceae, sterile polyploid (Julien et al. 2009)
	1.6. Ecosystems	Ecosystem alteration	5	3	1	3	Coetzee et al. 2011a; Doeleman 1989; Julien et al. 2009
Subtotal	-	-	22	2.67	2	3	
2. Economic impacts	2.1. Agricultural production	Reduced irrigation	4	3	1	3	Coetzee et al. 2011a; Doeleman 1989; Julien et al. 2009
	2.2. Animal production	Reduced irrigation	3	3	0	3	Coetzee et al. 2011a; Doeleman 1989; Julien et al. 2009
	2.3. Forestry production	None	0	3	0	3	NA
	2.4. Human infrastructure	Bridges, weirs, hydropower, pumps	3	3	1	3	Hill 2003
	2.5. Human health	Disease vectors	2	2	1	2	Bennet 1966; Coetzee et al. 2011a; Thomas and Room 1986
	2.6. Human social life	Reduced water-based recreation	4	3	1	3	Coetzee et al. 2011a; Hill 2003; Julien et al. 2009
Subtotal	-	-	16	2.83	4	2.83	-
Total	-	-	38	2.75	6	2.92	-
Azolla filiculoides							
1. Environmental impacts	1.1. On vegetation	Habitat alteration	5	3	0	3	Hill and McConnachie 2009; McConnachie et al. 2003, 2004
	1.2. On animals	Habitat alteration	5	3	0	3	Hill and McConnachie 2009; McConnachie et al. 2003, 2004
	1.3. On competition	Space, light, nutrients	3	3	0	3	Hill and McConnachie 2009; McConnachie et al. 2003, 2004
	1.4. Transmission of disease	Provide habitat for vectors of disease	2	2	0	3	Hill and McConnachie 2009; McConnachie et al. 2003, 2004
	1.5. Hybridisation	None	0	3	0	3	Madeira et al. 2013, 2016
	1.6. Ecosystems	Ecosystem alteration	5	3	0	3	Hill and McConnachie 2009; McConnachie et al. 2003, 2004
Subtotal	-	-	20	2.83	0	3	
2. Economic impacts	2.1. Agricultural production	Reduced irrigation	5	3	0	3	Ashton and Walmsley 1984; Hill and McConnachie 2009; McConnachie et al. 2003, 2004
	2.2. Animal production	Reduced irrigation	5	3	0	3	Ashton and Walmsley 1984; Hill and McConnachie 2009; McConnachie et al. 2003, 2004
	2.3. Forestry production	None	0	3	0	3	NA
	2.4. Human infrastructure	Bridges, weirs, hydropower, pumps	3	3	0	3	Ashton and Walmsley 1984; Hill and McConnachie 2009; McConnachie et al. 2003, 2004
	2.5. Human health	Disease vectors	2	2	0	3	Hill and McConnachie 2009
	2.6. Human social life	Reduced water-based recreation	5	3	0	3	Ashton and Walmsley 1984; Hill and McConnachie 2009; McConnachie et al. 2003, 2004
Subtotal	-	-	20	2.83	0	3	-
Total	-	-	40	2.83	0	3	-
Myriophyllum aquaticum							
1. Environmental impacts	1.1. On vegetation	Habitat alteration	5	3	2	3	Cilliers 1999; Coetzee et al. 2011a
	1.2. On animals	Habitat alteration	3	3	1	3	Cilliers 1999; Coetzee et al. 2011a
	1.3. On competition	Space, light, nutrients	4	3	2	3	Cilliers 1999; Coetzee et al. 2011a
	1.4. Transmission of disease	Provide habitat for vectors of disease	2	2	1	2	-
	1.5. Hybridisation	None	0	3	0	3	Only female plants in SA (Henderson and Cilliers 2002)
	1.6. Ecosystems	Ecosystem alteration	4	3	2	3	Cilliers 1999; Coetzee et al. 2011a
Subtotal			18	2.83	8	2.83	
2. Economic impacts	2.1. Agricultural production	Reduced irrigation	5	3	2	3	Cilliers 1999; Coetzee et al. 2011a
	2.2. Animal production	Reduced irrigation	4	3	1	3	Cilliers 1999; Coetzee et al. 2011a
	2.3. Forestry production	None	0	3	0	3	NA
	2.4. Human infrastructure	Bridges, weirs, hydropower, pumps	4	3	1	3	Cilliers 1999; Coetzee et al. 2011a
	2.5. Human health	Disease vectors	3	2	1	2	Cilliers 1999; Coetzee et al. 2011a

Table 1-A1 continues on next page →

TABLE 1-A1 (Continues...): Summary of GISS detailed impact levels (Nentwig et al. 2016), for eight freshwater invasive alien aquatic weeds, in South Africa. Scoring was conducted in both the absence of biological control (BC) (worst-case scenario – no BC) and the presence of biological control (current status – BC), where applicable. References are included. Where there is a paucity on impact data and no implementation of biological control to date in South Africa, impacts realised elsewhere have been considered (i.e. for *Sagittaria platyphylla*, *Egeria densa* and *Hydrilla verticillata*).

Impact	Species and impact category	Description	Impact level, no BC	Confidence[a], no BC	Impact level, BC	Confidence[a], BC	References
	2.6. Human social life	Reduced water-based recreation	4	3	2	3	Cilliers 1999; Coetzee et al. 2011a
Subtotal	-	-	20	2.83	7	2.83	-
Total	-	-	38	2.83	15	2.83	-
Sagittaria platyphylla							
1. Environmental impacts	1.1. On vegetation	Habitat alteration	4	1	-	-	Adair et al. 2012; Chapman and Dore 2009
	1.2. On animals	Habitat alteration	4	1	-	-	Adair et al. 2012; Chapman and Dore 2009
	1.3. On competition	Space, light, nutrients	4	1	-	-	Adair et al. 2012; Chapman and Dore 2009
	1.4. Transmission of disease	Provide habitat for vectors of disease	2	1	-	-	Adair et al. 2012; Chapman and Dore 2009
	1.5. Hybridisation	None	0	3	-	-	Adair et al. 2012; Chapman and Dore 2009
	1.6. Ecosystems	Ecosystem alteration	4	1	-	-	Adair et al. 2012; Chapman and Dore 2009
Subtotal	-	-	18	1.33	-	-	-
2. Economic impacts	2.1. Agricultural production	Reduced irrigation	4	1	-	-	Adair et al. 2012; Chapman and Dore 2009
	2.2. Animal production	Reduced irrigation	4	1	-	-	Adair et al. 2012; Chapman and Dore 2009
	2.3. Forestry production	None	0	3	-	-	Adair et al. 2012; Chapman and Dore 2009
	2.4. Human infrastructure	Bridges, weirs, hydropower, pumps	3	1	-	-	Adair et al. 2012; Chapman and Dore 2009
	2.5. Human health	Disease vectors	2	1	-	-	Adair et al. 2012; Chapman and Dore 2009
	2.6. Human social life	Reduced water-based recreation	4	1	-	-	Adair et al. 2012; Chapman and Dore 2009
Subtotal	-	-	17	1.33	-	-	-
Total	-	-	35	1.33	-	-	-
Egeria densa							
1. Environmental impacts	1.1. On vegetation	Habitat alteration	5	1	-	-	Cabrera Walsh et al. 2013; Coetzee et al. 2011b; Yarrow et al. 2009
	1.2. On animals	Habitat alteration	4	1	-	-	Cabrera Walsh et al. 2013; Coetzee et al. 2011b; Yarrow et al. 2009
	1.3. On competition	Space, light, nutrients	5	1	-	-	Cabrera Walsh et al. 2013; Coetzee et al. 2011b; Yarrow et al. 2009
	1.4. Transmission of disease	Provide habitat for vectors of disease	2	1	-	-	Cabrera Walsh et al. 2013; Coetzee et al. 2011b; Yarrow et al. 2009
	1.5. Hybridisation	None	0	3	-	-	Lambertini et al. 2010
	1.6. Ecosystems	Ecosystem alteration	5	1	-	-	Cabrera Walsh et al. 2013; Coetzee et al. 2011b; Yarrow et al. 2009
Subtotal	-	-	21	1.33	-	-	-
2. Economic impacts	2.1. Agricultural production	Reduced irrigation	5	1	-	-	Cabrera Walsh et al. 2013; Coetzee et al. 2011b; Yarrow et al. 2009
	2.2. Animal production	Reduced irrigation	4	1	-	-	Cabrera Walsh et al. 2013; Coetzee et al. 2011b; Yarrow et al. 2009
	2.3. Forestry production	None	0	3	-	-	Cabrera Walsh et al. 2013; Coetzee et al. 2011b; Yarrow et al. 2009
	2.4. Human infrastructure	Bridges, weirs, hydropower, pumps	5	1	-	-	Cabrera Walsh et al. 2013; Coetzee et al. 2011b; Yarrow et al. 2009;
	2.5. Human health	Disease vectors	2	1	-	-	Cabrera Walsh et al. 2013; Coetzee et al. 2011b; Yarrow et al. 2009
	2.6. Human social life	Reduced water-based recreation	5	1	-	-	Cabrera Walsh et al. 2013; Coetzee et al. 2011b; Yarrow et al. 2009
Subtotal			21	1.33	-	-	-
Total			42	1.33	-	-	-

Table 1-A1 continues on next page →

TABLE 1-A1 (Continues...): Summary of GISS detailed impact levels (Nentwig et al. 2016), for eight freshwater invasive alien aquatic weeds, in South Africa. Scoring was conducted in both the absence of biological control (BC) (worst-case scenario – no BC) and the presence of biological control (current status – BC), where applicable. References are included. Where there is a paucity on impact data and no implementation of biological control to date in South Africa, impacts realised elsewhere have been considered (i.e. for *Sagittaria platyphylla, Egeria densa* and *Hydrilla verticillata*).

Impact	Species and impact category	Description	Impact level, no BC	Confidence[a], no BC	Impact level, BC	Confidence[a], BC	References
Hydrilla verticillata							
1. Environmental impacts	1.1. On vegetation	Habitat alteration	5	1	-	-	Balciunas et al. 2002; Langeland 1996; Pieterse 1981; Souza 2011
	1.2. On animals	Habitat alteration	5	1	-	-	Balciunas et al. 2002; Langeland 1996; Pieterse 1981; Souza 2011
	1.3. On competition	Space, light, nutrients	5	1	-	-	Balciunas et al. 2002; Langeland 1996; Pieterse 1981; Souza 2011
	1.4. Transmission of disease	Provide habitat for vectors of disease	2	1	-	-	Balciunas et al. 2002; Langeland 1996; Pieterse 1981; Souza 2011
	1.5. Hybridisation	None	0	3	-	-	Balciunas et al. 2002; Langeland 1996; Pieterse 1981; Souza 2011
	1.6. Ecosystems	Ecosystem alteration	5	1	-	-	Balciunas et al. 2002; Langeland 1996; Pieterse 1981; Souza 2011
Subtotal			22	1.33	-	-	
2. Economic impacts	2.1. Agricultural production	Reduced irrigation	5	1	-	-	Balciunas et al. 2002; Langeland 1996; Pieterse 1981; Souza 2011
	2.2. Animal production	Reduced irrigation	5	1	-	-	Balciunas et al. 2002; Langeland 1996; Pieterse 1981; Souza 2011
	2.3. Forestry production	None	0	3	-	-	Balciunas et al. 2002; Langeland 1996; Pieterse 1981; Souza 2011
	2.4. Human infrastructure	Bridges, weirs, hydropower, pumps	5	1	-	-	Balciunas et al. 2002; Langeland 1996; Pieterse 1981; Souza 2011
	2.5. Human health	Disease vectors	2	1	-	-	Balciunas et al. 2002; Langeland 1996; Pieterse 1981; Souza 2011
	2.6. Human social life	Reduced water-based recreation	5	1	-	-	Balciunas et al. 2002; Langeland 1996; Pieterse 1981; Souza 2011
Subtotal	-	-	22	1.33	-	-	-
Total	-	-	44	1.33	-	-	-

[a], Confidence limits based on Nentwig et al. (2016) where 1 = low confidence – no empirical data or literature to support the impact score; 2 = medium confidence – no empirical data from South Africa, but literature from elsewhere to support the impact score; 3 = high confidence – empirical and published data from South Africa support the impact score.

6

Recommendations for municipalities to become compliant with national legislation on biological invasions

Authors:
Ulrike M. Irlich[1,2] ⓘ
Luke Potgieter[1,2] ⓘ
Louise Stafford[1]
Mirijam Gaertner[1,2]

Affiliations:
[1]Environmental Resource
Management Department
(ERMD), City of Cape Town,
Westlake Conservation
Office, South Africa

[2]Centre for Invasion Biology,
Department of Botany &
Zoology, Stellenbosch
University, South Africa

Corresponding author:
Ulrike Irlich,
irlich@gmail.com

Background: *The South African National Environmental Management: Biodiversity Act* (No. 10 of 2004) (NEM:BA) requires all Organs of State at all spheres of government to develop invasive species monitoring, control and eradication plans. Municipalities across South Africa are required to comply with the Alien and Invasive Species Regulations under NEM:BA but are faced with myriad challenges, making compliance difficult.

Objective: This paper unpacks some of the challenges municipalities face and provides guidance on how to overcome these in order to achieve NEM:BA compliance. Through a strategic, municipal-wide approach involving different landowners, compliance can be achieved and many of the associated challenges can be overcome. For example, lack of awareness and capacity within municipal structures can be addressed through various platforms that have proven successful in some areas.

Conclusions: Using the City of Cape Town as a case study, we highlight some of the notable successes in overcoming some of these challenges. For example, the City's Invasive Species Strategy has resulted in municipal buy-in, departmental collaboration and a city-wide invasive plant tender, allowing for streamlined invasive plant control across the city. We present a framework as a first step towards measuring compliance and how the national status report can measure the level of compliance by Organs of State.

Introduction

Biological invasions are a large and growing threat to ecosystem integrity in many parts of the world and have been identified as a priority for management, both nationally (Simberloff, Parker & Windle 2005; van Wilgen et al. 2012) and internationally (McNeely et al. 2001). The International Convention on Biological Diversity (CBD) Strategic Plan for Biodiversity (2011–2020), with the Aichi Biodiversity Target Nr. 9, states that invasive species with their associated pathways need to be identified and subsequent measures be put in place to minimise their spread (McGeoch et al. 2010). Furthermore, it stipulates that priority invasive species are to be controlled or eradicated (Caffrey et al. 2014). Legislation, regulations and strategies have been put in place at a global level (Global Strategy, McNeely et al. 2001) as well as for larger regions (e.g. EU Regulation 1143/2014 on Invasive Alien Species). Numerous countries, signatories as well as non-signatories to the CBD, have taken it upon themselves to follow suit [e.g. Mexico (National Advisory Committee on Invasive Species 2010), Great Britain (Great Britain Non-native Species Secretariat 2015)]. Similar approaches have been adopted at subnational levels, such as regional (Virginia, USA [Virginia Invasive Species Working Group 2012]) or specific areas, such as cities (Brisbane, Australia [Brisbane City Council 2013]) or nature reserves (Maunakea, Hawaii [Vanderwoude et al. 2015]).

Globally, urbanisation is on the rise, with an estimated 50% of the world's population currently living in cities. This trend is expected to increase drastically in the next few decades (Faeth, Saari & Bang 2012; Grimm et al. 2008). Increased urbanisation results in increased introductions of potentially invasive species to these human-dominated landscapes.

Biological invasions in urban areas are of concern as they can have considerable impacts on urban biodiversity and ecosystem services (Kowarik 2011). Cities are often points of introduction of non-native species (Pyšek 1998; Vitousek et al. 1997), and the associated large variety and frequency of pathways and vectors aids in the movement of species within an urban environment and surrounding areas (Alston & Richardson 2006; Hawthorne et al. 2015; von der Lippe & Kowarik 2008). In cities, non-native species encounter climatic conditions, habitats, hydrology and soils that have been profoundly altered by human activity, amplifying the establishment and spread of these species (Klotz & Kühn 2010; Kowarik 2011; Pickett et al. 2001).

Urbanisation and the associated introductions of non-native species present a significant challenge to people and landscapes in South Africa (van Wilgen 2012). The trade in ornamental plants and pets, and other enterprises that rely on non-native taxa, continues to introduce new species into urban areas, many of which remain undetected or unregulated, or both (Cronin et al. 2017). Invasive species management in urban areas is challenging for a number of reasons. Numerous entry-points, vectors and pathways within urban areas lead to high propagule pressure of invasive species (Kowarik & von der Lippe 2007; Pyšek 1998). Stakeholders in municipalities are numerous and often have strongly divergent views about the impacts and benefits of particular invasive species, and as a result, significant conflicts arise over the management of such species (Dickie et al. 2014; Gaertner et al. 2016; Zengeya et al. 2017).

The National Environmental Management: Biodiversity Act (No. 10 of 2004) (NEM:BA, hereafter referred to as the NEM:BA Act) covers all aspects of South Africa's biodiversity conservation and management at a country level and makes provision for the control and management of invasive species nationally (Alien and Invasive Species regulations under NEM:BA, hereafter referred to as NEM:BA regulations). Achieving NEM:BA compliance would require to meet the terms of the NEM:BA regulations, the specific actions it outlines and adhering to the timeframes stipulated, namely, submitting invasive species monitoring, control and eradication plans (from here onward referred to as 'area management plans') within 1 year from September 2016, after the guidelines for management plans were published (Section 5.2) (Figure 1).

A national strategy aimed at addressing biological invasions in South Africa (DEA 2014) has been drafted. Although the document has not been formally released, it is readily available. The strategy provides guidelines for Organs of State (Box 1) for managing invasive species, areas and pathways of introduction and movement against the background of the four stages of invasion (initial introduction, establishment, expansion and dominance).

Management of invasive species in South African municipalities is limited, with the City of Cape Town and eThekwini (metropolitan municipalities in the Western Cape and KwaZulu-Natal, respectively) being exceptions. For example, in Cape Town, a dedicated Invasive Species

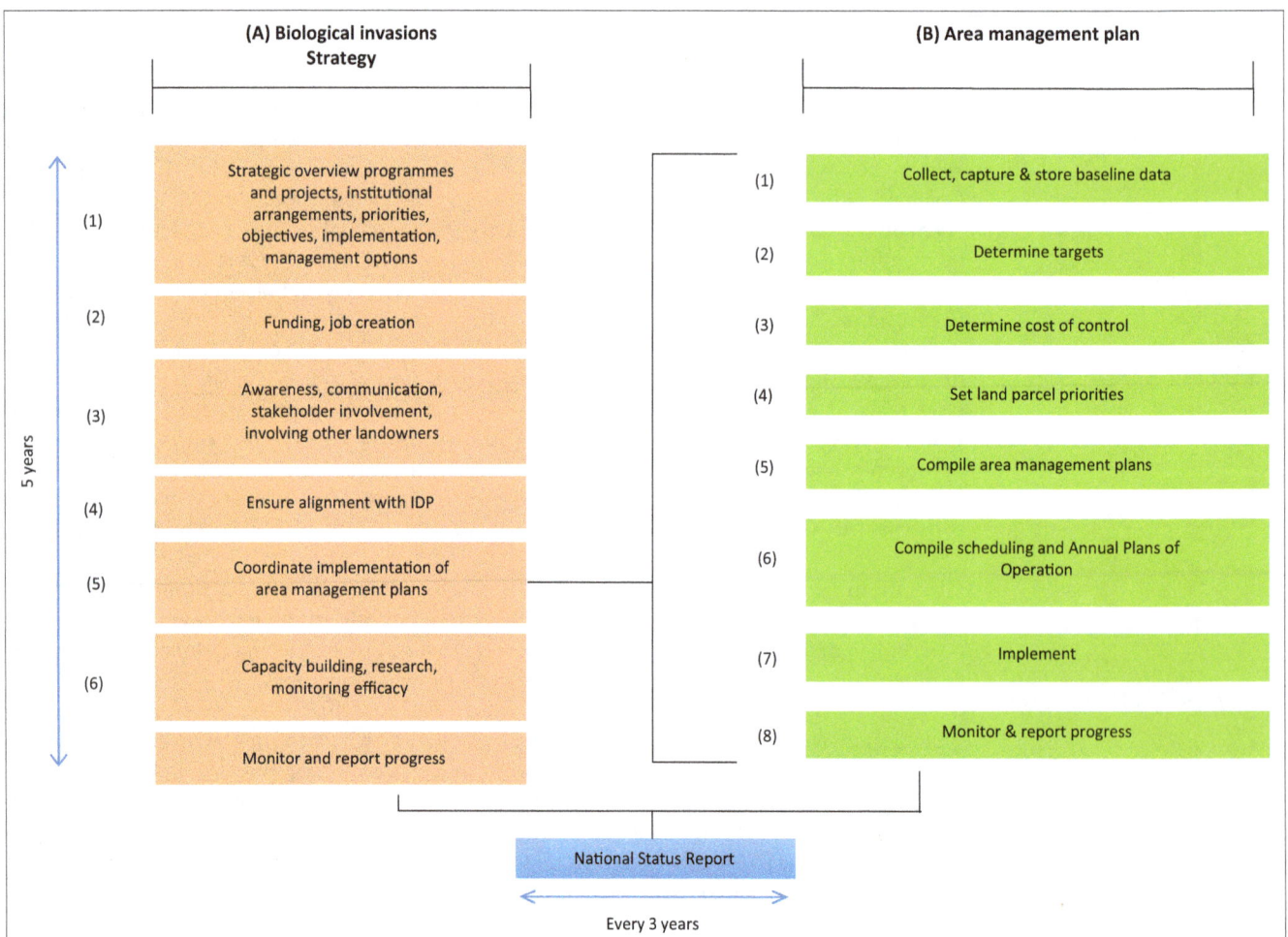

FIGURE 1: Overall framework of NEM:BA requirements (IDP: Integrated Development Plans).

BOX 1: Definitions and explanations.

Competent Authority: Any organ of state, delegated by DEA, that has the legally delegated or invested authority, capacity, or power to perform a designated function. Once an authority is delegated to perform a certain act, only the competent authority is entitled to take accounts therefrom and no one else. In terms of NEM:BA, a competent authority can be either (1) the Minister; (2) an organ of state in the national, provincial or local sphere of government or (3) any other organ of state.

District Municipality: Is a municipality which executes some of the functions of local government for a district. District municipalities are comprised of several local municipalities.

Integrated Development Plan (IDP): This is an overall strategy document for the municipality.

Invasive Species Monitoring, Control and Eradication Plan: A plan contemplated in section 76 of the NEM:BA Act and in Regulation 8.

Land under the control of Organs of State: There is uncertainty surrounding the interpretation of this clause – it may refer only to land parcels owned by a municipality (the stance taken by the City of Cape Town and adopted for the remainder of the paper) or it may refer to all parcels of land within a municipal boundary. The latter however, may prove impractical, as the municipality does not have authority over privately owned land and activities thereon. This matter needs to be clarified by DEA to ensure sound understanding and subsequent NEM:BA compliance by municipalities.

Land parcels: Land, or properties, owned and managed by municipalities can be protected areas, public open spaces, a river corridor, office buildings and road verges.

Legislative competence: Legal authority to carry out an activity.

Mandates: An official order or commission to do something.

NEM:BA compliance: Adhere to all actions stipulated within the legislation, within the timeframes given

Organs of State: Any department of state or administration in the national, provincial or local sphere of government.

Status Report: A national status report, tracking progress to compliance across the country needs to be compiled by SANBI as per Section 11 of the NEM:BA Regulations, not to be confused with the status report to be submitted by managing authorities of protected areas (as per Section 77 (1) and (2) of the NEM:BA Act). Thirdly, as part of the area management plans submitted by Organs of State a 'status report on the efficacy of previous control and eradication measures' needs to be submitted (as per Section 76(3d) of the NEM:BA Act).

Strategy: A plan of action or policy designed to achieve a major or overall aim.

TABLE 1: Measuring compliance and an indication on the level of awareness: Number of municipalities in each province (according to 2016 demarcations) (Part A) with a comparison of the number of municipalities that attended the awareness raising (NEM:BA roadshow) (Part B) and training events (South African Green Industries Council [SAGIC] training) (Part C). Number of plans submitted by September 2016 by municipalities within the different provinces (Part D) (data provided by DEA).

Province	Part A			Part B			Part C			Part D		
	Total number of municipalities			Municipalities attended NEM:BA roadshow			Municipalities attended SAGIC training			Number of submitted control plans		
	Metro	District	Local	Metro	District	Local	Metro	District	Local	Metro	District[a]	Local
Eastern Cape	2	6	31	2	1	6	1	2	5	-	-	-
Free State	1	4	18	1	1	3	1	1	3	-	-	-
Gauteng	3	2	6	3	-	5	3	-	3	-	-	-
KwaZulu-Natal	1	10	43	1	-	4	1	-	4	1	-	-
Limpopo	-	5	22	-	2	6	-	-	5	-	-	1
Mpumalanga	-	3	17	-	2	4	-	2	5	-	-	-
Northern Cape	-	5	26	-	2	2	-	2	3	-	-	-
North West	-	4	18	-	2	7	-	2	5	-	-	-
Western Cape	1	5	24	1	3	9	1	4	7	1	1 (1[b])	5 (3[c])
Total	**8**	**44**	**205**	**8**	**13**	**46**	**7**	**13**	**40**	**2**	**1**	**6**

Source: Authors' own work using data supplied by DEA and SAGIC

[a], Not all district municipalities manage or own land; thus, some district municipalities are not required to submit plans as per the current regulations. Hence the number of plans submitted will not equal the number of district municipalities once 100% compliance is achieved.

[b], Annexures missing, thus still viewed as incomplete.

[c], Letters were submitted to state (1) no budget is available to complete plans, (2) plans are complete but awaiting council approval and (3) the plans are being developed and will be submitted at a later stage.

Unit has been integrated into the municipal structure, aimed at streamlining and facilitating invasive species management across the metro (Gaertner et al. 2016). Some municipalities (e.g. Mbombela Local Municipality in Mpumalanga and Eden District Municipality in the Western Cape) have initiated actions to comply with the NEM:BA regulations since they were promulgated (SALGA 2016; and pers. comm. with municipalities by the authors, 2016). However, the majority of the remaining municipalities have not met the set timeframes, as seen from the number of submitted plans (Part D of Table 1). They are faced with multiple challenges such as a lack of capacity to develop area management plans and to implement, monitor and report on control programmes (K. Montgomery pers. comm., 2016). By using Cape Town as a case study, the challenges and complexities around invasive species management in urban areas are discussed with the intention of providing some guidance on how to overcome these challenges.

The aims of this paper are to (1) outline the requirements for municipalities to become NEM:BA compliant, (2) highlight the challenges faced by municipalities, (3) provide guidance on how to overcome such challenges, (4) outline the process for compiling area management plans and (5) discuss some indicators that can be used to measure progress towards compliance.

The City of Cape Town

The City of Cape Town (hereafter referred to as the City) is situated in the Cape Floristic Region, a biodiversity hotspot with high levels of endemism (Cowling et al. 1996), and is thus of high conservation priority (Holmes et al. 2012). The Cape Town municipality covers an area of 2460 km^2, of which over 61% has been transformed for urban development or agriculture (Holmes et al. 2012). Cape Town is the economic and social hub of the Western Cape, and the population has increased by almost 30% over a 10-year period from 2001 to

2011 (City of Cape Town 2012). Key pressures on the biodiversity surrounding the City include urban sprawl, agriculture, development for tourism (Holmes et al. 2012), exploitation through illegal harvesting (Petersen et al. 2012), changing of fire regimes through either suppressing or accelerating fire patterns (van Wilgen & Scott 2001) and invasive species (Rebelo et al. 2011).

Introduction of non-native species to Cape Town started with the first settlers in the 1600s, which brought in woody plant species for timber and dune stabilisation (Wilson et al. 2014). In Cape Town, invasive species not only negatively impact native biodiversity by outcompeting indigenous species (McKinney 2006), aquatic invasive species such as water hyacinth (*Eichhornia crassipes*) also cause flooding by clogging water ways (Richardson & van Wilgen 2004). Dense invasive plant stands pose serious risks to human settlements; for example, invasive pines and wattles increase the severity of wildfires near residential areas (van Wilgen & Scott 2001), provide shelter for criminal activities (Gaertner et al. 2016), pose human health risks (Taylor et al. 2008) and decrease river flows (Le Maitre et al. 2011).

NEM:BA requirements

NEM:BA places a 'Duty of Care' (Section 73(2) [as amended]) on all landowners, whether private or public, to control invasive species on their land. Section 76(2a) determines that all Organs of State at all spheres of government (from National through to Local Government) must compile area management plans for land under their control; Section 76(4 a–f) of the Act states the requirements of these plans (see Figure 1 for more detail on the Regulations). For Organs of State to become compliant with the NEM:BA regulations, they need to develop, submit for approval and implement area management plans, report back (Section 76[4][d]) and provide measurable indicators showing progress and timeframes for completion to national government (Department of Environmental Affairs [DEA]) (Figure 1). The guidelines for the development of these plans have been published (DEA 2015) and are available on DEA's website (https://www.environment.gov.za/sites/default/files/legislations/nemba_invasivespecies_controlguideline.pdf). The completed area management plans were required to be submitted by the end of September 2016 (1 year after the publication of the guidelines for management plans (NEM:BA Regulations [2] [b]) (Figures 1 and 2). Plans must be drawn up for all land under the control of Organs of State (Box 1; Figure 2; see Guidelines provided by DEA 2015).

Area management plans must include a description of the land parcels (Box 1) in question, detailed lists and descriptions of all the listed species found on each of the land parcels, the extent of invasion and the efficacy of previous control and eradication measures. These plans should be included into the municipal Integrated Development Plans (IDPs) (Section 76 [2][b]), to ensure subsequent implementation and budget allocation (Ruwanza & Shackleton 2016). Furthermore, the NEM:BA Act (Section 77[1]) states that all Organs of State managing

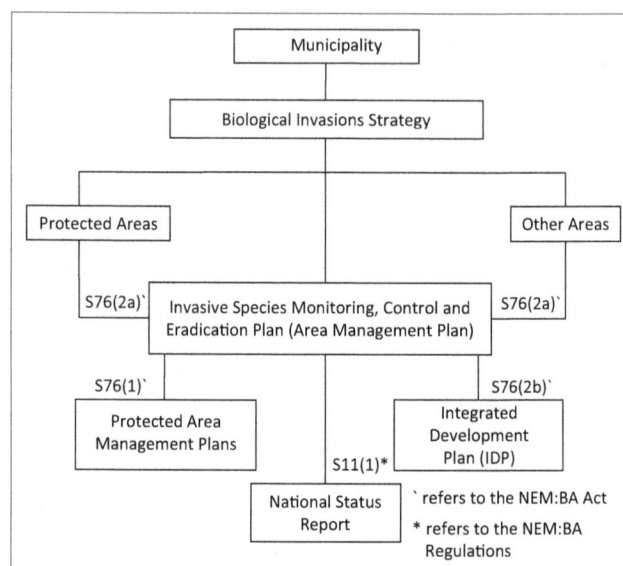

FIGURE 2: Framework for Biological Invasions Strategy and the steps required for developing area management plans for various parcels of land. Detailed guidelines on compiling area management plans are provided by DEA (2015).

protected areas are required to submit a status report (Box 1) 'at regular intervals', reporting on the progress made towards achieving the set targets. Smaller municipalities can develop a single plan for the entire municipal area. For larger municipalities, it is advisable to divide the municipality into more strategic areas (e.g. catchments or suburbs). Management plans should also make provision for invasive fauna. Collaboration with multiple landowners is required to assist with and ensure that plans are also developed for other land parcels within the municipality. Guidelines for private landowner area management plans are available on the City of Cape Town Invasive Species website (https//www.capetowninvasives.org.za).

Challenges faced by municipalities limiting NEM:BA compliance and recommendations to overcome these challenges

South African municipalities are facing a multitude of challenges, ranging from budget constraints to limited awareness and capacity. In this section, we firstly review these challenges, which have been identified through engagements with council officials, stakeholders, practitioners and scientists, and secondly present guidance on how to address the situation and to leverage invasive species management to benefit municipalities. Table 2 summarises the challenges and recommendations drawing on specific examples from the City of Cape Town.

Strategic planning and municipal buy-in

The Constitution of South Africa of 1993 regulates the responsibilities and legislative competence of each sphere of government. Municipal responsibilities include the delivery of a range of basic services such as access to water and sanitation (Section 73 of the *Municipal Systems Act*) to

TABLE 2: Challenges and proposed solutions in achieving NEM:BA compliance and managing invasive species management across municipalities.

Challenges	Solutions			Example	Sources
	Biological Invasion Strategy	Control plans	Communication/media	City of Cape Town: successes and challenges in invasive species management	
Land ownership	Determine land ownership and actions for addressing invasion across different landowners	Stakeholder involvement (e.g. Departments) for coproduction and coordination	Knowledge and information exchange; Interdepartmental and Institutional engagements; Landowner engagements; Creating of common vision and goal;	Cape Town's Invasive Species Strategy (2008) adopted by Council (Table 3), provided a platform for departmental alignment and resulted in city-wide invasive plant tender, resulting in streamlined clearing and better planning. This strategy is reviewed and updated every 5 years. Management plan development: The City was divided into four geographic regions. These regions were sub-divided into departmental land parcels, rivers, wetlands and protected areas. This sub-division strengthened the implementation of the strategy and management plans and helped to coordinate the invasive species responsibilities of the different departments.	Ruwanza & Shackleton 2016
Lack of awareness/ knowledge	Determine different audiences, means of communication and messages	Stakeholder involvement (private, business, governmental	Invasive Species Forums; Invasive species training (e.g. South African Green Industries Council [SAGIC]) Social media; Websites (e.g. www.invasives.org.za which municipalities can use to host their information and projects; currently used by three municipalities); Citizen science & citizen groups (e.g. garden clubs, friends' groups and ratepayer's associations) Knowledge and information exchange	Cape Town Invasive Species Forums (assisted with National roadshows and setting up of other forums); Partner with NGOs (e.g. Cape Town Environmental Education Trust); Partner with businesses (e.g. NCC Environmental Services; nurseries); Cape Town Invasive Species Facebook page; Spotter Network and Website providing information (www.capetowninvasives.org.za); Friends Groups (e.g. Friends of the Liesbeek; Friends of Constantia Valley Greenbelts); Volunteer Hack Groups; Garden Clubs, Ratepayers Associations; Media exposure (radio, TV, newspapers)	Crall et al. 2012 Cronin et al. 2017 Novoa et al. 2016 Sitas et al. 2016
Lack of capacity	Determine capacity needs; conduct needs analysis; actions to address	Dedicated environmental staff/ management; Provide necessary training; Outsource the development of control plans	Establish partnerships with different stakeholder groups; Collaboration with academic institutions; Identify 'champions' for invasive species management; Collaborate with different landowners within the municipal boundaries; Invasive species training (e.g SAGIC)	Establishment of Invasive Species Unit; Cape Town's Invasive Species Strategy (2008); Collaboration with the Research Institutes (CIB & Rhodes University); Partner with NGOs (e.g. Cape Town Environmental Education Trust); Mentorship of staff; Accommodating interns and volunteers to assist	Gaertner et al. 2016 Ruwanza & Shackleton 2016 Sitas et al. 2016
Limited and unpredictable budget	Determine long term strategic budget requirements; Prioritise; Establish partnerships; Job creation opportunities	Determine cost of control Annual Plans of Operation; Prioritise areas	Communicate with decision-makers, illustrate return on investment	EPWP allocation for invasive species management reduced by 50% in the 2016/17 financial year due to general budget cuts; Effective interdepartmental collaboration; Prioritisation workshop conducted following methodology of Forsyth et al. (2012)	Forsyth et al. 2012 Gaertner et al. 2016
Complexities/ conflicts of interest	Incorporate into IDP; Stakeholder involvement	Stakeholder involvement;	Research; Involve academic institutions Raise awareness (as stated above)	Conducted prioritisation workshop to identify priority areas for management; prioritisation process underway.	Gaertner et al. 2016

residents in a sustainable manner, promoting economic development and safe, healthy environments (Koma 2010). Environment is placed at the National and Provincial level of legislative competence, and thus local government prioritises service delivery over environmental aspects, such as invasive species control (Ruwanza & Shackleton 2016). However, municipalities play an important role in environmental planning and management but are not structured or mandated to perform their environmental responsibilities. Furthermore, many municipalities, particularly local municipalities, do not have dedicated environmental departments or staff, adding to a lack of environmental management at municipal levels. Currently, the NEM:BA delegations are not devolved to municipalities.

To encourage and assist local government to address environmental issues, we recommend that municipalities develop a biological invasions strategy in collaboration with their stakeholders (Figure 2). We further recommend that municipal strategies be aligned with the National Strategy (DEA 2014) for dealing with biological invasions by addressing the stages of invasion, priorities and management

approaches for species, areas, pathways of introduction and movement of species. Such a strategy can help achieve political buy-in and aid in delegating responsibilities across departments. Furthermore, it allows for more streamlined processes, ultimately resulting in more efficient expenditure and accountability. It also addresses the issue of multiple landowners within municipal boundaries, as further discussed below. DEA should consider the possibilities and processes of appointing municipalities as competent authorities to assist with invasive species management across municipalities (including privately owned land).

We recommend that a template and guidelines for municipal strategies be developed and made readily available to all municipalities. The City of Cape Town developed such a strategy in 2008, which was adopted by Council (Tables 2 and 3). Table 3 outlines some of the aims and indicators for success as per the City of Cape Town Invasive Species Strategy.

District municipalities play a coordinating role for several local municipalities; hence, we recommend they should

TABLE 3: The City of Cape Town's Invasive Species Strategy (City of Cape Town 2008).

Aims	Indicators for success
Obtain high level buy-in and support for the implementation of the biological invasions strategy	Achieved. The strategic framework was approved by the council.
Establish a management and coordination scenario for effective and integrated management of IAS within the City's boundaries	Achieved. An Invasive Species Unit was established coordinating invasive species functions across different line departments. Regular meetings with departments to iron out issues, plan and report back are conducted.
Develop an Invasive Alien Species education, communication and awareness strategy for the City of Cape Town	Achieved. Resulted in outreaches in schools, communities, visits to the biological control facility on environmental days, for example, World Wetlands Day and Invasive Species Week. Facebook page, Invasive species website, Spotter network and establishment of invasive species forums to facilitate public participation.
Develop and implement a legal and policy framework for IAS management	Achieved. Framework produced and recommendations are in process of being implemented. Risk assessment conducted.
Develop funding mechanisms to support IAS management	Achieved. Different funding mechanisms ensured implementation of control plans. Funding sources include EPWP, departmental, Working for Water, Working for Wetlands, ward allocations.
Establish priorities based on given resources and appropriate weighting of desired outcomes	Achieved. Although implementation is challenging because of the different dynamics in urban areas and inconsistent budgetary allocations.
Develop integrated control plans based on identified priorities, with clear timelines and required resources	Partly achieved in the absence of guidelines for developing the control plans; the City relied on annual plans of operation and long term control schedules. Process to develop control plans according to guidelines commenced in October 2015.
Monitor effectiveness of the IAS management in the City of Cape Town	Partly achieved because of capacity constraints.

IAS, Invasive Alien Species.

facilitate the development of district-wide strategies (in collaboration with their respective local municipalities) as well as the various municipal invasive species area management plans. The role of the South African Local Government Association (SALGA) is to ensure that municipalities are aware of the new legislation relevant to them and provide assistance by unblocking compliance challenges (N. Mtsewu pers. comm., 2016). According to NEM:BA Section 76(3), the minister may appoint the South African National Biodiversity Institute (SANBI) to assist municipalities with compiling management plans and status reports that report back on the efficacy of control measures (Wilson et al. 2017). As such, SALGA acts as an important link between DEA, SANBI and municipalities to assist and guide municipalities.

Consolidating the management plans under district-wide or metro-wide strategies will ensure consistency and higher standards of plans and reduce the number of plans to be submitted to DEA substantially (44 district and 8 metro plans would have to be submitted from local government authorities, instead of a total of 257 [205 Local, 44 District and 8 Metropolitan municipalities] plans [Part A of Table 1]).

Issues around promulgation of the NEM:BA regulations

Several shortcomings have been identified regarding the process of how the NEM:BA regulations were promulgated: NEM:BA was first promulgated in 2004, but the NEM:BA regulations were only promulgated in 2014; therefore, the determinations of NEM:BA still need to be institutionalised by municipalities, who in the absence of expertise are still not fully aware of their obligations. The timeframes and requirements set by the NEM:BA regulations therefore pose a challenge to municipalities (SALGA 2016). Furthermore, the institutions delegated to assist other Organs of State (the DEA and SANBI) have been criticised by municipalities for their lack of guidance (K. Montgomery pers. comm., 2016).

An additional concern is that the contents of the guidelines as well as the regulations are not easily interpretable by those having to apply these on the ground. Hence, simplification of the management plan guidelines and interpretation of the NEM:BA regulations should be considered by DEA and communicated by SALGA. Tools and templates should also be developed to assist with writing management plans, and all material should be made readily available to support municipalities in becoming NEM:BA compliant.

Multiple landowners within municipal boundaries

Within municipal boundaries, a mosaic of different landownerships co-exists, namely, national and provincial governments, residential, agricultural, industrial and communal (Table 2). The portions of land managed or owned by these different landowners vary in size, land use and levels of invasive species infestation as well as potential introduction and spread of invasive species. A lack of synergy and collaboration between municipal authorities and the different landowners can be problematic and counter-productive when managing invasive species in urban areas. Holistic management approaches require private landownership buy-in and cooperation. However, achieving such cooperation is complex and requires significant resource capacity, with few success stories to date (e.g. Sitas et al. 2016).

In an analysis conducted by SALGA (2016), several municipalities were found to be unsure of the number of properties registered under their name. We recommend municipalities conduct an audit of land parcels known to be under the control of the municipality and start developing area management plans for those land parcels. Furthermore, we recommend a register (or database) and a map of known municipal land parcels be kept and updated as and when new information becomes available. The strategy should make provision for dealing with land ownership and liaison between the municipality and other land owners within the municipal boundaries to ensure synergy.

Lack of awareness/knowledge

Awareness of invasive species impacts is generally poor and knowledge of the requirements set out by the NEM:BA regulations (under Chapter 5) is lacking. This applies to the public (Shackleton & Shackleton 2016) and professionals (e.g. nursery owners; Cronin et al. 2017; Table 2). Knowledge regarding invasive species-related matters within municipalities varies extensively. Although some municipalities are aware of the obligations placed on them by the NEM:BA regulations (100% of metros, 29.5% of districts and 22% of local municipalities attended NEM:BA-specific roadshows, Part B of Table 1), they often do not have the capacity or knowledge to address these requirements given the timeframes and are consequently stalled in their attempts to move forward. Others are unaware of the regulations and thus compliance cannot occur (SALGA 2016).

To raise awareness within municipalities, several initiatives are recommended (aimed at the public as well as the municipal staff). Municipal Invasive Species Forums are a useful platform for raising awareness about the impacts of invasive species, addressing municipal and landowner responsibilities, and allowing stakeholder and public input as well as obtaining buy-in. Other effective means of advocacy in engaging the public include social media and citizen science projects (such as spotter networks; e.g. Crall et al. 2012; Hawthorne et al. 2015).

Showcasing the negative impacts of invasive species (e.g. fire threat due to increased fuel loads), as well as the success of invasive species clearing projects, in enhancing ecosystem service delivery can be useful for raising awareness among the general public (van Wilgen et al. 2011). Furthermore, knowledge and information exchange with other municipalities (through informal discussions or inter-municipal workshops) or other relevant stakeholders is critical in bridging the knowledge gap (Sitas et al. 2016; Table 2).

Involving and collaborating with established interest groups (e.g. garden clubs, Table 2) provides an opportunity for municipalities to harness the interests and expertise within these stakeholder groups to achieve the collective purpose of reducing impacts of invasive species in urban areas (Table 2). Although this is a time-consuming activity, it has proven successful in mitigating potential conflicts and creating a common goal and understanding of the situation at hand. Table 2 lists additional examples of initiatives that could assist in raising awareness drawing on examples implemented by the City of Cape Town.

Lack of capacity

Municipalities are comprised of urban centres (within the urban edge, usually consisting of mixed use: residential, industrial and commercial) and peri-urban areas (generally consisting of a matrix of residential, agricultural and natural areas). Some municipalities (e.g. the City of Cape Town) own and manage protected areas within their boundaries, requiring an additional status report in terms of the NEM:BA Act (Section 77[1] and [2]). Depending on the municipal structures, different line functions or departments are responsible for managing the different parcels of land (e.g. the parks department manages public open spaces and roads department manages the road verges). This split in functions complicates invasive species management, as multiple departments have different mandates, access to resources and varying expertise in managing biological invasions.

The lack of capacity in terms of institutional and human resources limits municipal performance (Koma 2010). Numerous municipalities do not have dedicated environmental staff or departments (Ruwanza & Shackleton 2016). Understaffing because of budget constraints regularly results in staff having to double up on their responsibilities. Furthermore, the skills capacity gaps in some municipalities are a major challenge, where staff are placed in positions for which they are not adequately trained or experienced (Koma 2010).

In addition to these constraints, managing invasive species is not traditionally part of cities' or towns' mandates (Box 1); therefore, they are not institutionally geared for this task (Ruwanza & Shackleton 2016). Municipalities generally do not have the correct equipment, expertise, capacity or budget to address the issue of invasive species in addition to meeting everyday service delivery requirements (Table 2).

Faced with limited management capacity, as discussed above, municipalities are unable to achieve NEM:BA compliance. Several approaches can be adopted to aid in developing capacity. A starting point is to (1) raise awareness and involve multiple landowners within the municipal boundaries, (2) obtain high level municipal management buy-in and support for ensuring NEM:BA compliance, (3) create an understanding of what the requirements for compliance are for different stakeholders and landowners, (4) identify the resource and capacity requirements to achieve compliance, (5) determine what capacity and resources are available nationally for building capacity and assisting municipalities in collaboration with SALGA, (6) implement a programme to develop capacity and increase synergy and collaboration across different municipalities to ensure effective use of limited resources, (7) increase access to information (invasive species information and associated control methods) and finally (8) development and access to a central database for tracking clearing operations and guide planning processes. Options to build capacity can be addressed through different strategies and collaboration with multiple landowners, communities and business (see Table 2). An alternative approach would be the outsourcing of the management plan development and subsequent implementation. However, to ensure this is executed properly, in-house expertise is required to oversee and guide the process. Accrediting service providers in the invasive species realm will further ensure that competent service providers are appointed.

Limited and unpredictable budget

Municipal income varies according to the size of the municipality (Figure 3) and its ability to generate revenues (Ramakhula 2010; Ruwanza & Shackleton 2016). Property tax, one of the main sources of revenue for municipalities, is heavily reliant on privately owned land and as such is generally proportional to population size. This results in large differences across municipalities. Large, sparsely populated areas usually generate relatively poor revenues, while densely populated metropolitan municipalities generally generate much higher revenues (Figure 3). Municipal budgets are prepared every year (Nyalunga 2006) and applied to meet mandated service delivery requirements. Due to the pressures on service delivery, infrastructure and health, amongst others, municipalities are often not able to meet their mandates. Faced with backlogs in service delivery, municipalities can be further crippled through subsequent violent protests, which increase the pressure on resources and capacity, as municipalities have to restore damaged property and infrastructure.

The NEM:BA requirements do not make provision for additional financial resources to assist municipalities with data collection, compilation of the area management plans, implementation, monitoring and reporting (SALGA 2016). Municipalities further lack funding to appoint appropriately skilled service providers to compile management plans (SALGA 2016).

SALGA compiled an internal report on municipalities of the Western Cape and their level of compliance on the NEM:BA regulations (SALGA 2016). The report identified several challenges and found that one of the key challenges faced by municipalities is the fact that the NEM:BA regulations do not come with implementation budget. Furthermore, it found that most municipalities do not have the necessary capacity to perform the related environmental functions. As a result of the lack in capacity, components of the environmental management function are allocated to different departments within the municipality. 'This raises a serious concern, as it proves the lack of proper capacitation of local government to adequately perform the environmental management functions' (SALGA 2016).

The sum of all environmental budgets stipulated in municipal IDPs was found to consist of less than 1% of total municipal budgets (Ruwanza & Shackleton 2016). Ruwanza and Shackleton (2016) further found that budget allocation to environmental issues varies greatly between metropolitan, district and local municipalities. District municipalities generally allocate more budget to environmental issues than local and metropolitan municipalities. However, under environmental projects, invasive species management is allocated the lowest budget.

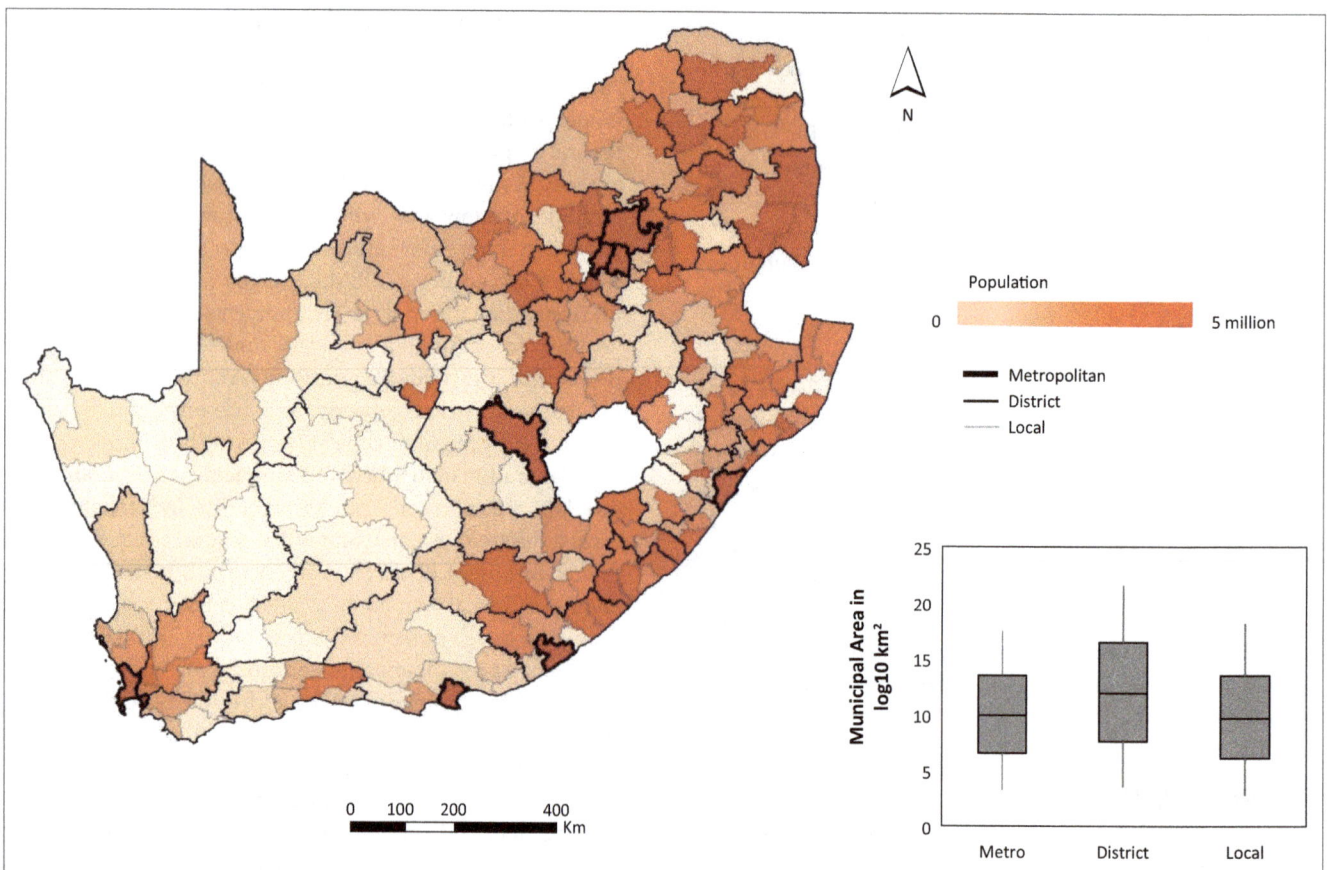

Source: StatsSA 2016 Community Survey (CS) data (http://cs2016.statssa.gov.za/ accessed 17 November 2016)

FIGURE 3: Large variation in municipal sizes and population count across the country. Differentiation is made between the different levels of municipalities, with metropolitan municipalities generally being small and highly populated. Variation in municipality sizes is also shown across the different municipal types (insert: area in km² on $\log_{10}$ scale).

Invasive species management is not separately funded but is mostly dependant on available operational funding (short-term) from the different departments responsible for managing land parcels. Municipalities across the country have the option to access Expanded Public Works Programme (EPWP) funding; however, this funding is primarily used for short-term job creation opportunities rather than invasive species management (Table 2). However, operational funds as well as access to funding through EPWP fluctuate widely between financial years, making it difficult to plan adequately. Allocating operational funding for managing invasive species is challenging because of competing priorities, the absence of a long term strategy, priorities and area management plans. If invasive species management is not regarded as a core function by the municipality, control mainly focusses on aesthetics (public open spaces and road verges) or addressing public complaints about security issues related to 'overgrown' land.

We recommend municipalities strengthen collaboration between different owners and managers of land parcels within a municipality to help leverage resources for achieving common goals and objectives. A Biological Invasions Strategy (see Figure 2; Table 3) can enable municipalities to improve budgeting processes and specifically allocate funding for invasive species management. Furthermore, it supports applications for resources and capacity (e.g. from Working for Water and EPWP) to fund priority invasive species interventions.

Prioritising land parcels for invasive species control assists with funds being appropriately allocated and utilised, resulting in the highest return on investment being achieved. It also assists municipalities in addressing competing issues characteristic of dynamic urban environments.

Measuring progress towards NEM:BA compliance

Municipalities are faced with a multitude of challenges, limiting their ability to comply with the recently promulgated NEM:BA regulations. This paper is not aimed at providing the solutions to all the challenges municipalities are facing; rather it is aimed at providing commentary on how municipalities could go about addressing limitations to ensure compliance with the NEM:BA regulations.

To measure the level of compliance, DEA should determine the number of municipalities that have submitted their plans according to the timeframes stipulated (Figure 4[1]).

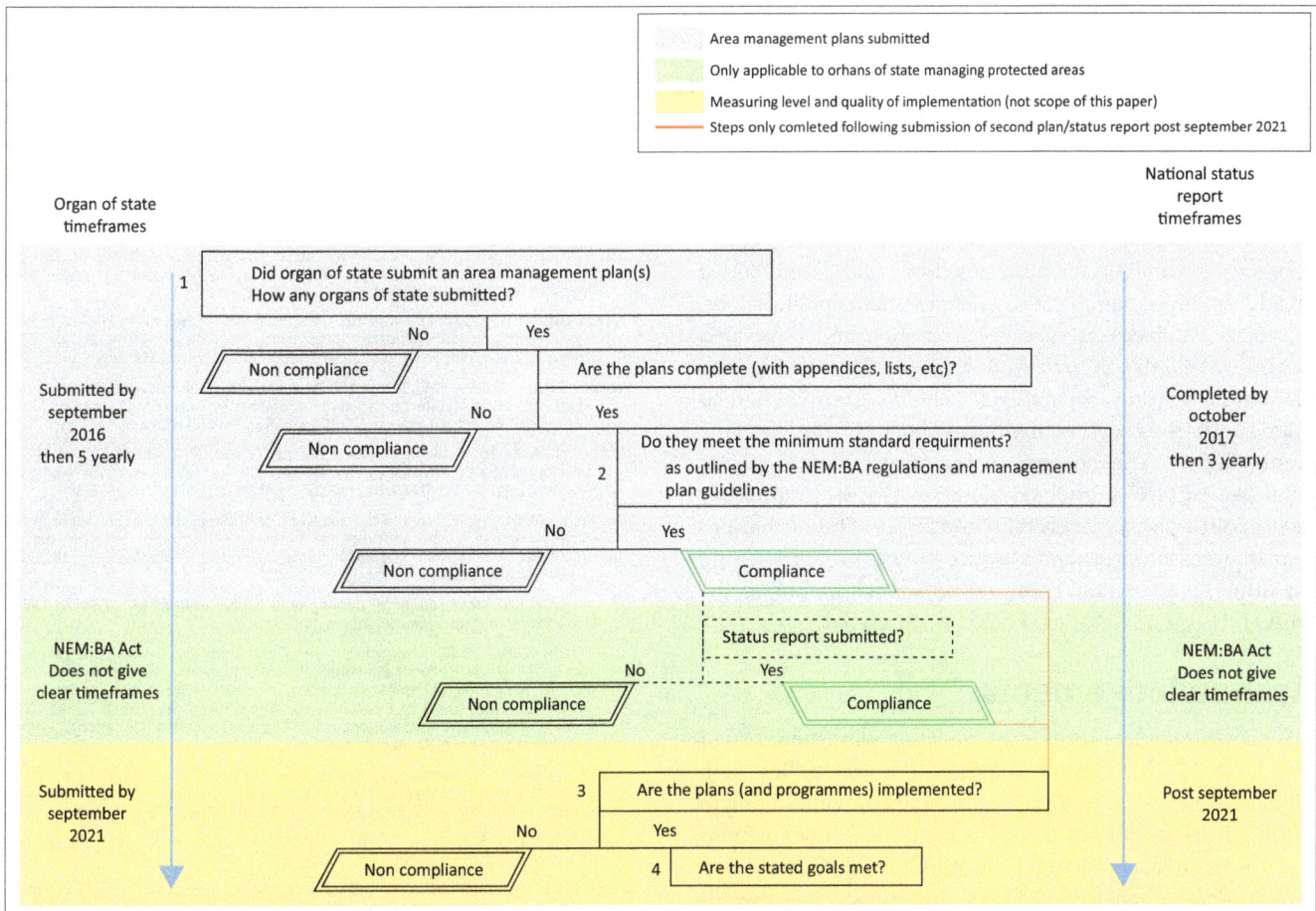

The first section (grey) refers to area management plans submitted (starting in September 2016 and to be repeated every 5 years).

The second section (green) only applies to nature reserves and the third section (orange) can only be completed once plans have submitted their second, revised plan, post September 2021. Numbers 1 to 4 refer to key indicators (narrative in text).

FIGURE 4: Measuring compliance: A simplified framework for measuring compliance of Organs of State, in line with the timeframes given to Organs of State and the National Status Report (Wilson et al. 2017).

The reasons for non-compliance should be determined to enable DEA to put measures in place in collaboration with SALGA to assist those municipalities who require additional support, resources and capacity. Appropriate actions should be taken by DEA to address these gaps to ensure those municipalities comply by the time the next national status report is compiled. Management plan standards will need to be assessed, using the guidelines as a baseline [Figure 4(2)]. The second national status report would then analyse the data to monitor the levels of change. An increase in the number of plans (and standard of plans) should indicate successful progress towards compliance (revisit Figure 4[1]).

Monitoring the implementation of area management plans (Figure 4[3]) requires municipalities to submit their updated plans and reports as required by the NEM:BA regulations. Alternatively, the uptake of control activities into municipal processes can be measured through the analysis of IDPs (Ruwanza & Shackleton 2016). The progress towards meeting set targets and goals (Figure 4[4]) can be determined through analysing updated management plans and reports submitted by Organs of State.

However, to increase the level of compliance for all Organs of State, steps need to be taken to overcome some of the challenges municipalities are faced with. Through coordinated national efforts by DEA, SANBI and SALGA working together with municipalities, nation-wide compliance can be achieved. Municipalities require guidance as to how best to bridge some of the challenges they are faced with when it comes to working towards NEM:BA compliance. Starting with increasing the level of awareness and capacity within municipalities should be one of the first steps undertaken by the said stakeholders. Ensuring easier access to information for municipalities could greatly assist in addressing some of the challenges municipalities face. A central database on species information and associated clearing methods, as well as a database for tracking and planning clearing operations, could greatly benefit municipalities as well as the development of the National Status Report. The hosting of a central invasive species database requires careful consideration as it will ultimately determine the usage, accessibility and ability of municipalities to plan, track progress and produce progress reports. Using the http://www.invasives.org.za website with a link to the DEA and SALGA websites should be considered.

Acknowledgements

Invited stakeholders for the prioritisation workshop: David Le Maitre (CSIR), Brian van Wilgen (Centre for Invasion Biology), Nicola van Wilgen (SANParks), Greg Forsyth (CSIR), Patricia Holmes (City of Cape Town), Luca Afonso (Centre for Invasion Biology), Chandre Rhoda (City of Cape Town), Chad Cheney (SANParks), Karen Esler (Centre for Invasion Biology) and Leighan Mossop (City of Cape Town).

Kay Montgomery for the contribution of the roadshow and South African Green Industries Council (SAGIC) training data as well as including insights into some of the points discussed. Ross Shackleton and Jana Fried for their contribution and insights into the paper.

We would like to thank David Le Maitre and three anonymous reviewers for their useful comments and input on previous versions of this manuscript.

Funding Information L.P. and M.G. acknowledge the funding provided by the DST-NRF Centre of Excellence for Invasion Biology and Working for Water Programme through their collaborative research project on 'Integrated Management of Invasive Alien Species in South Africa'.

Competing interests

The authors declare that they have no financial or personal relationship(s) that may have inappropriately influenced them in writing this article.

Authors' contributions

The paper was a joint product of all the authors. The writing of the paper was led by U.I., with input from all co-authors on an equal basis.

References

Alston, K.P. & Richardson, D.M., 2006, 'The roles of habitat features, disturbance, and distance from putative source populations in structuring alien plant invasions at the urban/wildland interface on the Cape Peninsula, South Africa', *Biological Conservation* 132(2), 183–198. https://doi.org/10.1016/j.biocon.2006.03.023

Brisbane City Council, 2013, *Brisbane Invasive Species Management Plan 2013–2017*, Brisbane City Council, Queensland, Australia.

Caffrey, J.M., Baars, J.-R., Barbour, J.H., Boets, P., Boon, P., Davenport, K. et al., 2014, 'Tackling invasive alien species in Europe: The top 20 issues', *Management of Biological Invasions* 5(1), 1–20. https://doi.org/10.3391/mbi.2014.5.1.01

City of Cape Town, 2008, *Framework for a strategy and action plan for the management of invasive alien species in the City of Cape Town*, Environmental Resource Management Department, Cape Town.

City of Cape Town, 2012, *City statistics and population census using 2011 and 2001 census data supplied by Statistics South Africa, South Africa*, viewed 15 July 2016, from http://www.capetown.gov.za/en/stats/Pages/Census2011.aspx

Cowling, R.M., Rundel, P.W., Lamont, B.B., Arroyo, M.K. & Arianoutsou, M., 1996, 'Plant diversity in Mediterranean-climate regions', *Trends in Ecology and Evolution* 11(9), 362–366. https://doi.org/10.1016/0169-5347(96)10044-6

Crall, A.W., Renz, M., Panke, B.J., Newman, G.J., Chapin, C., Graham, J. et al., 2012, 'Developing cost-effective early detection networks for regional invasions', *Biological Invasions* 14(12), 2461–2469. https://doi.org/0.1007/s10530-012-0256-3

Cronin, K., Kaplan, H., Gaertner, M., Irlich, U.M. & Hoffman, M.T., 2017, 'Aliens in the nursery: Assessing the attitudes of nursery managers to invasive species regulations', *Biological Invasions* 19(3), 925–937. https://doi.org/10.1007/s10530-016-1363-3

DEA, 2014, *A national strategy for dealing with biological invasions in South Africa*, Department of Environmental Affairs, Pretoria.

DEA, 2015, *Guidelines for monitoring, control and eradication plans as required by Section 76 of the National Environmental Management: Biodiversity Act, 2004 (Act No. 10 of 2004) (NEM:BA) for species listed as invasive in terms of Section 70 of this Act*, Department of Environmental Affairs, Pretoria, viewed 7 May 2016, from http//www.environment.gov.za/sites/default/files/legislations/nemba_invasivespecies_controlguideline.pdf

Dickie, I., Bennett, B., Burrows, L., Nuñez, M.A., Peltzer, D.A., Porté, A. et al., 2014, 'Conflicting values: Ecosystem services and invasive tree management', *Biological Invasions* 16(3), 705–719. https://doi.org/10.1007/s10530-013-0609-6

Faeth, S.H., Saari, S. & Bang, C., 2012, *Urban biodiversity: Patterns, processes and implications for conservation*, eLS 2012, Wiley, Chichester.

Forsyth, G.G., Le Maitre, D.C., O'Farrell, P.J. & van Wilgen, B.W., 2012, 'The prioritisation of invasive alien plant control projects using a multi-criteria decision model informed by stakeholder input and spatial data', *Journal of Environmental Management* 103, 51–57. https://doi.org/10.1016/j.jenvman.2012.01.034

Gaertner, M., Larson, B.M.H., Irlich, U.M., Holmes, P.M., Stafford, L., van Wilgen, B.W. et al., 2016, 'Managing invasive species in cities: A framework from Cape Town, South Africa', *Landscape and Urban Planning* 151, 1–9. https://doi.org/10.1016/j.landurbplan.2016.03.010

Great Britain Non-native Species Secretariat, 2015, *The Great Britain invasive non-native species strategy*, Sand Hutton, York, viewed 4 July 2016, from http://www.gov.uk/government/uploads/system/uploads/attachment_data/file/455526/gb-non-native-species-strategy-pb14324.pdf

Grimm, N.B., Faeth, S.H., Golubiewski, N.E., Redman, C.L., Wu, J., Bai, X. et al., 2008, 'Global change and the ecology of cities', *Science* 319(5864), 756–760. https://doi.org/10.1126/science.1150195

Hawthorne, T.L., Elmore, V., Strong, A., Bennett-Martin, P., Finnie, J., Parkman, J. et al., 2015, 'Mapping non-native invasive species and accessibility in an urban forest: A case study of participatory mapping and citizen science in', *Applied Geography* 56, 187–198. https://doi.org/10.1016/j.apgeog.2014.10.005

Holmes, P.M., Rebelo, A.G., Dorse, C. & Wood, J., 2012, 'Can Cape Town's unique biodiversity be saved? Balancing conservation imperatives and development needs', *Ecology and Society* 17(2), 28. https://doi.org/10.5751/ES-04552-170228

Klotz, S. & Kühn, I., 2010, 'Urbanisation and alien invasion', in K.J. Gaston (ed.), *Urban ecology*, pp. 120–133, Cambridge University Press, Cambridge.

Koma, S.B., 2010, 'The state of local government in South Africa: Issues, trends and option', *Journal of Public Administration* 45(1.1), 111–120.

Kowarik, I., 2011, 'Novel urban ecosystems, biodiversity, and conservation', *Environmental Pollution* 159(8/9), 1974–1983. https://doi.org/10.1016/j.envpol.2011.02.022

Kowarik, I. & von der Lippe, M., 2007, 'Pathways in plant invasions', in W. Nentwig (ed.), *Biological invasions. Ecological studies*, vol. 193, pp. 29–47, Springer, Berlin, New York.

Le Maitre, D.C., Gaertner, M., Marchante, E., Ens, E.-J., Holmes, P.M., Pauchard, A. et al., 2011, 'Impacts of invasive Australian acacias: Implications for management and restoration', *Diversity and Distributions* 17, 1015–1029. https://doi.org/10.1111/j.1472-4642.2011.00816.x

McGeoch, M.A., Butchart, S.H.M., Spear, D., Marais, E., Kleynhans, E.J., Symes, A. et al., 2010, 'Global indicators of biological invasion: Species numbers, biodiversity impact and policy responses', *Diversity and Distributions* 16, 95–108. https://doi.org/10.1111/j.1472-4642.2009.00633.x

McKinney, M.L., 2006, 'Urbanization as a major cause of biotic homogenization', *Biological Conservation*, 127(3), 247–260. https://doi.org/10.1016/j.biocon.2005.09.005

McNeely, J.A., Mooney, H.A., Neville, L.E., Schei, P. & Waage, J.K., 2001, *A global strategy on invasive species*, IUCN Gland, Switzerland.

National Advisory Committee on Invasive Species, 2010, *National strategy on invasive species in Mexico, prevention, control and eradication*, Comisión Nacional para el Conocimiento y Uso de la Biodiversidad, Comisión Nacional de Áreas Protegidas, Secretaría de Medio Ambiente y Recursos Naturales, México.

Novoa, A., Kaplan, H., Wilson, J.R. & Richardson, D.M., 2016, 'Resolving a prickly situation: Involving stakeholders in invasive cactus management in South Africa', *Environmental Management* 57, 998–1008. https://doi.org/10.1007/s00267-015-0645-3

Nyalunga, D., 2006, 'The revitalization of local government in South Africa', *International NGO Journal* 1(1), 1–6.

Petersen, L.M., Moll, E.J., Collins, R. & Hockings, M.T., 2012, 'Development of a compendium of local, wild-harvested species used in the informal economy trade, Cape Town, South Africa', *Ecology and Society* 17(2), 26. https://doi.org/10.5751/ES-04537-170226

Pickett, S.T.A., Cadenasso, M.L., Grove, M.J., Nilon, C.H., Pouyat, C.V., Zipperer, W.C. et al., 2001, 'Urban ecological systems: Linking terrestrial ecological, physical, and socioeconomic components of metropolitan areas', *Annual Review of Ecology and Systematics* 32, 127–157. https://doi.org/10.1146/annurev.ecolsys.32.081501.114012

Pyšek, P., 1998, 'Alien and native species in Central European urban floras: A quantitative comparison', *Journal of Biogeography* 25(1), 155–163. https://doi.org/10.1046/j.1365-2699.1998.251177.x

Ramakhula, M., 2010, 'Implications of the Municipal Property Rates Act (No: 6 of 2004) on Municipal Valuations', MSc Thesis, Faculty of Engineering and the Built Environment, University of the Witwatersrand.

Rebelo, A.G., Holmes, P.M., Dorse, C. & Wood, J., 2011, 'Impacts of urbanization in a biodiversity hotspot: Conservation challenges in Metropolitan Cape Town', *South African Journal of Botany* 77(1), 20–35. https://doi.org/10.1016/j.sajb.2010.04.006

Richardson, D.M. & van Wilgen, B.W., 2004, 'Invasive alien plants in South Africa: How well do we understand the ecological impacts?', *South African Journal of Science* 100(1/2), 45–52.

Ruwanza, S. & Shackleton, C.M., 2016, 'Incorporation of environmental issues in South Africa's municipal integrated development plans', *Journal of Sustainable Development & World Ecology* 23(1), 28–39. https://doi.org/10.1080/13504509.2015.1062161

SALGA, 2016, 'Status quo report on progress and challenges experienced by municipalities to comply with the NEMBA', Internal Report, City of Cape Town, South Africa.

Shackleton, C.M. & Shackleton, R.T., 2016, 'Knowledge, perceptions and willingness to control designated invasive tree species in urban household gardens in South Africa', *Biological Invasions 18(6), 1599–1609.* https://doi.org/10.1007/s10530-016-1104-7

Simberloff, D., Parker, I.M. & Windle, P.N., 2005, 'Introduced species policy, management, and future research needs', *Frontiers in Ecology and the Environment 3, 12–20.* https://doi.org10.1890/1540-9295(2005)003[0012:ISPMAF]2.0.CO;2

Sitas, N., Reyers, B., Cundill, G., Prozesky, H.E., Nel, J.L. & Esler, K.J., 2016, 'Fostering collaboration for knowledge and action in disaster management in South Africa', *Current Opinion in Environmental Sustainability* 19, 94–102. https://doi.org/10.1016/j.cosust.2015.12.007

Taylor, P.J., Arntzen, L., Hayter, M., Iles, M., Frean, J. & Belmain, S.R., 2008, 'Understanding and managing sanitary risks due to rodent zoonoses in an African city: Beyond the Boston model', *Integrative Zoology* 3, 38–50. https://doi.org/10.1111/j.1749-4877.2008.00072.x

van Wilgen, B.W., 2012, 'Evidence, perceptions, and trade-offs associated with invasive plant control in Table Mountain National Park, South Africa', *Ecology and Society* 17(2), 23. https://doi.org/10.5751/ES-04590-170223

van Wilgen, B.W., Cowling, R.M., Marais, C., Esler, K.J., McConnachie, M. & Sharp, D., 2012, 'Challenges in invasive alien plant control in South Africa', *South African Journal of Science* 108, 8–11. https://doi.org/10.4102/sajs.v108i11/12.1445

van Wilgen, B.W., Dyer, C., Hoffmann, J.H., Ivey, P., Le Maitre, D.C., Richardson, D.M. et al., 2011, 'National-scale strategic approaches for managing introduced plants: Insights from Australian acacias in South Africa', *Diversity and Distributions* 17, 1060–1075. https://doi.org/10.1111/j.1472-4642.2011.00785.x

van Wilgen, B.W. & Scott, D.F., 2001, 'Managing fires on the Cape Peninsula: Dealing with the inevitable', *Journal of Mediterranean Ecology* 2, 197–208.

Vanderwoude, C., Klasner, F., Kirkpatrick, J. & Kaye, S., 2015, *Maunakea Invasive Species Management Plan*, Technical Report No. 191, Pacific Cooperative Studies Unit, University of Hawai'i, Honolulu, Hawai'I, 84 pp.

Virginia Invasive Species Working Group, 2012, *Virginia Invasive Species Management Plan*, Natural Heritage Technical Document 12–13, Richmond, VA, 55 p.

Vitousek, P.M., D'Antonio, C.M., Loope, L.L., Rejmánek, M. & Westbrooks, R., 1997, 'Introduced species: A significant component of human-caused global change', *New Zealand Journal of Ecology* 21, 1–16.

von der Lippe, M. & Kowarik, I., 2008, 'Do cities export biodiversity? Traffic as dispersal vector across urban-rural gradients', *Diversity and Distributions* 14(1), 18–25. http://doi.wiley.com/10.1111/j.1472-4642.2007.00401.x

Wilson, J.R., Gaertner, M., Griffiths, C.L., Kotze, I., Le Maitre, D.C., Marr, S.M. et al., 2014, 'Biological invasions in the Cape Floristic Region: History, current patterns, impacts, and management challenges', in N. Allsopp, J.F. Colville & G.A. Verboom, (eds.), *Fynbos: Ecology, evolution, and conservation of a megadiverse region*, pp. 273–298, Oxford University Press, Oxford.

Wilson, J.R.U., Gaertner, M., Richardson, D.M. & van Wilgen, B.W. 2017, 'Contributions to the national status report on biological invasions in South Africa', *Bothalia* 47(2), a2207. https://doi.org/10.4102/abc.v47i2.2207

Zengeya, T., Ivey, P., Woodford, D.J., Weyl, O., Novoa, A., Shackleton, R. et al., 2017, 'Managing conflict-generating invasive species in South Africa: Challenges and trade-offs', *Bothalia* 47(2), a2160. https://doi.org/10.4102/abc.v47i2.2160

7

The balance of trade in alien species between South Africa and the rest of Africa

Authors:
Katelyn T. Faulkner[1,2]
Brett P. Hurley[3,4]
Mark P. Robertson[2]
Mathieu Rouget[5]
John R.U. Wilson[1,6]

Affiliations:
[1]Invasive Species Programme, South African National Biodiversity Institute, Kirstenbosch Research Centre, South Africa

[2]Centre for Invasion Biology, Department of Zoology and Entomology, University of Pretoria, South Africa

[3]Forestry and Agricultural Biotechnology Institute (FABI), University of Pretoria, South Africa

[4]Department of Zoology and Entomology, University of Pretoria, South Africa

[5]Centre for Invasion Biology, School of Agricultural, Earth and Environmental Sciences, University of KwaZulu-Natal, South Africa

[6]Centre for Invasion Biology, Department of Botany and Zoology, Stellenbosch University, South Africa

Corresponding author:
Katelyn Faulkner,
katelynfaulkner@gmail.com

Background: Alien organisms are not only introduced from one biogeographical region to another but also spread within regions. As South Africa shares land borders with six countries, multiple opportunities exist for the transfer of alien species between South Africa and other African countries; however, the direction and importance of intra-regional spread is unclear.

Objectives: The aim of this study was to gain a greater understanding of the introduction of alien species into Africa and the spread of species between South Africa and other African countries.

Method: We developed scenarios that describe the routes by which alien species are introduced to and spread within Africa and present case studies for each. Using data from literature sources and databases, the relative importance of each scenario for alien birds and insect pests of eucalypts was determined, and the direction and importance of intra-regional spread was assessed.

Results: Alien species from many taxonomic groups have, through various routes, been introduced to and spread within Africa. For birds and eucalypt insect pests, the number of species spreading in the region has recently increased, with South Africa being a major recipient of birds (14 species received and 5 donated) and a major donor of eucalypt insect pests (1 species received and 10 donated). For both groups, many introduced species have not yet spread in the region.

Conclusion: The intra-regional spread of alien species in Africa represents an important and possibly increasing threat to biosecurity. To address this threat, we propose a framework that details how African countries could cooperate and develop a coordinated response to alien species introductions.

Introduction

The movement of goods and people around the world is facilitating the introduction of organisms to regions where they are not native. Although many alien organisms are introduced directly from one biogeographical region to another ('inter-regional introduction'), the spread of species within biogeographical regions also contributes to biological invasions ('intra-regional spread') (Chiron, Shirley & Kark 2010; Hurley et al. 2016; Jaksic et al. 2002; Roques et al. 2016 in this article, the biogeographic region of interest is continental Africa). Relatively high propagule pressure [i.e. the number of individuals introduced and the number of introduction events for a specific species (Lockwood, Cassey & Blackburn 2005)] and short geographical distances mean that once an organism has been introduced to a region, further natural or human-aided spread is likely (Garnas et al. 2016; Hurley et al. 2016; Jaksic et al. 2002; Roques et al. 2016). Furthermore, organisms that are native to a biogeographical region might spread within the region, either naturally or with the aid of humans, to areas where they are not native (Chiron et al. 2010).

The intra-regional spread of species is often asymmetrical [i.e. one country donates more species than it receives (Ferus et al. 2015; Jaksic et al. 2002)] and, under some circumstances, introductions through intra-regional spread may be more common than those that occur through inter-regional introduction (Chiron et al. 2010). However, such patterns are the result of historical economic and socio-political processes and so can vary over time (Chiron et al. 2010; Essl et al. 2011; Roques et al. 2016). Furthermore, as properties linked with invasion success (e.g. likelihood of enemy release and propagule pressure) vary across dispersal pathways (e.g. extreme long-distance or leading-edge dispersal; see Wilson et al. 2009), whether an organism is introduced through inter-regional introduction or intra-regional spread can have consequences for its invasion success. To develop and improve efforts aimed at preventing or mitigating the introduction of invasive species, it is therefore important to identify the types of introduction within a region and determine their relative importance and direction [also see the Convention on Biological Diversity's Aichi Target 9 (UNEP 2011)].

Many organisms have been directly introduced to South Africa from other continents [e.g. the Sirex woodwasp (*Sirex noctilio*), which is native to Eurasia and northern Africa, was introduced to South Africa from Oceania and South America (Boissin et al. 2012), and the harlequin ladybird (*Harmonia axyridis*), which is native to Asia, was introduced to South Africa from North America (Lombaert et al. 2010)]. Given that South Africa shares land borders with six other African countries, multiple opportunities exist for species to spread between South Africa and other African countries, either through natural dispersal or with the aid of various human-related transport vectors (e.g. air, sea and land transport vectors). However, the relative importance of inter-regional introduction and intra-regional spread is currently not clear, and whether South Africa is primarily a donor or recipient of alien species is also unknown.

In an effort to gain a greater understanding of the movements of alien species into and within Africa, we aimed to (1) develop introduction route scenarios that describe how alien species might have been introduced to the region and spread between South Africa and elsewhere in Africa; (2) demonstrate these scenarios using case studies; (3) use the scenarios to quantify, for selected groups, the importance and direction of intra-regional spread; (4) determine if these patterns have changed through time; and (5) propose a framework for trans-boundary collaboration in biosecurity that could address the threat posed to Africa by the intra-regional spread of alien species.

Research method and design
Introduction route scenarios

Six introduction route scenarios that describe how alien species might have been introduced to the region and spread between continental South Africa (SA) and elsewhere in continental Africa (AF) were developed and examples identified. In these scenarios, both natural dispersal and the human-aided movement of species within the continent were considered as intra-regional spread. Furthermore, although introduction and spread through all human-related transport vectors were considered (i.e. land, air and sea transport vectors), these vectors were not discriminated in the scenarios. The scenarios describe introduction routes where species have been introduced to SA or AF from other regions, and consider whether subsequent intra-regional spread between SA and AF occurred. Also described are instances where species that are native to either SA or AF have spread between the subregions to areas where they are not native. Details on the introduction route scenarios and examples of species for each scenario are shown in Figure 1,

and the details for each example are provided in Appendix 2, Figure 1-A2 to Figure 6-A2.

Importance of introductions into Africa versus spread within Africa

For alien species in SA and AF, information on their native and introduced range (in Africa and elsewhere in the world), descriptions of introduction and spread as well as introduction data (introduction source, number of introductions, pathway of introduction and date of introduction or first record) were used to categorise species in terms of the most likely introduction route scenario that resulted in introduction. As detailed species-level introduction data are often lacking (Faulkner et al. 2015), we focused on two groups for which these data could be obtained: birds and insect pests of *Eucalyptus* trees (see Tables 1-A1 and 2-A1 for species lists). For birds, data were extracted from South African (i.e. Dean 2000; Peacock, van Rensburg & Robertson 2007; Picker & Griffiths 2011; van Rensburg et al. 2011) and global sources (i.e. CAB International 2016; Lever 1987, 2005; Long 1981). For each eucalypt insect pest in South Africa, date of first record for southern African countries was obtained from Bush et al. (2016), Wingfield et al. (2008) and local authorities, and other data were extracted from South African (i.e. Picker & Griffiths 2011) and global information sources (i.e. CAB International 2016). Re-introductions [e.g. helmeted guineafowl (*Numida meleagris*) in South Africa (Lever 1987; Long 1981)] and extralimital populations [species that have been translocated within the subregion where they are native to parts of that subregion where they are not native; e.g. red-eyed dove (*Streptopelia semitorquata*) in South Africa (Lever 1987; Long 1981)] were not included. For species that have been introduced to the region multiple times, all applicable scenarios were recorded (and therefore the total count of species for the scenarios can be larger than the total number of species investigated). Furthermore, this means that although a species might not have spread between the two subregions, as a result of independent introductions it might still occur in both SA and AF. For some introductions, it was clear which scenario was applicable [e.g. scenario 5 is clearly applicable for the common starling (*Sturnus vulgaris*) which was introduced in the 1800s from the United Kingdom to SA and then spread into neighbouring countries, i.e. AF], but for others this was not the case [e.g. the Indian subspecies of the rose-ringed parakeet (*Psittacula krameri*) was introduced to SA in the 1800s and then later to Egypt (i.e. AF), it is likely that these introductions were independent (i.e. both scenarios 1 and 4 are applicable); however, it is possible that the birds in Egypt came from SA (i.e. scenario 5)]. In an effort to account for this uncertainty, we categorised each introduction according to the most likely scenario (i.e. scenario 5 for *S. vulgaris* and scenarios 1 and 4 for *P. krameri*) and then rated our confidence in each designation as high or low. A high confidence rating was assigned when the scenario was clear (e.g. for *S. vulgaris*), and low confidence was assigned when more than one scenario was possible (e.g. for *P. krameri*). For some species, there were insufficient data to make a designation, for example, the common pigeon (*Columba livia*) is native to parts of North Africa and has also been introduced

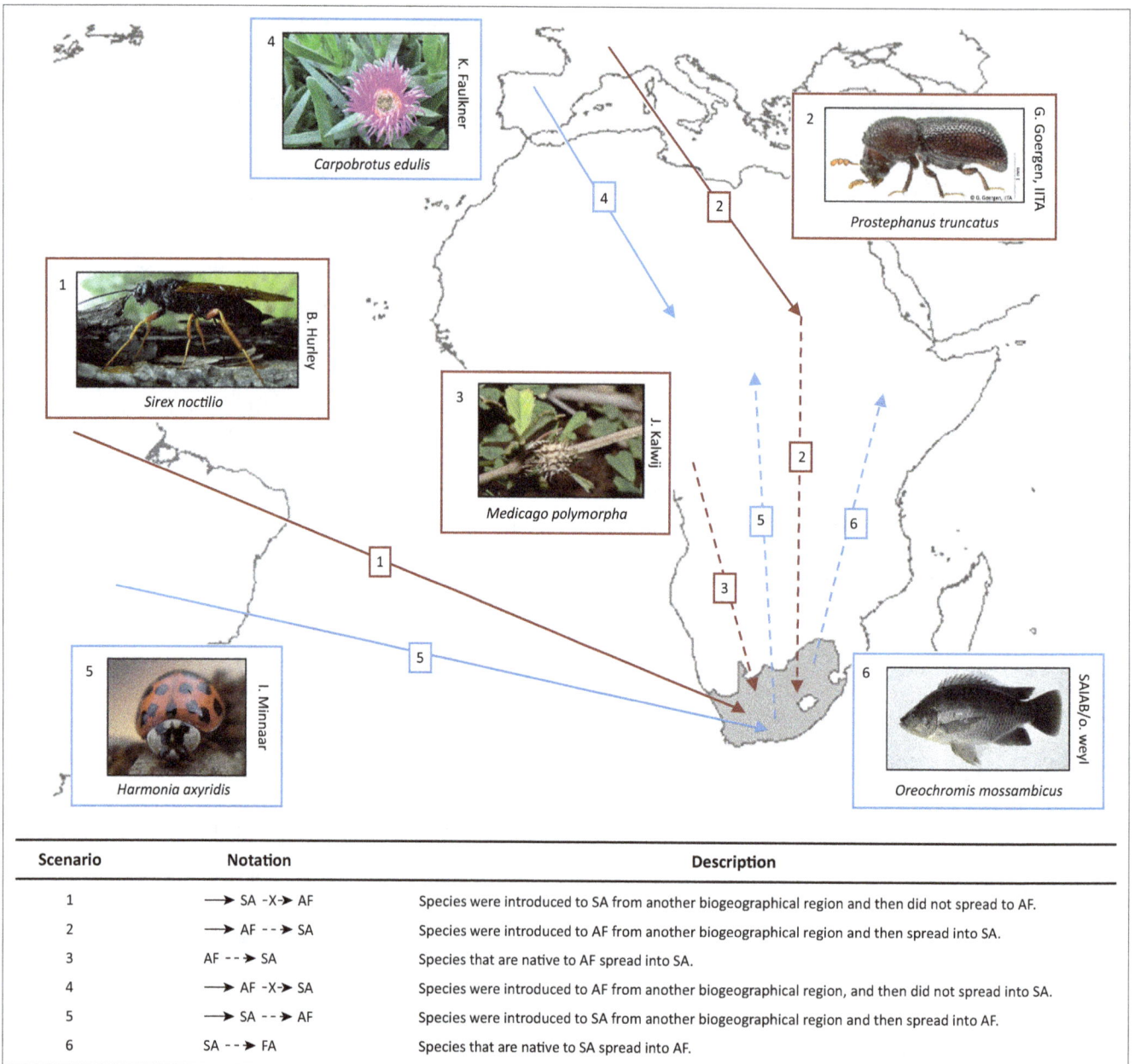

Scenario	Notation	Description
1	→ SA –X→ AF	Species were introduced to SA from another biogeographical region and then did not spread to AF.
2	→ AF – –→ SA	Species were introduced to AF from another biogeographical region and then spread into SA.
3	AF – –→ SA	Species that are native to AF spread into SA.
4	→ AF –X→ SA	Species were introduced to AF from another biogeographical region, and then did not spread into SA.
5	→ SA – –→ AF	Species were introduced to SA from another biogeographical region and then spread into AF.
6	SA – –→ FA	Species that are native to SA spread into AF.

South Africa is shown in grey, arrows with solid lines indicate introductions from other biogeographical regions (inter-regional introductions) and arrows with dashed lines indicate spread within the region (intra-regional spread). In dark red are scenarios where SA is the final recipient of the alien species and in light blue are scenarios where AF is the final recipient. Introduction or spread may be facilitated intentionally or unintentionally by humans, or may be as a result of natural dispersal by the organism.

FIGURE 1: Introduction route scenarios (indicated using numbers) for alien species in South Africa (SA) and in other parts of Africa (AF), and examples of species which conform to each scenario.

widely on the continent; however, details are imprecise for many of these introductions and thus multiple, equally likely scenarios are possible (Lever 2005). These species were recorded as having insufficient data, and they were not assigned an introduction route scenario.

While some of the scenarios involve only inter-regional introductions or intra-regional spread, others involve combinations of the two introduction types (i.e. a species is introduced to AF or SA and also spreads between SA and AF; see Figure 3). Therefore, the results for a number of the scenarios had to be combined to determine the relative importance of the two types of introduction. For instance, to determine the total number of species introduced from another biogeographical region to SA, results for scenarios

1 and 5 had to be combined (Figure 3). However, as the six scenarios provide useful details that are lost when combined (e.g. whether the species is native or alien to the region), we evaluated alien species movements in terms of both the scenarios and the types of introduction.

A generalised linear model (Poisson error distribution and log link) was used to analyse a two-way contingency table of species counts (Crawley 2007) and test the association between organism type (birds and eucalypt insect pests) and scenario. The relative importance of inter-regional introduction and intra-regional spread between SA and AF was assessed by calculating the number of bird and eucalypt insect pest species directly introduced from another region to SA (sum of the counts for scenarios 1 and 5) and AF (sum of the counts for scenarios 2 and

4), and the number of species that spread into SA from AF (sum of the counts for scenarios 2 and 3) and vice versa (sum of the counts for scenarios 5 and 6). A generalised linear model (Poisson error distribution and log link) was used to analyse a three-way contingency table of species counts and test the association between organism type (birds and eucalypt insect pests), recipient subregion (SA and AF) and introduction type (inter-regional introduction and intra-regional spread). To evaluate whether the relative importance of inter-regional introduction and intra-regional spread has varied over time, date of introduction or first record data were used to designate introductions into 50-year time periods, and for each period, the number of birds and eucalypt insect pests introduced to SA and AF through inter-regional introduction and intra-regional spread was determined. All generalised linear models were checked for overdispersion (Crawley 2007; Zuur et al. 2009), but no instances were noted. Counts that were significantly different from what might be expected based on chance alone were identified by calculating the standardised adjusted residuals and comparing these values to the critical values of the normal distribution (Bewick, Cheek & Ball 2004; Everitt 1977). In an effort to determine the influence of uncertainty on results, analyses were performed twice: using all the data and using a subset with only designations made with high certainty.

Results

Importance of introductions into Africa versus spread within Africa

Birds and eucalypt insect pests have been introduced through various introduction routes (Tables 3-A1 and 4-A1), but different scenarios were common for the two groups (significant association between scenario and organism type: $\chi^2 = 30.6$, $d.f. = 5$, $p < 0.001$). Many bird species that have been introduced to SA from another biogeographical region did not spread to AF (scenario 1, see Figure 2); however, four of these species have been independently introduced to AF (scenario 4). While the number of bird species for which scenario 1 was applicable was significantly higher than expected by chance, so too was the number of bird species that are native to AF that spread into SA (scenario 3, see Figure 2). For eucalypt insect pests, most species, and a significantly higher number than expected, were

introduced to SA from another region and then subsequently spread into AF (scenario 5, see Figure 2). A significantly higher number of eucalypt insect pests than expected by chance were also directly introduced to AF from another region and then did not spread into SA (scenario 4, see Figure 2). However, all of these species have also been introduced to SA (scenario 1). Multiple scenarios were applicable for eight bird and eight eucalypt insect pest species. There were few instances of insufficient data (3% for birds and 4% for eucalypt insect pests, see Figure 2). For birds, 68% of the scenario designations were made with high certainty, but for eucalypt insect pests, this was only the case for 26% of designations. Consequently, the results of the statistical analysis differed when only scenario designations with high certainty were included (the association between scenario and organism type was no longer significant), but for birds, the identified pattern (i.e. scenario 1 dominated, but for many species scenario 3 was applicable) remained the same (Figure 1-A1).

The relative importance of inter-regional introduction and intra-regional spread differed for birds and eucalypt insect pests and also varied based on the recipient subregion (Figure 3; significant association between introduction type, organism type and recipient subregion: $\chi^2 = 4.3$, $d.f. = 1$, $p = 0.04$). Based on the species for which the date of introduction data were available (75% for birds and 89% for eucalypt insect pests), it appears that for alien birds and eucalypt insect pests in SA and AF, the relative importance of inter-regional introduction and intra-regional spread changed over time (Figure 4). Although most alien bird species in SA were introduced from other regions, this number was significantly lower than expected by chance, and the number that spread from AF into SA was significantly higher than expected (Figure 3). Additionally, since 2000, more species have spread from AF to SA than have been introduced from other regions (Figure 4). For eucalypt insect pests in SA, a significantly higher number than expected have been introduced from other regions (Figure 3), but since

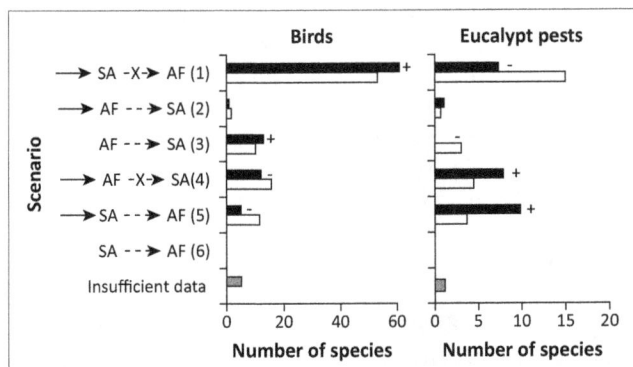

FIGURE 2: The number of alien bird and eucalypt insect pest species for which each introduction route scenario was applicable (in black).

Details of the scenarios are provided in Figure 1. In white are the expected values for each scenario. Species with insufficient data (shown in grey) were not included in the statistical analysis. Plus and minus signs indicate species counts that were significantly higher or lower than what was expected by chance.

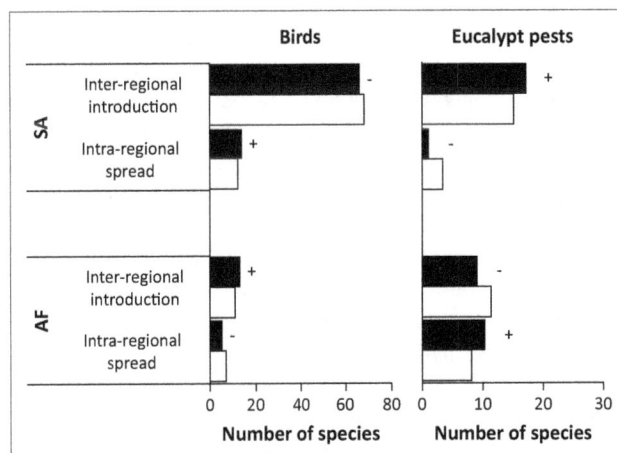

Expected values are shown in white. Plus and minus signs indicate numbers of species that were significantly higher or lower than what was expected by chance. Although the observed and expected values did not differ greatly, these differences were significant.

FIGURE 3: The number of alien bird and eucalypt insect pest species (in black) in South Africa (SA) and elsewhere in Africa (AF) that were introduced through a direct introduction from another region (inter-regional introduction) or through spread between the two subregions (intra-regional spread).

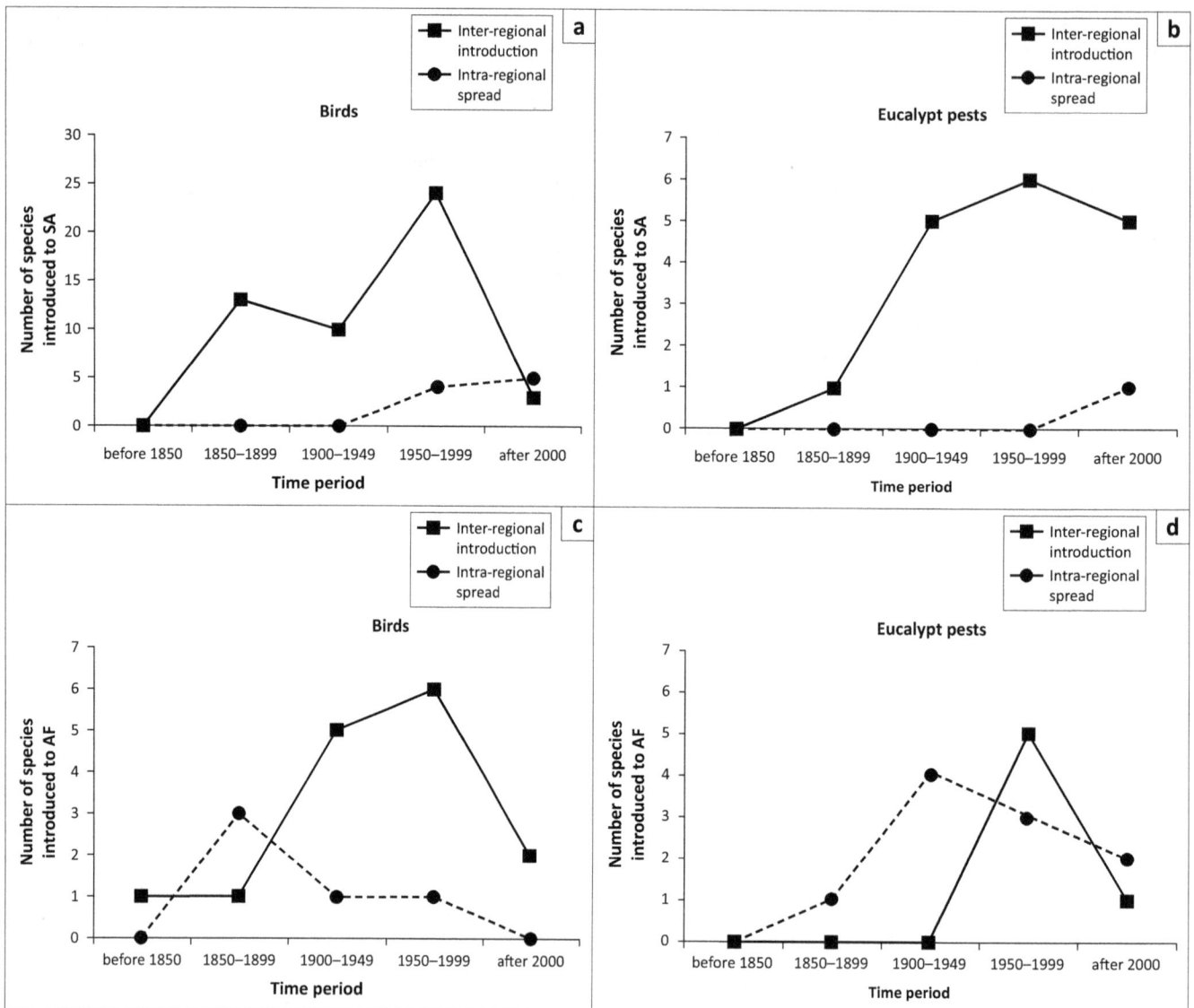

Please note the differing scales of the y-axes.

FIGURE 4: Temporal changes in the number of alien bird and eucalypt insect pest species in South Africa (a and b) and elsewhere in Africa (c and d) that were introduced through a direct introduction from another region (inter-regional introduction) or through spread between the two subregions (intra-regional spread).

2000, one species has also spread from AF to SA (Figure 4). A significantly higher number of bird species were introduced from other regions to AF than was expected (Figure 3), with this number being higher, for all time periods since 1900, than the number that have spread from SA into AF (Figure 4). The number of eucalypt insect pests that spread from SA to AF was significantly higher than that expected by chance (Figure 3), with more species spreading from SA to AF since 2000 than the number introduced from other regions (Figure 4). In the spread of alien species between SA and AF, SA was the major recipient of birds but the major donor of eucalypt insect pests (Figure 3). When only designations with high certainty were included in the analysis, the statistical results changed (the association between introduction type, organism type and recipient subregion was no longer significant), but the identified patterns were the same (i.e. most birds and eucalypt insect pests in SA and birds in AF were inter-regional introductions, but most eucalypt insect pests in AF spread in from SA, and SA was a major recipient of birds but a major donor of eucalypt insect pests; Figure 2-A1).

Discussion

The patterns of movement of alien species are often complex (Hurley et al. 2016) and, in line with this, our case studies and results show that alien organisms in Africa have been introduced through various introduction routes (e.g. see Measey et al. 2017 for a discussion on amphibians; and Visser et al. 2017 for grasses). Although many species are introduced to the continent directly from other regions, species are also spreading within Africa, with the relative importance and direction of spread varying across organisms and over time. This poses a challenge to biosecurity that needs to be addressed (see Keller & Kumschick 2017).

Importance of introductions to Africa versus spread within Africa

Many birds and eucalypt insect pests have been introduced to either SA or AF but have not yet spread between the two subregions. Similarly, many alien species in other regions [e.g. plants (Lambdon et al. 2008), birds (Chiron et al. 2010) and insects (Roques et al. 2016) in Europe] have not spread from the country where they were introduced. The spread of these species might be limited by a variety of factors, including the environment and their dispersal capabilities (Roques et al. 2016), but in many cases, it might simply be a matter of time before they spread across national boundaries. Most bird species that were introduced to SA and subsequently spread into AF were introduced over 100 years ago, while those species that have not yet spread tend to have been introduced during or after the 1970s. The future spread of these species likely represents a major invasion debt (Rouget et al. 2016).

Although introductions from other regions dominated in most cases, the spread of species between SA and AF has recently increased in importance. This trend might be driven by recent growth in trade between these two subregions (Figure 3-A1). The link between socio-economic factors and the introduction and spread of alien species is well documented (Essl et al. 2011) and, for example, political and economic changes in Europe (e.g. the cold war and the later opening of borders to movement and trade) have influenced the spread of alien bird and insect species in the region (Chiron et al. 2010; Roques et al. 2016).

South Africa is one of a few countries that serve as major introduction points for eucalypt insect pests (Hurley et al. 2016). Additionally, South Africa currently exports more goods to other African countries than it imports (Figure 4-A1). Thus, it is not surprising that South Africa appears to be a major donor in the intra-regional spread of eucalypt insect pests. Contrary to the dominant direction of trade, South Africa is the major recipient in the intra-regional spread of birds. However, as birds are often introduced intentionally, their movement patterns might be less likely than those of eucalypt insect pests (usually introduced accidentally), to reflect coarse trends in trade (but, see Seebens et al. 2015 for predictors of global flows of naturalised plants).

The results discussed above are based on only two groups for which some historical data could be obtained. Because of data quality issues, using historical data to determine the introduction routes of alien species can lead to imprecise inferences (for information on genetic techniques, see Estoup & Guillemaud 2010). For example, countries differ with regard to their surveillance and monitoring activities (Latombe et al. in press), and as a consequence, species that have been recorded in South Africa first might not have been introduced directly from another region, but might instead have spread into the country from elsewhere in Africa where their introduction was not detected. To get an indication of how data quality impacted our results, we conducted the analysis on the full dataset and on a subset in which we had high confidence. Although the results of these analyses differed, the overall conclusions were the same (i.e. that inter-regional introductions dominate but that intra-regional spread is important, and that South Africa is a major donor for eucalypt insect pests but a major

recipient for birds). Furthermore, by focusing on birds and eucalypt insect pests, we were able to highlight that organisms are being introduced to and are spreading within Africa, a pattern which, we show using our case studies, is true for a wide variety of organisms.

Intra-African spread as a biosecurity threat

Although inter-regional introductions and intra-regional spread may, in general, be increasing as a result of increased global travel and trade (Hurley et al. 2016), an increase in the number of species spreading within Africa (as shown here for birds and eucalypt insect pests) might pose a particularly high biosecurity threat. Shorter geographical distances, and higher propagule pressure and environmental similarity mean that the chances of naturalisation might be higher for species spreading within a region than for those introduced directly from other regions (van Kleunen et al. 2015). In keeping with this, for many regions (including Africa), a higher number of plant species than expected are native to a part of the region but have been introduced to and have become naturalised in other parts of the region where they are not native (van Kleunen et al. 2015). Furthermore, as many alien species present in Africa may still establish and spread (i.e. establishment and spread debt, see Rouget et al. 2016), it is likely that the biosecurity threat posed by intra-African spread will continue to increase.

The increasing number of species that have spread between South Africa and the rest of Africa highlights that current efforts in South Africa to prevent or reduce introductions from other African countries are likely insufficient. This potential weakness in South Africa's biosecurity deserves consideration in the first National Status Report on Biological Invasions, and in any future plans to manage South Africa's pathways of introduction (Wilson et al. 2017).

Unfortunately, preventing intra-regional spread is particularly difficult. Organisms from outside Africa can only be transported directly to South Africa by air or sea, and thus to prevent their introduction, border control only needs to be implemented at 18 official ports of entry (Table 1). In contrast, species in Africa can spread into South Africa through natural dispersal or with the aid of land, sea or air

TABLE 1: The modes of transportation, the types of introduction they could facilitate, the number of ports of entry for South Africa and the relative ease of managing introductions.

Mode of transport	Type of introduction	Ports of entry	Management ease
Sea	Intra-regional spread and inter-regional introduction	8	High
Air	Intra-regional spread and inter-regional introduction	10	Medium
Land	Intra-regional spread	54	Low
Natural dispersal	Intra-regional spread	Anywhere along the 4862-km-long land border	Extremely low

Only ports of entry where individuals may officially enter or exit the country were considered. Details on the ports of entry were obtained from the website of the South African Department of Home Affairs.

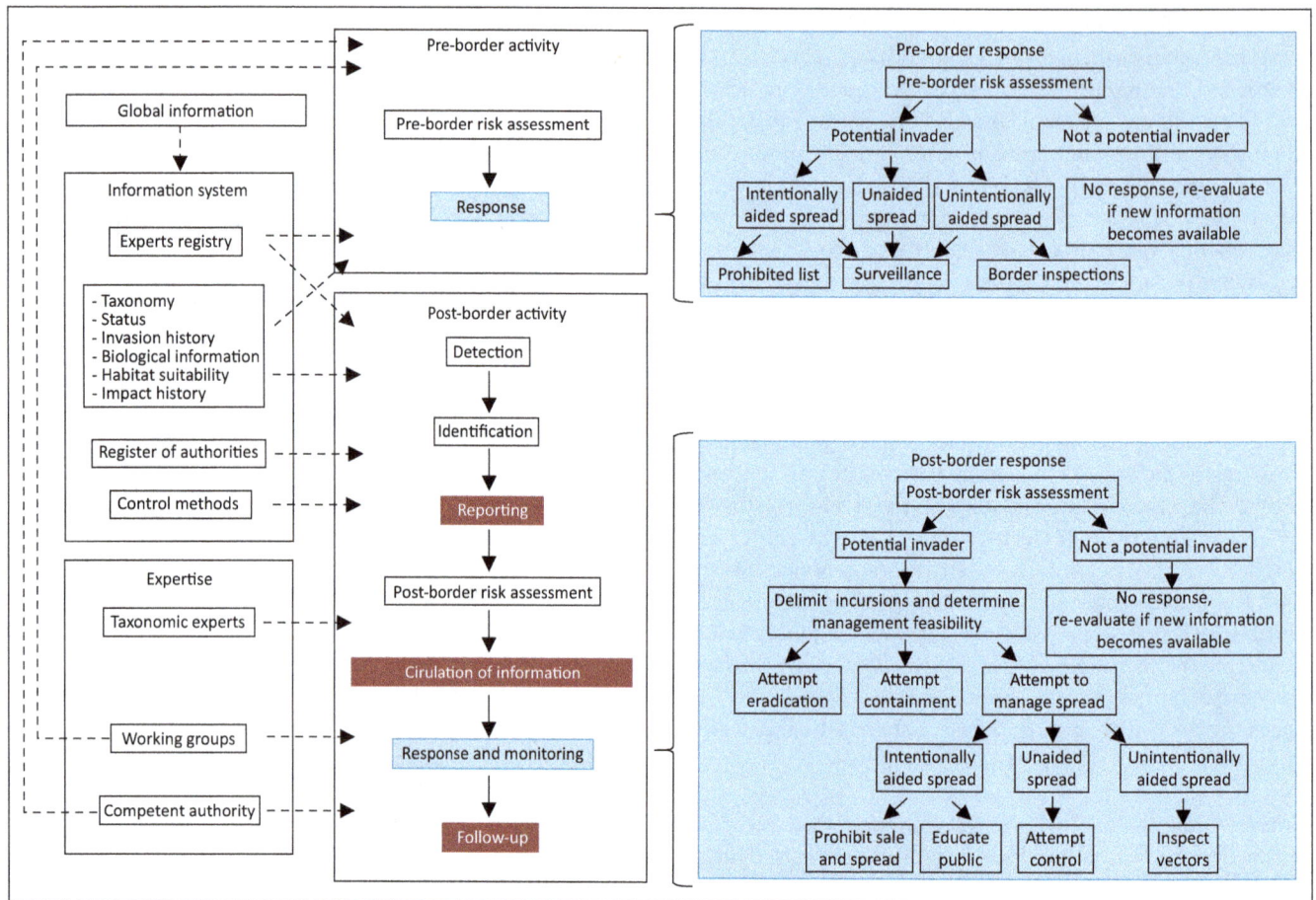

Source: Based on a framework for Europe (Genovesi et al. 2010)

Pre- and post-border response activities are in light blue, and details of these responses are presented in the light blue boxes, activities involving the sharing of information between countries are in dark red, and all other activities are in white. The stippled arrows demonstrate the type of information and expertise required for pre- or post-border activities.

FIGURE 5: Framework for a coordinated response to alien species introductions in Africa.

transportation (Table 1). Thus, to prevent organisms from spreading into the country, not only does border control need to be implemented at 73 official ports of entry, but South Africa's 4862-km-long land boundary (Central Intelligence Agency 2015) also needs to be managed (Table 1).

To overcome this problem, and better manage biological invasions in the region, a coordinated regional response is needed. Attempts have been made in other regions to achieve this and, for example, the Chilean and Argentinian governments have developed joint research and control programmes for alien mammals (Jaksic et al. 2002), while an early warning and response framework has been developed for Europe (Genovesi et al. 2010). Based on the European system, we have developed a framework, shown in Figure 5, to coordinate the response of African countries to alien species introductions. The framework details the activities that countries should perform before (i.e. pre-border activities) and after (i.e. post-border activities) an alien species is detected; highlights when communication and information sharing between countries is required (e.g. report the detection of an alien species); and details how countries should respond to the introduction of a potential invader based on its type of spread (i.e. unaided or intentionally or unintentionally aided by humans). To achieve these actions, various types of information (e.g. the status of a species and its

invasion history) and expertise (e.g. taxonomic experts for identifications) are required. As data availability on alien species varies across countries (McGeoch et al. 2010), we recommend that the required data should be maintained in a regional information system, which is regularly updated and to which information from global databases contributes [e.g. CABI's Invasive Species Compendium and the Global Invasive Species Information Network (GISIN), also see Lucy et al. 2016]. Such a database would improve not only the availability of alien species data but also, if standards are put in place, the quality of the data. Finally, it is unlikely that all of the required expertise will be available in every African country, for example, no country will have taxonomic experts for every taxonomic group (Klopper, Smith & Chikuni 2002). Regional cooperation, particularly in the training of personnel and the exchange of experts, is therefore required. In an effort to achieve this for taxonomy, regional networks [e.g. SABONET (Willis & Huntley 2001), which came to a close in 2005] and international initiatives [e.g. the Global Taxonomy Initiative (Secretariat of the Convention on Biological Diversity 2010)] have already been established.

Conclusion

A wide variety of alien species have been introduced to Africa and have spread within the region, with the movement patterns of these species varying across organisms and over time. Although direct introductions from other regions remain a concern, the number of species spreading within the region appears to be increasing, and these species probably pose a particularly high biosecurity threat. As preventing the intra-African spread of species is at best difficult, African countries need to cooperate and coordinate their responses. Achieving this requires communication, the development and implementation of standardised methods and systems, and political will. As the efficacy of a country's biosecurity greatly influences that of its neighbours, such an endeavour would benefit all of the countries involved.

Acknowledgements

We thank Georg Goergen, Jesse Kalwij, Ingrid Minnaar and Olaf Weyl for providing photos of example species. We also thank Donald Chungu (Copperbelt University, Zambia), Tembani Mduduzi (Forest Research Centre, Zimbabwe), Peter Kiwuso (Forest Research Institute, Uganda), Gerald Meke (FRIM, Malawi) and Eston Mutitu (KEFRI, Kenya) for providing information on the detection date of eucalypt insect pests in their respective countries.

This work was supported by the South African National Department of Environment Affairs through its funding of the South African National Biodiversity Institute's Invasive Species Programme. Additional funding was provided by the DST-NRF Centre for Invasion Biology. M.R. acknowledges funding from the South African Research Chairs Initiative of the Department of Science and Technology and National Research Foundation of South Africa.

Competing interests

The authors declare that they have no financial or personal relationship(s) that may have inappropriately influenced them in writing this article.

Author(s) contributions

K.T.F., M.P.R. and J.R.U.W. developed the scenarios. K.T.F. identified the examples. K.T.F. and B.P.H. collected the data. K.T.F. analysed the data. K.T.F., B.P.H., M.P.R., M.R. and J.R.U.W. wrote the manuscript.

References

Bewick, V., Cheek, L. & Ball, J., 2004, 'Statistics review 8: Qualitative data – Tests of association', *Critical Care* 8(1), 46–53. https://doi.org/10.1186/cc2428

Boissin, E., Hurley, B., Wingfield, M.J., Vasaitis, R., Stenlid, J., Davis, C. et al., 2012, 'Retracing the routes of introduction of invasive species: The case of the *Sirex noctilio* woodwasp', *Molecular Ecology* 21(23), 5728–5744. https://doi.org/10.1111/mec.12065

Bourgeois, K., Suehs, C.M., Vidal, E. & Médail, F., 2005, 'Invasional meltdown potential: Facilitation between introduced plants and mammals on French Mediterranean islands', *Écoscience* 12(2), 248–256. https://doi.org/10.2980/i1195-6860-12-2-248.1

Brandes, D., 2001, *Urban flora of Sousse (Tunisia)*, Braunschweig University Library Digitale Bibliothek.

Bromilow, C., 2010, *Problem plants and alien weeds of South Africa*, 3rd edn., BRIZA Publications, Pretoria.

Brown, P.M.J., Thomas, C.E., Lombaert, E., Jeffries, D.L., Estoup, A. & Lawson Handley, L.-J., 2011, 'The global spread of *Harmonia axyridis* (Coleoptera: Coccinellidae): Distribution, dispersal and routes of invasion', *BioControl* 56, 623–641. https://doi.org/10.1007/s10526-011-9379-1

Bush, S.J., Slippers, B., Neser, S., Harney, M., Dittrich-Schröder, G. & Hurley, B.P., 2016, 'Six recently recorded Australian insects associated with eucalyptus in South Africa', *African Entomology* 24(2), 539–544. https://doi.org/10.4001/003.024.0539

CAB International, 2000a, *Prostephanus truncatus, invasive species compendium*, viewed 10 February 2016, from http://www.cabi.org/isc/datasheet/44524.

CAB International, 2000b, *Medicago polymorpha, invasive species compendium*, viewed 22 February 2016, from http://www.cabi.org/isc/datasheet/33031

CAB International, 2000c, *Carpobrotus edulis, invasive species compendium*, viewed 17 February 2016, from http://www.cabi.org/isc/datasheet/10648

CAB International, 2016, *Invasive species compendium*, viewed 5 January 2016, from http://www.cabi.org/isc

Canonico, G.C., Arthington, A., McCrary, J.K. & Thieme, M.L., 2005, 'The effects of introduced tilapias on native biodiversity', *Aquatic Conservation: Marine and Freshwater Ecosystems* 15, 463–483. https://doi.org/10.1002/aqc.699

Carnegie, A.J., Matsuki, M., Haugen, D.A., Hurley, B.P., Ahumada, R., Klasmer, P. et al., 2006, 'Predicting the potential distribution of *Sirex noctilio* (Hymenoptera: Siricidae), a significant exotic pest of *Pinus* plantations', *Annals of Forest Science* 63, 119–128. https://doi.org/10.1051/forest:2005104

Central Intelligence Agency, 2015, *The CIA world factbook 2016*, Skyhorse Publishing, New York.

Chiron, F., Shirley, S.M. & Kark, S., 2010, 'Behind the Iron Curtain: Socio-economic and political factors shaped exotic bird introductions into Europe', *Biological Conservation* 143, 351–356. https://doi.org/10.1016/j.biocon.2009.10.021

Crawley, M.J., 2007, *The R book*, Wiley, Chichester.

D'Antonio, C.M., 1990, 'Seed production and dispersal in the non-native, invasive succulent *Carpobrotus edulis* (Aizoaceae) in coastal strand communities of central California', *Journal of Applied Ecology* 27(2), 693–702. https://doi.org/10.2307/2404312

de Moor, I.J. & Bruton, M.N., 1988, *Atlas of alien and translocated indigenous aquatic animals in southern Africa*, Council for Scientific and Industrial Research, Pretoria.

Deacon, J., 1986, 'Human settlement in South Africa and archaeological evidence for alien plants and animals', in I.A.W. Macdonald, F.J. Kruger & A.A. Ferrar (eds.), *The ecology and management of biological invasions in southern Africa*, pp. 3–19, Oxford University Press, Cape Town.

Dean, W.R.J., 2000, 'Alien birds in southern Africa: What factors determine success?', *South African Journal of Science* 96, 9–14.

Department of Home Affairs, 2016, *South African ports of entry*, viewed 9 June 2016, from http://www.dha.gov.za/index.php/immigration-services/south-african-ports-of-entry.

Essl, F., Dullinger, S., Rabitsch, W., Hulme, P.E., Hülber, K., Jarošík, V. et al., 2011, 'Socioeconomic legacy yields an invasion debt', *Proceedings of the National Academy of Sciences of the United States of America* 108(1), 203–207. https://doi.org/10.1073/pnas.1011728108

Estoup, A. & Guillemaud, T., 2010, 'Reconstructing routes of invasion using genetic data: Why, how and so what?', *Molecular Ecology* 19, 4113–4130. https://doi.org/10.1111/j.1365-294X.2010.04773.x

Everitt, B.S., 1977, *The analysis of contingency tables*, Chapman and Hall Ltd, London.

Falleh, H., Ksouri, R., Boulaaba, M., Guyot, S., Abdelly, C. & Magné, C., 2012, 'Phenolic nature, occurrence and polymerization degree as marker of environmental adaptation in the edible halophyte Mesembryanthemum edule', *South African Journal of Botany* 79, 117–124. https://doi.org/10.1016/j.sajb.2011.10.001

Faulkner, K.T., Spear, D., Robertson, M.P., Rouget, M. & Wilson, J.R.U., 2015, 'An assessment of the information content of South African alien species databases', *Bothalia: African Biodiversity and Conservation* 45(1). https://doi.org/10.4102/abc.v45i1.1103

Ferus, P., Culiţă, S., Eliáš, P., Konôpková, J., Ďurišová, L., Samuil, C. et al., 2015, 'Reciprocal contamination by invasive plants: Analysis of trade exchange between Slovakia and Romania', *Biologia* 70(7), 893–904. https://doi.org/10.1515/biolog-2015-0102

Garnas, J.R., Auger-Rozenberg, M.-A., Roques, A., Bertelsmeier, C., Wingfield, M.J., Saccaggi, D.L. et al., 2016, 'Complex patterns of global spread in invasive insects: Eco-evolutionary and management consequences', *Biological Invasions* 18(4), 921–933. https://doi.org/10.1007/s10530-016-1082-9

Genovesi, P., Scalera, R., Brunel, S., Roy, D. & Solarz, W., 2010, *Towards an early warning and information system for invasive alien species (IAS) threatening biodiversity in Europe*, European Environment Agency Technical Report, European Environmental Agency, Copenhagen.

Greuter, W. & Domina, G., 2015, 'Checklist of vascular plants collected during the 12th "Iter Mediterraneum" in Tunisia, 24 March–4 April 2014', *Bocconea* 27(1), 21–61.

Hodges, R.J., 1986, 'The biology and control of *Prostephanus truncatus* (Horn) (Coleoptera: Bostrichidae) – A destrutive storage pest with an increasing range', *Journal of Stored Products Research* 22(1), 1–14. https://doi.org/10.1016/0022-474X(86)90040-8

Hurley, B.P., Croft, P., Verleur, M., Wingfield, M.J. & Slippers, B., 2012, 'The control of the Sirex woodwasp in diverse environments: The South African experience', in B. Slippers, P. de Groot & M.J. Wingfield (eds.), *The Sirex woodwasp and its fungal symbiont: Research and management of a worldwide invasive pest*, pp. 247–264, Springer, Dordrecht, The Netherlands.

Hurley, B.P., Garnas, J., Wingfield, M.J., Branco, M., Richardson, D.M. & Slippers, B., 2016, 'Increasing numbers and intercontinental spread of invasive insects on eucalypts', *Biological Invasions* 18(4), 921–933. https://doi.org/10.1007/s10530-016-1081-x

Hurley, B.P., Slippers, B. & Wingfield, M.J., 2007, 'A comparison of control results for the alien invasive woodwasp, *Sirex noctilio*, in the Southern Hemisphere', *Agricultural and Forest Entomology* 9, 159–171. https://doi.org/10.1111/j.1461-9563.2007.00340.x

Jaksic, F.M., Iriarte, J.A., Jiménez, J.E. & Martínez, D.R., 2002, 'Invaders without frontiers: Cross-border invasions of exotic mammals', *Biological Invasions* 4, 157–173. https://doi.org/10.1023/A:1020576709964

Keller, R.P. & Kumschick, S., 2017, 'Promise and challenges of risk assessment as an approach for preventing the arrival of harmful alien species', *Bothalia* 47(2), a2136. https://doi.org/10.4102/abc.v47i2.2136

Kenis, M., Roy, H.E., Zindel, R. & Majerus, M.E.N., 2008, 'Current and potential management strategies against *Harmonia axyridis*', *BioControl* 53(1), 235–252. https://doi.org/10.1007/s10526-007-9136-7.

Klopper, R.R., Smith, G.F. & Chikuni, A.C., 2002, 'The global taxonomy initiative in Africa', *Taxon* 51, 159–165. https://doi.org/10.2307/1554974

Lambdon, P.W., Pyšek, P., Basnou, C., Hejda, M., Arianoutsou, M., Essl, F. et al., 2008, 'Alien flora of Europe: Species diversity, temporal trends, geographical patterns and research needs', *Preslia* 80, 101–149.

Lantschner, M.V., Villacide, J.M., Garnas, J.R., Croft, P., Carnegie, A.J., Liebhold, A.M. et al., 2014, 'Temperature explains variable spread rates of the invasive woodwasp *Sirex noctilio* in the Southern Hemisphere', *Biological Invasions* 16, 329–339. https://doi.org/10.1007/s10530-013-0521-0

Latombe, G., Pyšek, P., Jeschke, J.M., Blackburn, T.M., Bacher, S., Capinha, C. et al., in press, 'A vision for global monitoring of biological invasions', *Biological Conservation*. https://doi.org/10.1016/j.biocon.2016.06.013

Lever, C., 1987, *Naturalized birds of the world*, Longman Scientific and Technical, Essex.

Lever, C., 2005, *Naturalised birds of the world*, T & A D Poyser, London.

Lockwood, J.L., Cassey, P. & Blackburn, T.M., 2005, 'The role of propagule pressure in explaining species invasions', *Trends in Ecology and Evolution* 20(5), 223–228. https://doi.org/10.1016/j.tree.2005.02.004

Lombaert, E., Guillemaud, T., Cornuet, J.-M., Malausa, T., Facon, B. & Estoup, A., 2010, 'Bridgehead effect in the worldwide invasion of the biocontrol harlequin ladybird', *PLoS One* 5(3), e9743. https://doi.org/10.1371/journal.pone.0009743

Long, J.L., 1981, *Introduced birds of the world*, David & Charles, London.

Lucy, F.E., Roy, H.E., Simpson, A., Carlton, J.T., Hanson, J.M., Magellan, K. et al., 2016, 'INVASIVESNET towards an international association for open knowledge on invasive alien species', *Management of Biological Invasions* 7(2), 131–139. https://doi.org/10.3391/mbi.2016.7.2.01

Matthews, S. & Brand, K., 2004, *Africa invaded: The growing danger of invasive alien species*, The Global Invasive Species Programme, Cape Town.

McGeoch, M.A., Butchart, S.H.M., Spear, D., Marais, E., Kleynhans, E.J., Symes, A. et al., 2010, 'Global indicators of biological invasion: Species numbers, biodiversity impact and policy responses', *Diversity and Distributions* 16, 95–108. https://doi.org/10.1111/j.1472-4642.2009.00633.x.

Measey, J., Davies, S., Vimercati, G., Rebelo, A., Schmidt, W. & Turner, A., 2017, 'Invasive amphibians in southern Africa: A review of invasion pathways', *Bothalia* 47(2), a2117. https://doi.org/10.4102/abc.v47i2.2117

Muatinte, B.L., van den Berg, J. & Santos, L.A., 2014, '*Prostephanus truncatus* in Africa: A review of biological trends and perspectives on future pest management strategies', *African Crop Science Journal* 22(3), 237–256.

Peacock, D.S., van Rensburg, B.J. & Robertson, M.P., 2007, 'The distribution and spread of the invasive alien common myna, *Acridotheres tritis* L (Aves: Sturnidae), in southern Africa', *South African Journal of Science* 103, 465–473.

Picker, M. & Griffiths, C.L., 2011, *Alien and invasive animals: A South African perspective*, Struik Nature, Cape Town.

Poutsma, J., Loomans, A.J.M., Aukema, B. & Heijerman, T., 2008, 'Predicting the potential geographical distribution of the harlequin ladybird, *Harmonia axyridis*, using the CLIMEX model', *BioControl* 53, 103–125. https://doi.org/10.1007/s10526-007-9140-y

Roques, A., Auger-Rozenberg, M.-A., Blackburn, T.M., Garnas, J., Pyšek, P., Rabitsch, W. et al., 2016, 'Temporal and interspecific variation in rates of spread for insect species invading Europe during the last 200 years', *Biological Invasions* 18(4), 907–920. https://doi.org/10.1007/s10530-016-1080-y

Rouget, M., Robertson, M.P., Wilson, J.R.U., Hui, C., Essl, F., Renteria, J.L. et al., 2016, 'Invasion debt – Quantifying future biological invasions', *Diversity and Distributions* 22, 445–456. https://doi.org/10.1111/ddi.12408

Roy, H.E., Brown, P.M.J., Adriaens, T., Berkvens, N., Borges, I., Clusella-Trullas, S. et al., 2016, 'The harlequin ladybird, *Harmonia axyridis*: Global perspectives on invasion history and ecology', *Biological Invasions* 18(4), 997–1044. https://doi.org/10.1007/s10530-016-1077-6

Secretariat of the Convention on Biological Diversity, 2010, 'Guide to the global taxonomy initiative', CBD Technical Series, viewed 29 June 2016, from https://www.cbd.int/doc/publications/cbd-ts-30.pdf

Seebens, H., Essl, F., Dawson, W., Fuentes, N., Moser, D., Pergl, J. et al., 2015, 'Global trade will accelerate plant invasions in emerging economies under climate change', *Global Change Biology* 21(11), 4128–4140. https://doi.org/10.1111/gcb.13021.

Skelton, P., 1993, *A complete guide to the freshwater fishes of Southern Africa*, Southern Book Publishers, Harare.

Slippers, B., Wingfield, M.J., Coutinho, T.A. & Wingfield, B.D., 2001, 'Population structure and possible origin of *Amylostereum areolatum* in South Africa', *Plant Pathology* 50, 206–210. https://doi.org/10.1046/j.1365-3059.2001.00552.x

Slippers, B., Wingfield, B.D., Coutinho, T.A. & Wingfield, M.J., 2002, 'DNA sequence and RFLP data reflect geographical spread and relationships of *Amylostereum areolatum* and its insect vectors', *Molecular Ecology* 11, 1845–1854. https://doi.org/10.1046/j.1365-294X.2002.01572.x

Small, E., 2011, *Alfalfa and relatives: Evolution and classification of Medicago*, NRC Research Press, Ottawa.

Stals, R. & Prinsloo, G., 2007, 'Discovery of an alien invasive, predatory insect in South Africa: The multicoloured Asian ladybird beetle, *Harmonia axyridis* (Pallas) (Coleoptera: Coccinellidae)', *South African Journal of Science* 103, 123–126.

Stals, R., 2010, 'The establishment and rapid spread of an alien invasive lady beetle: *Harmonia axyridis* (Coleoptera: Coccinellidae) in southern Africa, 2001–2009', *IOBC/wprs Bulletin* 58, 125–132.

Taylor, J.S., 1962, '*Sirex noctilio* F., a recent introduction in South Africa', *Entomologist's Record* 74, 273–274.

Tribe, G.D., 1995, 'The woodwasp *Sirex noctilio* Fabricius (Hymenoptera: Siricidae), a pest of Pinus species, now established in South Africa', *African Entomology* 3(2), 215–217.

Tribe, G.D. & Cilliè, J.J., 2004, 'The spread of *Sirex noctilio* Fabricius (Hymenoptera: Siricidae) in South African pine plantations and the introduction and establishment of its biological control agents', *African Entomology* 12(1), 9–17.

UNEP, 2011, *Report of the tenth meeting of the conference of parties to the Convention on Biological Diversity*, Nagoya, Japan, 18–29 October 2010.

van Kleunen, M., Dawson, W., Essl, F., Pergl, J., Winter, M., Weber, E. et al., 2015, 'Global exchange and accumulation of non-native plants', *Nature* 525, 100–103. https://doi.org/10.1038/nature14910

van Rensburg, B.J., Weyl, O.L.F., Davies, S.J., van Wilgen, N.J., Spear, D., Chimimba, C.T. et al., 2011, 'Invasive vertebrates of South Africa', in D. Pimentel (ed.), *Biological invasions: Economic and environmental costs of alien plant, animal, and microbe species*, pp. 325–378, CRC Press, Boca Raton, FL.

Visser, V., Wilson, J.R.U., Canavan, S., Fish, L., Le Maitre, D.C., Nänni, I. et al., 2017, 'Grasses as invasive plants in South Africa revisited: Patterns, pathways and management', *Bothalia* 47(2), a2169. https://doi.org/10.4102/abc.v47i2.2169

Wells, M.J., Balsinhas, A.A., Joffe, H., Engelbrecht, V.M., Harding, G. & Stirton, C.H., 1986, 'A catalogue of problem plants in Southern Africa', *Memoirs of the Botanical Survey of South Africa* 53, 1–658.

Willis, C.K. & Huntley, B.J., 2001, 'Developing capacity within southern Africa's herbaria and botanical gardens', *Systematics and Geography of Plants* 71(2), 247–258. https://doi.org/10.2307/3668671

Wilson, J.R.U., Dormontt, E.E., Prentis, P.J., Lowe, A.J. & Richardson, D.M., 2009, 'Something in the way you move: Dispersal pathways affect invasion success', *Trends in Ecology and Evolution* 24(3), 136–144. https://doi.org/10.1016/j.tree.2008.10.007

Wilson, J.R.U., Gaertner, M., Richardson, D.M. & van Wilgen, B.W., 2017, 'Contributions to the National Status Report on Biological Invasions in South Africa', *Bothalia* 47(2), a2207. https://doi.org/10.4102/abc.v47i2.2207

Wingfield, M.J., Slippers, B., Hurley, B.P., Coutinho, T.A., Wingfield, B.D. & Roux, J., 2008, 'Eucalypt pests and diseases: Growing threats to plantation productivity', *Southern Forests* 70(2), 139–144. https://doi.org/10.2989/SOUTH.FOR.2008.70.2.9.537

Zuur, A.F., Ieno, E.N., Walker, N.J., Saveliev, A.A. & Smith, G.M., 2009, *Mixed effects models and extensions in ecology with R*, Springer, New York.

Appendix 1

Table 1-A1: Bird species introduced to South Africa and/or other parts of Africa.

No	Family	Species	Synonym	Common name
1	Alaudidae	Melanocorypha bimaculata		Bimaculated lark
2	Anatidae	Aix galericulata		Mandarin duck
3	Anatidae	Aix sponsa		Wood duck
4	Anatidae	Anas acuta		Northern pintail
5	Anatidae	Anas clypeata		Northern shoveler
6	Anatidae	Anas discors		Blue-winged teal
7	Anatidae	Anas platyrhynchos		Mallard
8	Anatidae	Anas querquedula		Garnaney
9	Anatidae	Anas rubripes	Anas obscura	American black duck
10	Anatidae	Aythya ferina		Common pochard
11	Anatidae	Aythya fuligula		Tufted duck
12	Anatidae	Aythya nyroca		Ferruginous duck
13	Anatidae	Cairina moschata		Muscovy duck
14	Anatidae	Callonetta leucophrys		Ringed teal
15	Anatidae	Cygnus atratus		Black swan
16	Anatidae	Cygnus olor		Mute swan
17	Anatidae	Dendrocygna autumnalis		Black-bellied whistling duck
18	Anatidae	Netta rufina		Red-crested pochard
19	Anatidae	Oxyura jamaicensis		Ruddy duck
20	Anatidae	Tadorna tadorna		European shelduck
21	Cacatuidae	Cacutua sulphurea		Yellow-crested cockatoo
22	Cacatuidae	Nymphicus hollandicus		Cockatiel
23	Columbidae	Columba livia		Common pigeon
24	Columbidae	Columbina inca	Scardafella inca	Inva dove
25	Columbidae	Geopelia cuneata		Diamond dove
26	Columbidae	Streptopelia decaoto		Eurasian collared dove
27	Columbidae	Streptopelia turtur		European turtle dove
28	Columbidae	Zenaida macroura		Mourning dove
29	Coraciidae	Coracias cyanogaster		Blue-bellied roller
30	Corvidae	Corvus frugilegus		Rook
31	Corvidae	Corvus monedula		Jackdaw
32	Corvidae	Corvus splendens		House Crow
33	Corvidae	Dendrocitta vagabunda		Rufous treepie
34	Emberizidae	Paroaria coronata	Paroaria dominicana	Red-crested Cardinal
35	Estrildidae	Amandava amandava	Estrilda amandava	Red Avadavat
36	Estrildidae	Estrilda melpoda		Orange-cheeked waxbill
37	Estrildidae	Euodice cantans	Lonchura cantans	African silverback
38	Estrildidae	Lonchura oryzivora	Padda oryzivora	Java Sparrow
39	Estrildidae	Taeniopygia guttata		Zebra finch
40	Falconidae	Falco columbarius		Merlin
41	Fringillidae	Carduelis carduelis		Goldfinch
42	Fringillidae	Fringilla coelebs		Common chaffinch
43	Leiothrichidae	Leiothrix argentauris		Silver-eared mesia
44	Meropidae	Merops malimbicus		Rosy beeater
45	Muscicapidae	Luscinia megarhynchos		Nightingale
46	Musophagidae	Crinifer piscator		Western gray plaintain-eater

Table 1-A1 (Continues): Bird species introduced to South Africa and/or other parts of Africa.

No	Family	Species	Synonym	Common name
47	Musophagidae	Criniferoides leucogaster		White-bellied-go-away-bird
48	Musophagidae	Musophaga violacea		Violet turaco
49	Passeridae	Passer domesticus		House sparrow
50	Phasianidae	Alectoris chukar	Alectoris graeca	Chukar partridge
51	Phasianidae	Alectoris melanocephalus		Arabian chukar
52	Phasianidae	Chrysolophus pictus		Golden pheasant
53	Phasianidae	Colinus virginianus		Bobwhite quail
54	Phasianidae	Coturnix chinensis		Asian blue quail
55	Phasianidae	Gallus gallus		Red jungle fowl
56	Phasianidae	Lophortyx californicus		California quail
57	Phasianidae	Lophura nycthemera		Silver pheasant
58	Phasianidae	Pavo cristatus		Common peacock
59	Phasianidae	Phasianus colchicus		Common pheasant
60	Ploceidae	Ploceus nigerrimus		Vieillot's black weaver
61	Psittacidae	Agapornis cana	Agapornis canus	Madagascar lovebird
62	Psittacidae	Amazona aestiva		Blue-fronted parrot
63	Psittacidae	Aratinga jandaya		Jandaya conure
64	Psittacidae	Aratinga pertinax		Brown-throated conure
65	Psittacidae	Aratinga weddellii		Dusky-headed conure
66	Psittacidae	Cyanoliseus patagonus		Patagonian conure
67	Psittacidae	Forpus passerinus		Blue-winged parrotlet
68	Psittacidae	Melopsittacus undulatus		Budgerigar
69	Psittacidae	Myiopsitta monachus		Monk parakeet
70	Psittacidae	Nandayus nenday		Black-hooded conure
71	Psittacidae	Poicephalus rueppellii		Ruppell's parrot
72	Psittacidae	Poicephalus rufiventris		African orange-bellied parrot
73	Psittacidae	Psittacula krameri		Rose-ringed parakeet
74	Psittacidae	Pyrrhura rupicola		Black-capped conure
75	Psittaculidae	Psittacula cyanocephala		Plum-headed parakeet
76	Pycnonotidae	Pycnonotus jocosus		Red-whiskered bulbul
77	Rallidae	Fulica Americana		American coot
78	Rallidae	Gallinula comeri		Gough moorhen
79	Rallidae	Gallinula nesiotis		Tristan moorhen
80	Sturnidae	Acridotheres tristis		Common myna
81	Sturnidae	Lamprotornis iris		Emerald starling
82	Sturnidae	Lamprotornis purpuropterus	Lamprotornis purpuroptera	Ruppells long-tailed starling
83	Sturnidae	Lamprotornis superbus		Superb starling
84	Sturnidae	Sturnus vulgaris		Common starling
85	Threskiornithidae	Eudocimus ruber		Scarlet ibis
86	Turdidae	Turdus merula		Blackbird
87	Turdidae	Turdus philomelos		Song thrush

Table 2-A1: Insect pests of *Eucalyptus* trees introduced to South Africa and other parts of Africa.

No	Family	Species	Common name
1	Adelgidae	*Pineus boerneri*	Pine woolly aphid
2	Aphididae	*Cinara cronartii*	Black pine aphid
3	Aphididae	*Eulachnus rileyi*	Pine needle aphid
4	Cerambycidae	*Phoracantha recurva*	Eucalyptus longhorn beetle
5	Cerambycidae	*Phoracantha semipunctata*	Eucalyptus longhorn beetle
6	Chrysomelidae	*Trachymela tincticollis*	Eucalyptus tortoise beetle
7	Curculionidae	*Gonipterus scutellatus*	Eucalyptus snout beetle
8	Curculionidae	*Pissodes nemorensis*	Pine weevil
9	Eulophidae	*Leptocybe invasa*	Bluegum chalcid
10	Eulophidae	*Ophelimus maskelli*	Eucalyptus gall wasp
11	Psyllidae	*Blastopyslla occidentalis*	Eucalyptus psyllid
12	Psyllidae	*Ctenarytaina eucalypti*	Bluegum psyllid
13	Psyllidae	*Glycaspis brimblecombei*	Redgum lerp psyllid
14	Psyllidae	*Spondyliaspis c.f. plicatuloides*	Shell lerp psyllid
15	Scolytidae	*Hylastes angustatus*	Pine bark beetle
16	Scolytidae	*Hylurgus ligniperda*	Red-haired pine bark beetle
17	Scolytidae	*Orthotomicus erosus*	Mediterranean pine engraver beetle
18	Siricidae	*Sirex noctilio*	Sirex woodwasp
19	Thaumastocoridae	*Thaumastocoris peregrinus*	Bronze bug

Table 3-A1: Bird species introduced to South Africa and/or other parts of Africa categorised in terms of the most likely introduction route scenario that resulted in introduction, and our confidence in each designation. For some species data were insufficient to make a designation.

Species	Common name	Scenario	Confidence
Cygnus olor	Mute swan	1	High
Lophortyx californicus	California quail	1	High
Gallus gallus	Red jungle fowl	1	High
Agapornis cana	Madagascar Lovebird	1	High
Melopsittacus undulatus	Budgerigar	1	High
Fringilla coelebs	Common Chaffinch	1	High
Paroaria coronata	Red-crested Cardinal	1	High
Pavo cristatus	Common peacock	1	High
Lophura nycthemera	Silver pheasant	1	High
Phasianus colchicus	Common pheasant	1	High
Coturnix chinensis	Asian blue quail	1	High
Dendrocygna autumnalis	Black-bellied whistling duck	1	High
Cygnus atratus	Black swan	1	High
Aix galericulata	Mandarin duck	1	High
Aix sponsa	Wood duck	1	High
Anas discors	Blue-winged teal	1	High
Cacutua sulphurea	Yellow-crested cockatoo	1	High
Nymphicus hollandicus	Cockatiel	1	High
Geopelia cuneata	Diamond dove	1	High
Zenaida macroura	Mourning dove	1	High
Taeniopygia guttata	Zebra finch	1	High
Chrysolophus pictus	Golden pheasant	1	High
Aratinga pertinax	Brown-throated conure	1	High
Nandayus nenday	Black-hooded conure	1	High
Forpus passerinus	Blue-winged parrotlet	1	High
Amazona aestiva	Blue-fronted parrot	1	High
Psittacula cyanocephala	Plum-headed Parakeet	1	High
Pycnonotus jocosus	Red-whiskered bulbul	1	High
Gallinula nesiotis	Tristan moorhen	1	High
Eudocimus ruber	Scarlet ibis	1	High
Tadorna tadorna	European shelduck	1	High
Callonetta leucophrys	Ringed teal	1	High
Anas rubripes	American black duck	1	High
Aythya nyroca	Ferruginous duck	1	High
Columbina inca	Inva dove	1	High
Dendrocitta vagabunda	Rufous treepie	1	High
Leiothrix argentauris	Silver-eared mesia	1	High
Pyrrhura rupicola	Black-capped conure	1	High
Aratinga jandaya	Jandaya conure	1	High
Aratinga weddellii	Dusky-headed conure	1	High
Cyanoliseus patagonus	Patagonian conure	1	High
Fulica Americana	American coot	1	High
Gallinula comeri	Gough moorhen	1	High
Psittacula krameri	Rose-ringed Parakeet	1	Low
Luscinia megarhynchos	Nightingale	1	Low
Turdus merula	Blackbird	1	Low
Turdus philomelos	Song Thrush	1	Low
Amandava amandava	Red Avadavat	1	Low
Lonchura oryzivora	Java Sparrow	1	Low

Table 3-A1 (Continues): Bird species introduced to South Africa and/or other parts of Africa categorised in terms of the most likely introduction route scenario that resulted in introduction, and our confidence in each designation. For some species data were insufficient to make a designation.

Species	Common name	Scenario	Confidence
Carduelis carduelis	Goldfinch	1	Low
Netta rufina	red-crested Pochard	1	Low
Aythya ferina	Common pochard	1	Low
Aythya fuligula	Tufted duck	1	Low
Cairina moschata	Muscovy duck	1	Low
Corvus frugilegus	Rook	1	Low
Streptopelia decaoto	Eurasian collared dove	1	Low
Melanocorypha bimaculata	Bimaculated lark	1	Low
Falco columbarius	Merlin	1	Low
Anas clypeata	Northern shoveler	1	Low
Anas acuta	Northern pintail	1	Low
Anas querquedula	Garnaney	1	Low
Corvus splendens	House Crow	2	High
Coracias cyanogaster	Blue-bellied roller	3	High
Criniferoides leucogaster	White-bellied-go-away-bird	3	High
Crinifer piscator	Western gray plaintain-eater	3	High
Musophaga violacea	Violet turaco	3	High
Ploceus nigerrimus	Vieillot's black weaver	3	High
Poicephalus rufiventris	African orange-bellied parrot	3	High
Poicephalus rueppellii	Ruppell's parrot	3	High
Lamprotornis iris	Emerald starling	3	High
Lamprotornis purpuropterus	Ruppells long-tailed starling	3	High
Lamprotornis superbus	Superb starling	3	High
Merops malimbicus	Rosy beeater	3	High
Estrilda melpoda	Orange-cheeked waxbill	3	Low
Euodice cantans	African silverback	3	Low
Alectoris melanocephalus	Arabian Chukar	4	High
Corvus splendens	House Crow	4	High
Corvus monedula	Jackdaw	4	High
Passer domesticus	House Sparrow	4	High
Lonchura oryzivora	Java Sparrow	4	High
Oxyura jamaicensis	Ruddy duck	4	High
Myiopsitta monachus	Monk parakeet	4	High
Anas platyrhynchos	Mallard	4	Low
Psittacula krameri	Rose-ringed Parakeet	4	Low
Acridotheres tristis	Common Myna	4	Low
Amandava amandava	Red Avadavat	4	Low
Cairina moschata	Muscovy duck	4	Low
Sturnus vulgaris	Common starling	5	High
Acridotheres tristis	Common Myna	5	High
Passer domesticus	House Sparrow	5	High
Anas platyrhynchos	Mallard	5	Low
Alectoris chukar	Chukar partridge	5	Low
Colinus virginianus	Bobwhite quail	Insufficient data	
Columba livia	Common pigeon	Insufficient data	
Streptopelia turtur	European turtle dove	Insufficient data	

Table 4-A1: Insect pests of *Eucalyptus* trees introduced to South Africa and other parts of Africa categorised in terms of the most likely introduction route scenario that resulted in introduction, and our confidence in each designation. For some species data were insufficient to make a designation.

Species	Common name	Scenario	Confidence
Pissodes nemorensis	Pine weevil	1	High
Trachymela tincticollis	Eucalyptus tortoise beetle	1	High
Sirex noctilio	Sirex woodwasp	1	High
Spondyliaspis c.f. plicatuloides	Shell lerp psyllid	1	High
Ctenarytaina eucalypti	Bluegum psyllid	1	Low
Blastopyslla occidentalis	Eucalyptus psyllid	1	Low
Ophelimus maskelli	Eucalyptus gall wasp	1	Low
Leptocybe invasa	Bluegum chalcid	2	High
Phoracantha recurva	Eucalyptus longhorn beetle	4	Low
Phoracantha semipunctata	Eucalyptus longhorn beetle	4	Low
Ctenarytaina eucalypti	Bluegum psyllid	4	Low
Cinara cronartii	Black pine aphid	4	Low
Pineus boerneri	Pine woolly aphid	4	Low
Blastopyslla occidentalis	Eucalyptus psyllid	4	Low
Glycaspis brimblecombei	Redgum lerp psyllid	4	Low
Ophelimus maskelli	Eucalyptus gall wasp	4	Low
Gonipterus scutellatus	Eucalyptus snout beetle	5	High
Hylastes angustatus	Pine bark beetle	5	High
Hylurgus ligniperda	Red-haired pine bark beetle	5	Low
Phoracantha recurva	Eucalyptus longhorn beetle	5	Low
Phoracantha semipunctata	Eucalyptus longhorn beetle	5	Low
Orthotomicus erosus	Mediterranean pine engraver beetle	5	Low
Cinara cronartii	Black pine aphid	5	Low
Pineus boerneri	Pine woolly aphid	5	Low
Thaumastocoris peregrinus	Bronze bug	5	Low
Glycaspis brimblecombei	Redgum lerp psyllid	5	Low
Eulachnus rileyi	Pine needle aphid	Insufficient data	

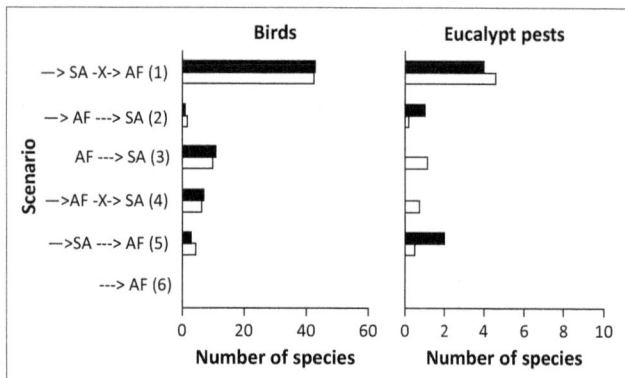

FIGURE 1-A1: The number of alien bird and eucalypt insect pest species for which each introduction route scenario was applicable (in black). Scenario designations with only high certainty were included. Details of the scenarios are provided in Figure 1. Expected values are shown in white. The association between scenario and organism type was not significant: $\chi^2 = 0.06$, $d.f. = 5$, $p = 0.1$.

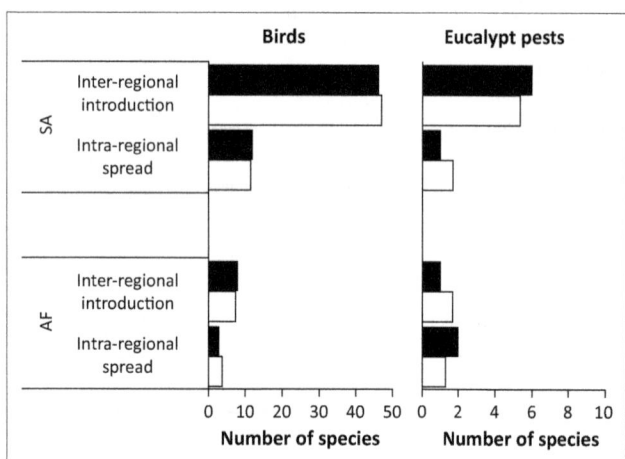

FIGURE 2-A1: The number of alien bird and eucalypt insect pest species (black) in South Africa (SA) and elsewhere in Africa (AF) that were introduced through a direct introduction from another region (inter-regional introduction) or through spread between the two subregions (intra-regional spread). Scenario designations with only high certainty were included and expected values are shown in white. The association between introduction type, organism type and recipient sub-region was not significant: $\chi^2 = 1.5$, $d.f. = 1$, $p = 0.2$.

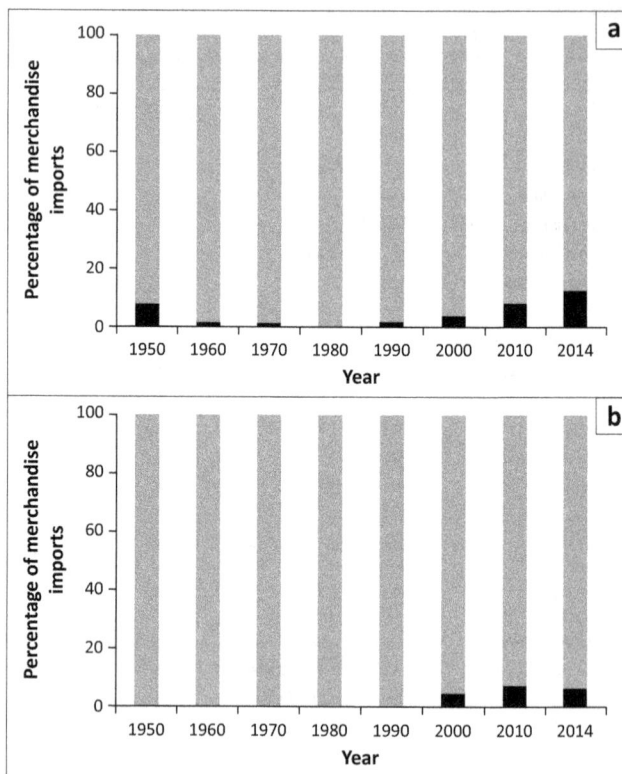

FIGURE 3-A1: Direction of trade statistics from the International Monetary Fund showing temporal changes in the contribution of (a) Africa (black) and other regions (grey) to South African imports and (b) South Africa (black) and other regions (grey) to African imports.

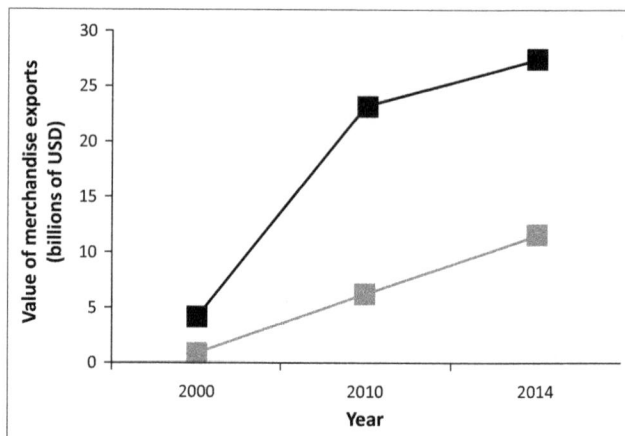

FIGURE 4-A1: Recent temporal changes in the value of merchandise exports from South Africa to elsewhere in Africa (black) and from elsewhere in Africa to South Africa (grey). Data were obtained from the International Monetary Fund.

Appendix 2

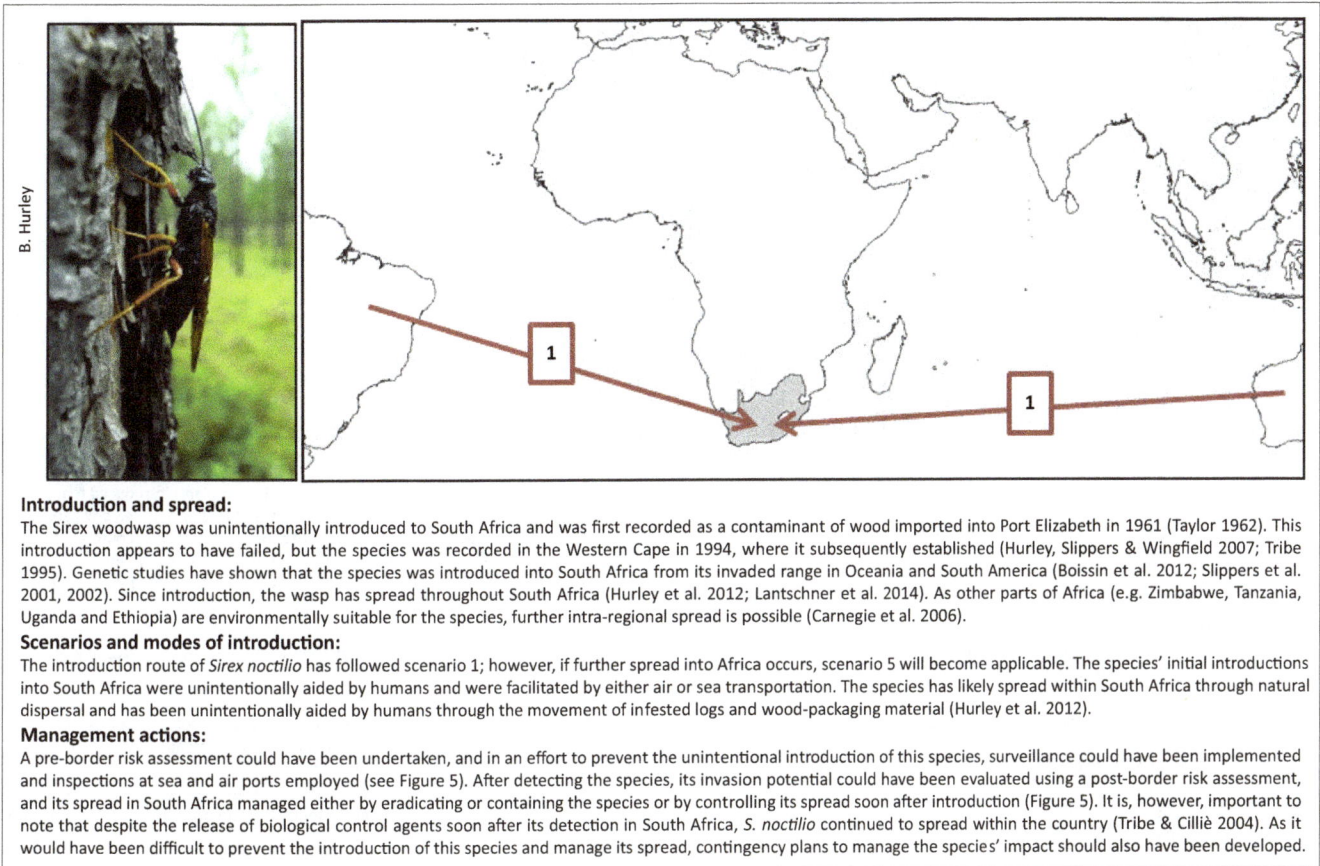

Introduction and spread:

The Sirex woodwasp was unintentionally introduced to South Africa and was first recorded as a contaminant of wood imported into Port Elizabeth in 1961 (Taylor 1962). This introduction appears to have failed, but the species was recorded in the Western Cape in 1994, where it subsequently established (Hurley, Slippers & Wingfield 2007; Tribe 1995). Genetic studies have shown that the species was introduced into South Africa from its invaded range in Oceania and South America (Boissin et al. 2012; Slippers et al. 2001, 2002). Since introduction, the wasp has spread throughout South Africa (Hurley et al. 2012; Lantschner et al. 2014). As other parts of Africa (e.g. Zimbabwe, Tanzania, Uganda and Ethiopia) are environmentally suitable for the species, further intra-regional spread is possible (Carnegie et al. 2006).

Scenarios and modes of introduction:

The introduction route of *Sirex noctilio* has followed scenario 1; however, if further spread into Africa occurs, scenario 5 will become applicable. The species' initial introductions into South Africa were unintentionally aided by humans and were facilitated by either air or sea transportation. The species has likely spread within South Africa through natural dispersal and has been unintentionally aided by humans through the movement of infested logs and wood-packaging material (Hurley et al. 2012).

Management actions:

A pre-border risk assessment could have been undertaken, and in an effort to prevent the unintentional introduction of this species, surveillance could have been implemented and inspections at sea and air ports employed (see Figure 5). After detecting the species, its invasion potential could have been evaluated using a post-border risk assessment, and its spread in South Africa managed either by eradicating or containing the species or by controlling its spread soon after introduction (Figure 5). It is, however, important to note that despite the release of biological control agents soon after its detection in South Africa, *S. noctilio* continued to spread within the country (Tribe & Cilliè 2004). As it would have been difficult to prevent the introduction of this species and manage its spread, contingency plans to manage the species' impact should also have been developed.

FIGURE 1-A2: Details on the introduction and spread of *Sirex noctilio*, the relevant introduction route scenario and modes of transport, and the management actions that could have prevented or mitigated the invasion.

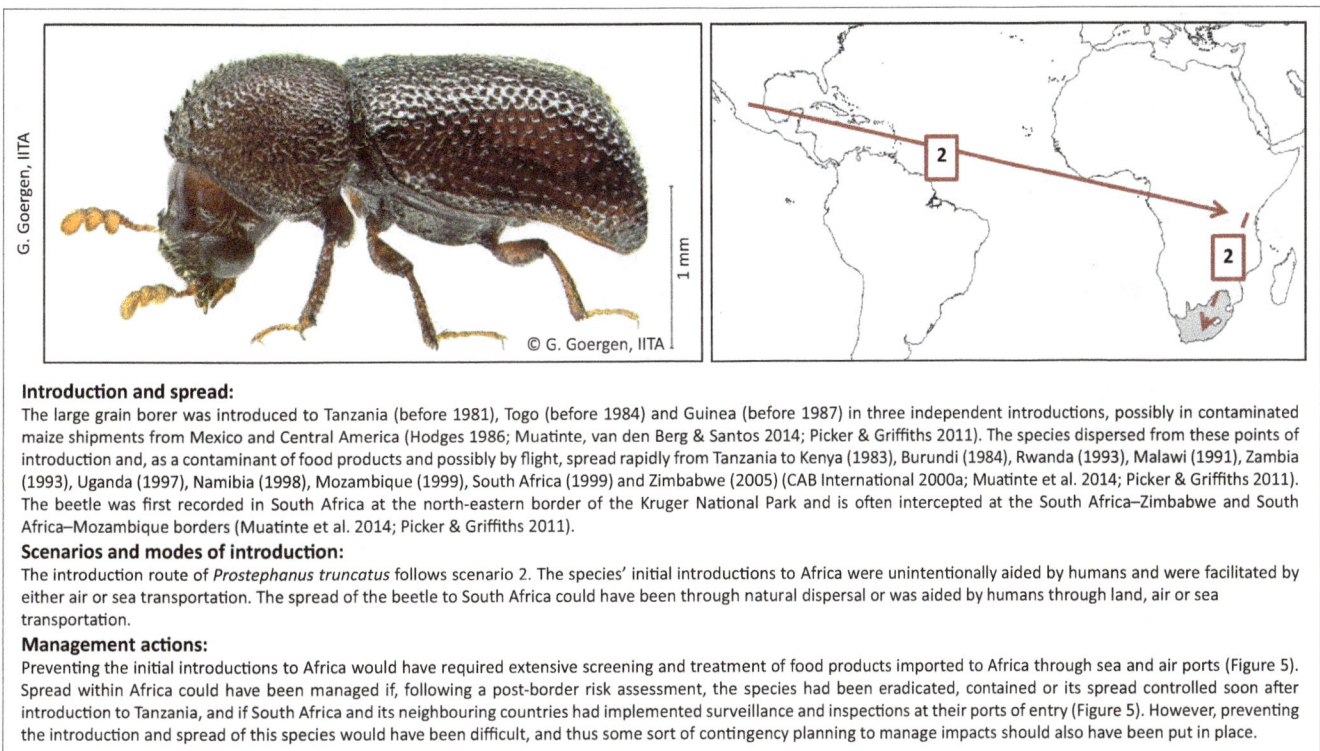

Introduction and spread:

The large grain borer was introduced to Tanzania (before 1981), Togo (before 1984) and Guinea (before 1987) in three independent introductions, possibly in contaminated maize shipments from Mexico and Central America (Hodges 1986; Muatinte, van den Berg & Santos 2014; Picker & Griffiths 2011). The species dispersed from these points of introduction and, as a contaminant of food products and possibly by flight, spread rapidly from Tanzania to Kenya (1983), Burundi (1984), Rwanda (1993), Malawi (1991), Zambia (1993), Uganda (1997), Namibia (1998), Mozambique (1999), South Africa (1999) and Zimbabwe (2005) (CAB International 2000a; Muatinte et al. 2014; Picker & Griffiths 2011). The beetle was first recorded in South Africa at the north-eastern border of the Kruger National Park and is often intercepted at the South Africa–Zimbabwe and South Africa–Mozambique borders (Muatinte et al. 2014; Picker & Griffiths 2011).

Scenarios and modes of introduction:

The introduction route of *Prostephanus truncatus* follows scenario 2. The species' initial introductions to Africa were unintentionally aided by humans and were facilitated by either air or sea transportation. The spread of the beetle to South Africa could have been through natural dispersal or was aided by humans through land, air or sea transportation.

Management actions:

Preventing the initial introductions to Africa would have required extensive screening and treatment of food products imported to Africa through sea and air ports (Figure 5). Spread within Africa could have been managed if, following a post-border risk assessment, the species had been eradicated, contained or its spread controlled soon after introduction to Tanzania, and if South Africa and its neighbouring countries had implemented surveillance and inspections at their ports of entry (Figure 5). However, preventing the introduction and spread of this species would have been difficult, and thus some sort of contingency planning to manage impacts should also have been put in place.

FIGURE 2-A2: Details on the introduction and spread of *Prostephanus truncatus*, the relevant introduction route scenario and modes of transport, and the management actions that could have prevented or mitigated the invasion.

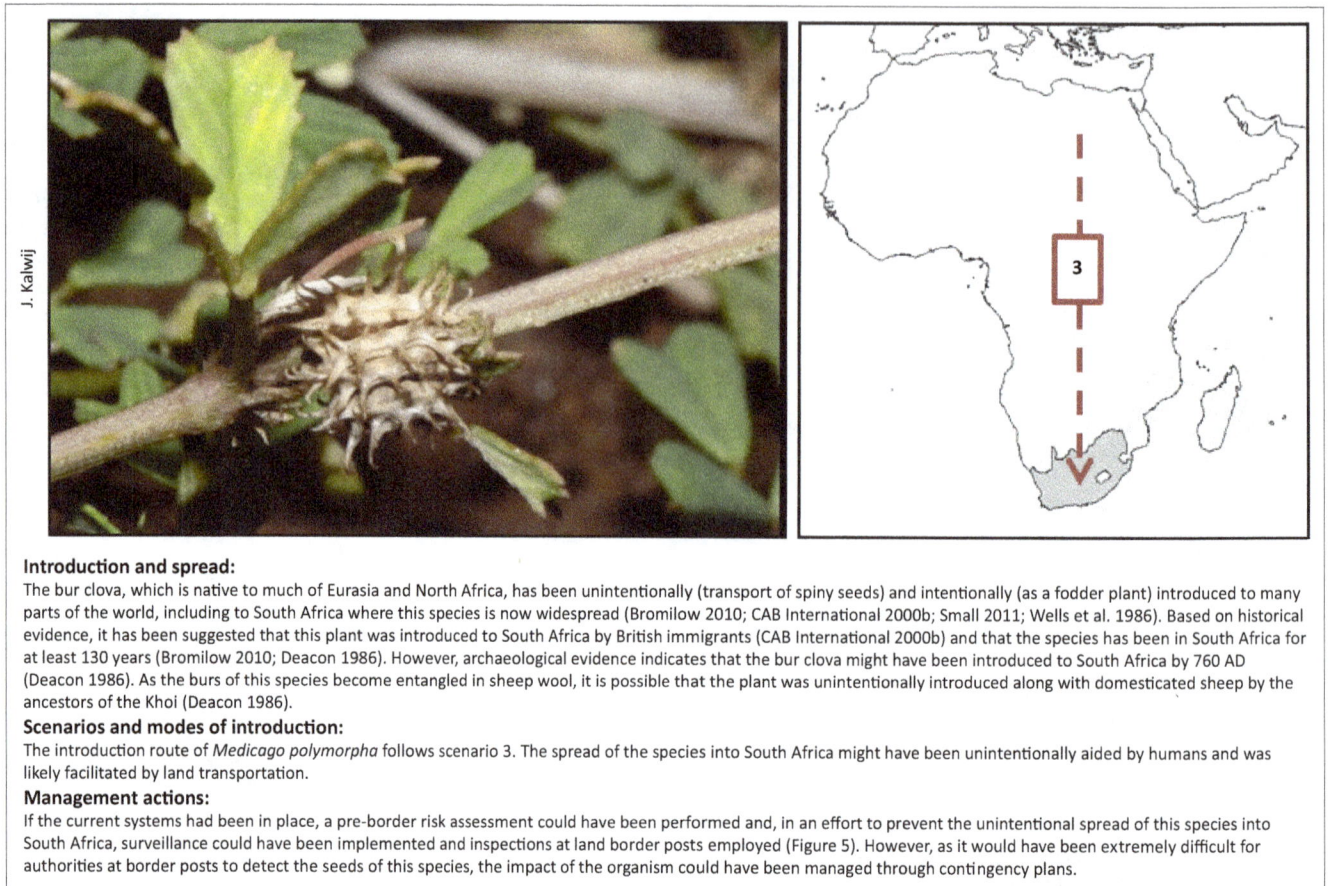

Introduction and spread:

The bur clova, which is native to much of Eurasia and North Africa, has been unintentionally (transport of spiny seeds) and intentionally (as a fodder plant) introduced to many parts of the world, including to South Africa where this species is now widespread (Bromilow 2010; CAB International 2000b; Small 2011; Wells et al. 1986). Based on historical evidence, it has been suggested that this plant was introduced to South Africa by British immigrants (CAB International 2000b) and that the species has been in South Africa for at least 130 years (Bromilow 2010; Deacon 1986). However, archaeological evidence indicates that the bur clova might have been introduced to South Africa by 760 AD (Deacon 1986). As the burs of this species become entangled in sheep wool, it is possible that the plant was unintentionally introduced along with domesticated sheep by the ancestors of the Khoi (Deacon 1986).

Scenarios and modes of introduction:

The introduction route of *Medicago polymorpha* follows scenario 3. The spread of the species into South Africa might have been unintentionally aided by humans and was likely facilitated by land transportation.

Management actions:

If the current systems had been in place, a pre-border risk assessment could have been performed and, in an effort to prevent the unintentional spread of this species into South Africa, surveillance could have been implemented and inspections at land border posts employed (Figure 5). However, as it would have been extremely difficult for authorities at border posts to detect the seeds of this species, the impact of the organism could have been managed through contingency plans.

FIGURE 3-A2: Details on the introduction and spread of *Medicago polymorpha*, the relevant introduction route scenario and modes of transport, and the management actions that could have prevented or mitigated the invasion.

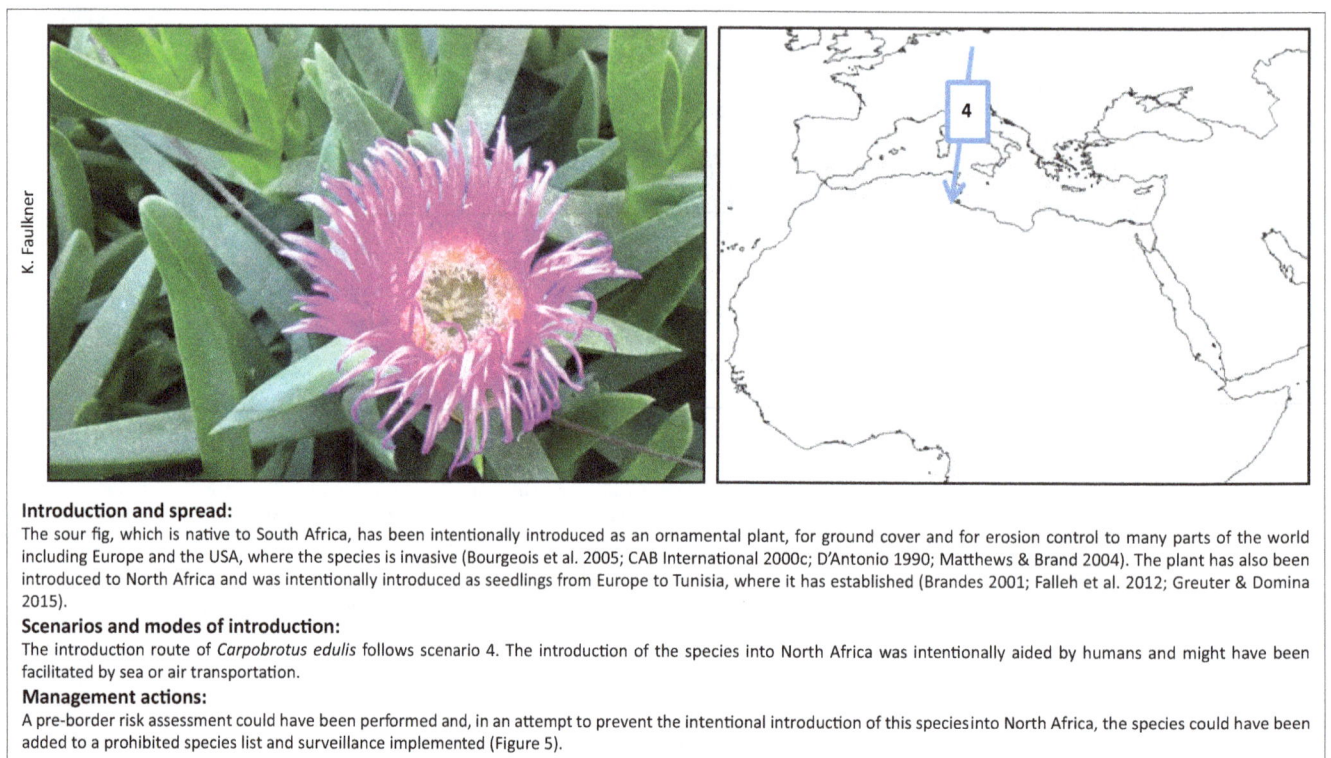

Introduction and spread:

The sour fig, which is native to South Africa, has been intentionally introduced as an ornamental plant, for ground cover and for erosion control to many parts of the world including Europe and the USA, where the species is invasive (Bourgeois et al. 2005; CAB International 2000c; D'Antonio 1990; Matthews & Brand 2004). The plant has also been introduced to North Africa and was intentionally introduced as seedlings from Europe to Tunisia, where it has established (Brandes 2001; Falleh et al. 2012; Greuter & Domina 2015).

Scenarios and modes of introduction:

The introduction route of *Carpobrotus edulis* follows scenario 4. The introduction of the species into North Africa was intentionally aided by humans and might have been facilitated by sea or air transportation.

Management actions:

A pre-border risk assessment could have been performed and, in an attempt to prevent the intentional introduction of this species into North Africa, the species could have been added to a prohibited species list and surveillance implemented (Figure 5).

FIGURE 4-A2: Details on the introduction and spread of *Carpobrotus edulis*, the relevant introduction route scenario and modes of transport, and the management actions that could have prevented or mitigated the invasion.

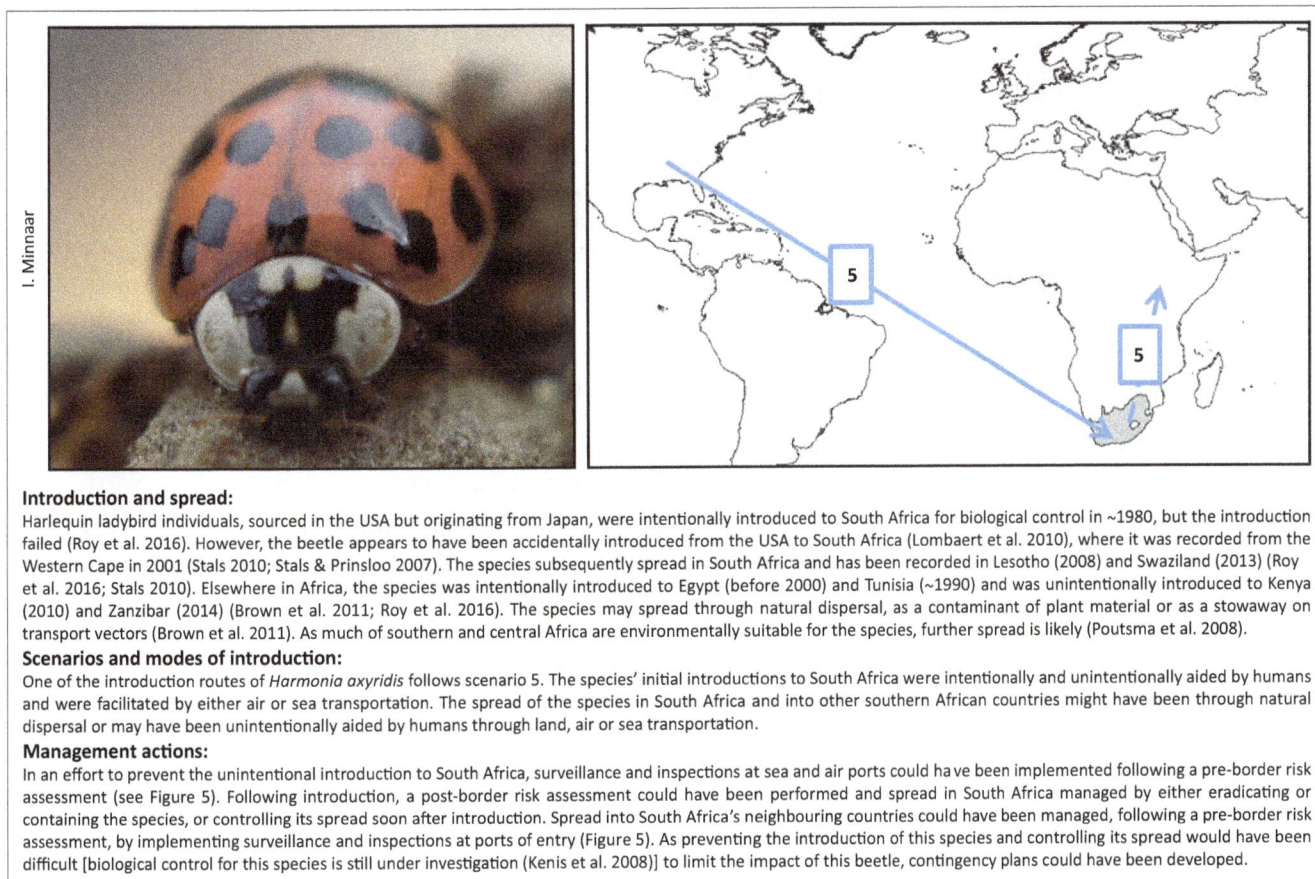

Introduction and spread:

Harlequin ladybird individuals, sourced in the USA but originating from Japan, were intentionally introduced to South Africa for biological control in ~1980, but the introduction failed (Roy et al. 2016). However, the beetle appears to have been accidentally introduced from the USA to South Africa (Lombaert et al. 2010), where it was recorded from the Western Cape in 2001 (Stals 2010; Stals & Prinsloo 2007). The species subsequently spread in South Africa and has been recorded in Lesotho (2008) and Swaziland (2013) (Roy et al. 2016; Stals 2010). Elsewhere in Africa, the species was intentionally introduced to Egypt (before 2000) and Tunisia (~1990) and was unintentionally introduced to Kenya (2010) and Zanzibar (2014) (Brown et al. 2011; Roy et al. 2016). The species may spread through natural dispersal, as a contaminant of plant material or as a stowaway on transport vectors (Brown et al. 2011). As much of southern and central Africa are environmentally suitable for the species, further spread is likely (Poutsma et al. 2008).

Scenarios and modes of introduction:

One of the introduction routes of *Harmonia axyridis* follows scenario 5. The species' initial introductions to South Africa were intentionally and unintentionally aided by humans and were facilitated by either air or sea transportation. The spread of the species in South Africa and into other southern African countries might have been through natural dispersal or may have been unintentionally aided by humans through land, air or sea transportation.

Management actions:

In an effort to prevent the unintentional introduction to South Africa, surveillance and inspections at sea and air ports could have been implemented following a pre-border risk assessment (see Figure 5). Following introduction, a post-border risk assessment could have been performed and spread in South Africa managed by either eradicating or containing the species, or controlling its spread soon after introduction. Spread into South Africa's neighbouring countries could have been managed, following a pre-border risk assessment, by implementing surveillance and inspections at ports of entry (Figure 5). As preventing the introduction of this species and controlling its spread would have been difficult [biological control for this species is still under investigation (Kenis et al. 2008)] to limit the impact of this beetle, contingency plans could have been developed.

FIGURE 5-A2: Details on the introduction and spread of *Harmonia axyridis*, the relevant introduction route scenario and modes of transport, and the management actions that could have prevented or mitigated the invasion.

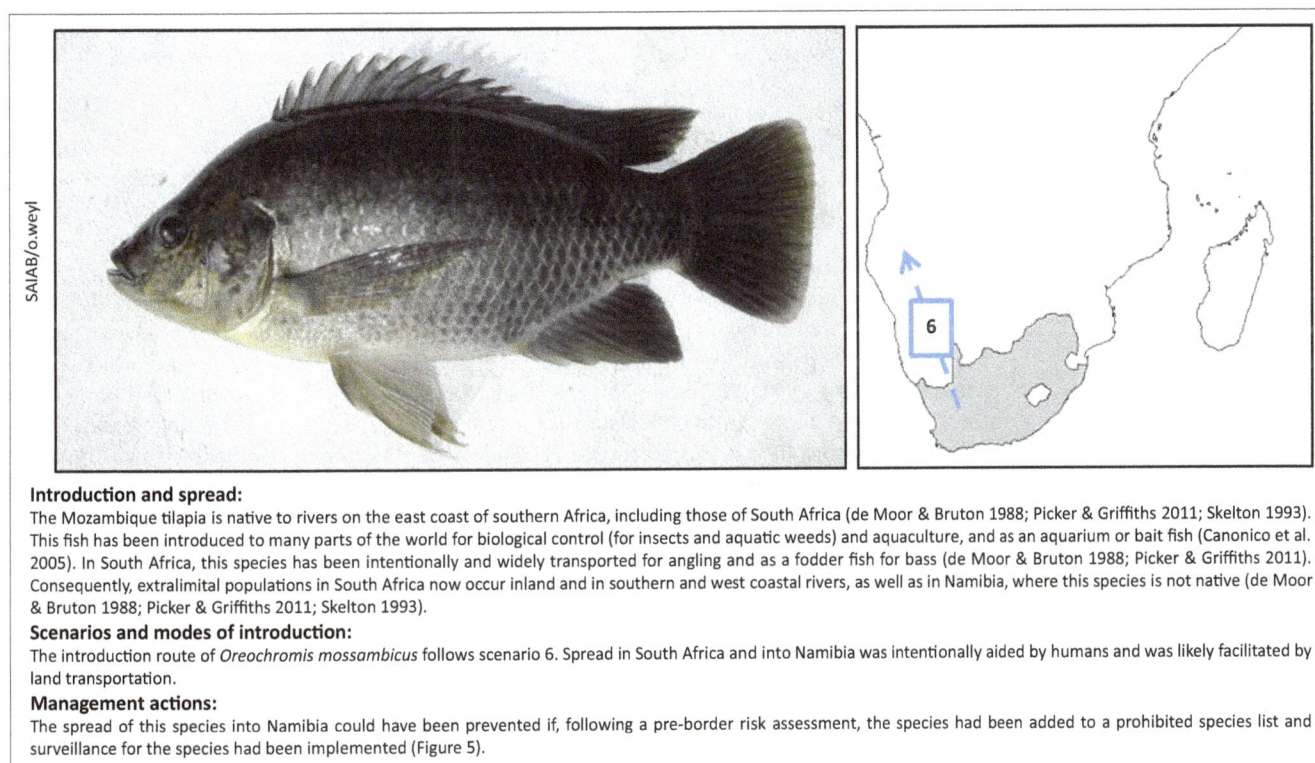

Introduction and spread:

The Mozambique tilapia is native to rivers on the east coast of southern Africa, including those of South Africa (de Moor & Bruton 1988; Picker & Griffiths 2011; Skelton 1993). This fish has been introduced to many parts of the world for biological control (for insects and aquatic weeds) and aquaculture, and as an aquarium or bait fish (Canonico et al. 2005). In South Africa, this species has been intentionally and widely transported for angling and as a fodder fish for bass (de Moor & Bruton 1988; Picker & Griffiths 2011). Consequently, extralimital populations in South Africa now occur inland and in southern and west coastal rivers, as well as in Namibia, where this species is not native (de Moor & Bruton 1988; Picker & Griffiths 2011; Skelton 1993).

Scenarios and modes of introduction:

The introduction route of *Oreochromis mossambicus* follows scenario 6. Spread in South Africa and into Namibia was intentionally aided by humans and was likely facilitated by land transportation.

Management actions:

The spread of this species into Namibia could have been prevented if, following a pre-border risk assessment, the species had been added to a prohibited species list and surveillance for the species had been implemented (Figure 5).

FIGURE 6-A2: Details on the introduction and spread of *Oreochromis mossambicus*, the relevant introduction route scenario and modes of transport, and the management actions that could have prevented or mitigated the invasion.

8

Biological invasions in South African National Parks

Authors:
Llewellyn C. Foxcroft[1,2] ⓘ
Nicola J. van Wilgen[1,2]
Johan A. Baard[1]
Nicolas S. Cole[1] ⓘ

Affiliations:
[1]South African National Parks (SANParks), South Africa

[2]Centre for Invasion Biology, Department of Botany and Zoology, Stellenbosch University, South Africa

Corresponding author:
Llewellyn Foxcroft,
Llewellyn.foxcroft@sanparks.org

Objectives: A core objective in South African National Parks (SANParks) is biodiversity conservation and the maintenance of functional ecosystems, which is compromised by alien species invasions. The 2016 Alien and Invasive Species Regulations of the *National Environmental Management: Biodiversity Act* (NEM:BA) requires landowners to develop management plans for alien and invasive species, as well as report on the status and efficacy of control.

Method: To compile the species list, we started with the 2011 SANParks alien species list. Name changes were updated and SANParks ecologists and park managers contacted to verify the species lists and add new records. Species reported by external experts were added in the same manner. The management programme costs and species controlled per park per year were extracted from SANParks' Working for Water programme database.

Results: SANParks has listed 869 alien and extra-limital species, including 752 plants and 117 animals, increasing from 781 alien species in 2011. About R 590 million has been spent by the Working for Water/Biodiversity Social Programmes since 2000/2001. Of the species recorded, 263 are listed by NEM:BA, including 12 Category 1a species, 184 Category 1b species, 28 Category 2 species and 39 Category 3 species.

Conclusion: While large clearing programmes have been maintained since at least 1998, improving prioritisation is necessary. We provide a short synopsis of (1) what alien species are present in SANParks, (2) the species and parks that management has focused on, (3) the implications of the NEM:BA Invasive Alien Species Regulations and (4) future developments in monitoring.

Introduction

The South African National Parks (SANParks) estate includes 19 national parks across South Africa, covering about 39 000 km², which includes fynbos, forest, arid and sub-tropical savanna (Figure 1). SANParks' primary mandate is biodiversity conservation and the maintenance of heritage assets and thereby providing human benefits (SANParks 2015). The role protected areas (PAs) are required to play in maintaining biodiversity and ecosystem services is becoming increasingly important as landscapes become progressively fragmented (Watson et al. 2014). Changes in land use types surrounding PAs lead to habitat transformation that is not always compatible with conservation. Higher human population density in areas surrounding SANParks' PAs has been shown to be a significant predictor of invasions (Spear et al. 2013). These source populations around urban centres drive continual input into the system, increase propagule pressure and ultimately heighten the risk of impacts to PAs.

The insidious nature of invasions, typical lag phase (Crooks 2011) and the difficulty of detecting the resulting ecological change mean that concerns are often only raised and actions taken once the invasion is well advanced. Moreover, more quantitative data are needed to show that the observed impact on response variables (e.g. on plant richness) manifests as an impact on ecosystem processes (Hulme et al. 2013). Nonetheless, there are numerous examples that can be used as indicators of how alien species impacts may effect PAs, which should be used to illustrate concerns and motivate for control in the early stages. Studies in PAs that are considered intact natural ecosystems show that invasive alien plants (IAPs) dominate and displace native species and communities, alter fire regimes, directly or indirectly alter biogeochemistry and nutrient cycles and can use significantly more water than native vegetation because of the densities they reach (Foxcroft et al. 2013; Le Maitre, Versfeld & Chapman 2000).

Chromolaena odorata in Hluhluwe-iMfolozi Game Reserve (South Africa), for example, has affected spiders and mammals. In invaded areas, native spider assemblages changed in abundance, diversity and estimated species richness, but these changes were reversed following clearing (Mgobozi, Somers & Dippenaar-Schoeman 2008). Both small and large mammals had higher

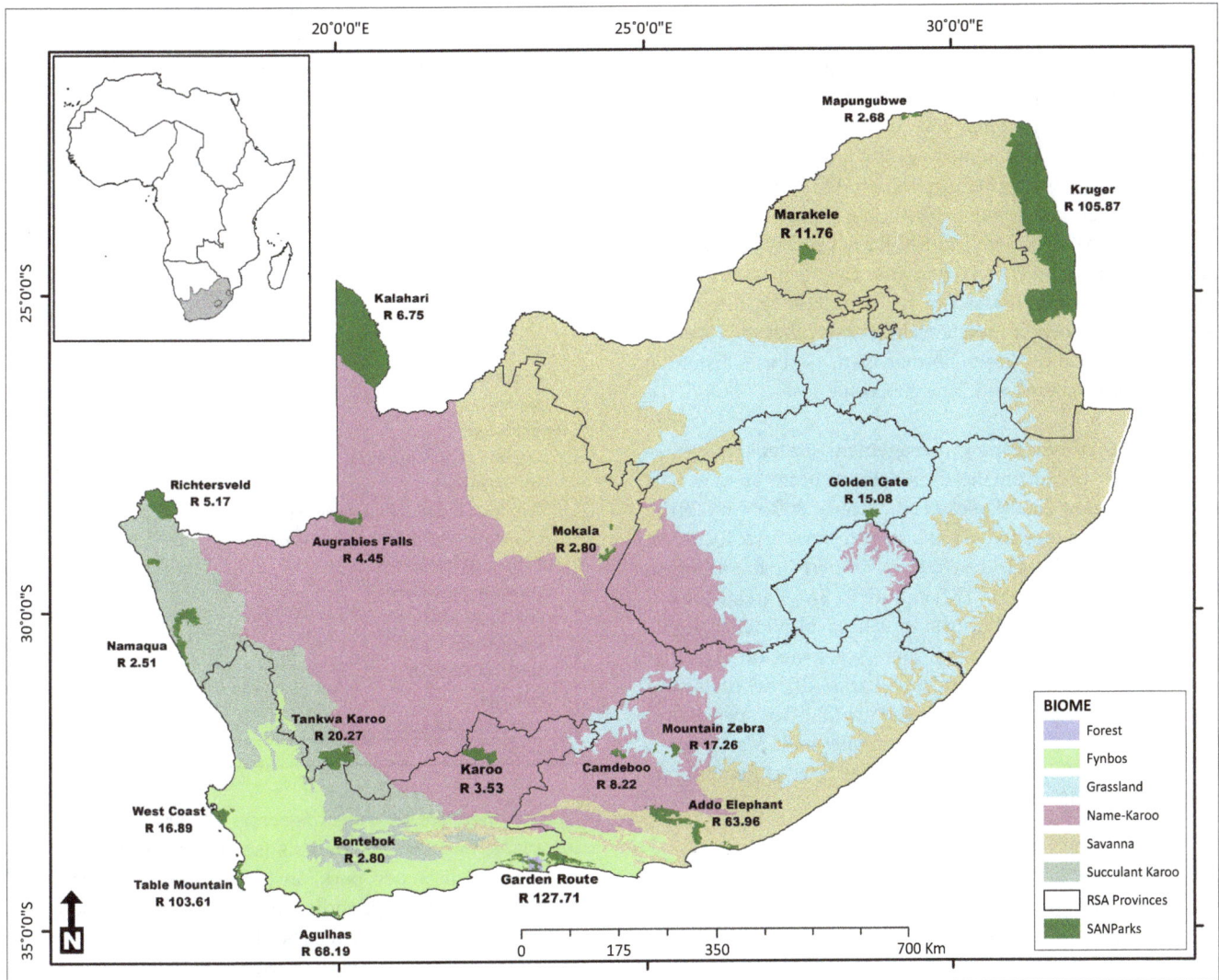

Source: Courtesy of S. MacFadyen
The values indicate annual funding spent by Working for Water (Biodiversity Social Projects) in SANParks between the 2002/2003 and 2015/2016 financial years, in ZAR millions.

FIGURE 1: Distribution of South African National Parks' 19 protected areas.

species richness and diversity in uninvaded sites compared with invaded sites (Dumalisile 2008). Similarly, *Opuntia stricta* in Kruger National Park (Kruger; all parks hereafter given by name) significantly altered beetle assemblages (Robertson et al. 2011). Although now successfully under biological control, the example illustrates the species level effects that *O. stricta* could have if not managed. In the fynbos biome, the presence of IAPs alters the fuel load as well as the horizontal and vertical connectivity of fuel. This can increase fire intensity and spread (Chamier et al. 2012) with serious implications for ecosystems and the ability to manage fires and human safety (Van Wilgen, Forsyth & Prins 2012). The effects of invasive alien fish are well documented, often resulting in irreversible change to native species communities and ecosystem function (see Ellender & Weyl 2014 for a review in South Africa) and, therefore, the presence of 17 alien fish species across at least 10 parks is of great concern. Some of these species include bass (*Micropterus* spp.), common carp (*Cyprinus carpio*), Nile tilapia (*Oreochromis niloticus*) as well as its hybridised form with the indigenous Mozambique tilapia (*Oreochromis mossambicus*) (Woodford

et al. 2017). Feral animals are problematic in almost all parks. Although currently with low incidence, the potential of feral cats (*Felis silvestris catus*) to hybridise with African wild cats (*Felis silvestris lybica*) (Le Roux et al. 2015) is concerning. These examples paint a worrying picture of how invasive-species-led habitat transformation, ecosystem function impairment, loss of native biodiversity or genetically pure species could undermine the ability of SANParks to achieve its objectives and compromise its status.

The highly complex biophysical context of SANParks makes managing invasive alien species (IAS) across its PAs difficult, for example, the vast area, number of parks, the distribution across South Africa's biomes and the degree to which they are invaded makes planning challenging. Decisions about when, where and how to implement actions, therefore, need to be prioritised in line with available resources (e.g. Forsyth et al. 2012; Roura-Pascual et al. 2009), across parks and within key areas within each park (Forsyth & Le Maitre 2011), although current funding provision processes complicate prioritisation. Management also needs to take into account the complexities of species' distribution, abundance, spread, and the multiple interacting environmental and socio-economic factors (Roura-Pascual et al. 2009). Difficult decisions need to be made to trade-off benefits against losses for different ecosystems and different species, thereby accepting the fact that some negative impacts are inevitable in some areas or parks. An initial Analytical Hierarchy Process-driven assessment highlighted important criteria that should be considered in SANParks and recommended species that should receive management (Forsyth & Le Maitre 2011). While plants can be controlled, because of the kinds of species and ecosystems inhabited other taxa such as fish cannot in most cases be managed, posing substantial threats to ecosystems and indigenous species.

The process of developing management strategies requires two sources of information, namely accurate species lists and distribution data (Pyšek et al. 2013; Tu & Robison 2013). These data are needed to assess priorities and focus on species posing the greatest threat. As part of the strategic adaptive management (SAM) culture in SANParks (Roux & Foxcroft 2011) assessing past practices provides insights for continuous improvement. Here we focus on (1) what alien species are present in SANParks, 2) the costs and parks that management has focused on, (3) listed species and implications of the *National Environmental Management: Biodiversity Act* (NEM:BA) Invasive Alien Species Regulations and (4) future developments.

Methods

To compile the species list, we used the list in 'Alien species in South Arica's national parks' (Spear et al. 2011) as a starting point (data collection methods are provided in Spear et al. 2011). All the species were checked for name changes and then verified. SANParks botanists, ecologists and park managers were contacted and new species that had been positively identified since 2011 were added. Species reported by external experts were verified and added in the same manner. The control costs were extracted from SANParks' Working for Water programme database, as well as the species controlled per park per year.

Alien species in South African National Parks

The first comprehensive account of alien and invasive species in SANParks documented 781 species (including extra-limital and feral species, but excluding biological control agents as 'alien', Spear et al. 2011). The list comprised 655 plants and 115 animals. Current revisions based on (1) new species introductions, (2) updated nomenclature and (3) correcting for misidentified species have increased the list to a total of 869 species (Table 1). Of these, 752 are plants and 117 are animals (Table 1). The number of mammals (26) and insects (13) has not changed substantially. Two of the three parks with the highest number of plants changed marginally, while the number of listed plants in Garden Route increased from 171 to 251. Kruger's numbers increased to 363 plants (from 348) and Table Mountain to 243 species (from 239).

TABLE 1: Total alien plants and animals recorded across SANParks estate, reported by class.

Species	Total
Plants	**752**
Dicots	568
Monocots	152
Pinophytes	18
Ferns (Pteridopsida)	11
Cycads	3
Animals	**117**
Vertebrates	**54**
Mammals	26
Fish	17
Birds	9
Amphibians	1
Reptiles	1
Invertebrates	**63**
Slugs and snails (Gastropoda)	19
Insects (Insecta)	13
Collembola and relatives (Entognatha)	11
Crustaceans (Maxillopoda and Malacostraca)	5
Earthworms (Oligochaeta)	4
Sea squirts (Ascidiacea)	3
Bivalves (Bivalvia)	2
Millipedes (Diplopoda)	2
Spiders (Arachnida)	2
Anthozoa	1
Centipedes (Chilopoda)	1
All species	**869**

In total, there are 1878 records across all parks, of which 1622 are plants and 256 are animals (see Online Appendix 1 for full species lists per park, including kingdom, class and family, and Online Appendix 2 for a list of species, indicating subspecies, common names and class). At least 18 plant species occur in 10 or more parks; fortunately only a third of these represent major concerns, for example, *Pennisetum setaceum*, *Arundo donax*, *Lantana camara*, *Melia azedarach* and *Schinus molle*. Of greater concern is that except for four parks (Kalahari Gemsbok [3], Richtersveld [6], Namaqua [8] and West Coast [8]), almost all parks have large numbers of 'transformer' species (sensu McGeoch, Chown & Kalwij 2006; Richardson et al. 2000). Furthermore, even in the parks with few transformer species, the species present are often highly invasive. For example, *Parkinsonia aculeata*, *Prosopis glandulosa* and *S. molle* in Kalahari Gemsbok; *P. glandulosa* and *S. molle* in Richtersveld; and *Acacia cyclops* and *Acacia saligna* in West Coast (predominantly a problem in new sections that are in the process of being added to the park) (Online Appendix 1).

Management

Effectiveness, costs and challenges of invasive alien plant control

As early as the 1940–1950s, there have been efforts to control plant invasions in some areas now falling within SANParks' estate, for example, Table Mountain and Kruger. In what is now part of Table Mountain, control was initiated in 1941 but by the 1970s efforts were still considered unsuccessful (Macdonald et al. 1988). By the mid-1980s, 40% of the Cape of Good Hope Nature Reserve's (now incorporated into Table Mountain) annual budget was being used for IAP control, but the distribution continued to expand (Macdonald et al. 1988). In 2015/2016, the budget for alien plant clearing in Table Mountain totalled about R 22.7 million (Figure 2). As Table Mountain falls in a species rich region with about 2285 indigenous plant species, of which 158 are endemic and 141 appear on the Red Data List (SANParks 2016), ongoing efforts attest to the importance placed on bringing the IAPs to maintenance control levels, which has had substantial success in certain areas. For example, the current density of IAPs at Cape Point in Table Mountain (uninvaded to scattered individuals; Appendix 1, Figure 1-A1, TMNP Management Plan, SANParks 2016) are lower than previous decades, where up to 25% of the area was densely invaded with *A. cyclops* and related species (Taylor & Macdonald 1985; Taylor, Macdonald & Macdonald 1985). However, the inflexibility of clearing programmes to respond quickly to changing priorities (e.g. following fires) undermines attempts to reduce the density of IAPs (Van Wilgen & Wannenburgh 2016).

In Kruger, small-scale efforts date back to the mid-1950s, focusing on the control of *Melia azedarach*. In the early 1980s, Kruger created an Alien Plant Control Officer post and a team of 10 people to control the IAPs. However, the size of the problem proved too large and species such as *Lantana camara* and *O. stricta* continued to invade (Foxcroft & Freitag-Ronaldson 2007). The programme started expanding with funding from the Royal Netherlands Embassy who provided R 3 million between 1997 and 2000, and then with the initiation of the Working for Water programme in Kruger in 1997.

While individual parks provide resources from their own operational funds for the control of IAPs, the overriding majority of funding comes from the Expanded Public Works Program, through the Department of Environmental Affairs Natural Resource Management Program (Working for Water). Since the start of the 2002/2003 financial year, from when more detailed records have been kept, about R 590 million has been spent in SANParks on IAP control (Figure 2). Between 2011/2012 and 2014/2015, the annual budget fluctuated between R 60 million and R 75 million, with a R 42 million increase in 2015/2016 to a total of R 114 million (Figure 2). Between 2002/2003 and 2015/2016, about 80% of the total funding was spent across five parks, namely Garden Route (R 127 million), Kruger (R 105 million), Table Mountain (R 103 million), Agulhas (R 68 million) and Addo Elephant (R 64 million) (Figure 3). For the last 6 years (since 2009/2010), Garden Route has accounted for about 20% of the annual budget, increasing to 29% in 2015/2016. The IAP control programme for Kruger constituted about 46% of the total budget for 2002–2004, which decreased to 13.5% in 2015/2016 (Figure 4).

It would be disingenuous not to acknowledge the problems that have arisen and in a programme operating at such a large-scale are inevitable, but need to be addressed promptly. For example, a recent assessment of the costs of controlling IAPs in 25 PAs (not only SANParks) in the Cape Floristic Region argued that without careful prioritisation and substantial increases in funding, the likelihood of achieving successful control is low (Van Wilgen et al. 2016; Van Wilgen & Wannenburgh 2016). In addition, evidence from the Garden Route suggests that significant management intervention is required to increase the impact and effectiveness of funds that are available (Kraaij et al. 2017). Additional challenges arise from the numerous parcels of land being added to national parks as part of the PA expansion strategy. In many instances, alien species have not previously been managed on the new land, which is also often transferred without accompanying financial resources for IAP management or at best a once-off payment

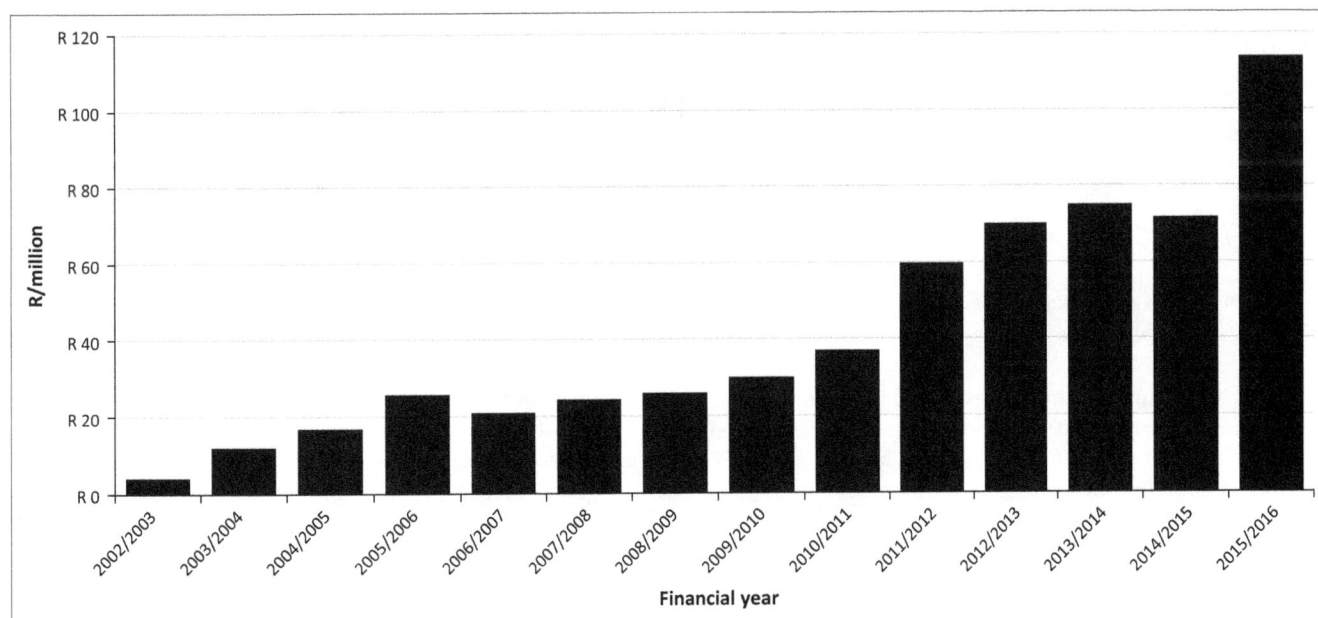

FIGURE 2: Annual funding spent by Working for Water (Biodiversity Social Projects) in SANParks between the 2002/2003 and 2015/2016 financial years.

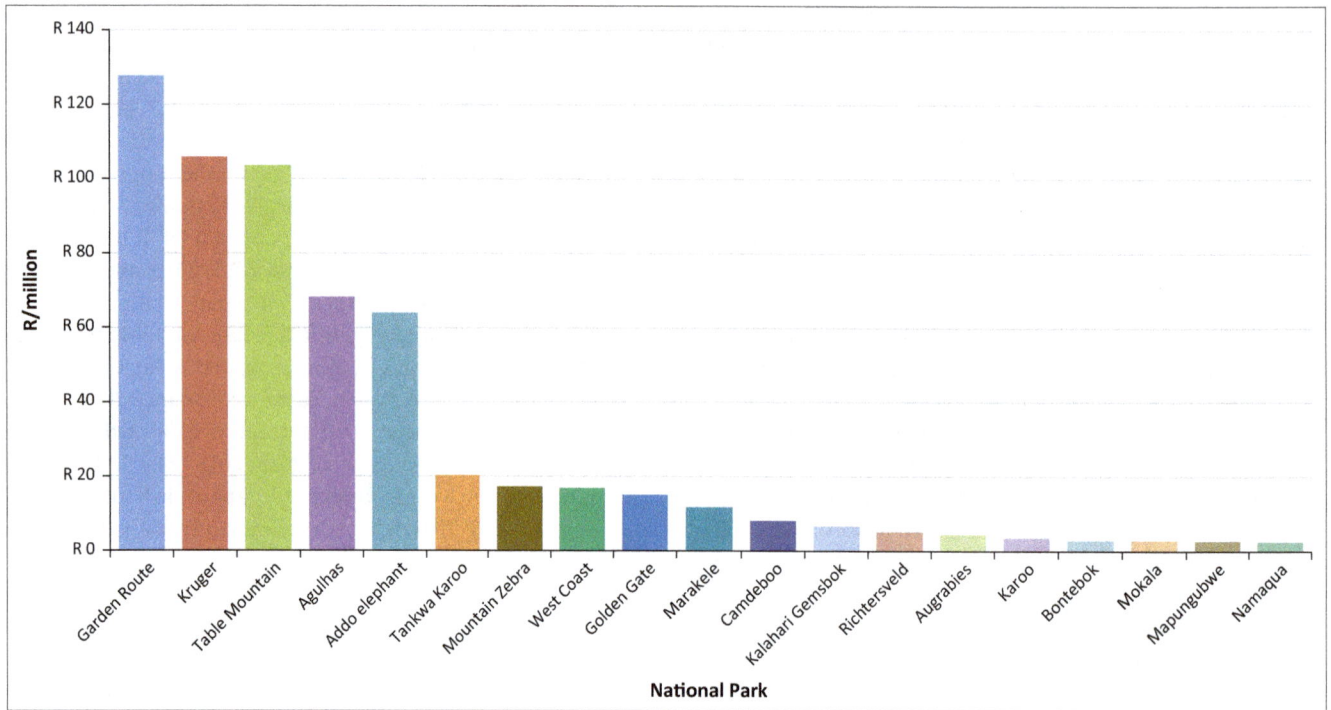

FIGURE 3: Total funding spent between 2002/2003 and 2015/2016 by Working for Water (Biodiversity Social Projects) in SANParks per national park.

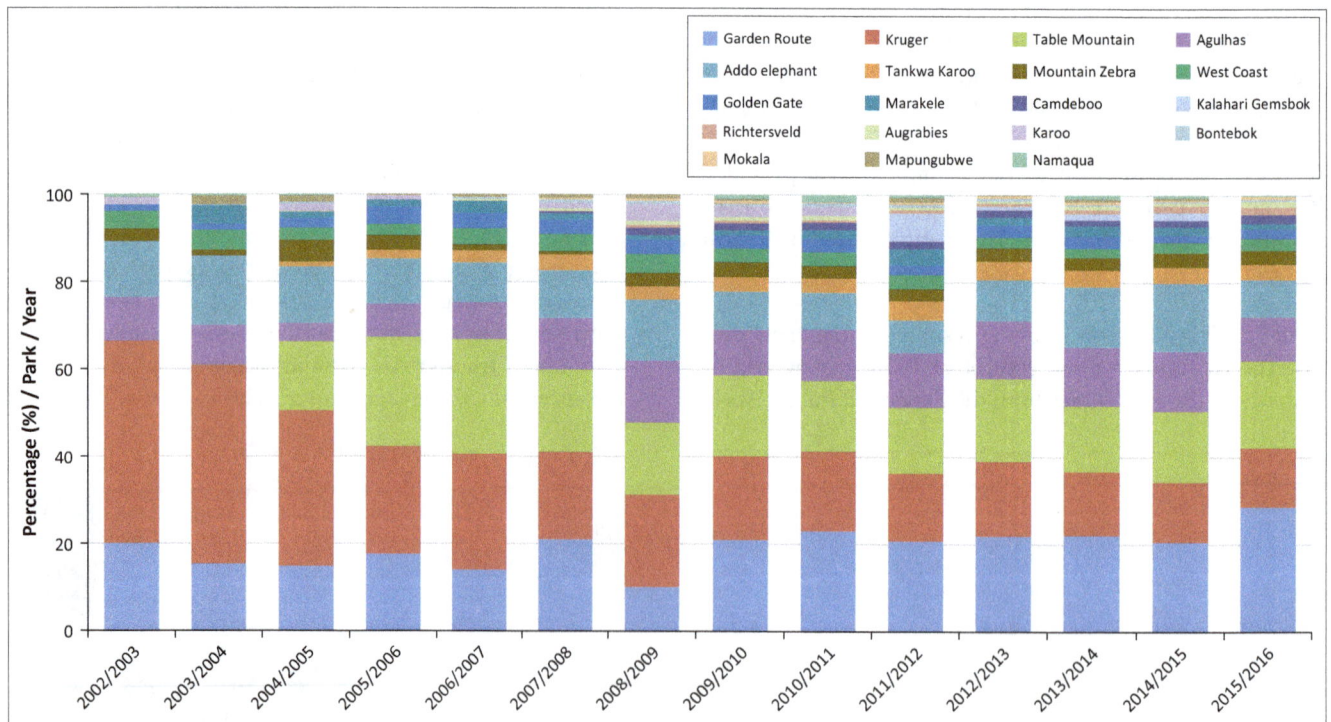

The colours shown in the legend for each park correspond to Figure 3

FIGURE 4: The percent of the total funding spent in each park per financial year.

for short-term management. For example, in the large areas where commercial forestry is withdrawing from the Garden Route, a once-off amount of R 5.335 million (plus R 4.438 million outstanding) was received from the landowners when transferred to SANParks; however, no additional funds for long-term management have been made available. A similar situation exists in Tokai, Table Mountain, where the problem had been exacerbated by fires in 2015, which burnt most of the remaining plantation areas. Funding is

also required to take advantage of these unexpected events as fire stimulates the germination of seeds and burning after 1 to 2 years kills seedlings before they mature (Van Wilgen, Forsyth & Prins 2012). Despite numerous challenges, the need to revise and align management plans with the NEM:BA regulations, along with improved species lists, increasing distribution data on key species and assessments of past programmes provides the opportunity to strategically plan future directions.

Biological control of invasive alien plants – Introduction and efficacy

Biological control is an essential component of any long-term IAP management programme and perhaps more so in PAs where there is resistance to the use of herbicides (Van Driesche & Center 2013). In addition manual or chemical control of IAPs over extensive areas may not be feasible even where large amounts of funding are provided. There are also numerous successes around the world and especially in South Africa (Moran, Hoffmann & Zimmermann 2005; Zachariades et al. 2017). In light of this, biological control of IAPs is potentially the only control option available to many PAs. In SANParks, 38 biological control agents (35 invertebrates and 3 fungi), totalling 47 records across all parks, have been released for the management of IAPs (Online Appendix 3). Of these, 23 have been released in Kruger, 12 in Garden Route and 5 recorded in Camdeboo. In some parks, for example, Kruger, biological control has been used since the mid-1980s, where it has been highly successful in the management of *Pistia stratiotes* (water lettuce) (Foxcroft & Freitag-Ronaldson 2007; see also Hill & Coetzee 2017). Additionally, biological control using *Dactylopius opuntiae* (cochineal) together with *Cactoblastis cactorum* (cactoblastis moth) is the primary control method for *O. stricta* (Paterson et al. 2011), with chemical control used in new foci outside of the core management zones (cf. Kaplan et al. 2017). Similarly in Camdeboo and Addo Elephant, biocontrol using *Dactylopius* spp. is one of the core management approaches for the cacti. Biocontrol agents have been released on invasive *Acacia* spp. in Addo Elephant, Agulhas, Garden Route and Table Mountain.

Management of extra-limital, alien and invasive animals

Of the alien animals in SANParks, 23 are extra-limital species (which we defined as species that are indigenous to South Africa but that have been introduced into national parks outside their historical ranges, Spear et al. 2011). Programmes are underway to remove species, for example, nyala (*Tragelaphus angasii*) and blesbuck (*Damaliscus pygargus*) have been removed from Marakele (SANParks 2014). Blue wildebeest (*Connochaetes taurinus*), springbuck (*Antidorcas marsupialis*) and gemsbok (*Oryx gazella*) are planned for removal from West Coast, while annual culling of warthogs (*Phacochoerus africanus*) takes place in Addo Elephant. As animals are removed from the parks, ongoing surveillance is required to ensure that no animals were missed and that reinvasions do not occur. However, in many instances eradication of invasive alien animals is likely to be impossible, for example, invertebrate species such as collembola (springtails) that occur in the soil and which are further complicated as native taxa are often poorly described. Some species require specialised training to assist in control operations, for example, in the case of highly aggressive German wasps (*Vespula germanica*) where Table Mountain has collaborated with specialists from The City of Cape Town in the removal of nests (L. Stafford [Environmental Resource Management Department, City of Cape Town] pers. comm., May 2016). Few alien animals have been introduced into SANParks, with some present prior to park proclamation,

often for many decades, for example, Himalayan tahr (*Hemitragus jemlahicus*), sambar (*Rusa unicolor*) and grey squirrel (*Sciurus carolinensis*) in Table Mountain (Online Appendix 1). Vertebrates have been shown to have a high probability of becoming invasive once introduced (Jeschke & Strayer 2005). Once they have become invasive, the ability to control them is usually extremely difficult, labour intensive (e.g. fish; Ellender & Weyl 2014) and often compounded by fierce public resistance [e.g. the case of mallard ducks (*Anas platyrhynchos*) in Cape Town, Erasmus 2013].

NEM:BA listed species and implications for SANParks

The management of IAS in SANParks is governed by two primary policy instruments: the *National Environmental Management Protected Areas Act* (No. 57 of 2003), which requires alien species to be included in park management plans, and the *National Environmental Management: Biodiversity Act* (Act No. 10 of 2003, hereafter NEM:BA), which through its associated Alien and Invasive Species Regulations (2016) requires landowners to develop specific plans for the control, eradication and monitoring of alien and invasive species.

Of the 869 species in SANParks (including extra-limital and feral species), 263 are included in the NEM:BA alien and invasive species regulations (Table 2). This poses significant challenges for the management of IAS in the organisation (1) in the complexity of developing strategic plans for the numerous listed species across 19 national parks and (2) because of these extensive species lists, implementation thereof, even where the best available strategies have been developed. The NEM:BA regulations include four categories that aim to prevent introduction, manage existing species populations and regulate the use of commercially important but potentially IAS. Specifically, Category 1a includes 'Invasive species which must be combatted and eradicated' and 1b includes 'Invasive species which must be controlled and wherever possible, removed and destroyed'. The SANParks list includes 12 Category 1a species and 184 Category 1b species (Table 2). Table Mountain, Kruger and Garden Route each have more than 75 Category 1b species present. Since 2001/2002, these parks have however only focused on a small number of NEM:BA listed species, for example, 47 species (60%) in Table Mountain and 37 species each in Kruger (40%) and Garden Route (51%); typically less than half this number of listed species are worked on in any given year. For example, 15 (21%) and 29 (37%) Category 1b species were worked on in Garden Route and Table Mountain in 2015/2016, respectively. Category 2 species are 'Invasive species, or species deemed to be potentially invasive, in which a permit is required to carry out a restricted activity' and Category 3 species may only be allowed under specific terms. Category 2 species are generally used in commercial plantations and being granted exemptions under Category 3 is, as a PA agency, highly unlikely. This in effect adds 28 and 39 species to Category 1b, which then requires control in the same manner as Category 1b. Therefore, with the current funding of about R 110 million per year (Figure 2), SANParks will not be able to expand the current programme to eradicate or actively control

TABLE 2: The total number of alien species per park and the number of these species listed in each Category of the NEM:BA regulations.

Park	Total species per park				NEM:BA listed species										
	# species	# plants	# animals	# biocontrol	Total spp.	Total plants	1a plants	1b plants	2 plants	3 plants	Total animals	1a animals	1b animals	2 animals	3 animals
Addo Elephant	149	124	24	1	69	63	2	52	6	3	6	0	3	1	2
Agulhas	95	82	10	3	55	50	0	38	6	6	5	0	3	2	0
Augrabies Falls	59	54	5	0	31	29	0	24	3	2	2	0	0	0	2
Bontebok	95	80	15	0	37	28	0	23	5	0	9	0	4	2	3
Camdeboo	56	45	6	5	35	34	0	28	5	1	1	0	1	0	0
Garden Route	283	251	20	12	110	98	2	72	11	13	12	0	3	4	5
Golden Gate Highlands	89	77	12	0	63	57	0	44	9	4	6	0	2	1	3
Kalahari Gemsbok	21	18	3	0	11	9	0	8	0	1	2	0	0	0	2
Karoo	35	22	13	0	23	19	1	14	4	0	4	0	0	1	3
Kruger	415	363	28	22	130	118	1	93	7	17	12	0	9	0	3
Mapungubwe	49	40	9	0	33	29	0	27	2	0	4	0	1	0	3
Marakele	28	20	8	0	24	18	0	14	2	2	6	0	2	1	3
Mokala	34	27	6	1	22	21	0	16	2	3	1	0	0	0	1
Mountain Zebra	111	101	8	2	49	45	1	36	6	2	4	0	1	0	3
Namaqua	23	18	5	0	14	13	0	10	2	1	1	0	0	0	1
Richtersveld	21	15	6	0	15	12	1	9	1	1	3	0	0	0	3
Table Mountain	295	243	51	1	126	114	8	78	13	15	12	0	4	4	4
Tankwa Karoo	33	27	6	0	20	18	0	13	4	1	2	0	0	0	2
West Coast	36	15	21	0	19	11	0	8	1	2	8	0	0	2	6

Total spp., species per park.

many listed species (Table 2 and Online Appendix 2) and some form of prioritisation and triage will be necessary. However, strategies at the corporate level and park management plans have been developed and are being revised to determine the best approach for each park or group of listed species.

An additional point in the NEM:BA regulations, which states that any 'form of trade, propagation or planting is strictly prohibited', also applies to SANParks, especially with regard to nurseries selling native plants in some parks or the use of ornamental species in tourist facilities and staff accommodation. Ornamental species are well known to be an important pathway of invasion into PAs (e.g. in Kruger; Foxcroft, Richardson & Wilson 2008). These pathways can, however, be managed and nurseries are restricted to indigenous species only. For example, ornamental alien plants and landscaping in Kruger are strictly regulated and should be followed by the other national parks. The revised standard operating procedure allows only those indigenous species naturally occurring within a particular landscape for use in the tourist camps, while other non-invasive but alien ornamental species are being phased out (Kruger National Park [KNP] 2015). Follow-up control and awareness will have to remain a key part of the programme in the long term. However, from 1999 when the first comprehensive ornamental plant survey was conducted in Kruger, significant progress in managing the species used in camps and staff gardens has been achieved (Foxcroft, Richardson & Wilson 2008).

Future developments – Monitoring and indicators

Two key processes that have been lacking or only partially developed in some parks are outcomes or ecologically based monitoring and standardised operating procedures or guiding frameworks. A core element of the SAM approach

that SANParks has adopted is monitoring and the concept of thresholds of potential concern (TPC) (Roux & Foxcroft 2011). The TPC approach has guided management interventions and drawn attention to important potentially invasive species, highlighting new or potential introductions and new foci of a species in a park (e.g. Foxcroft 2009). For example, where the TPCs were implemented in Kruger, a new introduction or increase in distribution of a species would breach a pre-defined threshold. This would trigger a process whereby the Kruger Conservation Management department was officially notified, the most appropriate course of action determined, implemented, and feedback given to the department until satisfactorily dealt with (Foxcroft & Downey 2008). In addition, as part of SANParks' biodiversity monitoring programme (McGeoch et al. 2011) an IAS monitoring programme was developed (Foxcroft & McGeoch 2011). This programme provides seven headline indicators against which progress in management of invasive species is measured over time, frequently a 3- or 5-year period. These include (1) the number of alien species in a park, (2) the number of populations, (3) the coverage or density of each species per park, (4) the total area of park invaded, (5) the number of species of special concern threatened by IAS, (6) the percentage of invasive species being actively controlled and (7) the percentage area controlled with abundance maintained at an acceptable threshold. These indicators can be disaggregated into finer level indicators, for example, for number of alien species in a park, temporal trends in changes to species can be listed by taxon, status, transformer or extra-limital species (Foxcroft & McGeoch 2011). Detailed monitoring of change and response to IAS on ecosystems will likely be implemented in the form of focused scientific studies, for example, using indicator species at a fine spatial scale. Successful implementation of the indicators is, however, contingent on extensive and detailed monitoring and will not be possible without the requisite resources.

The collection of baseline data on species distribution and abundance at a fine scale using stratified sampling methods has been initiated in some parks (e.g. Table Mountain, Bontebok, Agulhas and West Coast). The data include species, age class, abundance and control status, for example, in Table Mountain and Bontebok (e.g. Figure 5a–d). Species-specific distribution monitoring for priority species is also conducted, for example, *Parthenium hysterophorus* (Figure 5e) and *O. stricta* (Figure 5f) in Kruger, which is used to inform preparation of management plans. This form of data is highly valuable not only in planning but also in monitoring the impact of management interventions. Baseline data collection beyond the parks mentioned and repeated monitoring will, however, require sustainable funding and dedicated human capital, which should be accounted for in future budgeting and funding applications.

Conclusion

Alien species management in SANParks is at an important junction, providing not only many opportunities but also substantial challenges and threats. The NEM:BA regulations provide an opportunity for much needed reorganisation and prioritisation of key targets, whether priority species or areas. However, increases in the number of alien species that are

mandated to receive attention mean that difficult decisions are required to determine optimal allocation of funds. It is likely that additional funding streams will be required to maintain the status of areas currently being managed, as well as resources for the new species and areas that will be prioritised. The threat however is that should funding be reduced or reallocated, some areas that were under control will return to an invaded state without some level of follow-up or maintenance control.

Protected areas form a nexus between conservation and society. With a broad constituency across the 5.2 million annual visitors (SANParks 2015), SANParks should play a large role in creating awareness of alien species invasions. For South African visitors, SANParks should also be a source of awareness and information not only of the threat of IAS but also the NEM:BA regulations to gain additional support for implementation.

Globally, PA organisations are recognising the risk posed to the biodiversity entrusted to their care and the severe state of invasions already found in some PAs (see examples in Foxcroft et al. 2013). While there are many challenges that PAs may face in trying to implement management programmes (Tu & Robison 2013), there are numerous tools

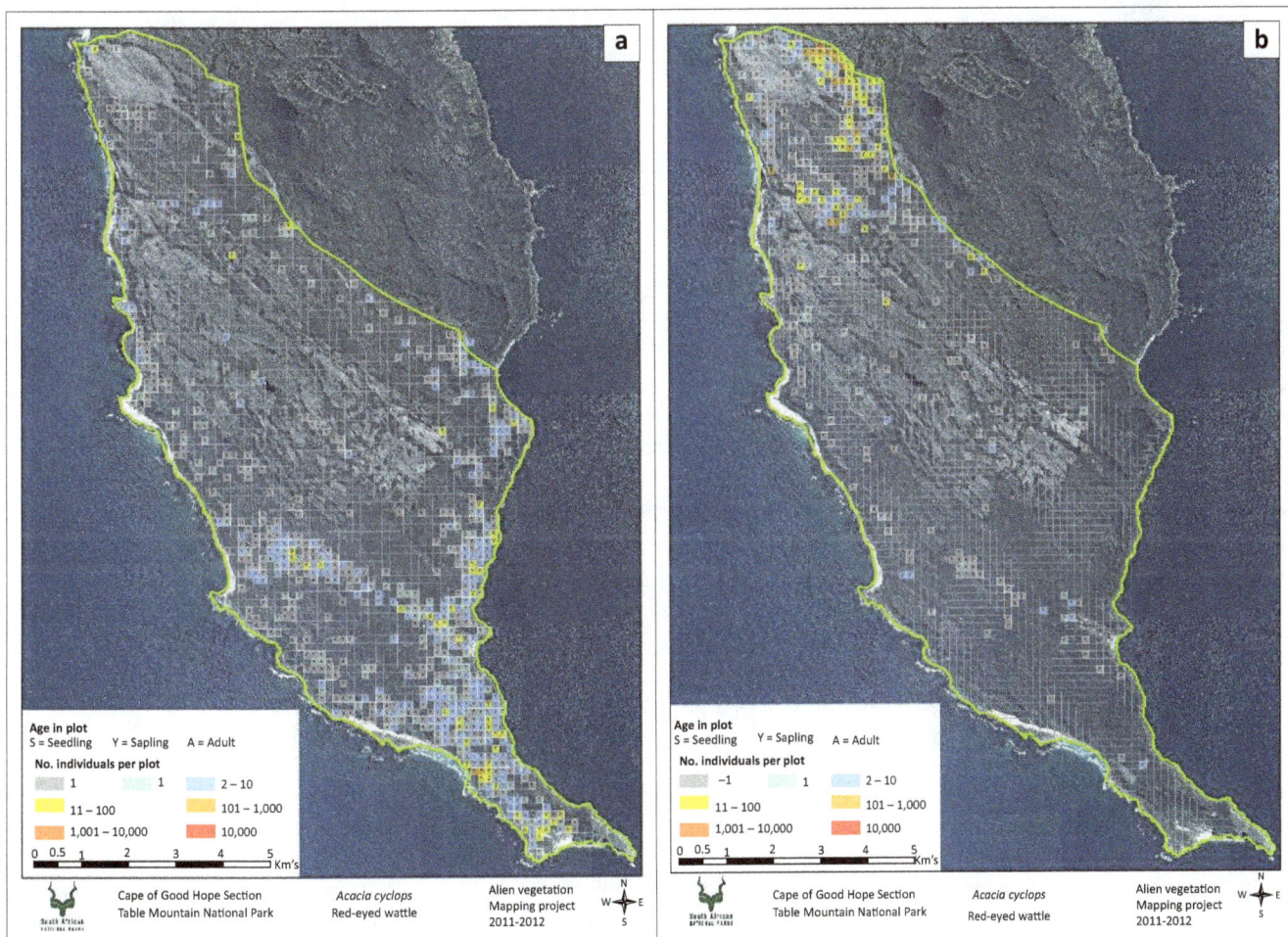

Source: Figure a–d courtesy of C. Cheney; e–f courtesy of S. MacFadyen

(a) *Acacia cyclops* and (b) *A. saligna* in Table Mountain, (c) *A. cyclops* and (d) *A. saligna* in Bontebok, (e) *Parthenium hysterophorus* and (f) *Opuntia stricta* in Kruger.

FIGURE 5: Invasive alien plant distribution surveys.

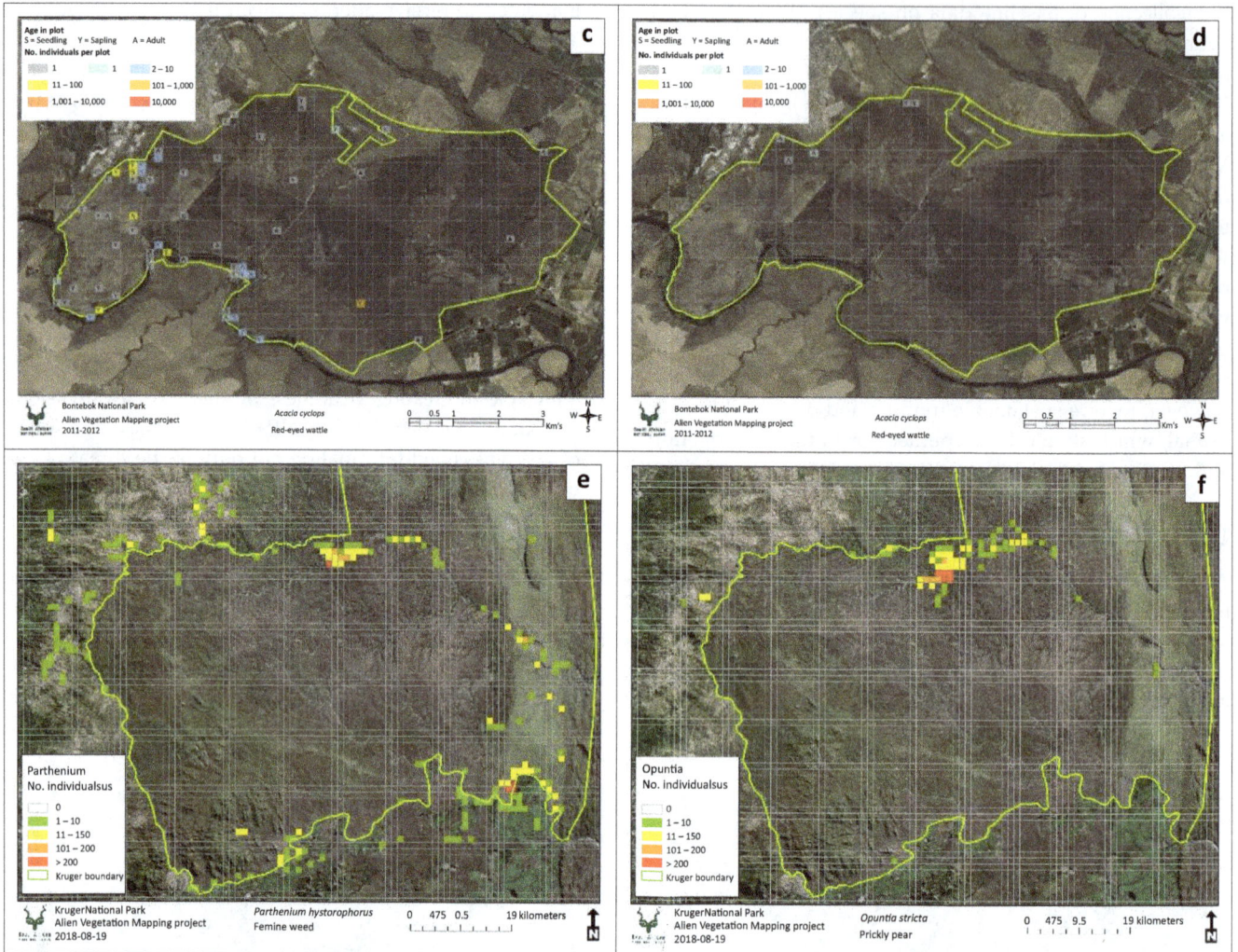

Source: Figure a–d courtesy of C. Cheney; e–f courtesy of S. MacFadyen

(a) *Acacia cyclops* and (b) *A. saligna* in Table Mountain, (c) *A. cyclops* and (d) *A. saligna* in Bontebok, (e) *Parthenium hysterophorus* and (f) *Opuntia stricta* in Kruger.

FIGURE 5 (Continues...): Invasive alien plant distribution surveys.

and examples that can be used to assist managers in developing approaches to managing IAS. However, even where management approaches are in place, an ubiquitous problem is the lack of monitoring and the basic data, such as species lists and distribution data, required to inform programmes and assess progress. Unless larger proportions of funding are allocated to formal monitoring programmes, with long-term commitments, the sustainability of large control programmes may be in jeopardy.

SANParks is in the unenviable position of having recorded 869 alien species, with extensive alien plant lists such as 251 species in Garden Route and 363 in Kruger. Moreover, 263 species found in SANParks are listed in the NEM:BA alien and invasive species regulations. However, SANParks is acutely aware of the status and has been implementing a large-scale management programme in a bid to minimise the potential impacts to biodiversity. The organisation is instituting processes and frameworks to assist in improving planning and implementing monitoring programmes to determine trends in future progress. This review can therefore be used in various ways by providing an updated status and species list against which indicators can be assessed for

detecting trends in invasion, providing the information required as part of the National Status Report and providing a basis for evaluating management implementation with a view to ongoing improvements (Wilson et al. 2017).

Acknowledgements

L.C.F. thanks South African National Parks, the DST-NRF Centre of Excellence (C•I•B) for Invasion Biology and Stellenbosch University, and the National Research Foundation of South Africa (project numbers IFR2010 041400019 and IFR160215158271). We thank T. Thwala for assisting in checking the KNP species list, C. Cheney for commenting on the Bontebok species list, S. Engel for comments on the Agulhas list, H. Malgas for comments on the Tankwa list and E. van Wyk for providing information on species that South African National Biodiversity Institute recently detected within SANParks. We sincerely thank 'Working for Water' for the long-term support of invasive alien species control in SANParks. We thank two anonymous reviewers and the editor, M. Gaertner, for their constructive comments that helped improve the article.

Competing interests

The authors declare that they have no financial or personal relationships that may have inappropriately influenced them in writing this article.

Authors' contributions

L.C.F. was the project leader, collected and checked species data and led the writing of the manuscript. N.J.v.W. collated and corrected the species lists, analysed the lists and wrote the manuscript. J.A.B. collated data and checked the species lists. N.S.C. provided data on costs of control and list of species controlled by Biodiversity Social Projects.

References

Chamier, J., Schachtschneider, K., Le Maitre, D.C., Ashton, P.J. & van Wilgen, B.W., 2012, 'Impacts of invasive alien plants on water quality, with particular emphasis on South Africa', Water SA 38, 345–356. https://doi.org/10.4314/wsa.v38i2.19

Crooks, J.A., 2011, 'Lag times', in D. Simberloff & M. Rejmánek (eds.), Encyclopedia of biological invasions, pp. 404–410, University of California Press, Berkeley, CA.

Dumalisile, L., 2008, 'The effects of Chromolaena odorata on mammalian biodiversity in Hluhluwe-iMfolozi Park', MSc thesis, University of Pretoria.

Ellender, B.R. & Weyl, O.L.F., 2014, 'A review of current knowledge, risk and ecological impacts associated with non-native freshwater fish introductions in South Africa', Aquatic Invasions 9, 117–132. https://doi.org/10.3391/ai.2014.9.2.01

Erasmus, E., 2013, 'Invasive mallards must take a duck, says city', The Tygerburger, Durbanville, 23 October, p. 26.

Forsyth, G.G., Le Maitre, D.C., O'Farrell, P.J. & van Wilgen, B.W., 2012, 'The prioritisation of invasive alien plant control projects using a multi-criteria decision model informed by stakeholder input and spatial data', Journal of Environmental Management 103, 51–57. https://doi.org/10.1016/j.jenvman.2012.01.034

Forsyth, G.G. & Le Maitre, D.C., 2011, Prioritising national parks for the management of invasive alien plants: Report on the development of models to prioritise invasive alien plant control operations, CSIR Natural Resources and the Environment Report number: CSIR/NRE/ECO/ER/2011/0036/B, CSIR, Stellenbosch.

Foxcroft, L.C., 2009, 'Developing thresholds of potential concern for invasive alien species: Hypotheses and concepts', Koedoe 50(1), Art. #157, 1–6. https://doi.org/10.4102/koedoe.v51i1.157

Foxcroft, L.C. & Downey, P.O. 2008. 'Protecting biodiversity by managing alien plants in national parks: Perspectives from South Africa and Australia', in B. Tokarska-Guzik, J.H. Brock, G. Brundu, L. Child, C.C. Daehler & P. Pyšek (eds.), Plant invasions: Human perception, ecological impacts and management, pp. 387–403, Backhuys Publishers, Leiden.

Foxcroft, L.C. & Freitag-Ronaldson, S., 2007, 'Seven decades of institutional learning: Managing alien plant invasions in the Kruger National Park, South Africa', Oryx 41, 160–167. https://doi.org/10.1017/S0030605307001871

Foxcroft, L.C. & McGeoch, M.A., 2011, South African National Parks Biodiversity Monitoring Programme: Alien and invasive species, Scientific Report 09/2011, Kruger National Park, Skukuza.

Foxcroft, L.C., Pyšek, P. Richardson, D.M., Pergl, J. & Hulme, P.E., 2013, 'The bottom line: Impacts of alien plant invasions in protected areas', in L.C. Foxcroft, P. Pyšek, D.M. Richardson & P. Genovesi (eds.), Plant invasions in protected areas. Patterns, problems and challenges, pp. 19–41, Springer, Dordrecht.

Foxcroft, L.C., Richardson, D.M. & Wilson, J.R.U., 2008, 'Ornamental plants as invasive aliens: Problems and solutions in Kruger National Park, South Africa', Environmental Management 41, 32–51. https://doi.org/10.1007/s00267-007-9027-9

Hill, M.P. & Coetzee, J.A., 2017, 'The biological control of aquatic weeds in South Africa: Current status and future challenges', Bothalia 47(2), a2152. https://doi.org/10.4102/abc.v47i2.2152

Hulme, P.E., Pyšek, P., Jarošík, V., Pergl, J., Schaffner, U. & Vilá, M., 2013, 'Bias and error in understanding plant invasion impacts', Trends in Ecology and Evolution 28, 212–218. https://doi.org/10.1016/j.tree.2012.10.010

Jeschke, J.M. & Strayer, D.L., 2005, 'Invasion success of vertebrates in Europe and North America', Proceedings of the National Academy of Sciences 102, 7198–7202. https://doi.org/10.1073/pnas.0501271102

Kaplan, H., Wilson, J.R.U., Klein, H., Henderson, L., Zimmermann, H.G., Manyama, P. et al., 2017, 'A proposed national strategic framework for the management of Cactaceae in South Africa', Bothalia 47(2), a2149. https://doi.org/10.4102/abc.v47i2.2149

Kraaij, T., Baard, J.A., Rikhotso, D.R., Cole, N.S. & Van Wilgen, B.W., 2017, 'Assessing the efficiency of invasive alien plant management in a large fynbos protected area', Bothalia 47(2), a2105. https://doi.org/10.4102/abc.v47i2.2105

Kruger National Park (KNP), 2015, Protocol for the management of ornamental alien plants and landscaping in all developed areas of the Kruger National Park, South African National Parks Reference Number: 16/Pr-KNP Management, Kruger National Park, Skukuza.

Le Maitre, D.C., Versfeld, D.B. & Chapman, R.A., 2000, 'The impact of invading alien plants on surface water resources in South Africa: A preliminary assessment', Water SA 26, 397–408.

Le Roux, J.J., Foxcroft, L.C., Herbst, M. & MacFadyen, S., 2015, 'Genetic status of the African wildcat (Felis silvestris lybica) in South Africa: The role of protected areas in conserving genetic purity', Ecology and Evolution 5, 288–299. https://doi.org/10.1002/ece3.1275

Macdonald, I.A.W., Graber, D.M., Debenedetti, S., Groves, R.H., Fuentes, E.R., 1988, 'Introduced species in nature reserves in Mediterranean-type climatic regions of the world', Biological Conservation 44, 37–66.

McGeoch, M.A., Chown, S.L. & Kalwij, J.M., 2006, 'A global Indicator for biological invasion', Conservation Biology 20, 1635–1646. https://doi.org/10.1111/j.1523-1739.2006.00579.x.

McGeoch, M.A., Dopolo, M., Novellie, P., Hendriks, H., Freitag–Ronaldson, S., Ferreira, S., et al., 2011, 'A strategic framework for biodiversity monitoring in SANParks', Koedoe 53(2), Art. #991, 1–10. https://doi.org/10.4102/koedoe.v53i2.991

Mgobozi, M.P., Somers, M.J. & Dippenaar-Schoeman, A.S., 2008, 'Spider responses to alien plant invasion: The effect of short- and long-term Chromolaena odorata invasion and management', Journal of Applied Ecology 45, 1189–1197.

Moran, V.C., Hoffmann, J.H. & Zimmermann, H.G., 2005, 'Biological control of invasive alien plants in South Africa: Necessity, circumspection, and success', Frontiers in Ecology and the Environment 3, 71–77. https://doi.org/10.1890/1540-9295(2005)003[0071:BCOIAP]2.0.CO;2

Paterson, I.D., Hoffmann, J.H., Klein, H., Mathenge, C.W., Neser, S. & Zimmermann, H.G., 2011, 'Biological control of Cactaceae in South Africa', African Entomology 19, 230–246. https://doi.org/10.4001/003.019.0221

Pyšek, P., Genovesi, P., Pergl, J., Monaco, A. & Wild, J., 2013, 'Invasion of protected areas in Europe: An old continent facing new problems', in L.C. Foxcroft, P. Pyšek, D.M. Richardson & P. Genovesi (eds.), Plant invasions in protected areas. Patterns, problems and challenges, pp. 209–240, Springer, Dordrecht.

Richardson, D.M., Pyšek, P., Rejmánek, M., Barbour, M.G., Panetta, F.D. & West, C.J., 2000, 'Naturalization and invasion of alien plants: Concepts and definitions', Diversity and Distributions 6, 93–107. https://doi.org/10.1046/j.1472-4642.2000.00083.x

Robertson, M.P., Harris, K.R., Coetzee, J., Foxcroft, L.C., Dippenaar-Schoeman, A.S. & van Rensburg, B.J., 2011, 'Assessing local scale impacts of Opuntia stricta (Cactaceae) invasion on beetle and spider diversity in Kruger National Park, South Africa', African Zoology 46, 205–223. https://doi.org/10.3377/004.046.0202

Roura-Pascual, N., Richardson, D.M., Krug, R.M., Brown, A., Chapman, R.A., Forsyth, G.G., et al. 2009, 'Ecology and management of alien plant invasions in South African fynbos: Accommodating key complexities in objective decision making', Biological Conservation 142, 1595–1604. https://doi.org/10.1016/j.biocon.2009.02.029

Roux, D.J. & Foxcroft, L.C., 2011, 'The development and application of strategic adaptive management within South African National Parks', Koedoe 53(2), Art. #1049, 1–5. https://doi.org/10.4102/Koedoe.v53i2.1049

SANParks, 2014, Marakele National Park. Park management plan 2014–2024, South African National Parks, Pretoria.

SANParks, 2015, South African National Parks strategic plan for 2016/17–2019/20, South African National Parks, Groenkloof, Pretoria.

SANParks, 2016, Table Mountain National Park. Park management plan: 2015–2025, South Africa National Parks, Table Mountain National Park, Constantia.

Spear, D., Foxcroft, L.C., Bezuidenhout, H. & McGeoch M.A., 2013, 'Human population density explains alien species richness in protected areas', Biological Conservation 159, 137–147. https://doi.org/10.1016/j.biocon.2012.11.022

Spear, D., McGeoch, M.A., Foxcroft, L.C. & Bezuidenhout, H., 2011, 'Alien species in South Africa's National Parks (SANParks)', Koedoe 53(1), Art. #1032, 1–4. https://doi.org/10.4102/koedoe.v53i1.1032

Taylor, H.C. & Macdonald, S.A., 1985, 'Invasive alien woody plants in the Cape of Good Hope Nature Reserve. I. Results of a first survey in 1966', South African Journal of Botany 51, 14–20. https://doi.org/10.1016/S0254-6299(16)31696-9

Taylor, H.C., Macdonald, S.A. & Macdonald, I.A.W., 1985, 'Invasive alien woody plants in the Cape of Good Hope Nature Reserve. II. Results of a second survey from 1976–1980', South African Journal of Botany 51, 21–29. https://doi.org/10.1016/S0254-6299(16)31697-0

Tu, M. & Robison, R.A., 2013, 'Overcoming barriers to the prevention and management of alien plant invasions in protected areas: A practical approach', in L.C. Foxcroft, P. Pyšek, D.M. Richardson & P. Genovesi (eds.), Plant invasions in protected areas. Patterns, problems and challenges, pp. 529–547, Springer, Dordrecht. https://doi.org/10.1007/978-94-007-7750-7_24

Van Driesche, R. & Center, T., 2013, 'Biological control of invasive plants in protected areas', in L.C. Foxcroft, P. Pyšek, D.M. Richardson & P. Genovesi (eds.), Plant invasions in protected areas. Patterns, problems and challenges, pp. 561–597, Springer, Dordrecht.

Van Wilgen, B.W., Fill, J.M., Baard, J., Cheney, C., Forsyth, A.T. & Kraaij, T., 2016, 'Historical costs and projected future scenarios for the management of invasive alien plants in protected areas in the Cape Floristic Region', Biological Conservation 200, 168–177. https://doi.org/10.1016/j.biocon.2016.06.008

Van Wilgen, B.W., Forsyth, G.G. & Prins, P., 2012, 'The management of fire-adapted ecosystems in an urban setting: The case of Table Mountain National Park, South Africa', Ecology and Society 17, 8. https://doi.org/10.5751/ES-04526-170108

Van Wilgen, B.W. & Wannenburgh, A., 2016, 'Co-facilitating invasive species control, water conservation and poverty relief: Achievements and challenges in South Africa's Working for Water programme', Current Opinion in Environmental Sustainability 19, 7–17. https://doi.org/10.1016/j.cosust.2015.08.012

Watson, J.E.M, Dudley, N., Segan, D.B. & Hockings, M., 2014, 'The performance and potential of protected areas', *Nature* 515, 67–73. https://doi.org/10.1038/nature13947

Wilson, J.R.U., Gaertner, M., Richardson, D.M. & Van Wilgen, B.W., 2017, 'Contributions to the National Status Report on biological invasions in South Africa', *Bothalia* 47(2), a2207. https://doi.org/10.4102/abc.v47i2.2207

Woodford, D.J., Ivey, P., Jordaan, M.S., Kimberg, P.K., Zengeya, T. & Weyl, O.L.F., 2017, 'Optimising invasive fish management in the context of invasive species legislation in South Africa', *Bothalia* 47(2), a2138. https://doi.org/10.4102/abc.v47i2.2138

Zachariades, C., Paterson, I.D., Strathie, L.W., Hill, M.P. & Van Wilgen, B.W., 2017, 'Assessing the status of biological control as a management tool for suppression of invasive alien plants in South Africa', *Bothalia* 47(2), a2142. https://doi.org/10.4102/abc.v47i2.2142

Appendix 1

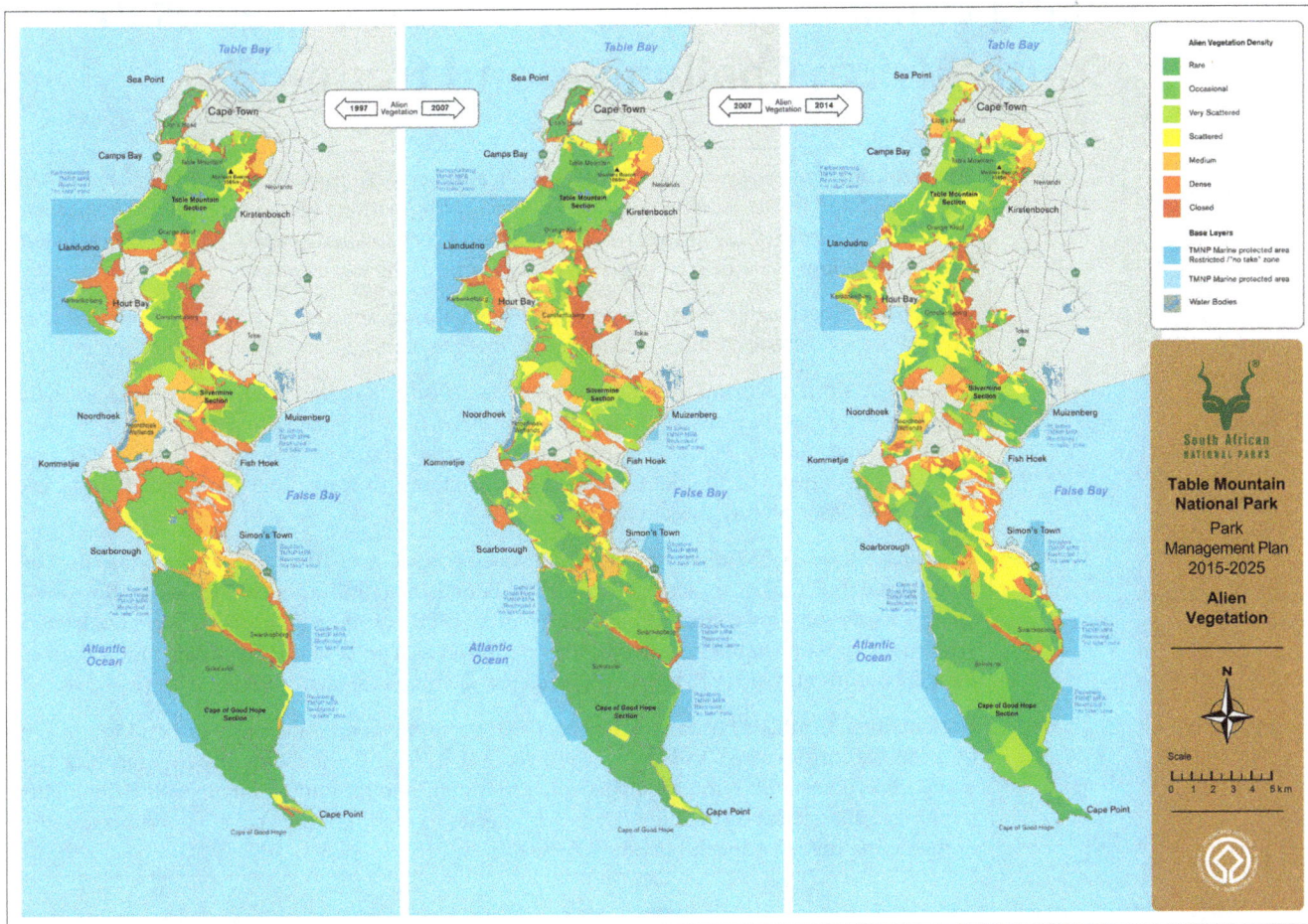

Source: TMNP Management Plan, SANParks 2016

FIGURE 1-A1: Alien vegetation density of Table Mountain National Park over the past decades.

9

Managing conflict-generating invasive species in South Africa: Challenges and trade-offs

Authors:
Tsungai Zengeya[1] ⓘ
Philip Ivey[2]
Darragh J. Woodford[3,4]
Olaf Weyl[4] ⓘ
Ana Novoa[5] ⓘ
Ross Shackleton[5] ⓘ
David Richardson[5] ⓘ
Brian van Wilgen[5] ⓘ

Affiliations:
[1]South African National
Biodiversity Institute,
Kirstenbosch Research
Centre, South Africa

[2]Invasive Species Programme,
South African National
Biodiversity Institute,
Kirstenbosch Research
Centre, South Africa

[3]Centre for Invasion Biology,
Animal, Plant and
Environmental Sciences,
University of the
Witwatersrand, South Africa

[4]Centre for Invasion Biology,
South African Institute for
Aquatic Biodiversity (SAIAB),
South Africa

[5]Centre for Invasion Biology,
Department of Botany and
Zoology, Stellenbosch
University, South Africa

Corresponding author:
Tsungai Zengeya,
T.Zengeya@sanbi.org.za

Background: This paper reviewed the benefits and negative impacts of alien species that are currently listed in the Alien and Invasive Species Regulations of the *National Environmental Management: Biodiversity Act* (Act no 10 of 2004) and certain alien species that are not yet listed in the regulations for which conflicts of interest complicate management.

Objectives: Specifically, it identified conflict-generating species, evaluated the causes and driving forces of these conflicts and assessed how the conflicts have affected management.

Method: A simple scoring system was used to classify the alien species according to their relative degree of benefits and negative impacts. Conflict-generating species were then identified and further evaluated using an integrated cognitive hierarchy theory and risk perception framework to identify the value systems (intrinsic and economic) and risk perceptions associated with each conflict.

Results: A total of 552 alien species were assessed. Most of the species were classified as inconsequential (55%) or destructive (29%). Beneficial (10%) and conflict-generating (6%) species made a minor contribution. The majority (46%) of the conflict cases were associated with more than one value system or both values and risk perception. The other conflicts cases were based on intrinsic (40%) and utilitarian (14%) value systems.

Conclusions: Conflicts based on value and risk perceptions are inherently difficult to resolve because authorities need to balance the needs of different stakeholders while meeting the mandate of conserving the environment, ecosystem services and human well-being. This paper uses the identified conflict-generating species to highlight the challenges and trade-offs of managing invasive species in South Africa.

Introduction

South Africa has a long history of alien species introductions and interventions for managing biological invasions (Richardson et al. 2011). The primary reasons for introductions of alien species were to provide food and raw materials, for recreational ornamentation and as pets, and for erosion control and dune stabilisation. In addition, many species were introduced accidently (Richardson et al. 2003). These introductions have included commercially important trees such as many species of acacias, eucalypts and pines for forestry (van Wilgen & Richardson 2014), fish species such as salmonids and black bass for aquaculture and recreational fishing (Ellender & Weyl 2014) and mammals for the game industry (Brooke, Lloyd & de Villiers 1986; van Rensburg et al. 2011). Moreover, numerous species of birds, fishes, mammals and plants were introduced for ornamentation and as pets (Brooke et al. 1986; Foxcroft, Richardson & Wilson 2008; Picker & Griffiths 2017; Richardson et al. 2003). Such alien species with a high societal value have been widely disseminated across the country and in some areas they are now conspicuous components of natural ecosystems. Although considerable socio-economic benefits have been derived from many alien species, some have become invasive and have caused adverse ecological and socio-economic impacts in recipient areas (De Wit, Crookes & van Wilgen 2001; Ellender et al. 2014; Le Maitre et al. 2011; van Rensburg et al. 2011; van Wilgen et al. 2011). Impacts caused by alien species include loss of biodiversity (Powell, Chase & Knight 2013), changes to ecosystem functioning, economic losses (Holmes et al. 2009) or impacts on human health (Hulme 2014).

The main legislative instrument that guides the management of alien species in South Africa is the *National Environmental Management: Biodiversity Act* (NEM:BA) (Act 10 of 2004) and the regulations relating to this Act (Republic of South Africa [RSA] 2004). Management measures include interventions directed at restricting the importation of high-risk alien species, regulating the movement and utilisation of alien species, and interventions aimed at eradicating species that occur in low numbers over limited areas, containing invasions, and reducing the extent and impact of well-established invaders (RSA 2014). Management actions directed at species that have both benefits and negative impacts are, however, complicated. For such species (hereafter 'conflict-generating species'), there is current or potential disagreement between stakeholders regarding their benefits and the damage that they inflict on the ecosystems where they occur. Issues pertaining to such species are increasingly complicating the management of biological invasions worldwide (Woodford et al. 2016).

The benefits and negative impacts of alien species vary widely in type and magnitude and are dependent on the species, their invasive potential, the extent to which they have invaded, the nature of the invaded environment and socio-economic contexts (Kueffer 2013; Kueffer & Kull 2017; Kull et al. 2011; Shackleton et al. 2007; van Wilgen & Richardson 2014). Nevertheless, the relative degree of negative impacts of alien species and the benefits associated with their utilisation can be used to place them in a conceptual framework that divides species into four broad categories (van Wilgen & Richardson 2014) (Figure 1). In the first two categories, alien species have neither substantial negative impact nor benefit (these are termed 'inconsequential species') or are 'beneficial species' that have relatively low negative impact but provide significant benefit. The third category comprises species that have no substantial benefits but high negative impacts ('destructive species'). Species in these three categories usually have less complex dimensions to their management and the degree of social contestation regarding ways to control, eradicate or otherwise manage them is low. However, species in the fourth category – that is species with high negative impacts and benefits ('conflict-generating species') – generate most controversy because of the polarised perceptions of their impacts and benefits between different stakeholders and the options for managing them. Human attitudes and behaviours towards the use and management of conflict-generating species are largely influenced by individual or group demographics and knowledge and by properties of the species itself (Rotherham & Lambert 2011; van Wilgen & Richardson 2014). Furthermore, the complexity of societal issues around conflict-generating

species can also vary from simple issues centred on one stakeholder's perception of the problem to complex issues that involve many conflicting stakeholders' perspectives (Novoa et al. 2016; Woodford et al. 2016). Therefore, the dimensions of conflicts that arise and the options that exist for resolving these conflicts can be highly taxon and region-specific.

This paper, focusing on conflict-generating species, is directly aligned to one of the tenets of this special issue of *Bothalia – African Biodiversity and Conservation* on efficacy of interventions for managing biological invasions in South Africa (Wilson et al. 2017). This topic is, however, also relevant to global audiences as conflicts of interest around invasive species are a global issue (Pyšek & Richardson 2010). We review the benefits and negative impacts of alien species that are currently, or may be, listed in the Alien and Invasive Species (A&IS) Regulations (RSA 2014) of the NEM:BA (Act no 10 of 2004) (RSA 2004). Specifically, we aimed to identify conflict-generating species, evaluate the causes and driving forces of these conflicts and assess how the conflicts have affected management.

Methods

The 549 alien species that are currently listed in the A&IS Regulations (2014) were classified as inconsequential, beneficial, destructive or conflict-generating species according to their relative degree of benefits and negative impacts (Online Appendix). Three additional contentious species that are either not listed or were removed from the A&IS Regulations (2014) to avoid conflicts were also classified. Species were classified through a simple scoring system (Table 1) that had two categories each for negative impacts [ecological and socio-economic (including intrinsic impacts under 'socio' part)] and benefits (economic and intrinsic). For each category, the negative impacts and benefits of each species were quantified by a three-level scale that ranged from 1 to 3 (no or little evidence of negative impacts or benefits), 4 to 6 (localised negative impacts or benefits) and 7 to 10 (widespread negative impacts or benefits). The final score from impact scoring systems can be obtained in several ways (see Nentwig et al. 2016) depending on the focus of the assessment. For this study, the overall scores of a species were obtained by taking the maximum score for any of the two categories of negative impacts and benefits. The scoring was done by the co-authors, all of whom are active researchers in the field of invasion biology in South Africa. The authors were divided into two groups (plants and animals) according to their expertise. Within each group, each author was allocated a number of species to independently evaluate them based on the available literature. Subsequently, all evaluations were discussed among all group members. We acknowledge that scoring could be improved if all the authors evaluated all the assessed species and through consultations with a broader stakeholder group and therefore recommend that this study should be expanded in future to get a more balanced assessment.

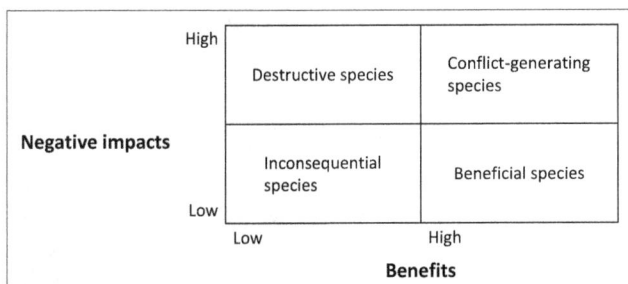

		Low	High
Negative impacts	High	Destructive species	Conflict-generating species
	Low	Inconsequential species	Beneficial species
		Benefits	

Source: van Wilgen and Richardson 2014.

FIGURE 1: A conceptual framework to categorise alien and invasive species based on their relative environmental and socio-economic negative impacts and benefits.

TABLE 1: The scoring system used to identify conflict-generating species on the list of the alien species in the Alien and Invasive Species Regulations of 2014.

Impacts	Category	Score levels
Negative impacts	Ecological impacts	*1–3:* No or little evidence of impact on other species and/or causes little to no changes to the supply of ecosystem services (provisioning, regulating, supporting and cultural): Low consequences – negligible impacts on ecological systems, pattern and processes and/or impacts arising are easily restored *4–6:* Localised impact on a few species and/or reduces the supply one or more ecosystem service but does not result in the total loss of these services: Moderate consequences – impacts on ecological systems, pattern and processes and impacts arising are potentially restorable *7–10:* Widespread impacts on multiple species and/or results in the loss of one or more ecosystem service, and has, or threatens to cause, a total loss in the services supplied: Substantial consequences – impacts on ecological systems, pattern and processes and impacts arising are difficult to restore/reverse
	Socio-economic impacts	*1–3:* Has no or very low economic impact on a few communities or stakeholders and/or has very little to no impact local human well-being at local levels: Low consequences – negligible impacts on social-economic systems, pattern and processes and impacts arising are easily restored *4–6:* Has multiple economic impacts for several communities or stakeholders at localised levels and can hinder livelihoods and reduce economic returns and/or disrupts local livelihoods and well-being in some manner but does not prevent economic activity entirely: Moderate consequences – impacts on social-economic systems, pattern and processes and impacts arising are easily restored or reversed *7–10:* Has multiple economic impacts for numerous communities or stakeholders at a national level that threatens livelihoods and economic returns and/or disrupts livelihoods and well-being on a large scale and considerably increases the vulnerability of local communities in a variety of ways: Substantial consequences – impacts on economic activities are difficult to restore or reverse
Benefits	Economic benefits	*1–3:* Has no economic benefits – it is not harvested or traded and provides no employment *4–6:* Is harvested or traded by local communities on a small scale (informal and localised) (localised areas) – provides low employment *7–10:* Is harvested or traded at a national scale – it contributes to national gross domestic product (GDP) and provides jobs to large numbers of people
	Intrinsic benefits	*1–3:* Provides little to no intrinsic (aesthetic appeal, moralistic appeal or recreation value) value to local communities. Local would not notice or care if the species was removed based on intrinsic reasons *4–6:* Provides some intrinsic value (aesthetic appeal, moralistic appeal or recreation value) to local communities (small part of the population) and/or one particular stakeholders group. There might be resistance to removal because of intrinsic reasons *7–10:* Provides high intrinsic value (aesthetic appeal, moralistic appeal or recreation value) for a large proportion of the population and numerous communities and/or is beneficial to multiple stakeholder groups: resistance to removal based on intrinsic reasons expected

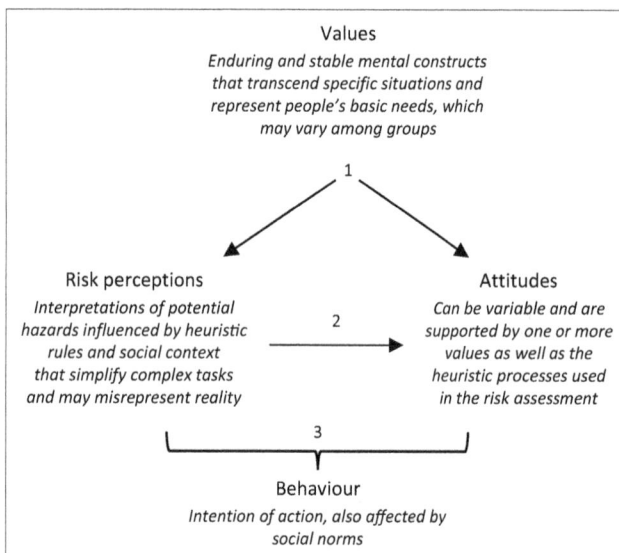

Source: Estévez et al. 2015

FIGURE 2: A conceptual framework created by integrating cognitive hierarchy theory (CHT) and risk perception theories.

The identified conflict-generating species were then evaluated using an integrated cognitive hierarchy theory – risk perception framework (Estévez et al. 2015; Figure 2), to identify the conflict type and cognitive level at which the conflicts occurred. Two conflict types were identified – conflicts based on specific values or heuristic rules. Heuristic rules are simple and efficient ways in which people reduce complex mental tasks to simpler ones to form judgements and make decisions, for example, focusing on one aspect of a complex problem and ignoring others (Slovic 1999). Cognitive levels were defined as subjective classifications of conflicts based on whether there are based on values or perceptions. The conceptual framework is a tiered system of values, risk perceptions, attitudes and behaviour that has been applied to

the management of conflict invasive species (Estévez et al. 2015). Values are the basis of the hierarchal framework and are defined as enduring and fundamental beliefs that represent people's needs and may vary among groups. For the purposes of this paper, we used Kellert's (1993) classification that describes eight fundamental values that humans associate with nature (Table 2). These vary from intrinsic values (aesthetic, dominionistic, humanistic, moralistic and naturalistic) that represent some form of emotional relationship between society and nature to values that represent a practical value or material benefit of nature (scientific and utilitarian). Risk perceptions are interpretations of potential hazards that are influenced by mental strategies or heuristic rules and social context (cultural backgrounds and personal values) that reduce complex mental task to simpler ones (Slovic 1999). Risk perceptions may either represent reality or generate substantial and persistent biases that lead to attitudes that misrepresent the magnitude or severity of risks (Burgman 2005). Attitudes are flexible constructs that are supported by one or several values and are affected by risk perception (Fulton, Manfredo & Lipscom 1996). Behaviour is defined as an intention of action that is directly influenced by attitudes and risk perception (Rokeach 1973).

Results

Negative impacts and benefits framework

Most of the 552 assessed alien species were classified as either inconsequential (55%) or destructive (29%). Far fewer species were classified as beneficial (10%) and conflict-generating species (6%) (Figure 3). The results for animals and plants showed similar general trends, where inconsequential and destructive species predominate and conflict and beneficial species make up the remainder. Inconsequential species

TABLE 2: Eight fundamental values that humans associate with nature.

Conflict level	Value or heuristic rule	Definition
Value system	Naturalistic	Exploration of nature and outdoor recreation
	Aesthetic	Physical attraction and appeal of nature
	Dominionistic	Mastery and control over nature
	Humanistic	Emotional, spiritual, or symbolic affection for nature
	Moralistic	Moral concern about the right and treatment of nature
	Negativistic	Fear or aversion towards nature
	Scientific	Systematic and empirical study of nature
	Utilitarian	Practical value or material benefit of nature
Risk perception	Evaluation of potentials	Differences in evaluations of potential hazards
	Lack of institutional trust	Lack of trust between stakeholders and government agencies could result from lack of community engagement and transparencies in decision making processes, differences in evaluations of potential hazards, and lack of confidence in government authorities

Source: Kellert 1993

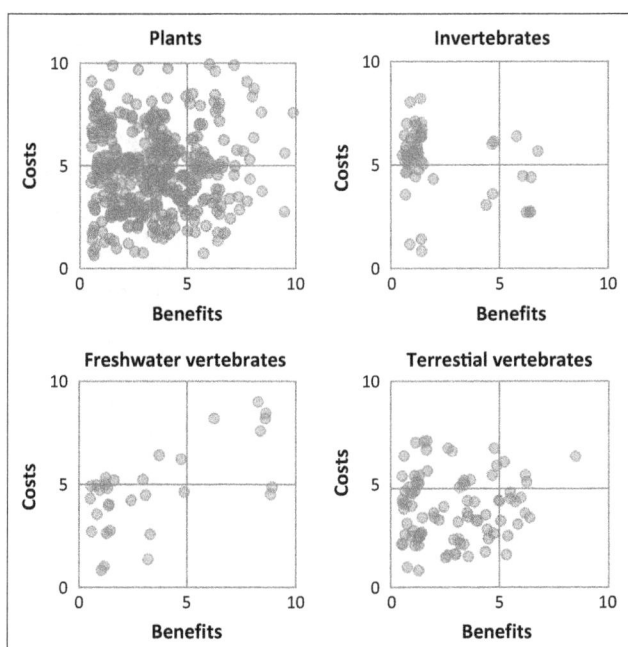

FIGURE 3: Categorisation of alien species listed under the A&IS Regulation (2014) based on the degree of their negative impacts (ecological and socio-economic) and benefits (economic and intrinsic). Jitter was used to indicate the density of dots where there is overlap.

consisted of 99 animal and 203 plant species that had a score ≤ 5 for both negative impacts and benefits. Nearly half of the animal (48 out of 100 species) and 30% (61/203) of plant species in this category recorded the lowest possible scores (< 3) for both negative impact and benefits. Furthermore, almost all the listed amphibian species (5 out of 7), bird (18/24) and reptiles (33/35) species were placed in this category.

Beneficial species comprised 13 animal and 42 plant species that had a score ≤ 5 for negative impacts but > 5 for benefits (Figure 3). The highest economic (≥ 8) and intrinsic (≥ 8) benefit scores were allocated to three *Eucalyptus* species [*E. cladocalyx* (sugar gum); *E. diversicolor* (karri gum) and *E. grandis* (saligna gum)]. The 'destructive species' category had 54 animal and 107 plant species that scored > 5 for negative impacts but ≤ 5 for benefits (Figure 3). Among the animals, all the listed microbes and terrestrial insects (except stick insects; species in the order Phasmatodea) were classified as destructive species and

the highest negative impact score of 8 was reached by *Phytophthora cinnamomi* (microbe) and *Trogoderma granarium* (khapra beetle). Among the plant species, 60 species were classified as highly destructive (score range 7–10) and the highest negative impact scores (> 9) were reached by lantana (*Lantana* spp.), three *Acacia* species (*A. decurrens*, *A. longifolia* and *A. saligna*), *Dolichandra unguiscati* (cat's claw creeper), *Echium plantagineum* (Patterson's curse), *Chromolaena odorata* (triffid/siam weed), *Eucalyptus camaldulensis* (river red gum) and *Hakea sericea* (silky hakea).

Conflict-generating species consisted of 9 animal and 25 plant species that had a score > 5 for either negative impacts or benefits (Figure 3). The animal species assigned the highest scores (≥ 8) each for negative impacts and benefits were five fish species [*Micropterus salmoides* (largemouth bass), *Micropterus dolomieu* (smallmouth bass), *Oreochromis niloticus* (Nile tilapia), *Oncorhynchus mykiss* (rainbow trout) and *Salmo trutta* (brown trout)]. The plant species that reached the highest score include *Acacia* species (*A. cyclops*, *A. dealbata*, *A. mearnsii* and *A. melanoxylon*), *Pinus* species (*P. elliottii*, *P. patula* and *P. radiata*), *Prosopis* species and cacti (*Cylindropuntia fulgida* var. *mamillata*, *C. imbricata*, *C. leptocaulis*, *C. pallida* and *C. spinosior*). See Woodford et al. (2017) for a review of the conflicts that have arisen with introduced fish and Kaplan et al. (2017) for conflicts with introduced cacti.

Conflict types and cognitive levels

The majority (46%) of the conflict cases could be explained by more than one conflict type and cognitive level (Table 3). Different conflict types occurred when a species was associated with both intrinsic and economic values. This was observed for two plant species: *Pinus radiata* (Monterey pine) and *Psidium guajava* (guava). Different cognitive levels were observed when a conflict was based on values and risk perceptions. Risk perceptions were mainly derived from a fear or aversion of possible adverse impacts associated with the species and control methods used in its management. Plant species in this category include mesquite (*P. glandulosa* and *P. velutina*), acacias (*A. cyclops*, *A. dealbata*, *A. mearnsii* and *A. melanoxylon*) and pines (*P. elliottii* and *P. patula*). The remainder comprised animal

TABLE 3: Conflict-generating invasive species showing the negative impacts (ecological and social-economic), benefits (economic and intrinsic), cognitive level (VS = value system and PB = perceptions based) and value system or heuristic rule at which conflicts occurred.

Descriptive category	Common name	Species	Taxon	Costs		Benefits		Conflict level	Value or heuristic rule
				Ecological	Social-economic	Economic	Intrinsic		
Agricultural initiatives	Leucaena	*Leucaena leucocephala*	P	7	7	6	5	VS	Utilitarian
	Mission prickly pear	*Opuntia ficus-indica*	P	6	6	8	4	VS	Utilitarian
	Blue-leaf cactus	*Opuntia robusta*	P	6	6	7	4	VS	Utilitarian
	Honey mesquite	*Prosopis glandulosa*	P	9	9	8	6	VS, PB	Utilitarian, evaluation of potential hazard
	Velvet mesquite	*Prosopis velutina*	P	9	9	8	6	VS, PB	Utilitarian, evaluation of potential hazard
	Guava	*Psidium guajava*	P	6	4	9	6	VS	Aesthetic, utilitarian
Angling	Smallmouth bass	*Micropterus dolomieu*	FV	9	2	9	9	VS, PB	Utilitarian, evaluation of potential hazard
	Largemouth bass	*Micropterus salmoides*	FV	8	2	9	9	VS, PB	Utilitarian, evaluation of potential hazard
	Brown trout	*Salmo trutta*	FV	8	2	9	1	VS, PB	Utilitarian, evaluation of potential hazard
Aquaculture	Marron	*Cherax tenuimanus*	I	6	3	6	1	VS, PB	Utilitarian, evaluation of potential hazard
	Japanese oyster	*Crassostrea gigas*	I	6	4	7	1	VS	Utilitarian
	Nile tilapia	*Oreochromis niloticus*	FV	8	2	7	1	VS, PB	Utilitarian, evaluation of potential hazard
Aquaculture, angling	Rainbow trout	*Oncorhynchus mykiss*	FV	8	2	9	1	VS, PB	Utilitarian, evaluation of potential hazard
Commercial important trees	Golden wattle	*Acacia pycnantha*	P	6	3	1	6	VS	Aesthetic, naturalistic
	Red eye/rooikrans	*Acacia cyclops*	P	10	7	6	3	VS, PB	Utilitarian, evaluation of potential hazard
	Silver wattle	*Acacia dealbata*	P	10	7	6	3	VS, PB	Utilitarian, evaluation of potential hazard
	Black wattle	*Acacia mearnsii*	P	10	7	9	4	VS, PB	Utilitarian, evaluation of potential hazard
	Australian blackwood	*Acacia melanoxylon*	P	10	7	7	5	VS, PB	Utilitarian, evaluation of potential hazard
	Slash pine	*Pinus elliottii*	P	8	4	8	5	VS, PB	Utilitarian, evaluation of potential hazard
	Patula pine	*Pinus patula*	P	8	3	10	4	VS, PB	Utilitarian, evaluation of potential hazard
	Monterey pine	*Pinus radiata*	P	8	3	8	6	VS	Aesthetic, naturalistic, utilitarian
Erosion control, sand dune stabilisation	Marram grass	*Ammophila arenaria*	P	7	6	6	2	VS	Utilitarian
Escapees	Domestic cat	*Felis catus*	TV	7	5	1	8	VS	Aesthetic, moralistic
	Himalayan tahr	*Hemitragus jemlahicus*	TV	6	3	1	6	VS	Humanistic
Ornamentals	Chain-fruit cholla	*Cylindropuntia fulgida* var. *fulgida*	P	7	8	4	6	VS	Aesthetic
	Boxing-glove cactus	*Cylindropuntia fulgida* var. *mamillata*	P	7	8	4	6	VS	Aesthetic
	Imbricate cactus	*Cylindropuntia imbricate*	P	7	8	4	6	VS	Aesthetic
	Pencil cactus	*Cylindropuntia leptocaulis*	P	7	8	4	6	VS	Aesthetic
	Pink-flowered sheathed cholla	*Cylindropuntia pallida*	P	7	8	4	6	VS	Aesthetic
	Cane cholla	*Cylindropuntia spinosior*	P	7	8	4	6	VS	Aesthetic
	Syringa	*Melia azedarach*	P	6	1	2	8	VS	Aesthetic
	Yellow bunny-ears	*Opuntia microdasys*	P	6	6	3	7	VS	Aesthetic
	Jerusalem thorn	*Parkinsonia aculeata*	P	7	7	3	6	VS	Aesthetic
	Belhambra	*Phytolacca dioica*	P	8	4	2	6	VS	Aesthetic
	Pink tamarisk	*Tamarix ramosissima*	P	7	7	2	6	VS	Aesthetic

Species were grouped into four descriptive taxon categories were *P* = plants, I = invertebrates, TV = terrestrial vertebrates, FV = freshwater vertebrates.

species such as *M. dolomieu, M. salmoides, O. niloticus, O. mykiss, S. trutta* and *Cherax tenuimanus* (marron).

Conflicts centred on intrinsic value systems collectively accounted for 40% of all the examined cases of conflict-generating species (Table 3). The detected value systems included naturalistic, humanistic, aesthetic and moralistic values that were associated with 12 ornamental plants and 2 vertebrate species [*Hemitragus jemlahicus* (Himalayan tahr) and *Felis catus* (domestic cat)]. The ornamental plants include

A. pycnantha (golden wattle), cacti (*C. fulgida* var. *fulgida, C. fulgida* var. *mamillata, C. imbricata, C. leptocaulis, C. pallida, C. spinosior* and *Opuntia microdasys*), *Melia azedarach* (syringa), *Parkinsonia aculeate* (Jerusalem thorn), *Phytolacca dioica* (ombu) and *Tamarix ramosissima* (pink tamarisk). Conflicts based on only utilitarian values made up 14% of observed conflicts and were identified mainly from commercial important species such as *Leucaena leucocephala* (leucaena), *Ammophila arenaria* (marram grass), *Opuntia ficus-indica* (prickly pear), *Opuntia robusta* (blue-leaf cactus) and *Crassostrea gigas* (Japanese oyster).

Discussion

Inconsequential species

The negative impacts and benefits framework indicated that almost half of the 552 assessed species could be classified as 'inconsequential species' that have neither substantial benefits nor negative impacts. Most of these inconsequential species have either limited distribution or no known impacts in South Africa or elsewhere in the world. For example, freshwater fish can only spread within a river system and their spread across the country is facilitated by human movement. Some of the species such as *Perca fluviatilis* (European perch) are of little interest to the mainstream angling and aquaculture fraternity (major pathways of fish introductions in South Africa; Richardson et al. 2011). They are therefore not moved as much as commercially important and widely used introduced species such as trout, carp and black basses (Ellender et al. 2014). Moreover, most inconsequential mammal and bird species were introduced for novelty, and as ornamentals for private collections and game ranching. Some of these species have not established self-sustaining populations and are either no longer present in the country or actively maintained in captive facilities such as botanical and zoological gardens or in private properties where they have not caused any documented impact on the environment. For example, there is no evidence that the rare *Elaphurus davidianus* (Père David's Deer) that is endemic to China is invasive anywhere in the world, and known populations outside its native range are in captivity (Long 2003). It is important to note that because of the limited data on the potential impact and benefits associated with species in this category, reassessments might be necessary as more information becomes available. For example, the eastern grey squirrel (*Sciurus carolinensis*) has relatively minor impacts on biodiversity although some potentially major impacts such as predation on native birds have not yet been thoroughly assessed and any future management plans of the species needs an increased appreciation of its potential impacts on native biota. Similarly, some plant species have long lag phases (low initial population growth rate) after introduction and as a result they may become widely utilised and accepted as either inconsequential or beneficial species. The benefits are, however, often surpassed by negative impacts when the species become invasive (Geerts et al. 2013; Shackleton et al. 2007).

Destructive species

Species that were classified as destructive (no substantial benefits but high negative impact) made up the second largest proportion of all the listed species on the A&IS Regulations. Species in this category are largely regarded as pests and weeds because of the deleterious effect they have on society and the environment. Many of these species were also accidental introductions. As a result, the degree of social contestation regarding ways to control or eradicate them is low. For example, there has been little public resistance to the management of invasive rodents (*Rattus* spp.) in several townships in Gauteng because the species have been implicated in causing damage to infrastructure and transmission of zoonotic diseases (Jassat et al. 2013). The same is true for most of the listed microbes and terrestrial insects that are known pests and pathogens of agricultural crops (Burgess et al. 2007; Durán et al. 2009; Hunter et al. 2011; Picker & Griffiths 2017; Wingfield et al. 2008a, 2008b, 2010). Several animal species that were classified as destructive have been introduced globally as part of the pet trade and game ranching industry (Brooke et al. 1986; Picker & Griffiths 2017; Richardson et al. 2003; van Rensburg et al. 2011). Many of these species escaped captivity through direct or accidental releases by the public into the wild. In the wild, the species have been linked to a variety of impacts on native biota (see Long 2003). However, because we are concerned only with populations in the wild, species that are important in the pet trade or game ranching industry might not have much value once they escape captivity. While eradicating such species from the wild would therefore cause little conflict, banning them from the trade would result in conflict. Examples of destructive plants – that is plants that have little use, high negative impacts and a low degree of social contestation regarding ways to control them – include *Lantana* spp., which outcompetes and replaces indigenous species and forms dense thickets that obstruct access to land, and *Parthenium hysterophorus*, which reduces grazing potential and causes human health related problems (McConnachie et al. 2011; Strathie 2015; Vardien et al. 2012; Zachariades et al. 2009).

Beneficial species

There are also some species that were categorised as beneficial but not harmful. This suggests that active management is not necessary or should only be done in particular cases. Some of these perceived beneficial species still create conflict even though there is not much evidence of negative impacts and represent some unique cases. Previous attempts to manage these unique cases have led to some controversy where either proposed management actions have been completely put aside or there have been some trade-offs and compromises. For example, *Jacaranda mimosifolia* (jacaranda) is an iconic tree species in the city of Pretoria, where the species is regarded as part of the identity and 'sense of place' of the city. There was huge public resistance to their removal and to regulations preventing replanting (Dickie et al. 2014; Kasrils 2001). The conflicts were resolved through active stakeholder engagement and through compromises, which involved regulating the species by area (management interventions that are area specific). As a result, jacarandas are not listed as invasive species in urban areas and around farm houses in several provinces where they occur (Gauteng, KwaZulu-Natal, Limpopo, Mpumalanga and North-West). It is envisaged that in these areas, trees will be gradually phased out by preventing further replanting of jacaranda. The seed source is, however, likely to remain for years (see Irlich et al. 2017 for a review of such issues facing municipalities). Similarly, several individual alien trees and groups of alien trees comprising 41 species were classified as champion trees and given protected status under Section 12 of the *National*

Forests Act of 1998 by the Department of Agriculture, Forestry and Fisheries (DAFF) because of their remarkable size, age and aesthetic, cultural, historic or tourism value (DAFF 2016). An example is the large *Cinnamomum camphora* (camphor trees) that were planted in the Vergelegen Estate in Somerset West more than three centuries ago by Governor Willem Adriaan van der Stel (DAFF 2016). These trees were classified as national heritage or champion trees and are not listed on the A&IS Regulations. The species as a whole is, however, regulated by area, and camphor trees that are not listed as heritage/champion trees have to be controlled in the Eastern Cape, KwaZulu-Natal, Limpopo and Mpumalanga, but utilisation is allowed in the Western Cape, subject to certain prohibitions (e.g. no selling or distribution). Camphor trees are not listed in A&IS Regulations in other provinces such as the North-West, Free State, Gauteng and Northern Cape.

Similar compromises have also been made for other listed plants that have high economic and intrinsic values, such as eucalypts (*E. cladocalyx, E. diversicolor, E. grandis*) (RSA 2014) which is also regulated by area. For all biomes, eucalypt trees must be removed from riparian areas, protected areas (nature reserves and national parks) and ecosystems of conservation concern. Eucalypt trees that occur in the Nama-Karoo, Succulent Karoo and Desert biomes are exempt from removal, but trees in all other biomes are to be controlled except when they occur in a formal plantation, in urban areas, close to farm homesteads and within cultivated land.

Examples of beneficial animal species where compromises had to be made for their management include rock doves (*Columba livia*) and ungulate mammal species. Feral rock doves are regarded as a health risk to humans and indigenous bird species and as a result were initially listed in Category 1b ('Invasive species requiring compulsory control as part of an invasive species control programme. These species are deemed to have such a high invasive potential that infestations can qualify to be placed under a government sponsored invasive species management program. No permits will be issued to introduce or use this species in South Africa'). This caused some consternation among the pigeon racing community. The regulations were amended to accommodate their concerns in the A&IS list that was published in July 2016 (RSA 2016). The rock doves are now listed in Category 2 where utilisation for all restricted activities related to racing and showing of pigeons is permitted subject to certain prohibitions. On the contrary, some alien ungulate species have been promoted by the game industry because of their economic importance for hunting. These species are now regulated by area as a compromise between conservation authorities that advocated for control and the game industry that preferred unrestricted movement. Among animal species, *Anas platyrhynchos* (European mallard) represents an extreme example. Proposed management actions of removal (i.e. killing) of the species were met with fierce resistance and were stopped, because the perceived value was considered to outweigh the impacts (Gaertner et al. 2016). The mallard hybridises with the native *A. undulata* (yellow-billed duck),

but impacts on the genetic integrity of indigenous congeneric species are insidious and more difficult to explain to the public and local policy-makers than other more evident ecological impacts of other invasive species such as predation and overgrazing. The current management option in the City of Cape Town is to tolerate the species and apply no management (Gaertner et al. 2016).

Conflict-generating species

Actual and potential conflict-generating species made up the smallest proportion (6%) of all listed species. Species in this category had both benefits and negative impacts. The majority of these conflicts could be explained by more than one cognitive level, such as utilitarian values based on practical or material benefits and risk perceptions of possible impacts from invasive species. Examples include *Prosopis* species that were introduced to provide fodder, firewood and shade in arid parts of the country (Shackleton et al. 2014). However, they have also been implicated in adverse social and ecological impacts (Shackleton, Le Maitre & Richardson 2015a, 2015c; Shackleton et al. 2015b). This has led to conflicts between communities who see the species as a resource [some farmers and non-governmental organisations (NGOs)] and others who are concerned about its negative impacts (conservation managers and some farmers) (Shackleton et al. 2015c). The use of the species is still promoted by NGOs in many countries and as a result it has spread widely. Millions of Dollars are being spent on control of this species in South Africa annually (van Wilgen 2012). Further conflicts have been observed in the use of biological control agents (lethal vs. non-lethal agents). The biological control of *Prosopis* using seed-attacking insects has been ineffective and there is a need to use more lethal options. However, risk perception over potential loss of benefits has previously prevented the use of the latter option (Shackleton et al. 2016; Wise, van Wilgen & Le Maitre 2012; Zachariades, Hoffmann & Roberts 2011). However, as negative impacts increase and managing interventions lag behind invasion rates the willingness to use alternative and more lethal agents is increasing (Shackleton et al. 2016; Wise et al. 2012).

Similarly, *Acacia* species such as *A. mearnsii* (black wattle) are commercially important and contribute to livelihoods but are also aggressive invaders that have significant ecological impacts and have caused major conflicts of interests (De Wit et al. 2001; Shackleton et al. 2007). Research on biological control for these *Acacia* species was blocked for many years to protect the interests of the wattle growers (Pieterse & Boucher 1997). Eventually, there was reluctant acceptance of biological control to reduce seed output, but the use of lethal biological control remains blocked (Impson et al. 2011; Stubbings 1977; van Wilgen et al. 2011). Another example is the use of *Pissodes validirostris* (a cone-feeding weevil) for biological control of invasive *P. pinaster*. The forestry industry was concerned that adult weevils fed on shoots of pines, thereby potentially facilitating infection by *Fusarium* fungi which pose a possible risk to commercially important *Pinus* species. This risk led to the discontinuation of research on the use of biological control agents on pines in South Africa (Hoffmann, Moran & van Wilgen 2011).

The aversion of possible impacts of invasive species with utilitarian values is clearly illustrated by conflict around the proposed regulation of *S. trutta* and *O. mykiss* that are utilised for recreational angling and commercial aquaculture (Ellender et al. 2014; Woodford et al. 2017). Angling is dependent on introductions into the wild where trout reduce native biota (see Ellender & Weyl 2014; Jackson et al. 2015; Shelton, Samways & Day 2015). Management of trout has been contentious because of conflicting values of stakeholders. The differences are mainly centred on risk perception (evaluation of potential hazards and lack of institutional trust) and benefits derived from the trout industry. The trout fraternity (concerned stakeholders) refused to acknowledge that trout are invasive species and highlighted the lack of scientific evidence of the risks posed by trout to biodiversity (Cox 2013). There is also a lack of institutional trust because the trout fraternity view A&IS legislation as an instrument for destroying the trout industry. Because of the strong opposition to the inclusion of trout in the A&IS legislation, both *S. trutta* and *O. mykiss* were removed from the national lists of regulated invasive species until a consensus on options for their management could be reached (Woodford et al. 2017). Consultations are continuing in the hope of reaching consensus among stakeholders; issues remain highly polarised among stakeholders and it is unclear whether an end to the deadlock is in sight (Woodford et al. 2016). Similar conflicts could emerge for other angling species such as black bass species (e.g. *M. salmoides* and *M. dolomieu*) that have been implicated in causing adverse ecological impacts on native biota but also contribute to local and regional economies through industries that provide goods and services to anglers (e.g. fishing tackle, boats and bed-nights) (Ellender et al. 2014). The proposed management of these species has, however, been less contentious than that of trout because concerned stakeholders have seen little threat to their industry from the legislation. This is because black bass are fully established in many reservoirs and the fisheries that they support are therefore less dependent on stocking than those based on trout.

The utilisation of *O. niloticus* in aquaculture is also likely to create major conflicts soon because of contrasting values among stakeholders. Its introduction into river systems in southern Africa is a cause for concern for the conservation of indigenous congeneric species that risk extirpation through hybridisation and competition arising from habitat and trophic overlaps (Zengeya, Booth & Chimimba 2015). In areas where *O. niloticus* has established, it has rapidly replaced indigenous congeneric to the extent that some populations have become extirpated (Firmat et al. 2013; Weyl 2008; Zengeya & Marshall 2007). Despite these well-documented adverse ecological effects, it remains one of the most widely cultured and propagated fish species in aquaculture and stock enhancements in the southern Africa sub-region (Denies et al. 2016). Decisions on its management

will be based on the trade-offs between socio-economic benefits and potential adverse ecological effects.

In some cases, species were associated with different conflict types that could be either intrinsic or utilitarian values. For example, the removal of invasive trees in urban or peri-urban environments in the City of Cape Town created conflicts because of the intrinsic and utilitarian values attached to certain species such as acacias, eucalypts and pines (Gaertner et al. 2016; van Wilgen 2012). The intrinsic (naturalistic and aesthetic) values were derived from the physical attraction and appeal of nature, while utilitarian values were centred on derived practical or material benefits from the invasive trees species such as carbon sequestration, economic value of timber and honey production (Allsopp & Cherry 2004; van Wilgen 2012). Some of these concerns could not be substantiated based on current knowledge and were set aside. Trade-offs were proposed for supported concerns. For example, some plantations of *E. diversicolor* that are less invasive than pines (Forsyth et al. 2004), were retained to maintain aesthetic value, for recreational purposes and honey production (Gaertner et al. 2016; van Wilgen 2012). Concerns continue to be raised periodically but, despite opposition, the policy promoting alien plant removal has remained in place, and considerable progress has been made towards clearing alien plants from the park. This is largely attributed to political support, arising largely from job creation, and a strong body of scientific evidence that could be cited in support of the programme (van Wilgen 2012).

Conflicts centred on intrinsic values represented some form of emotional relationship between society and nature. The detected value systems included naturalistic, humanistic, aesthetic and moralistic values systems. For example, moralistic values are centred on the right of invasive animals to live and not to be abused. Control measures often involve culling which is strongly opposed by some sectors of society such as animal rights organisations (Bremner & Park 2007; Ford-Thompson et al. 2012). For example, the introduced *H. jemlahicus* in Table Mountain National Park has been the focus of eradication attempts, despite strong opposition to control – in this case because the perceived impacts (overgrazing) are clearly evident and outweigh the benefits (recreational values) (Gaertner et al. 2016). Active engagement was needed to offset opposition through equal but opposite support for the eradication from government conservation agencies, NGOs and leading academics. In contrast *F. catus* represents a case where proposed management actions (control and or eradication) have either been completely put aside because they would be too controversial or there has been some trade-offs and compromises. Feral cats have been introduced worldwide as pets, for biological control of rodents and accidentally via shipping (Brooke et al. 1986). Conversely, feral cats have also been implicated in causing adverse impacts on biodiversity through predation, hybridisation, competition for resources and transmission of diseases (Tennent, Downs & Bodasing 2009; but see Le Roux et al. 2015). Because the perceived benefits outweigh negative impacts the species is to be tolerated, and is not listed on the A&IS Regulations on the South African mainland. The species has, however, been successfully eradicated from Marion Island (Nogales et al. 2004).

Conflicts based on utilitarian values were observed for species that are economically important because they provide food and raw materials for industry and local communities. For example, many cactus species were introduced as part of agricultural initiatives to improve fruit production for human consumption, fodder for livestock and ornamental purposes (Novoa et al. 2015a). Unfortunately, some of the introduced cactus species have also been implicated in causing adverse ecological impacts and this dichotomy has caused conflicts (Novoa et al. 2015). Cactus species associated with utilitarian benefits include *O. ficus-indica* and *O. robusta.* The other seven listed species of cactus (*C. fulgida* var. *fulgida*, *C. fulgida* var. *mamillata*, *C. imbricate*, *C. leptocaulis*, *C. pallida*, *C. spinosior* and *O. microdasys*) are mainly associated with intrinsic values (aesthetic and naturalistic) as they are utilised for ornamental purposes. Active stakeholder engagement is ongoing to try and resolve the conflicts and the results will be used to advise and develop a national cactus management strategy for South Africa (Kaplan et al. 2017; Novoa & Shackleton 2015; Novoa et al. 2016). There are many approaches to enable stakeholders and managers to find common ground in such contentious situations. One useful method is the deliberative multi-criteria evaluation approach (Liu et al. 2011). In this method, participants assess the different risk factors associated with managing an invasive species, and by assigning risk weighting to different management strategies chart the management approach that will cause the least conflict among the stakeholders (Liu et al. 2011; Woodford et al. 2016).

Species that are on the margins (i.e. scoring high on the benefits) but medium on the negative impacts (and vice versa) should be prioritised for directed research as they represent areas where new conflicts might emerge. For example, *Cyprinus carpio* and *Micropterus floridanus* were classified as beneficial but are likely to have substantial negative impacts. The medium score for negative impacts reflects a lack of research effort on the species in South Africa (see Ellender & Weyl 2014). The situation is similar for species such as *Cherax quadricarinatus* (redclaw crayfish) that scored high on negative impacts and medium on benefits because the benefits derived from the species have not yet been quantified.

Conclusions

Most conflicts around the management of invasive species in South Africa could be explained by more than one value system (intrinsic vs. utilitarian) and cognitive level (values systems vs. risk perception). Value-based conflicts are inherently difficult to resolve because management authorities have to balance the needs of different stakeholders while still conserving the environment. An ideal management plan is where parties with different value systems agree on a

win-win solution where invasive species can still deliver benefits, but adverse impacts are reduced. This is potentially possible through open dialogue among stakeholders, trade-offs and compromises. In cases where the perceived benefits outweigh impacts such as those observed for most of the intrinsic-based conflicts, the management approach has generally been to tolerate the species and monitor whether they potentially cause impacts in the future. In contrast, when the impacts outweigh perceived benefits, management options have involved trade-offs and compromises that have minimised impact of the invasive species but retained a large proportion of their amenity values. In extreme cases, control efforts have proceeded despite opposition because of a strong body of scientific evidence and political support. Conflicts based on risk perception were mainly centred on the fear and aversion of impacts of the invasive species or the control methods proposed for it management. In some cases, such as the use of biological agents to control invasive plants species, management authorities have employed strategies to try and effectively communicate the risks through open dialogue among stakeholders and this has resulted in trade-offs and compromises (Zachariades et al. 2017). Only one case (trout species) identified a lack of trust between stakeholders and government agencies that could have resulted from lack of community engagement and transparencies in decision-making processes, differences in evaluations of potential hazards and lack of confidence in government authorities. Consultations are continuing in an attempt to reach consensus on issues among stakeholders; these issues, however, remain highly polarised, and no obvious solution is in sight. This might be a case where a formal process of scientific assessment is required (Scholes, Schreiner & Snyman-Van der Walt 2017).

The majority of invasive alien species listed in the A&IS Regulations were not considered to be conflict-generating. However, the small proportion of species identified as conflict-generating hold the potential to negatively impact the future efficiency of conservation management in South Africa by forcing regulators and managers to spend great amounts of time and resources addressing stakeholder complaints and concerns instead of discharging their duties in dealing with the species that do not generate controversy. The initial delay in promulgating the lists because of the objections of the trout lobby, and the subsequent amount of time and energy spent by the Department of Environmental Affairs (DEA) staff in negotiating the relisting of trout (Woodford et al. 2017), is a stark reminder to managers to anticipate potential management conflicts before they have a chance to disrupt problem-solving. When assessing the best strategy to deal with conflict-generating species, it is critical to identify all stakeholders at the outset and to recognise that they might hold severely divergent perceptions on the problem posed by the invasive species (Woodford et al. 2016). When these issues are addressed from the start of the development of management plans, and when stakeholders are directly engaged to determine their perceptions of the risks posed by these species, the chance of arriving at practical, equitable and non-controversial management strategies is greatly increased.

Acknowledgements

A.N., D.M.R., R.T.S. and B.W.vW. acknowledge funding from the DST-NRF Centre of Excellence for Invasion Biology. D.M.R. (grant 85417), B.W.vW (84512), O.L.F.W. (77444), D.J.W. (103581) and T.Z. (103602) received support from the National Research Foundation. A.N. and P.I. acknowledge funding provided by the Working for Water Programme of the South African Department of Environmental Affairs, through the South African National Biodiversity Institute Invasive Species Programme.

Competing interests

The authors declare that they have no financial or personal relationships which may have inappropriately influenced them in writing this article.

Authors' contributions

All authors collected data (scored species) and T.Z. analysed the data. R.T.S. developed the scoring framework. T.Z. drafted the manuscript with input from all other authors.

References

Allsopp, M. & Cherry, M., 2004, *Assessment of the economic impact on the bee and agricultural industries in the Western Cape of the clearing of certain Eucalyptus species*, Agricultural Research Council, Stellenbosch, South Africa.

Bremner, A. & Park, K., 2007, 'Public attitudes to the management of invasive non-native species in Scotland', *Biological Conservation* 139, 306–314. https://doi.org/10.1016/j.biocon.2007.07.005

Brooke, R.K., Lloyd, P.H. & de Villiers, A.L., 1986, 'Alien and translocated terrestrial vertebrates in South Africa', in A.W. Macdonald, F.J. Kruger & A.A. Ferrar (eds.), *The ecology and management of biological invasions in Southern Africa*, pp. 63–74, Oxford University Press, Cape Town.

Burgess, T.I., Andjic, V., Wingfield, M.J. & Hardy, G.E.S.J., 2007, 'The eucalypt leaf blight pathogen *Kirramyces destructans* discovered in Australia', *Australasian Plant Disease Notes* 2, 141–144. https://doi.org/10.1071/DN07056

Burgman, M.A., 2005, *Risks and decisions for conservation and environmental management*, Cambridge University Press, Cambridge.

Cox, I., 2013, 'Is this the end of the line for freshwater fishing?', *Farmers Weekly* 13036, 8–9.

De Wit, M., Crookes, D. & van Wilgen, B.W., 2001, 'Conflicts of interest in environmental management: Estimating the costs and benefits of a tree invasion', *Biological Invasions* 3, 167–178. https://doi.org/10.1023/A:1014563702261

Denies, A.M., Wittmann, M.E., Denies, J.M. & Lodge, D.M., 2016, 'Tradeoffs among ecosystems services with global tilapia introductions', *Reviews in Fisheries & Aquaculture* 24, 178–191. https://doi.org/10.1080/23308249.2015.1115466

Department of Agriculture, Forestry and Fisheries (DAFF), 2016, *Champion tress*, viewed 11 April 2016, from http://www.daff.gov.za/daffweb3/Branches/Forestry-Natural-Resources-Management/Forestry-Regulation-Oversight/Sustainable-Forestry/Champion-Trees

Dickie, I.A., Bennett, B.M., Burrows, L.E., Nuñez, M.A., Peltzer, D.A., Porté, A. et al., 2014, 'Conflicting values: Ecosystem services and invasive tree management', *Biological Invasions* 16, 705–719. https://doi.org/10.1007/s10530-013-0609-6

Durán, A., Slippers, B., Gryzenhout, M., Ahumada, R., Drenth, A., Wingfield, B.D. et al., 2009, 'DNA-based method for rapid identification of the pine pathogen, *Phytophthora pinifolia*', *FEMS Microbiology Letters* 298, 99–104. https://doi.org/10.1111/j.1574-6968.2009.01700.x

Ellender, B.R. & Weyl, O.L.F., 2014, 'A review of current knowledge, risk and ecological impacts associated with non-native freshwater fish introductions in South Africa', *Aquatic Invasions* 9, 117–132. https://doi.org/10.3391/ai.2014.9.2.01

Ellender, B.R., Woodford, D.J., Weyl, O.L.F. & Cowx, I.G., 2014, 'Managing conflicts arising from fisheries enhancements based on non-native fishes in southern Africa', *Journal of Fish Biology* 85, 1890–1906. https://doi.org/10.1111/jfb.12512

Estévez, R.A., Anderson, C.B., Pizarro, J.C. & Burgman, M.A., 2015, 'Clarifying values, risk perceptions, and attitudes to resolve or avoid social conflicts in invasive species management', *Conservation Biology* 29, 19–30. https://doi.org/10.1111/cobi.12359

Firmat, C., Alibert, P., Losseau, M., Baroiller, J.F. & Schliewen, U.K., 2013, 'Successive invasion-mediated interspecific hybridizations and population structure in the endangered cichlid *Oreochromis mossambicus*', *PLoS One* 8, e63880. https://doi.org/10.1371/journal.pone.0063880

Ford-Thompson, A., Snell, C., Saunders, G. & White, P.C.L., 2012, 'Stakeholder participation in the management of invasive vertebrates', *Conservation Biology* 26, 345–356. https://doi.org/10.1111/j.1523-1739.2011.01819.x

Forsyth, G.G., Richardson, D.M., Brown, P.J. & van Wilgen, B.W., 2004, 'A rapid assessment of the invasive status of *Eucalyptus* species in two South African provinces', *South African Journal of Science* 100, 75–77.

Foxcroft, L.C., Richardson, D.M. & Wilson, J.R.U., 2008, 'Ornamental plants as invasive aliens: Problems and solutions in Kruger National Park, South Africa', *Environmental Management* 41, 32–51. https://doi.org/10.1007/s00267-007-9027-9

Fulton, D.C., Manfredo, M.J. & Lipscom, J., 1996, 'Wildlife value orientations: A conceptual and measurement approach', *Human Dimensions of Wildlife* 1, 24–47. https://doi.org/10.1080/10871209609359060

Gaertner, M., Larson, B.M.H., Irlich, U.M., Holmes, P.M., Stafford, L., van Wilgen, B.W. et al., 2016, 'Managing invasive species in cities: A framework from Cape Town, South Africa', *Landscape and Urban Planning* 151, 1–9. https://doi.org/10.1016/j.landurbplan.2016.03.010

Geerts, S., Moodley, D., Gaertner, M., Le Roux, J.J., McGeoch, M.A., Muofhe, C. et al., 2013, 'The absence of fire can cause a lag phase: The invasion dynamics of *Banksia ericifolia* (Proteaceae)', *Austral Ecology* 38–931. https://doi.org/10.1111/aec.12035

Hoffmann, J.H., Moran, V.C. & van Wilgen, B.W., 2011, 'Prospects for the biological control of invasive *Pinus* species (Pinaceae) in South Africa', *African Entomology* 19, 393–401. https://doi.org/10.4001/003.019.0209

Holmes, T.P., Aukema, J.E., Von Holle, B., Liebhold A. & Sills, E., 2009, 'Economic impacts of invasive species in forests: Past, present, and future', *Annals of the New York Academy of Sciences* 1162, 18–38. https://doi.org/10.1111/j.1749-6632.2009.04446.x

Hulme, P.E., 2014, 'Invasive species challenge the global response to emerging diseases', *Trends in Parasitology* 30, 267–270. https://doi.org/10.1016/j.pt.2014.03.005

Hunter, G.C., Crous, P.W., Carnegie, A.J., Burgess, T.I. & Wingfield, M.J., 2011, '*Mycosphaerella* and *Teratosphaeria* diseases of Eucalyptus; easily confused and with serious consequences', *Fungal Diversity* 50, 145–166. https://doi.org/10.1007/s13225-011-0131-z

Impson, F.A.C., Kleinjan, C.A., Hoffmann, J.H., Post, J.A. & Wood, A.R., 2011, 'Biological control of Australian *Acacia* species and *Paraserianthes lophantha* (Willd.) Nielsen (Mimosaceae) in South Africa', *African Entomology* 19, 186–207. https://doi.org/10.4001/003.019.0210

Irlich, U.M., Potgieter, L., Stafford, L. & Gaertner, M., 2017, 'Recommendations for municipalities to become compliant with National legislation on biological invasions', *Bothalia* 47(2), a2156. https://doi.org/10.4102/abc.v47i2.2156

Jackson, M.C., Woodford, D.J., Bellingan, T.A., Weyl, O.L.F., Potgieter, M.J., Rivers-Moore, N.A. et al., 2015, 'Diet overlap between fish and riparian spiders: Potential impacts of an invasive fish on terrestrial consumers', *Ecology and Evolution* 6, 1745–1752. https://doi.org/10.1002/ece3.1893

Jassat, W., Naicker, N., Naidoo, S. & Mathee, A., 2013, 'Rodent control in urban communities in Johannesburg, South Africa: From research to action', *International Journal of Environmental Health Research* 23, 474–483. https://doi.org/10.1080/09603123.2012.755156

Kaplan, H., Wilson, J.R.U., Klein, H., Henderson, L., Zimmermann, H.G., Manyama, P. et al., 2017, 'A proposed national strategic framework for the management of Cactaceae in South Africa', *Bothalia* 47(2), a2149. https://doi.org/10.4102/abc.v47i2.2149

Kasrils, R., 2001, *Jacaranda-Xenophobia in the name of environment management?* viewed 13 April 2016, from http://www.stratek.co.za/.%5Carchive%5Cronniekasrils.html

Kellert, S.R., 1993, 'Values and perceptions of invertebrates', *Conservation Biology* 7, 845–855. https://doi.org/10.1046/j.1523-1739.1993.740845.x

Kueffer, C., 2013, 'Integrating natural and social sciences for understanding and managing plant invasions', in S. Larrue (ed.), *Biodiversity and societies in the Pacific Islands*, pp. 71–96, Collection 'Confluent des Sciences', Marseille, Presses Universitaires de Provence & ANU ePress, Canberra, Australia.

Kueffer, C. & Kull, C., 2017, 'Non-native species and the aesthetics of nature', in M. Vilà & P. Hulme (eds.), *Impact of biological invasions on ecosystem services*, Springer, Berlin.

Kull, C.A., Shackleton, C.M., Cunningham, P.S., Ducatillon, C., Dufour Dror, J.-M., Esler, K.J. et al., 2011, 'Adoption, use, and perception of Australian acacias around the world', *Diversity and Distributions* 17, 822–836. https://doi.org/10.1111/j.1472-4642.2011.00783.x

Le Maitre, D.C., de Lange, W.J., Richardson, D.M., Wise, R.M. & van Wilgen, B.W., 2011, 'The economic consequences of the environmental impacts of alien plant invasions in South Africa', in D. Pimentel (ed.), *Biological invasions: Economic and environmental costs of alien plant, animal, and microbe species*, pp. 295–323, CRC Press, Boca Raton, FL.

Le Roux, J.J., Foxcroft, L.C., Herbst, M. & MacFadyen, S., 2015, 'Genetic analysis shows low levels of hybridization between African wildcats (*Felis silvestris lybica*) and domestic cats (*F. s. catus*) in South Africa', *Ecology and Evolution* 5, 288–299. https://doi.org/10.1002/ece3.1275

Liu, S., Sheppard, A., Kriticos, D. & Cook, D., 2011, 'Incorporating uncertainty and social values in managing invasive alien species: A deliberative multi-criteria evaluation approach', *Biological Invasions* 13, 2323–2337. https://doi.org/10.1007/s10530-011-0045-4

Long, J.L., 2003, *Introduced mammals of the world. Their history distribution and influence*. CABI Publishing, Wallingford, U.K.

McConnachie, A.J., Strathie, L.W., Mersie, W., Gebrehiwot, L., Zewdie, K., Abdurehim, A. et al., 2011, 'Current and potential geographical distribution of the invasive plant *Parthenium hysterophorus* (Asteraceae) in eastern and southern Africa', *Weed Research* 51, 71–84. https://doi.org/10.1111/j.1365-3180.2010.00820.x

Nentwig, W., Bacher, S., Pyšek, P., Vilà, M. & Kumschick, S., 2016, 'The generic impact scoring system (GISS): A standardized tool to quantify the impacts of alien species', *Environmental Monitoring and Assessment* 188, 315. https://doi.org/10.1007/s10661-016-5321-4

Nogales, M., Martin, A., Tershy, B.R., Donlan, C.J., Veitch, D., Puerta, N. et al., 2004, 'A review of feral cat eradication on islands', *Conservation Biology* 18, 310–319. https://doi.org/10.1111/j.1523-1739.2004.00442.x

Novoa, A., Kaplan, H., Kumschick, S., Wilson, J.R.U. & Richardson, D.M., 2015, 'Soft touch or heavy hand? Legislative approaches for preventing invasions: Insights from cacti in South Africa', *Invasive Plant Science and Management* 8, 307–316. https://doi.org/10.1614/IPSM-D-14-00073.1

Novoa, A., Kaplan, H., Wilson, J.R.U. & Richardson, D.M., 2016, 'Resolving a prickly situation: Involving stakeholders in invasive cactus management in South Africa', *Environmental Management* 57, 998–1008. https://doi.org/10.1007/s00267-015-0645-3

Novoa, A., Le Roux, J.J., Robertson, M.P., Wilson, J.R.U. & Richardson, D.M., 2015a, 'Introduced and invasive cactus species: A global review', *AoB Plants* 7, plu078. https://doi.org/10.1093/aobpla/plu078

Novoa, A. & Shackleton R., 2015, 'Stakeholder involvement: Making strategies workable', *Quest* 11, 54–56.

Picker, M.D. & Griffiths, C.L., 2017, 'Alien animals in South Africa – composition, introduction history, origins and distribution patterns', *Bothalia: African Biodiversity and Conservation*.

Pieterse, P.J. & Boucher, C., 1997, 'A case against controlling introduced Acacias-19 years later', *South African Forestry Journal* 180, 37–44. https://doi.org/10.1080/10295925.1997.9631166

Powell, K.I., Chase, J.M. & Knight, T.M., 2013, 'Invasive plants have scale-dependent effects on diversity by altering species-area relationships', *Science* 339, 317–319. https://doi.org/10.1126/science.1226817

Pyšek, P. & Richardson, D.M., 2010, 'Invasive species, environmental change and management, and ecosystem health', *Annual Review of Environment and Resources* 35, 25–55. https://doi.org/10.1146/annurev-environ-033009-095548

Republic of South Africa (RSA), 2004, National Environmental Management: Biodiversity Act 10 of 2004, Proc. R47/Government Gazette No. 26887/20041008.

Republic of South Africa (RSA), 2014, Government Notice No. 37885, Vol. 590, Regulation Gazette No. 10244.

Republic of South Africa (RSA), 2016, Government Notice R864, Government Gazette No. 40166.

Richardson, D.M., Cambray, J.A., Chapman, R.A., Dean, W.R.J., Griffiths, C.L., Le Maitre, D.C. et al., 2003, 'Vectors and pathways of biological invasions in South Africa – past, future and present', in G. Ruiz & J. Carlton (eds.), *Invasive species: Vectors and management strategies*, pp. 292–349, Island Press, Washington, DC.

Richardson, D.M., Wilson, J.R.U., Weyl, O.L.F. & Griffiths, C.L., 2011, 'South Africa: Invasions', in D. Simberloff & M. Rejmánek (eds.), *Encyclopedia of biological invasions*, pp. 643–651, University of California Press, Berkeley, CA.

Rokeach, M., 1973, *The nature of human values*, The Free Press, New York.

Rotherham, I.D. & Lambert R.A., 2011, *Invasive and introduced plants and animals: Human perceptions, attitudes and approaches to managemen*, Earthscan, New York.

Scholes, R.J., Schreiner, G. & Snyman-Van der Walt, L., 2017, 'Scientific assessments: Matching the process to the problem', *Bothalia* 47(2), a2144. https://doi.org/10.4102/abc.v47i2.2144

Shackleton, C.M., McGarry, D., Fourie, S., Gambiza, J., Shackleton, S.E. & Fabricius, C., 2007, 'Assessing the effects of invasive alien species on rural livelihoods: Case examples and a framework from South Africa', *Human Ecology* 35, 113–127. https://doi.org/10.1007/s10745-006-9095-0

Shackleton, R.T., Le Maitre, D.C., Pasiecznik, N.M. & Richardson, D.M., 2014, '*Prosopis*: A global assessment of the biogeography, benefits, impacts and management of one of the world's worst woody invasive plant taxa', *AoB Plants* 6, plu027. https://doi.org/10.1093/aobpla/plu027

Shackleton, R.T., Le Maitre, D.C. & Richardson, D.M., 2015a, '*Prosopis* invasions in South Africa: Population structures and impacts on native tree population stability', *Journal of Arid Environments* 114, 70–78. https://doi.org/10.1016/j.jaridenv.2014.11.006

Shackleton, R.T., Le Maitre, D.C. & Richardson, D.M., 2015c, 'Stakeholder perceptions and practices regarding *Prosopis* (mesquite) invasions and management in South Africa', *Ambio* 44, 569–581. https://doi.org/10.1007/s13280-014-0597-5

Shackleton, R.T., Le Maitre, D.C., van Wilgen, B.W. & Richardson, D.M., 2015b, 'The impact of invasive alien *Prosopis* species (mesquite) on native plants in different environments in South Africa', *South African Journal of Botany* 97, 25–31. https://doi.org/10.1016/j.sajb.2014.12.008

Shackleton, R.T., Le Maitre, D.C., van Wilgen, B.W. & Richardson, D.M., 2016, 'Identifying barriers to effective management of widespread invasive alien trees: *Prosopis* species (mesquite) in South Africa as a case study', *Global Environmental Change* 38: 183–194. https://doi.org/10.1016/j.gloenvcha.2016.03.012

Shelton, J.M., Samways, M.J. & Day, J.A., 2015, 'Predatory impact of non-native rainbow trout on endemic fish populations in headwater streams in the Cape Floristic Region of South Africa', *Biological Invasions* 17, 365–379. https://doi.org/10.1007/s10530-014-0735-9

Slovic, P., 1999, 'Trust, emotion, sex, politics, and science: Surveying the risk-assessment battlefield', *Risk Analysis* 19, 689–701. https://doi.org/10.1111/j.1539-6924.1999.tb00439.x

Strathie, L., 2015, 'Managing the invasive alien plant *Parthenium hysterophorus* in South Africa', *Grassroots* 15, 15–21.

Stubbings, J.A., 1977, 'A case against controlling introduced acacias', in A.A. Balkema (eds.), *Proceedings of the Second National Weeds Conference of South Africa*, Stellenbosch, South Africa, pp. 89–107, Cape Town.

Tennent, J., Downs, C.T. & Bodasing, M., 2009, 'Management recommendations for feral cat (*Felis catus*) populations within an urban conservancy in KwaZulu-Natal, South Africa', *South African Journal of Wildlife Research* 39, 137–142. https://doi.org/10.3957/056.039.0211

van Rensburg, B.J., Weyl, O.L.F., Davies, S.J., van Wilgen, N.J., Spear, D., Chimimba, C.T. et al., 2011, 'Invasive vertebrates of South Africa', in D. Pimentel (ed.), *Biological invasions: Economic and environmental costs of alien plant, animal, and microbe species*, pp. 326–378, CRC Press, Boca Raton, FL.

van Wilgen, B.W., 2012, 'Evidence, perceptions, and trade-offs associated with invasive alien plant control in the Table Mountain National Park, South Africa', *Ecology and Society* 17(2), 23. https://doi.org/10.5751/ES-04590-170223

van Wilgen, B.W., Dyer, C., Hoffmann, J.H., Ivey, P., Le Maitre, D.C., Richardson, D.M. et al., 2011, 'National-scale strategic approaches for managing introduced plants: Insights from Australian acacias in South Africa', *Diversity and Distributions* 17, 1060–1075. https://doi.org/10.1111/j.1472-4642.2011.00785.x

van Wilgen, B.W. & Richardson, D.M., 2014, 'Challenges and trade-offs in the management of invasive alien trees', *Biological Invasions* 16, 721–734. https://doi.org/10.1007/s10530-013-0615-8

Vardien, W., Richardson, D.M., Foxcroft, L.C., Thompson, G.D., Wilson, J.R.U. & Le Roux, J.J., 2012, 'The introduction history, spread, and current distribution of *Lantana camara* in South Africa', *South African Journal of Botany* 81, 81–94. https://doi.org/10.1016/j.sajb.2012.06.002

Weyl, O.L.F., 2008, 'Rapid invasion of a subtropical lake fishery in central Mozambique by Nile tilapia, *Oreochromis niloticus* (Pisces: Cichlidae)', *Aquatic Conservation: Marine and Freshwater Ecosystems* 18, 839–851. https://doi.org/10.1002/aqc.897

Wilson, J.R.U., Gaertner, M., Richardson, D.M. & van Wilgen, B.W., 2017, 'Contributions to the national status report on biological invasions in South Africa', *Bothalia* 47(2), a2207. https://doi.org/10.4102/abc.v47i2.2207

Wingfield, M.J., Hammerbacher, A., Ganley, R.J., Steenkamp, E.T., Gordon, T.R., Wingfield, B.D. et al., 2008a, 'Pitch canker caused by *Fusarium circinatum* – A growing threat to pine plantations and forests worldwide', *Australasian Plant Pathology* 37, 319–334. https://doi.org/10.1071/AP08036

Wingfield, M.J., Slippers, B., Hurley, B.P., Coutinho, T.A., Wingfield, B.D. & Roux, J., 2008b, 'Eucalypt pests and diseases: Growing threats to plantation productivity', *Southern Forests* 70, 139–144. https://doi.org/10.2989/SOUTH.FOR.2008.70.2.9.537

Wingfield, M.J., Slippers, B. & Wingfield, B.D., 2010, 'Novel associations between pathogens, insects and tree species threaten world forests', *New Zealand Journal of Forestry Science* 40, S95–S103.

Wise, R.M., van Wilgen, B.W. & Le Maitre, D.C., 2012, 'Costs, benefits and management options for an invasive alien tree species: The case of mesquite in the Northern Cape, South Africa', *Journal of Arid Environments* 84, 80–90. https://doi.org/10.1016/j.jaridenv.2012.03.001

Woodford, D.J., Ivey, P., Jordaan, M.S., Kimberg, P.K., Zengeya, T. & Weyl, O.L.F., 2017, 'Optimising invasive fish management in the context of invasive species legislation in South Africa', *Bothalia* 47(2), a2138. https://doi.org/10.4102/abc.v47i2.2138

Woodford, D.J., Richardson, D.M., MacIsaac, H.J., Mandrak, N., van Wilgen, B.W., Wilson, J.R.U. et al., 2016, 'Confronting the wicked problem of managing invasive species', *Neobiota* 31, 63–86. https://doi.org/10.3897/neobiota.31.10038

Zachariades, C., Day, M., Muniappan, R. & Reddy, G.V.P. 2009, '*Chromolaena odorata* (L.) King and Robinson (Asteraceae) ', in R. Muniappan, G.V.P. Reddy & A. Raman (eds.), *Biological control of tropical weeds using arthropods*, pp. 130–162, Cambridge University Press, Cambridge, UK.

Zachariades, C., Hoffmann, J.H. & Roberts, A., 2011, 'Biological control of mesquite (*Prosopis* species) (Fabaceae) in South Africa', *African Entomology* 19, 402–415. https://doi.org/10.4001/003.019.0230

Zachariades, C., Paterson, I.D., Strathie, L.W., Hill, M.P. & Wilgen, B.W.V., 2017, 'Assessing the status of biological control as a management tool for suppression of invasive alien plants in South Africa', *Bothalia* 47(2), a2142. https://doi.org/10.4102/abc.v47i2.2142

Zengeya, T.A., Booth, A.J. & Chimimba, C.T., 2015, 'Broad niche overlap between invasive Nile tilapia *Oreochromis niloticus* and indigenous congenerics in southern Africa: Should we be concerned?', *Entropy* 17, 4959–4973. https://doi.org/10.3390/e17074959

Zengeya, T.A. & Marshall, B.E., 2007, 'Trophic interrelationships amongst cichlid fishes in a tropical African reservoir (Lake Chivero, Zimbabwe)', *Hydrobiologia* 592, 175–182. https://doi.org/10.1007/s10750-007-0790-7

Impacts of invasive plants on animal diversity in South Africa

Authors:
Susana Clusella-Trullas[1] ⓘ
Raquel A. Garcia[1] ⓘ

Affiliations:
[1]Centre for Invasion Biology, Department of Botany and Zoology, Stellenbosch University, South Africa

Corresponding author:
Susana Clusella-Trullas,
sct333@sun.ac.za

Background: Increasing numbers of invasive alien plant (IAP) species are establishing around the globe and can have negative effects on resident animal species function and diversity. These impacts depend on a variety of factors, including the extent of invasion, the region and the taxonomic group affected. These context dependencies make extrapolations of IAP impacts on resident biota from region to region a substantial challenge.

Objectives: Here, we synthesised data from studies that have examined the effects of IAPs on animal diversity in South Africa. Our focus is on ectothermic organisms (reptiles, amphibians and invertebrates).

Method: We sourced relevant articles using keywords relating to (1) the effects of IAPs on species diversity (abundance, richness and composition), (2) the IAP and (3) the native ectotherm. We extracted the taxonomic and spatial coverage of IAPs and affected native species and assessed the extent of information given on potential mechanisms driving IAP impacts.

Results: Across the 42 studies, IAPs had a decreasing or neutral effect on native animal abundance and richness and significantly changed species composition. This review highlighted the paucity of studies and the research deficits in taxonomic and geographic coverage and in the mechanisms underlying IAP impacts on ectotherms.

Conclusion: By assessing the status of knowledge regarding the impacts of IAPs on resident animal species in South Africa, this study identifies information gaps and research priorities at the country level with a view to informing monitoring and conservation efforts, such as alien plant removal and control programmes, and ensuring that endemic terrestrial animal diversity is maintained.

Introduction

Invasive alien species are considered a major pressure on the current state of biodiversity globally (Butchart et al. 2010). Invasive alien plants (IAPs), in particular, have spread rapidly and extensively in many regions of the world, impacting resident species diversity, ecosystem processes and people's livelihoods (Levine et al. 2003; Pyšek et al. 2012; Schirmel et al. 2016; Vilà et al. 2011). South Africa is no exception and nearly two million hectares of land have been invaded by alien plants (Van Wilgen et al. 2012), with well-known impacts on hydrology, nutrient cycling and fire regimes (Kraaij, Cowling & Van Wilgen 2011; Le Maitre, Gush & Dzikiti 2015; Richardson & Van Wilgen 2004). This estimate of alien plant coverage includes 27 species, without incorporating arid and transformed land except for *Prosopis* trees in the arid northwest of the country (Kotzé et al. 2010; Van den Berg, Kotze & Beukes 2013). Alien *Acacia* species cover most of the dense areas of invasion, followed by *Eucalyptus* and *Pinus* trees, *Opuntia* cacti and *Chromolaena odorata* shrubs. These invasions extend across the country, with higher concentrations in the southwestern, southern and particularly eastern coastal belts and the adjacent interior (Henderson 2007; Kotzé et al. 2010; Van Wilgen et al. 2012). Overall, there is reasonable knowledge of alien plant occurrence in South Africa, especially at a coarse spatial resolution. Whereas their effects on native plant diversity have been fairly well assessed (e.g. Gaertner et al. 2009; Richardson & Van Wilgen 2004), fewer studies have focused on the impacts of alien plants on native animal communities (Richardson et al. 2011, but see Breytenbach 1986).

Species population and community metrics such as abundance, richness and composition can provide useful baseline data as indicators of animal diversity change between invaded and uninvaded areas. The direction and magnitude of effects of alien plant invasions on animal communities can, however, depend on a variety of factors, including the scale of the plant invasion (extent and density), the stage of invasion, and the region and taxonomic group affected

(Kumschick et al. 2015; Ricciardi et al. 2013). These context dependencies make extrapolations of the effects of IAPs on resident biota from region to region a substantial challenge, especially for the development of generalised management frameworks across diverse habitats. The invasion of alien plants into natural or previously uninvaded habitats involves a number of significant changes to the habitat, often negatively affecting resident fauna and sometimes in counterintuitive or non-obvious ways (Figure 1). Alien plants may directly modify the structure and complexity of the physical environment and, thus, restrict the opportunities for the animal to thermoregulate or hydroregulate within its microenvironment or impose barriers to essential functions such as moving, creating nests or finding refuges. Alien plants can also directly or indirectly affect food resources for animal communities (e.g. Groot, Kleijn & Jogan 2007). For example, a change in plant composition will affect herbivores directly by reducing the amount or quality of plant hosts whereas a change in habitat structure, microenvironment and litter or soil properties can indirectly affect prey availability or predator abundance and alter trophic interactions (Figure 1; e.g. Pearson 2009).

Whereas reports of negative impacts of invasive plant species are pervasive in the literature, positive effects have also been reported, for example, via increases in suitable habitat or net resources to recipient fauna (e.g. Schlaepfer, Sax & Olden 2010). Whether the latter represent rare case studies or whether any general patterns can be drawn from these is presently unclear. Thus, knowledge of the underlying mechanisms should improve the understanding of the consequences of alien plant invasions on native diversity. More importantly, knowledge of proximate and ultimate causes of species declines may enhance our ability to predict responses of animal communities in newly invaded areas of South Africa with similar characteristics to those studied previously, or facing concomitant pressures such as global climate change or habitat transformation (Ricciardi et al. 2013). Together, these are essential elements of the scientific framework that will allow invasion science to robustly predict the impacts of new and developing invasions, and not simply be viewed as a series of unique invasion case studies.

In this study, we aim to synthesise studies that have examined the effects of IAPs on animal diversity in South Africa. We concentrate this review on ectotherms (reptiles, amphibians and invertebrates) for several reasons. Firstly, their energy budgets are more directly influenced by the environment compared with endotherms (Gates 1980). Consequently, environmental factors likely play a major role in determining a suite of physiological and behavioural attributes and, at

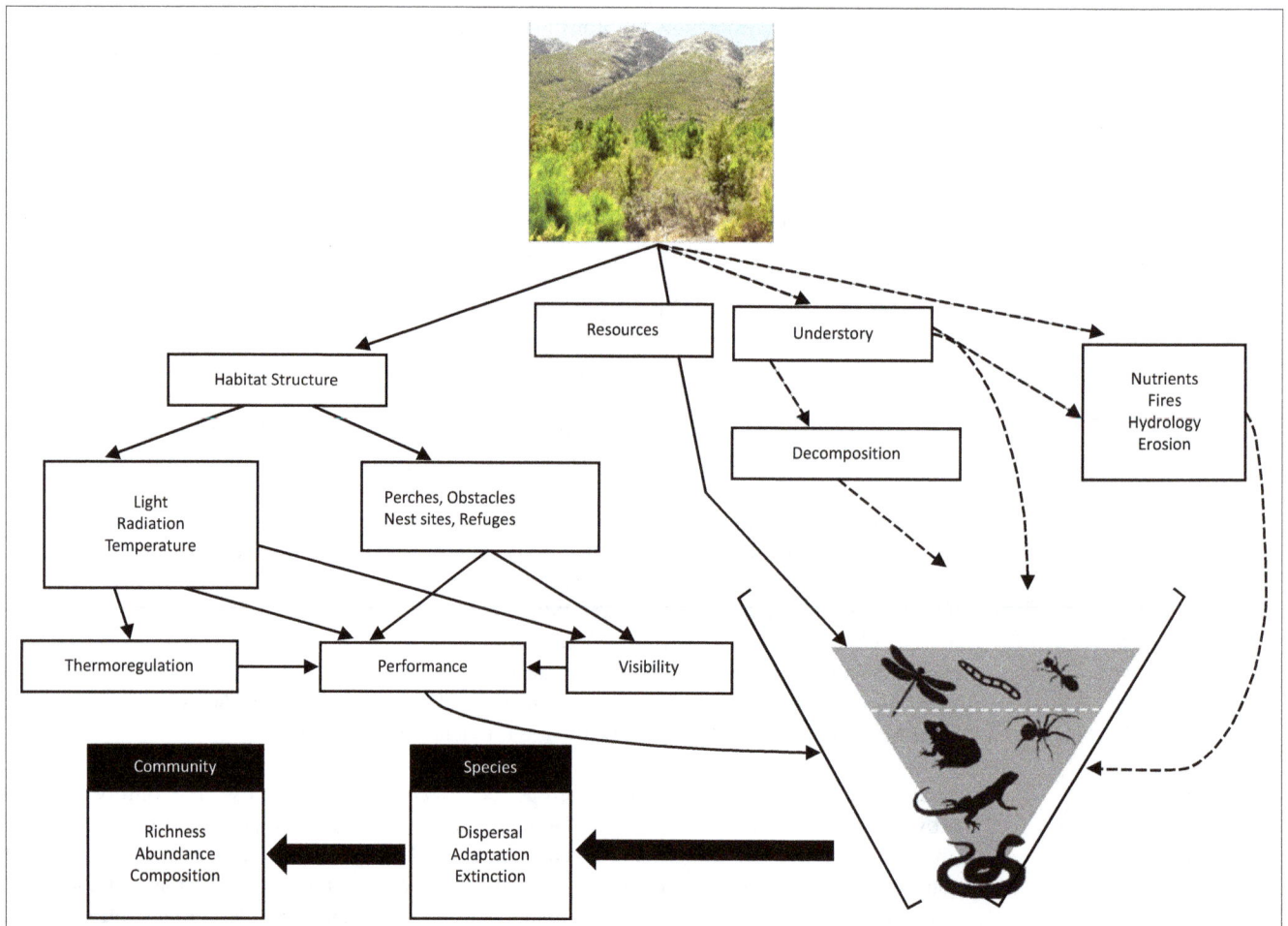

The picture depicts the invasion of an alien plant species (e.g. *Pinus sp.*) into native fynbos vegetation (e.g. *Protea sp.*). Solid and dashed arrows refer to direct and indirect effects of IAPs on native species, respectively.

FIGURE 1: Potential mechanisms through which invasive alien plants (IAPs) affect ectotherm diversity.

least partly, influence life history and timing of key phenological events (e.g. mating and reproduction) in the group. Secondly, they typically have smaller dispersal abilities than mammals and birds (Endler 1977), likely reducing their capacity to move away from disturbance or suboptimal conditions. Finally, they make a large contribution to overall animal diversity that is mostly explained by high insect diversity and abundance (Wilson & Peter 1988), playing a central role in food webs and ecosystem function. To present a comprehensive view of the status of knowledge of impacts of IAPs on resident animal species in South Africa and identify information gaps, we ask three key questions: (1) How many studies have addressed the effects of IAPs on terrestrial ectotherm diversity and what general patterns can we draw from these?, (2) What is the taxonomic and spatial coverage of IAPs and affected native species studied so far? and (3) How much is known about the mechanisms underlying these impacts? These questions are central to assess the status of knowledge of alien plant impacts on animal species diversity and inform invasive plant monitoring and control programmes in South Africa (Wilson et al. 2017).

Methods

We searched the ISI Web of Science for relevant studies comparing abundance, richness or composition of terrestrial ectotherms between invaded and uninvaded sites in South Africa. Our search combined terms for (1) invasive plants; (2) native reptiles, amphibians and terrestrial invertebrates; and (3) effects on species abundance, richness or composition (shown in detail in Appendix 1). We included studies comparing sites with native vegetation or cleared of alien vegetation to sites invaded by alien plants or with plantations of alien species. A second search targeted studies addressing only the mechanisms underlying the potential effects on ectotherm species, without necessarily quantifying changes in species diversity. This second search thus replaced the search terms for effects with terms for mechanisms such as altered thermoregulatory behaviour, prey availability or reproductive output (Appendix 1). We retrieved articles, reviews or book chapters for all years available on the Web of Science Core Collection on 10 June 2016. The articles considered relevant for our review were then screened for additional references.

We gathered information from the studies on the location of the field sites and respective biomes (Mucina et al. 2014), and on the native animal and IAP species included. We classified the studies according to the phyla of native animals potentially affected by plant invasions (Arthropoda, Annelida, Chordata and Onychophora). For the IAP species included, we used three classifications. Firstly, we assigned IAP species to six major growth forms: tree, shrub or bush, vine, forb, grass and aquatic plant. Secondly, we used the categories of the Alien and Invasive Species (A&IS) Lists that were published under the National Environmental Management: Biodiversity Act (NEM:BA) in 2014. The NEM:BA A&IS categories include eradication targets (category 1a), widespread invasive species where a national species management programme is required

(category 1b) and invasive species that can be kept under managed circumstances (categories 2 and 3). Thirdly, to assess the extent to which the existing studies focus on the invasive plants with the largest potential impacts on ectotherms, we also considered a published classification of the most prominent invasive plant species in South Africa, according to their estimated impacts on native biodiversity (Van Wilgen et al. 2008). For all classifications above, we assigned studies to multiple categories when they presented individual comparisons for more than one biome, native phylum or IAP category.

To assess the impacts of alien plants on native animal diversity, we focused on comparisons of species abundance or richness between sites with native and alien vegetation and classified the effects as positive (i.e. an increase in abundance or richness in sites with alien vegetation), negative (decrease) or neutral effect. Native vegetation sites included sites cleared of alien plants in two studies where authors indicated that sufficient time had elapsed for recovery of native vegetation. We incorporated comparisons based on original data, accumulation and rarefaction curves, and richness estimators (e.g. Chao1 and Chao2) but did not include diversity indices (e.g. Shannon's or Simpson's index) as their use was very variable across studies. For composition, comparisons were classified as alien vegetation resulting in a change or not in species composition. These comparisons were typically the result of ordination techniques based on similarity measures or cluster analyses. When studies presented the effects of alien plants on fauna for several functional or taxonomic groups (e.g. herbivores vs. predators and beetles vs. spiders) or for different habitat types (e.g. invaded by *Eucalyptus* vs. *Pinus*), we considered each comparison separately in the synthesis unless there were no statistical tests associated with these. Therefore, the number of comparisons was typically higher than the number of studies assessed. If available, we extracted information on the growth form, stand age and spatial coverage of the IAP and on mechanisms driving the impacts such as changes in habitat structure, thermal opportunities, food resources, predators and refuge or nest site availability.

Results

Our first search for studies addressing the effects of invasive plants on ectotherm diversity in South Africa yielded 358 studies. Of these, only 42 were relevant for this review (Appendix 2). Our second search for articles studying the mechanisms underlying the effects of alien plants on ectotherms in South Africa yielded 702 papers. Among these, we only retained six relevant studies (Appendix 2), partly because a large portion of the articles found investigated the viability of biological agents for invasive plant control which was outside the focus of this study.

The 42 papers reviewed were published between 1985 and 2016, with a slight increase in the annual number of publications since 2000 (Figure 2a). The vast majority of studies

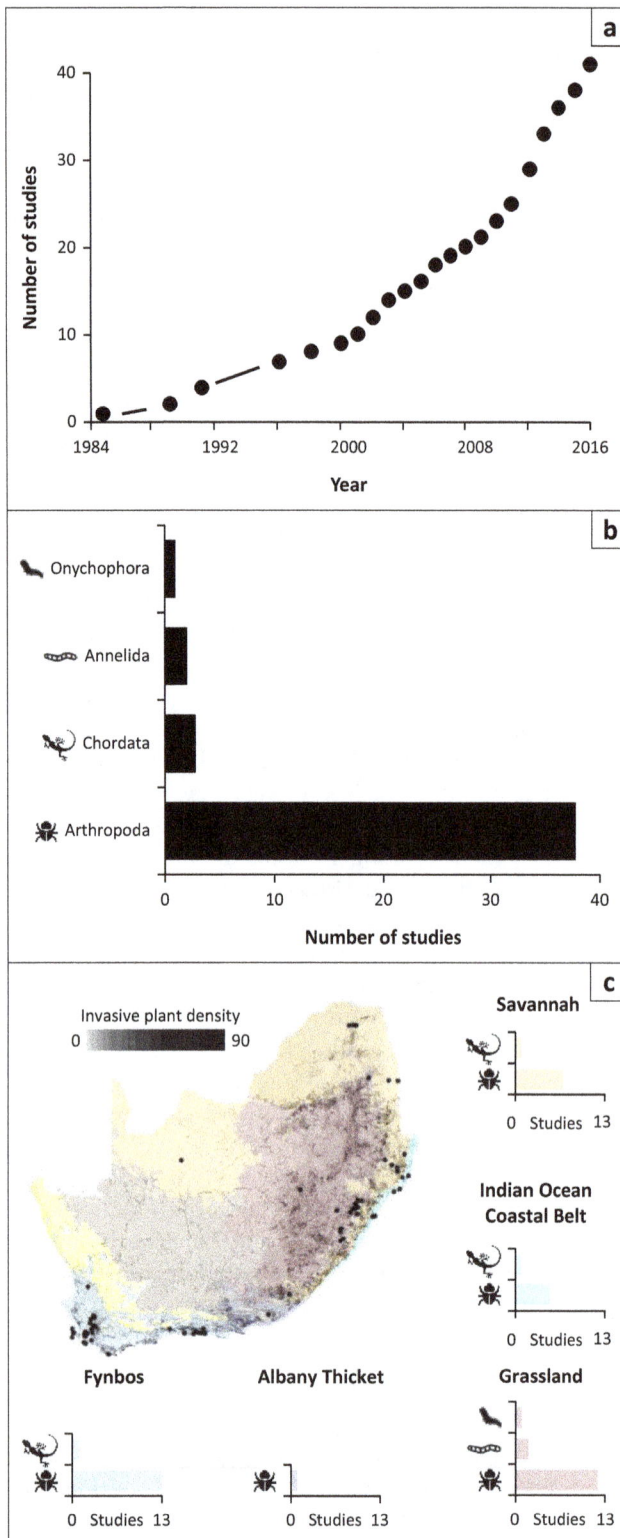

Source: (c) Mucina et al. 2014; Kotzé et al. 2010
There are no studies in the Succulent Karoo (light yellow) and Nama Karoo (brown) biomes. The forest and desert biomes are not represented in the map due to their small range.

FIGURE 2: Distribution of studies (a) per year of publication, (b) per taxonomic group (phylum) of native ectothermic organisms, and (c) per biome and taxonomic group of native ectothermic organisms. The map in (c) shows the distribution of study sites (black dots) in South Africa across seven biomes, with the shades of grey indicating the density of invasive plants.

A single study included Onychophora and two studies focused on earthworms. Only three studies addressed vertebrate ectotherms, covering amphibians (Order Anura), lizards and snakes (Order Squamata).

Most studies compared sites along the coastal belt and adjacent interior in the Western Cape and KwaZulu-Natal provinces (Figure 2c). The majority of studies took place in the Fynbos and Grassland biomes, whereas a few studies covered the Savanna, Indian Ocean Coastal Belt and Albany Thicket biomes (Figure 2c). Native arthropods remained the focal native class across biomes (Figure 2c). Trees, particularly those belonging to the genera *Acacia*, *Eucalyptus* and *Pinus*, were the most common growth form of invasive plants (Figure 3a), followed by shrubs such as *Chromolaena odorata*, *Hakea* species, *Lantana camara* and *Solanum mauritianum*. These alien trees and shrubs are listed as major invaders in the country (Henderson 2007; Le Maitre, Versfeld, & Chapman 2000; Van Wilgen et al. 2012). Many of them are species that need to be controlled according to the NEM:BA A&IS regulations (category 1b; Figure 3a) and are estimated to have high impacts on native biodiversity (Van Wilgen et al. 2008; Figure 3b).

Alien plants had a larger decreasing effect on native species abundance compared to species richness (Figure 4a and Table 1). Most studies found that the effects of alien plants were either neutral or decreasing for native animal species richness, but increasing effects of alien plants were rare for both species richness and abundance. Alien plants had a substantial impact on species composition (Figure 4b and Table 1). Among the 15 IAPs listed on NEM:BA, only 3 species (*Arundo donax*, *Chromolaena odorata* and *Passiflora edulis*) showed no negative effects on the arthropod species studied. However, these results refer to a single study conducted for each alien plant, highlighting the data deficiency for these species (Table 1). *Acacia mearnsii* was the IAP, most commonly studied, with negative or neutral effects on arthropod species found in four studies (Table 1). Although most studies incorporated standard metrics of species abundance and richness, accumulation and rarefaction curves were presented (or mentioned) in only 13 and 7 studies, respectively, and 10 studies provided comparisons of metrics between invaded and uninvaded sites without presenting accompanying statistical analyses.

Seventy percent of the studies investigating differences in species diversity between invaded and uninvaded habitats referred to potential mechanisms underlying the patterns found. These mechanisms included changes in habitat structure ($n = 15$ of 29 studies), microclimates ($n = 13$ of 29 studies) and food resources (including host specificity for herbivores, $n = 14$ of 29 studies), and the degree of species' ecological breadth (e.g. generalists vs. specialists, $n = 3$ of 29 studies). In the 6 publications that solely referred to mechanisms (Appendix 2), the authors highlighted habitat structure, microclimate and food resources as mechanisms affecting functional diversity (according to McGill et al.

focused on arthropods (Figure 2b), particularly in the Insecta (mostly Coleoptera, Hymenoptera, Diptera, Lepidoptera and Odonata) and Arachnida (mostly Araneae) classes.

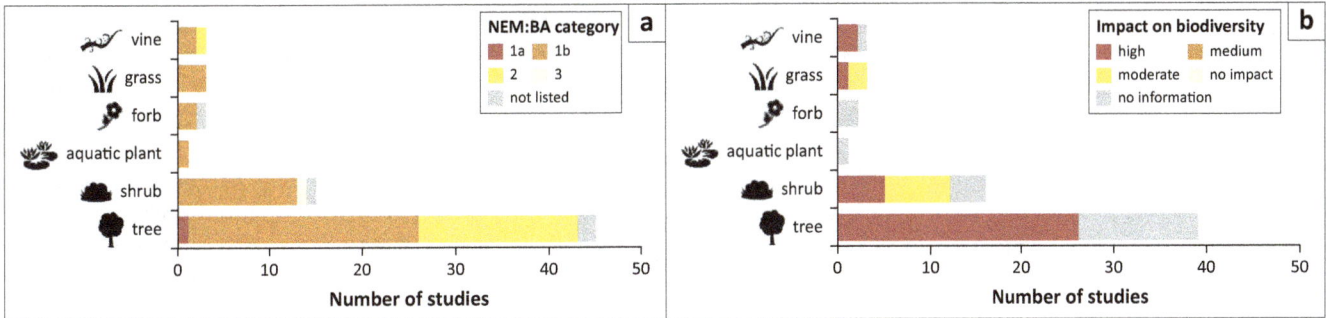

Source: (a) National Environmental Management: Biodiversity Act (NEM:BA), 2014; (b) van Wilgen et al. 2008

FIGURE 3: Distribution of studies per growth form of the invasive plant species studied and their classification according to (a) the NEM:BA regulations for control and management of invasive species and (b) their estimated impacts on biodiversity according to a published classification.

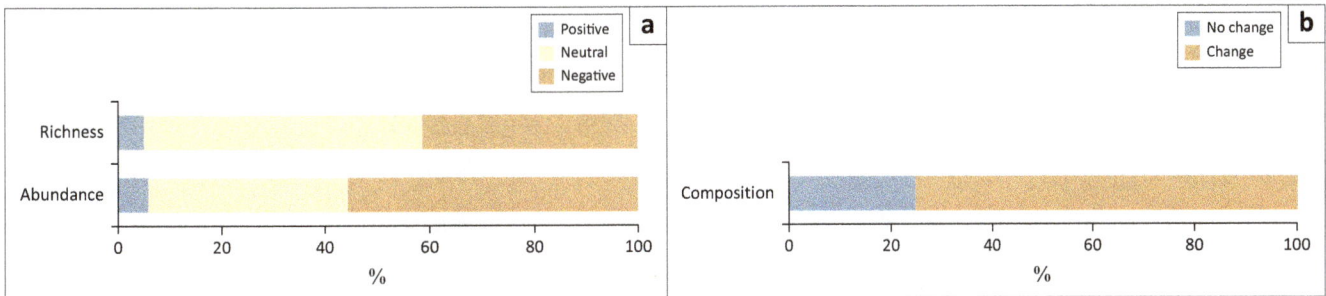

FIGURE 4: Percentage of comparisons performed in the 42 studies reviewed that found invaded sites to have (a) positive (increased diversity), neutral or negative (decreased diversity) effects on native ectotherm species richness (n = 80 comparisons) and abundance (n = 52), and (b) the same or altered species composition (n = 36) as uninvaded sites.

TABLE 1: Impacts of specific invasive alien plants on abundance, richness and composition of several taxonomic groups of ectotherms.

Invasive alien plant species	Impacts on native species		
	Abundance	Richness	Composition
Acacia dealbata	Negative: Insecta (1)	Negative: Insecta (1)	Different: Insecta (1)
Acacia mearnsii	Negative: Araneae, Lepidoptera, Formicidae (2) Neutral: Odonata (3), Coleoptera, Diptera, Hemiptera, Hymenoptera (2)	Negative: Odonata (3), Araneae and Formicidae (2), Araneae*, Coleoptera•, Diptera*, Hemiptera•, Hymenoptera• and Formicidae* (4) Neutral: Coleoptera, Diptera, Hemiptera, Hymenoptera, Lepidoptera (2), Lepidoptera• and Orthoptera• (4)	Different: Odonata (3, 5), Orthoptera*, Formicidae•, Hymenoptera• and Hemiptera• (4) Not different: Araneae•, Coleoptera•, Diptera•, Lepidoptera• (4)
Acacia saligna	Negative: Formicidae (6)	Neutral: Formicidae (6)	Different: Formicidae (6)
Arundo donax	-	Neutral: Insecta and Arachnida (7)	-
Chromolaena odorata	-	Neutral: Araneae (8)	-
Eucalyptus camaldulensis	Negative: Araneae, Coleoptera, Hymenoptera, Lepidoptera, Formicidae (9) Neutral: Diptera, Hemiptera, Orthoptera (9)	Negative: Araneae, Orthoptera, Hymenoptera, Lepidoptera, Formicidae (9) Neutral: Coleoptera, Diptera, Hemiptera (9)	Different: Araneae, Orthoptera, Hymenoptera, Lepidoptera, Formicidae, Coleoptera, Diptera, Hemiptera (9)
Eucalyptus grandis	-	-	Different: Formicidae (10) Not different: Lepidoptera, Orthoptera, Araneae and Scarabaeidae (10)
Eucalyptus lehmannii	-	Negative: Hymenoptera, Opiliones and Amphipoda (11)	Different: Hymenoptera, Opiliones and Amphipoda (11)
Opuntia stricta	Positive: Coleoptera (12) Neutral: Araneae (12)	Neutral: Coleoptera (12), Araneae (12)	Different: Coleoptera (12) Not different: Araneae (12)
Passiflora edulis	Neutral: Coleoptera (13)	Neutral: Coleoptera (13)	
Pinus patula	-	-	Different: Formicidae and Araneae (10) Not different: Orthoptera, Lepidoptera and Scarabaeidae (10)
Pinus radiata	Negative: Squamata (14) Positive: Collembola (15)	Negative: Squamata (14) Positive: Collembola (15)	Different: Collembola (15), Squamata (14)
Prosopis glandulosa	-	Negative: Scarabaeidae (16)	Different: Scarabaeidae (16)
Rubus cuneifolius	-	Positive: Anisoptera (17) Neutral: Zygoptera (17)	Not different: Odonata (17)
Solanum mauritianum	-	Neutral: Insecta (18)	Different: Insecta (18)

(1) Coetzee, van Rensburg & Robertson 2007; (2) van der Colff et al. 2015; (3) Samways & Sharratt 2010; (4) Maoela et al. 2016; (5) Samways & Grant 2006; (6) French & Major 2001; (7) Canavan et al. 2014; (8) Mgobozi et al. 2008; (9) Roets & Pryke 2012; (10) Pryke & Samways 2012; (11) Ratsirarson et al. 2002; (12) Robertson et al. 2011; (13) Padayachi, Proches & Ramsay 2014; (14) Schreuder & Clusella-Trullas in press; (15) Liu et al. 2012; (16) Steenkamp & Chown 1996; (17) Kietzka et al. 2015; (18) Olckers & Hulley 1989.

For invasive alien plants listed in South Africa's *National Environmental Management: Biodiversity Act* (NEM:BA) of 2014, the table summarises the negative, neutral or positive impacts on native reptiles, amphibians or arthropods that were found in the studies reviewed. When reported, the reversibility or irreversibility of negative impacts after alien plant clearing or restoration is indicated with an asterisk (*) or a filled circle (•), respectively. Only statistically significant results are included.

2006) and animal behaviours such as the ability to create nests, flight dynamics and flower visitation rates. Of a total of 35 studies that addressed mechanisms, 21 studies used an observational (correlative) approach to assess the link between impact and mechanism, 4 studies followed a manipulative experimental approach (including a study that incorporated the clearing history of the invasive plant) and 10 studies speculated on potential mechanisms.

Discussion

IAPs have been recognised as a threat to South Africa's native biodiversity for more than two decades, with efforts to manage invasions underway through the Working for Water Programme since 1995 (Van Wilgen et al. 2016). Our review underscores the importance of controlling IAPs to reduce their impacts on terrestrial ectotherm diversity in the country. Although not always statistically significant, reductions in abundance and richness of native ectothermic species were found in 56% and 41% of the comparisons presented in the studies reviewed, respectively, and changes in species composition were reported in almost 75% of the comparisons. These findings echo those of global reviews and meta-analyses, which have reported significant decreases in native animal species abundance, diversity and fitness as impacts of plant invasions (Pyšek et al. 2012; Schirmel et al. 2016; Van Hengstum et al. 2014; Vilà et al. 2011). In the following three sections, we discuss key directions for future research, the context-dependent nature of the problem and issues relating to methodologies used to assess impacts.

Research gaps

The studies reviewed in this synthesis covered some of the areas of South Africa most heavily invaded by alien plants such as *Acacia*, *Eucalyptus* and *Pinus* tree species, along a southern and eastern coastal belt (Kotzé et al. 2010; Figure 2c). The extent of this coverage was, however, very small relative to the distribution of invasive plants in the country. Study locations tend to be clumped in particular areas (Figure 2c), likely associated with the accessibility of sites, management prioritisation for particular problematic alien plant species or fields of interest of research institutes and invasion biologists. The arid regions of the country, including the Succulent Karoo, Nama Karoo and Desert biomes, and the arid parts of the Savanna and Grassland biomes have not been assessed. However, these areas are invaded by dense stands of *Prosopis* spp. (Van den Berg et al. 2013), listed as NEM:BA category 1b and estimated to have high impacts on natural ecosystems. Less attention has also been given to invasive forbs, aquatic plants, vines and grasses (Figure 3). Alien grasses, for example, are recognised as understudied in South Africa (Milton 2004; Richardson & Van Wilgen 2004; Visser et al. 2017), despite their prominence in invasion science in other regions of the world (Hulme et al. 2013; Visser et al. 2016). In South Africa, however, forbs, vines and grasses tend to be less prominent than invasive shrubs and trees, and introduced grasses are not

well adapted to local conditions, such as fire regimes (Van Wilgen et al. 2008, 2012; Visser et al. 2016). Overall, the 42 studies assessing alien plant impacts on native ectotherms addressed some of the major IAP species of South Africa in terms of potential impacts and control regulations in place (e.g. *Acacia*, *Eucalyptus* and *Pinus* species; Figure 3 and Table 1) but gave less or no attention to other important alien plants such as *Lantana camara* in the Savanna Biome and *Populus* species in the Grassland Biome (Henderson 2007). For each NEM:BA-listed IAP species examined individually, the evidence collated typically comes from a small number of studies and refers to a small number of native taxonomic groups (Table 1).

Our synthesis highlights a serious deficit in the knowledge of impacts of IAPs on terrestrial ectothermic groups other than arthropods, which mostly comprised insects and spiders. South Africa is known for its exceptional reptile and amphibian diversity and its high level of endemism; between 40% and 67% of its indigenous species of amphibians, chelonians, lizards and snakes are unique to the country (Bates et al. 2014; Measey 2011). Invasive alien vegetation has been highlighted as a major threat to reptile and amphibian diversity, both nationally and globally (Gibbons et al. 2000; Martin & Murray 2011; Measey 2011; Mokhatla et al. 2012), but despite this realisation, little focus has been given to the effects of alien vegetation on these organisms in South Africa. The three studies that examined the effects of alien vegetation on squamate and amphibian species found significant reductions in species richness and substantial changes in species composition (Russell & Downs 2012; Schreuder & Clusella-Trullas in press; Trimble & van Aarde 2014). A single study demonstrated that the encroachment of alien vegetation into nesting habitat of the Nile crocodile (*Crocodylus niloticus*) altered the temperatures in egg chambers, potentially affecting hatchling sex ratios, and clearing of the roots of the alien plant increased the number of females nesting (Leslie & Spotila 2001). To our knowledge, no study has been conducted to assess the impacts of alien vegetation on tortoises, a taxonomic group with high species diversity in South Africa and high endemism in the Cape region (Branch, Benn & Lombard 1995). Some of the life history characteristics of tortoises, such as herbivory, long lifespans and habitat structure needed for shelter, likely increase their vulnerability to alien plant invasions (Gray & Steidl 2015; Stewart, Austin & Bourne 1993). Similarly, only two studies (Russell & Downs 2012; Trimble & van Aarde 2014) considered amphibians, despite their potential vulnerability to the impacts of alien plants on water resources (Le Maitre et al. 2015).

Context dependencies

It is well recognised that the impacts of alien plants on fauna are context-dependent and are shaped by a variety of factors, including the abundance and distribution of the IAP, the time since its introduction and the invasion history (e.g. rate of spread and lag times), the spatial extent of the study area, the degree of contrast of the alien plant form and function to the native vegetation, the ecosystem type and climatic conditions (e.g. seasonality) and the habitat preference of the animal species assessed (Kumschick et al. 2015; Maron & Marler 2008; Pyšek et al. 2012; Schirmel et al. 2016; Vilà et al. 2011). Little knowledge of these factors and the extent to which they may interact makes the prediction of impacts of invasive alien vegetation on native animal diversity a difficult task. Furthermore, these variables are essential for incorporating moderators in meta-analyses that seek to explore the direction of effects of alien plants on community metrics and avoid spurious results. These difficulties were encountered in our synthesis in various ways. The 42 studies differed in alien vegetation abundance, ranging from mildly invaded sites (e.g. Robertson et al. 2011) to plantations of alien plants (e.g. Pryke & Samways 2009) but, in general, little information was provided about the extent of the invasion, the study site size and its landscape context (e.g. degree of fragmentation and edge effects), and the invasion history, with a few notable exceptions (e.g. Liu, Janion & Chown 2012; Mgobozi, Somers & Dippenaar-Schoeman 2008). In some cases, or regions, the lack of long-term vegetation surveys or a poor knowledge of the alien plant (e.g. site of origin and genetic lineage; Thompson et al. 2014) may explain the lack of such detailed accounts.

Our synthesis also showed that opposite effects of alien plants on the same taxonomic group are found and generally depend on the animal group investigated, the type of IAP and the occurrence of other environmental stresses (cattle and habitat alteration). For example, alien plants can have decreasing (Samways & Grant 2006), neutral (Kinvig & Samways 2000; Samways & Sharratt 2010) or increasing (Kietzka, Pryke & Samways 2015) effects on dragonfly species diversity. Kietzka et al. (2015) suggested that the invasive American bramble (*Rubus cuneifolius*) provided additional perching sites and protection from predators to dragonflies, but where the alien plant stands were very dense, the negative effects outweighed the positive. The direction of the impact is also influenced by the behaviour and physiological requirements of the species or groups investigated. For example, shade specialists such as some Odonata species can benefit from increased shade (Samways & Grant 2006), whereas basking species are negatively affected by closed canopies that can result from plant invasions. Similarly, herbivore arthropods were generally more affected than detritivores (e.g. van der Colff et al. 2015). Finally, novel or very dissimilar alien plant forms to those of native plants species are likely to impact local communities most (Martin & Murray 2011), and although some comparisons involving alien plants with similar growth form to the native vegetation found no differences in native species abundance or richness (e.g. Olckers & Hulley 1989; van der Colff et al. 2015; van der Merwe, Dippenaar-Schoeman & Scholtz 1996), this was not always the case (Pryke & Samways 2009; Ratsirarson et al. 2002).

Study approaches

This review further revealed that most studies examining the impacts of alien plants on ectothermic animals employed the same methodological approach, comparing invaded and non-invaded sites. A very small proportion of studies included a gradient of alien plant abundance (e.g. Schreuder & Clusella-Trullas in press; Robertson et al. 2011) or incorporated comparisons with restored sites that had been cleared from alien plants (e.g. Maoela et al. 2016) or among sites with different times since invasion (e.g. Mgobozi et al. 2008). Although all of these approaches have their strengths and caveats (Hulme et al. 2013; Kueffer, Pyšek & Richardson 2013), progress in the understanding of impacts of alien plants on animal diversity requires a multifaceted approach (Kumschick et al. 2015). For example, comparisons of invaded and uninvaded plots can be confounded by differences that are inherent to each site (altitude, topography and soil properties) and require increased replication to boost the power of the analyses by incorporating variation originating from each site's characteristics. It is often particularly difficult to find uninvaded reference sites because baseline data (prior to the invasion) are not available. Despite South Africa having a national-scale government-funded project to clear alien plants, the Working for Water Programme, relatively few studies have incorporated removal of aliens as part of their study design or assessed how rapidly fauna assemblages reflect pre-invasion reference sites.

Species abundance, richness and composition results provided useful data to compare key animal community changes between invaded and uninvaded habitats (albeit with the known problems associated with, e.g., comparison of species richness; Gotelli & Colwell 2001). However, this review also illustrated that in some cases, these metrics were insufficient to obtain an adequate understanding of the change detected or lack thereof. For example, Magoba & Samways (2012) showed that adult dragonfly species richness was not hampered by riparian alien vegetation, but assemblages changed drastically in sites cleared of IAPs. Although some generalist, widespread species (e.g. spiders and hymenopterans) can thrive in invaded areas, rare, endemic or sensitive species often disappear from alien-infested sites (e.g. Stewart & Samways 1998). These findings illustrate that additional metrics such as beta-diversity, species evenness, measures of commonness and rarity and functional richness (e.g. Magurran & McGill 2011) should be incorporated in assessments of impacts of alien vegetation on animal communities in South Africa.

Overall, these metrics alone give little insight into the mechanisms underpinning community changes. Less than 13% of the studies reviewed here incorporated an experimental approach for testing for mechanisms. These are essential for describing processes underpinning patterns, distinguishing direct and indirect effects of alien plants on ecosystems (Hulme et al. 2013) and, ultimately, enabling predictions of invasion outcomes in the context of future distributions of alien plants in South Africa (Rouget et al. 2004) and in the face of other environmental alterations and climate change. The ability to forecast impacts of IAPs and develop management strategies will rest on knowing these mechanisms.

Conclusion

The current state of knowledge of the impacts of IAPs on resident ectothermic animal species in South Africa relies heavily on a few key studies, with distinct biases in geographic locations and taxonomic groups. In cases where detailed information is available, it is nevertheless clear that there are pronounced negative impacts of IAPs on terrestrial animal (ectotherm) species diversity. The mechanisms underlying these impacts are unclear, but here we highlight a few key abiotic and biotic processes that could be examined in future, especially if microenvironments determine key behaviours and life-cycle timing that lead to changes in population abundance. Such an integrated approach to the question of IAPs and their impact on native animal species diversity would be of direct value to monitoring and conservation efforts, such as alien plant removal and control programmes. At present, it is wholly unclear whether the removal of IAPs will be sufficient to allow recovery of native ectotherm biodiversity.

Acknowledgements

S.C.-T. is supported by the Centre for Invasion Biology, Stellenbosch University and the Incentive Funding for Rated Researchers from the South African National Research Foundation. R.A.G. is supported by a post-doctoral fellowship from the Centre for Invasion Biology, Stellenbosch University.

Competing interests

The authors declare that they have no financial or personal relationships that may have inappropriately influenced them in writing this article.

Author(s) contributions

Both S.C.-T. and R.A.G. contributed equally to the article.

References

Bates, M.F., Branch, W.R., Bauer, A.M., Burger, M., Marais, J., Alexander, G.J. et al., 2014, *Atlas and Red List of the Reptiles of South Africa, Lesotho and Swaziland*. Suricata 1. South African National Biodiversity Institute, Pretoria.

Branch, W.R., Benn, G.A. & Lombard, A.T., 1995. 'The tortoises (Testudinidae) and terrapins (Pelomedusidae) of southern Africa: Their diversity, distribution and conservation', *South African Journal of Zoology* 30(3), 91–102. https://doi.org/10.1080/02541858.1995.11448377

Breytenbach, G.J., 1986, 'Impacts of alien organisms on terrestrial communities with emphasis on communities of the south-western Cape', in I.A.W. Macdonald, F.J. Kruger & A.A. Ferrar (eds.), *The ecology and management of biological invasions in southern Africa*, pp. 229–238, Oxford University Press, Cape Town.

Butchart, S.H.M., Walpole, M., Collen, B., van Strien, A., Scharlemann, J.P.W., Rosamunde, E.A.A. et al. 2010, 'Global biodiversity: Indicators of recent declines', *Science* 328, 1164–1168. https://doi.org/10.1126/science.1187512

Canavan, K., Paterson, I. & Hill, M.P., 2014, 'The herbivorous arthropods associated with the invasive alien plant, *Arundo donax*, and the native analogous plant, *Phragmites australis*, in the Free State Province, South Africa', *African Entomology* 22(2), 454–459. https://doi.org/10.4001/003.022.0204

Coetzee, B.W.T., van Rensburg, B.J. & Robertson, M.P., 2007, 'Invasion of grasslands by silver wattle, *Acacia dealbata* (Mimosaceae), alters beetle (Coleoptera) assemblage structure', *African Entomology* 15(2), 328–339. https://doi.org/10.4001/1021-3589-15.2.328

de Groot, M., Kleijn, D. & Jogan, N., 2007, 'Species groups occupying different trophic levels respond differently to the invasion of semi-natural vegetation by *Solidago canadensis*', *Biological Conservation* 136(4), 612–617. https://doi.org/10.1016/j.biocon.2007.01.005

Endler, J.A., 1977, 'Geographic variation, speciation, and clines', *Monographs in Population Biology* 10, 1–246.

French, K. & Major, R.E., 2001, 'Effect of an exotic *Acacia* (Fabaceae) on ant assemblages in South African fynbos', *Austral Ecology* 26(4), 303–310. https://doi.org/10.1046/j.1442-9993.2001.01115.x

Gaertner, M., Den Breeyen, A., Hui, C. & Richardson, D.M., 2009, 'Impacts of alien plant invasions on species richness in Mediterranean-type ecosystems: A meta-analysis', *Progress in Physical Geography* 33(3), 319–338. https://doi.org/10.1177/0309133309341607

Gates, D.M., 1980, *Biophysical ecology*, Springer-Verlag, New York.

Gibbons, J.W., Scott, D.E., Ryan, T.J., Buhlmann, K.A., Tuberville, T.D., Metts, B.S. et al., 2000, 'The global decline of reptiles, déjà vu amphibians', *BioScience* 50(8), 653–666. https://doi.org/10.1641/0006-3568(2000)050[0653:TGDORD]2.0.CO;2

Gotelli, N.J. & Colwell, R.K., 2001, 'Quantifying biodiversity: Procedures and pitfalls in the measurement and comparison of species richness', *Ecology Letters* 4(4), 379–391. https://doi.org/10.1046/j.1461-0248.2001.00230.x

Gray, K.M. & Steidl, R.J., 2015, 'A plant invasion affects condition but not density or population structure of a vulnerable reptile', *Biological Invasions* 17(7), 1979–1988. https://doi.org/10.1007/s10530-015-0851-1

Henderson, L., 2007, 'Invasive, naturalized and casual alien plants in southern Africa: A summary based on the Southern African Plant Invaders Atlas (SAPIA)', *Bothalia* 37(2), 215–248. https://doi.org/10.4102/abc.v37i2.322

Hulme, P.E., Pyšek, P., Jarošík, V., Pergl, J., Schaffner, U. & Vilà, M., 2013, 'Bias and error in understanding plant invasion impacts', *Trends in Ecology & Evolution* 28(4), 212–218. https://doi.org/10.1016/j.tree.2012.10.010

Kietzka, G.J., Pryke, J.S. & Samways, M.J., 2015, 'Landscape ecological networks are successful in supporting a diverse dragonfly assemblage', *Insect Conservation and Diversity* 8(3), 229–237. https://doi.org/10.1111/icad.12099

Kinvig, R.G. & Samways, M.J., 2000, 'Conserving dragonflies (Odonata) along streams running through commercial forestry', *Odonatologica* 29(3), 195–208.

Kotzé, J.D.F., Beukes, H., van der Beg, E. & Newby, T., 2010, *National Invasive Alien Plant Survey – Dataset*, Agricultural Research Council: Institute for Soil, Climate and Water, Pretoria.

Kraaij, T., Cowling, R.M. & Van Wilgen, B.W., 2011, 'Past approaches and future challenges to the management of fire and invasive alien plants in the new Garden Route National Park', *South African Journal of Science* 107(9–10), 16–26.

Kueffer, C., Pyšek, P. & Richardson, D.M., 2013, 'Integrative invasion science: Model systems, multi-site studies, focused meta-analysis and invasion syndromes', *New Phytologist* 200, 615–633. https://doi.org/10.1111/nph.12415

Kumschick, S., Gaertner, M., Vilà, M., Essl, F., Jeschke, J.M., Pyšek, P. et al., 2015, 'Ecological impacts of alien species: Quantification, scope, caveats, and recommendations', *Bioscience* 65(1), 55–63. https://doi.org/10.1093/biosci/biu193

Le Maitre, D.C., Gush, M.B. & Dzikiti, S., 2015, 'Impacts of invading alien plant species on water flows at stand and catchment scales', *AoB Plants* 7, plv043. https://doi.org/10.1093/aobpla/plv043

Le Maitre, D.C., Versfeld, D.B. & Chapman, R.A., 2000, 'The impact of invading alien plants on surface water resources in South Africa: A preliminary assessment', *Water SA* 26(3) 397–408. https://doi.org/10.1093/aobpla/plv043

Leslie, A.J. & Spotila, J.R., 2001, 'Alien plant threatens Nile crocodile (*Crocodylus niloticus*) breeding in Lake St. Lucia, South Africa', *Biological Conservation* 98, 347–355. https://doi.org/10.1016/S0006-3207(00)00177-4

Levine, J.M., Vila, M., Antonio, C.M., Dukes, J.S., Grigulis, K. & Lavorel, S., 2003, 'Mechanisms underlying the impacts of exotic plant invasions', *Proceedings of the Royal Society of London B: Biological Sciences* 270(1517), 775–781. https://doi.org/10.1098/rspb.2003.2327

Liu, W.P.A., Janion, C. & Chown, S.L., 2012, 'Collembola diversity in the critically endangered Cape Flats Sand Fynbos and adjacent pine plantations', *Pedobiologia – International Journal of Soil Biology* 55, 203–209.

Maoela, M.A., Roets, F., Jacobs, S.M. & Esler, K.J., 2016, 'Restoration of invaded Cape Floristic Region riparian systems leads to a recovery in foliage-active arthropod alpha- and beta-diversity', *Journal of Insect Conservation* 20(1), 85–97. https://doi.org/10.1007/s10841-015-9842-x

Maron, J.L. & Marler, M., 2008, 'Effects of native species diversity and resource additions on invader impact', *The American Naturalist* 172(suppl. 1), S18–S33. https://doi.org/10.1086/588303

Martin, L.J. & Murray, B.R., 2011, 'A predictive framework and review of the ecological impacts of exotic plant invasions on reptiles and amphibians', *Biological Reviews* 86(2), 407–419. https://doi.org/10.1111/j.1469-185X.2010.00152.x

McGill, B.J., Enquist, B., Weiher, E. & Westoby, M., 2006, 'Rebuilding community ecology from functional traits', *Trends in Ecology & Evolution* 21(4), 178–185. https://doi.org/10.1016/j.tree.2006.02.002

Magoba, R.N. & Samways, M.J., 2012, 'Comparative footprint of alien, agricultural and restored vegetation on surface-active arthropods', *Biological Invasions* 14(1), 165–177. https://doi.org/10.1007/s10530-011-9994-x

Magurran, A.E. & McGill, B.J. 2011, *Biological diversity: Frontiers in measurement and assessment*, Oxford University Press, Oxford, xvii + 345 p.

Measey, G.J., 2011, *Ensuring a future for South Africa's frogs: A strategy for conservation research*. SANBI Biodiversity Series 19. South African National Biodiversity Institute, Pretoria.

Mgobozi, M.P., Somers, M.J. & Dippenaar-Schoeman, A.S., 2008, 'Spider responses to alien plant invasion: The effect of short- and long-term *Chromolaena odorata* invasion and management', *Journal of Applied Ecology* 45(4), 1189–1197. https://doi.org/10.1111/j.1365-2664.2008.01486.x

Milton, S.J., 2004, 'Grasses as invasive alien plants in South Africa', *South African Journal of Science* 100(1/2), 69–75.

Mokhatla, M.M, Measey, G.J., Chimimba, C.T. & van Rensburg, B.J., 2012, 'A biogeographical assessment of anthropogenic threats to areas where different frog breeding groups occur in South Africa: Implications for anuran conservation', *Diversity and Distributions* 18, 470–480. https://doi.org/10.1111/j.1472-4642.2011.00870.x

Mucina, L., Rutherford, M.C., Powrie, L.W., van Niekerk, A. & van der Merwe, J.H., 2014, *Vegetation Field Atlas of Continental South Africa, Lesotho and Swaziland. In Strelitzia 33*. South African National Biodiversity Institute, Pretoria.

Olckers, T. & Hulley, P.E., 1989, 'Insect herbivore diversity on the exotic weed *Solanum mauritianum* Scop. and three other *Solanum* species in the eastern Cape Province', *Journal of the Entomological Society of Southern Africa* 52(1), 81–93.

Padayachi, Y., Proches, S. & Ramsay, L.F., 2014, 'Beetle assemblages of indigenous and alien decomposing fruit in subtropical Durban, South Africa', *Arthropod-Plant Interactions* 8, 135–142.

Pearson, D.E., 2009, 'Invasive plant architecture alters trophic interactions by changing predator abundance and behavior', *Oecologia* 159, 549–558. https://doi.org/10.1007/s00442-008-1241-5

Pryke, J.S. & Samways, M.J., 2009, 'Recovery of invertebrate diversity in a rehabilitated city landscape mosaic in the heart of a biodiversity hotspot', *Landscape and Urban Planning* 93(1), 54–62. https://doi.org/10.1016/j.landurbplan.2009.06.003

Pryke, J.S. & Samways, M.J., 2012, 'Conservation management of complex natural forest and plantation edge effects', *Landscape Ecology* 27(1), 73–85. https://doi.org/10.1007/s10980-011-9668-1

Pyšek, P., Jarošík, V., Hulme, P.E., Pergl, J., Hejda, M., Schaffner, U. et al., 2012, 'A global assessment of invasive plant impacts on resident species, communities and ecosystems: The interaction of impact measures, invading species' traits and environment', *Global Change Biology* 18(5), 1725–1737. https://doi.org/10.1111/j.1365-2486.2011.02636.x

Ratsirarson, H., Robertson, H.G., Picker, M.D. & van Noort, S., 2002, 'Indigenous forests versus exotic eucalypt and pine plantations: A comparison of leaf-litter invertebrate communities', *African Entomology* 10(1), 93–99.

Ricciardi, A., Hoopes, M.F., Marchetti, M.P. & Lockwood, J.L., 2013, 'Progress towards understanding the ecological impacts of nonnative species', *Ecological Monographs* 83(3), 263–282. https://doi.org/10.1890/13-0183.1

Richardson, D.M. & van Wilgen, B.W., 2004, 'Invasive alien plants in South Africa: How well do we understand the ecological impacts?', *South African Journal of Science* 100, 45–52.

Richardson, D.M., Wilson, J.R.U., Weyl, O.L.F. & Griffiths, C.L., 2011, 'South Africa: Invasions', in D. Simberloff & M. Rejmánek (eds.), *Encyclopedia of biological invasions*, pp. 643–651, University of California Press, Berkeley, CA.

Robertson, M.P., Harris, K.R., Coetzee, J.A., Foxcroft, L.C., Dippenaar-Schoeman, A.S. & van Rensburg, B.J., 2011, 'Assessing local scale impacts of *Opuntia stricta* (Cactaceae) invasion on beetle and spider diversity in Kruger National Park, South Africa', *African Zoology* 46(2), 205–223. https://doi.org/10.3377/004.046.0202

Roets, F. & Pryke, J.S., 2012, 'The rehabilitation value of a small culturally significant island based on the arthropod natural capital', *Journal of Insect Conservation* 17(1), 53–65. https://doi.org/10.1007/s10841-012-9485-0

Rouget, M., Richardson, D.M., Nel, J.L., Le Maitre, D.C., Egoh, B. & Mgidi, T., 2004, 'Mapping the potential ranges of major plant invaders in South Africa, Lesotho and Swaziland using climatic suitability', *Diversity and Distributions* 10(5–6), 475–484. https://doi.org/10.1111/j.1366-9516.2004.00118.x

Russell, C. & Downs, C.T., 2012, 'Effect of land use on anuran species composition in north-eastern KwaZulu-Natal, South Africa', *Applied Geography* 35(1–2), 247–256. https://doi.org/10.1016/j.apgeog.2012.07.003

Samways, M.J. & Grant, P.B.C., 2006, 'Regional response of Odonata to river systems impacted by and cleared of invasive alien trees', *Odonatologica* 35(3), 297–303.

Samways, M.J. & Sharratt, N.J., 2010, 'Recovery of endemic dragonflies after removal of invasive alien trees', *Conservation Biology* 24(1), 267–277. https://doi.org/10.1111/j.1523-1739.2009.01427.x

Schlaepfer, M.A., Sax, D.F. & Olden, J.D., 2010, 'The potential conservation value of non-native species', *Conservation Biology* 25(3), 428–437. https://doi.org/10.1111/j.1523-1739.2010.01646.x

Schirmel, J., Bundschuh, M., Entling, M.H., Kowarik, I. & Buchholz, S., 2016, 'Impacts of invasive plants on resident animals across ecosystems, taxa, and feeding types: A global assessment', *Global Change Biology* 22(2), 594–603. https://doi.org/10.1111/gcb.13093

Schreuder, E. & Clusella-Trullas, S., (in press), 'Exotic trees modify the thermal landscape and food resources for lizard communities', *Oecologia* 182(4), 1213–1225. https://doi.org/10.1007/s00442-016-3726-y

Steenkamp, H.E. & Chown, S.L., 1996, 'Influence of dense stands of an exotic tree, *Prosopis glandulosa* Benson, on a savanna dung beetle (Coleoptera: Scarabaeinae) assemblage in southern Africa', *Biological Conservation* 78(3), 305–311. https://doi.org/10.1016/S0006-3207(96)00047-X

Stewart, M.C., Austin, D.F. & Bourne, G.R., 1993, 'Habitat structure and the dispersion of gopher tortoises on a nature preserve', *Florida Scientist* 56, 70–81.

Stewart, D.A.B. & Samways, M.J., 1998, 'Conserving dragonfly (Odonata) assemblages relative to river dynamics in an African savanna game reserve', *Conservation Biology* 12(3), 683–692. https://doi.org/10.1111/j.1523-1739.1998.96465.x

Thompson, G.D., Bellstedt, D.U., Richardson, D.M., Wilson, J.R.U. & Le Roux, J.J., 2014, 'A tree well travelled: Global genetic structure of the invasive tree *Acacia saligna*', *Journal of Biogeography* 42(2), 305–314. https://doi.org/10.1111/jbi.1243

Trimble, M.J. & van Aarde, R.J., 2014, 'Amphibian and reptile communities and functional groups over a land-use gradient in a coastal tropical forest landscape of high richness and endemicity', *Animal Conservation* 17(5), 441–453. https://doi.org/10.1111/acv.12111

Van den Berg, E., Kotze, I. & Beukes, H., 2013, 'Detection, quantification and monitoring of *Prosopis* in the Northern Cape Province of South Africa using Remote Sensing and GIS', *South African Journal of Geomatics* 2(2), 68–81.

van der Colff, D., Dreyer, L.L., Valentine, A., & Roets, F., 2015, 'Invasive plant species may serve as a biological corridor for the invertebrate fauna of naturally isolated hosts', *Journal of Insect Conservation* 19(5), 863–875. https://doi.org/10.1007/s10841-015-9804-3

van der Merwe, M., Dippenaar-Schoeman, A.S. & Scholtz, C.H., 1996, 'Diversity of ground-living spiders at Ngome State Forest, Kwazulu/Natal: A comparative survey in indigenous forest and pine plantations', *African Journal of Ecology* 34(4), 342–350. https://doi.org/10.1111/j.1365-2028.1996.tb00630.x

van Hengstum, T., Hooftman, D.A.P., Oostermeijer, J.G.B. & van Tienderen, P.H., 2014, 'Impact of plant invasions on local arthropod communities: A meta-analysis', *Journal of Ecology* 102, 4–11. https://doi.org/10.1111/1365-2745.12176

Van Wilgen, B.W., Carruthers, J., Cowling, R.M., Esler, K.J., Forsyth, A.T., Gaertner, M. et al., 2016, 'Ecological research and conservation management in the Cape Floristic Region between 1945 and 2015: History, current understanding and future challenges', *Transactions of the Royal Society of South Africa* 71(3), 207–303. https://doi.org/10.1080/0035919X.2016.1225607

van Wilgen, B.W., Forsyth, G.G., Le Maitre, D.C., Wannenburgh, A., Kotzé, J.D.F. van der Berg, E. et al. 2012, 'An assessment of the effectiveness of a large, national-scale invasive alien plant control strategy in South Africa', *Biological Conservation* 148(1), 28–38. https://doi.org/10.1016/j.biocon.2011.12.035

van Wilgen, B.W., Reyers, B., Le Maitre, D.C., Richardson, D.M. & Schonegevel, L., 2008, 'A biome-scale assessment of the impact of invasive alien plants on ecosystem services in South Africa', *Journal of Environmental Management* 89(4), 336–349. https://doi.org/10.1016/j.jenvman.2007.06.015

Vilà, M., Espinar, J.L., Hejda, M., Hulme, P.E., Jarošík, V., Maron, J.L. et al. 2011, 'Ecological impacts of invasive alien plants: A meta-analysis of their effects on species, communities and ecosystems', *Ecology Letters* 14(7), 702–708. https://doi.org/10.1111/j.1461-0248.2011.01628.x

Visser, V., Wilson, J.R.U, Fish, L., Brown, C., Cook, G.D. & Richardson, D.M., 2016, 'Much more give than take: South Africa as a major donor but infrequent recipient of invasive non-native grasses', *Global Ecology and Biogeography* 25(6), 679–692. https://doi.org/10.1111/geb.12445

Visser, V., Wilson, J.R.U., Canavan, K., Canavan, S., Fish, L., Le Maitre, D. et al., 2017, Grasses as invasive plants in South Africa revisited: Patterns, pathways and management, *Bothalia* 47(2), a2169. https://doi.org/10.4102/abc.v47i2.2169

Wilson, E.O. & Peter, F., 1988, *Biodiversity*, National Academies, Washington, DC, 521 pp.

Wilson, J.R.U., Gaertner, M., Richardson, D.M. & van Wilgen, B.W., 2017, 'Contributions to the national status report on biological invasions in South Africa', *Bothalia* 47(2), a2207. https://doi.org/10.4102/abc.v47i2.2207

Appendix 1

Literature review search terms

TABLE 1-A1: Search terms used in the literature review on the ISI Web of Science.

Issue	Search terms
Location	"South Africa*"
Invasive plants	((invas* OR alien* OR non$nativ* OR exotic* OR introduced OR non$indigenous OR naturali?ed OR plantation*) AND (plant* OR vegetat* OR tree* OR shrub* OR grass* OR forest* OR forb* OR herb* OR vine* OR *weed*)) OR (invaded AND (habitat* OR site* OR plot*))
Reptiles	reptil* OR squamata OR snake* OR python* OR boa* OR cobra* OR mamba* OR viper* OR adder* OR colubrid* OR elapid* OR lizard* OR gecko* OR skink* OR chameleon * OR agama* OR monitor* OR lacertid* OR amphisbaenid* OR cordylid* OR testudine* OR chenolian* OR turtle* OR tortoise* OR terrapin* OR crocodylia OR crocodil*
Amphibians	amphibian* OR frog* OR anura* OR tadpole*
Terrestrial invertebrates	invertebrate* OR platyhelminthe* OR *worm* OR nematod* OR nematomorph* OR nemertea* OR acanthocephalan* OR annelid* OR oligochaet* OR leech* OR mollus* OR gastropod* OR snail* OR slug* OR tardigrad* OR onychophora* OR arthropod* OR crustacea* OR *lice OR "terrestrial crab*" OR amphipod* OR isopod* OR myriapod* OR centipede* OR millipede* OR chilopod* OR diplopod* OR chelicerat* OR Araneae OR arachnid* OR spider* OR Acari OR acarin* OR mite* OR tick* OR opiliones OR harvestm?n OR scorpion* OR hexapod* OR insect* OR apterygot* OR odonat* OR dragonfl* OR damselfl* OR orthoptera* OR grasshopper* OR cricket* OR isoptera* OR termite* OR mantodea* OR mantis* OR mantid* OR blattodea* OR cockroach* OR embioptera* OR webspinner* OR phasmid* OR phasmatodea* OR hemiptera* OR *bug* OR cicada* OR aphid* OR *hopper* OR thysanoptera* OR thrip* OR psocoptera* OR coleoptera* OR beetle* OR lepidoptera* OR butterfl* OR moth* OR diptera* OR *flies OR *fly OR mosquito* OR flea* OR hymenoptera* OR wasp* OR ant OR ants OR bee OR bees OR neuroptera* OR lacewing* OR antlion* OR pollinat*
Effect on diversity	((population* OR communit* OR assemblage* OR species) AND (abundan* OR richness OR ((population* OR communit* OR assemblage* OR species) AND (abundan* OR richness OR diversity OR composition OR evenness OR dominance OR equitability OR structure OR poor* OR impoverish*)) OR ((functional OR *genetic) AND diversity)
Effect mechanism	(habitat* NEAR/3 (quality OR structure OR heterogeneity)) OR shad* OR thermal* OR hydrolog* OR micro$site* OR micro$climate* OsR micro$habitat* OR refuge* OR prey* OR activity OR thermo$regulat* OR behavio$r* OR bask* OR predat* OR competit* OR herbivo$r* OR resource* OR nutrient* OR fire* OR soil* OR sediment* OR locomoti* OR host* OR reproducti* OR toxic* OR poison* OR hybrid* OR disease* OR parasit*

A first search combined terms for location, invasive plants, reptiles, amphibians and terrestrial invertebrates, and effects on diversity, whereas a second search replaced the terms for effects with terms for mechanisms.

Search 1: Effect on diversity (358 records)

TOPIC = "South Africa*" AND (((invas* OR alien* OR non$nativ* OR exotic* OR introduced OR non$indigenous OR naturali?ed OR plantation*) AND (plant* OR vegetat* OR tree* OR shrub* OR grass* OR forest* OR forb* OR herb* OR vine* OR *weed*)) OR (invaded AND (habitat* OR site* OR plot*))) AND ((reptil* OR squamata OR snake* OR python* OR boa* OR cobra* OR mamba* OR viper* OR adder* OR colubrid* OR elapid* OR lizard* OR gecko* OR skink* OR chameleon * OR agama* OR monitor* OR lacertid* OR amphisbaenid* OR cordylid* OR testudine* OR chenolian* OR turtle* OR tortoise* OR terrapin* OR crocodylia OR crocodil*) OR (amphibian* OR frog* OR anura* OR tadpole*) OR (invertebrate* OR platyhelminthe* OR *worm* OR nematod* OR nematomorph* OR nemertea* OR acanthocephalan* OR annelid* OR oligochaet* OR leech* OR mollus* OR gastropod* OR snail* OR slug* OR tardigrad* OR onychophora* OR arthropod* OR crustacea* OR *lice OR "terrestrial crab*" OR amphipod* OR isopod* OR myriapod* OR centipede* OR millipede* OR chilopod* OR diplopod* OR chelicerat* OR Araneae OR arachnid* OR spider* OR Acari OR acarin* OR mite* OR tick* OR opiliones OR harvestm?n OR scorpion* OR hexapod* OR insect* OR apterygot* OR odonat* OR dragonfl* OR damselfl* OR orthoptera* OR grasshopper* OR cricket* OR isoptera* OR termite* OR mantodea* OR mantis* OR mantid* OR blattodea* OR cockroach* OR embioptera* OR webspinner* OR phasmid* OR phasmatodea* OR hemiptera* OR *bug* OR cicada* OR aphid* OR *hopper* OR thysanoptera* OR thrip* OR psocoptera* OR coleoptera* OR beetle* OR lepidoptera* OR butterfl* OR moth* OR diptera* OR *flies OR *fly OR mosquito* OR flea* OR hymenoptera* OR wasp* OR ant OR ants OR bee OR bees OR neuroptera* OR lacewing* OR antlion* OR pollinat*)) AND (((population* OR communit* OR assemblage* OR species) AND (abundan* OR richness OR diversity OR composition OR evenness OR dominance OR equitability OR structure OR poor* OR impoverish*)) OR ((functional OR *genetic) AND diversity))

Refined by: DOCUMENT TYPES: (ARTICLE OR REVIEW OR BOOK CHAPTER)

Timespan: All years.

Search 2: Effect mechanisms (702 records)

TOPIC = "South Africa*" AND (((invas* OR alien* OR non$nativ* OR exotic* OR introduced OR non$indigenous OR naturali?ed OR plantation*) AND (plant* OR vegetat* OR tree* OR shrub* OR grass* OR forest* OR forb* OR herb* OR vine* OR *weed*)) OR (invaded AND (habitat* OR site* OR plot*))) AND ((reptil* OR squamata OR snake* OR python* OR boa* OR cobra* OR mamba* OR viper* OR adder* OR colubrid* OR elapid* OR lizard* OR gecko* OR skink* OR chameleon * OR agama* OR monitor* OR lacertid* OR amphisbaenid* OR cordylid* OR testudine* OR chenolian* OR turtle* OR tortoise* OR terrapin* OR crocodylia OR crocodil*) OR (amphibian* OR frog* OR anura* OR tadpole*) OR (invertebrate* OR platyhelminthe* OR *worm* OR nematod* OR nematomorph* OR nemertea* OR acanthocephalan* OR annelid* OR oligochaet* OR leech* OR mollus* OR gastropod* OR snail* OR slug* OR tardigrad* OR onychophora* OR arthropod* OR crustacea* OR *lice OR "terrestrial crab*" OR amphipod* OR isopod* OR myriapod* OR centipede* OR millipede* OR chilopod* OR diplopod* OR chelicerat* OR Araneae OR arachnid* OR spider* OR Acari OR acarin* OR mite* OR tick* OR opiliones OR harvestm?n OR scorpion* OR hexapod* OR insect* OR apterygot* OR odonat* OR dragonfl* OR damselfl* OR orthoptera* OR grasshopper* OR cricket* OR isoptera* OR termite* OR mantodea* OR mantis* OR mantid* OR blattodea* OR cockroach* OR embioptera* OR webspinner* OR phasmid* OR phasmatodea* OR hemiptera* OR *bug* OR cicada* OR aphid* OR *hopper* OR thysanoptera* OR thrip* OR psocoptera* OR coleoptera* OR beetle* OR lepidoptera* OR butterfl* OR moth* OR diptera* OR *flies OR *fly OR mosquito* OR flea* OR hymenoptera* OR wasp* OR ant OR ants OR bee OR bees OR neuroptera* OR lacewing* OR antilon* OR pollinat*)) AND ((habitat* NEAR/3 (quality OR structure OR heterogeneity)) OR shad* OR thermal* OR hydrolog* OR micro$site* OR micro$climate* OsR micro$habitat* OR refuge* OR prey* OR activity OR thermo$regulat* OR behavio$r* OR bask* OR predat* OR competit* OR herbivo$r* OR resource* OR nutrient* OR fire* OR soil* OR sediment* OR locomoti* OR host* OR reproducti* OR toxic* OR poison* OR hybrid* OR disease* OR parasit*)

Refined by: DOCUMENT TYPES: (ARTICLE OR REVIEW OR BOOK CHAPTER)

Timespan: All years.

Appendix 2

Studies reviewed

Search 1: Effects on diversity

Botzat, A., Fischer, L. & Farwig, N., 2013, 'Forest-fragment quality rather than matrix habitat shapes herbivory on tree recruits in South Africa', *Journal of Tropical Ecology*, 29(02), 111–122. https://doi.org/10.1017/S0266467413000102

Canavan, K., Paterson, I. & Hill, M.P., 2014, 'The Herbivorous Arthropods Associated with the Invasive Alien Plant, *Arundo donax*, and the Native Analogous Plant, *Phragmites australis*, in the Free State Province, South Africa', *African Entomology*, 22(2), 454–459. https://doi.org/10.4001/003.022.0204

Coetzee, B.W.T., van Rensburg, B.J. & Robertson, M.P., 2007, 'Invasion of grasslands by silver wattle, *Acacia dealbata* (Mimosaceae), alters beetle (Coleoptera) assemblage structure', *African Entomology*, 15(2), 328–339. https://doi.org/10.4001/1021-3589-15.2.328

Dlamini, T.C. & Haynes, R.J., 2004, 'Influence of agricultural land use on the size and composition of earthworm communities in northern KwaZulu-Natal, South Africa', *Applied Soil Ecology*, 27(1). https://77-88.10.1016/j.apsoil.2004.02.003

Donnelly, D. & Giliomee, J., 1985, 'Community structure of epigaeic ants in a pine plantation in newly burnt fynbos', *Journal of the Entomological Society of Southern Africa*, 48(2), 259–265.

Esterhuizen, J., Green, K.K., Marcotty, T. & van den Bossche, P., 2005, 'Abundance and distribution of the tsetse flies, *Glossina austeni* and *G. brevipalpis*, in different habitats in South Africa', *Medical and Veterinary Entomology*, 19(4), 367–371. https://doi.org/10.1111/j.1365-2915.2005.00582.x

French, K. & Major, R.E., 2001, 'Effect of an exotic *Acacia* (Fabaceae) on ant assemblages in South African fynbos', *Austral Ecology*, 26(4), 303–310. https://doi.org/10.1046/j.1442-9993.2001.01115.x

Grass, I., Berens, D.G., Peter, F. & Farwig, N., 2013, 'Additive effects of exotic plant abundance and land-use intensity on plant-pollinator interactions', *Oecologia*, 173(3), 913–923. https://doi.org/10.1007/s00442-013-2688-6

Haynes, R.J., Dominy, C.S. & Graham, M.H., 2003, 'Effect of agricultural land use on soil organic matter status and the composition of earthworm communities in KwaZulu-Natal, South Africa', *Agriculture, Ecosystems & Environment*, 95(2–3), 453–464. https://doi.org/10.1111/j.1747-0765.2006.00084.x

Kietzka, G.J., Pryke, J.S. & Samways, M.J., 2015, 'Landscape ecological networks are successful in supporting a diverse dragonfly assemblage', *Insect Conservation and Diversity*, 8(3), 229–237. https://doi.org/10.1111/icad.12099

Kinvig, R.G. & Samways, M.J., 2000, 'Conserving dragonflies (Odonata) along streams running through commercial forestry', *Odonatologica*, 29(3), 195–208.

Lawrence, J.M. & Samways, M.J., 2002, 'Influence of hilltop vegetation type on an African butterfly assemblage and its conservation', *Biodiversity and Conservation*, 11(7), 1163–1171. https://doi.org/10.1023/A:1016017114473

Liu, S. & Samways, M.J., 2002, 'Conservation management recommendations for the Karkloof blue butterfly, *Orachrysops ariadne* (Lepidoptera: Lycaenidae)', *African Entomology*, 10(1), 149–159.

Magoba, R.N. & Samways, M.J., 2012, 'Comparative footprint of alien, agricultural and restored vegetation on surface-active arthropods', *Biological Invasions*, 14(1), 165–177. https://doi.org/10.1007/s10530-011-9994-x

Magoba, R.N. & Samways, M.J., 2010, 'Recovery of benthic macroinvertebrate and dragonfly assemblages in response to large scale removal of riparian invasive alien trees', *Journal of Insect Conservation*, 14(6), 627–636. https://doi.org/10.1007/s10841-010-9291-5

Mata, M.A., Roets, F., Jacobs, S.M. & Esler, K.J., 2016, 'Restoration of invaded Cape Floristic Region riparian systems leads to a recovery in foliage-active arthropod alpha- and beta-diversity', *Journal of Insect Conservation*, 20(1), 85–97. https://doi.org/10.1007/s10841-015-9842-x

Mgobozi, M.P., Somers, M.J. & Dippenaar-Schoeman, A.S., 2008, 'Spider responses to alien plant invasion: the effect of short- and long-term *Chromolaena odorata* invasion and management', *Journal of Applied Ecology*, 45(4), 1189–1197. https://doi.org/10.1111/j.1365-2664.2008.01486.x

Oberprieler, T. & Hulley, P.E., 1991, 'Impoverished insect herbivore faunas on the exotic weed *Solanum mauritianum* Scop. relative to indigenous *Solanum* species in Natal/KwaZulu and the Transkei', *Journal of the Entomological Society of Southern Africa*, 54(1), 39–50.

Oberprieler, T. & Hulley, P.E., 1989, 'Insect herbivore diversity on the exotic weed *Solanum mauritianum* Scop. and three other *Solanum* species in the eastern Cape province', *Journal of the Entomological Society of Southern Africa*, 52(1), 81–93.

Ochi, Y., Proches, S. & Ramsay, L.F., 2014, 'Beetle assemblages of indigenous and alien decomposing fruit in subtropical Durban, South Africa', *Arthropod-Plant Interactions*, 8, 135–142.

Proches, S., Wilson, J.R.U., Richardson, D.M. & Chown, S.L., 2008, 'Herbivores, but not other insects, are scarce on alien plants', *Austral Ecology*, 33, 691–700. https://doi.org/10.1111/j.1442-9993.2008.01836.x

Pryke, J.S., Roets, F. & Samways, M.J., 2013, 'Importance of habitat heterogeneity in remnant patches for conserving dung beetles', *Biodiversity and Conservation*, 22(12), 2857–2873. https://doi.org/10.1007/s10531-013-0559-4

Pryke, J.S. & Samways, M.J., 2012, 'Conservation management of complex natural forest and plantation edge effects', *Landscape Ecology*, 27(1), 73–85. https://doi.org/10.1007/s10980-011-9668-1

Pryke, J.S. & Samways, M.J., 2009, 'Recovery of invertebrate diversity in a rehabilitated city landscape mosaic in the heart of a biodiversity hotspot', *Landscape and Urban Planning*, 93(1), 54–62. https://doi.org/10.1016/j.landurbplan.2009.06.003

Pryke, S.R. & Samways, M.J., 2003, 'Quality of remnant indigenous grassland linkages for adult butterflies (Lepidoptera) in an afforested African landscape', *Biodiversity and Conservation*, 12(10), 1985–2004. https://doi.org/10.1023/A:1024103527611

Ratsirarson, H., Robertson, H.G., Picker, M.D. & van Noort, S., 2002, 'Indigenous forests versus exotic eucalypt and pine plantations: a comparison of leaf-litter invertebrate communities', *African Entomology*, 10(1), 93–99.

Robertson, M.P., Harris, K.R., Coetzee, J.A., Foxcroft, L.C., Dippenaar-Schoeman & van Rensburg, B.J., 2011, 'Assessing local scale impacts of *Opuntia stricta* (Cactaceae) invasion on beetle and spider diversity in Kruger National Park, South Africa', *African Zoology*, 46(2), 205–223. https://doi.org/10.3377/004.046.0202

Roets, F. & Pryke, J.S., 2012, 'The rehabilitation value of a small culturally significant island based on the arthropod natural capital', *Journal of Insect Conservation*, 17(1), 53–65. https://doi.org/10.1007/s10841-012-9485-0

Russell, C. & Downs, C.T., 2012, 'Effect of land use on anuran species composition in north-eastern KwaZulu-Natal, South Africa', *Applied Geography*, 35(1–2), 247–256. https://doi.org/10.1016/j.apgeog.2012.07.003

Samways, M.J., Caldwell, P.M. & Osborn, R., 1996, 'Ground-living invertebrate assemblages in native, planted and invasive vegetation in South Africa', *Agriculture, Ecosystems & Environment*, 59(1), 19–32. https://doi.org/10.1016/0167-8809(96)01047-X

Samways, M.J. & Grant, P.B.C., 2006, 'Regional response of Odonata to river systems impacted by and cleared of invasive alien trees', *Odonatologica*, 35(3), 297–303.

Samways, M.J. & Moore, S.D., 1991, 'Influence of exotic conifer patches on grasshopper (Orthoptera) assemblages in a grassland matrix at a recreational resort, Natal, South Africa', *Biological Conservation*, 57(2), 117–137. https://doi.org/10.1016/0006-3207(91)90134-U

Samways, M.J. & Sharratt, N.J., 2010, 'Recovery of endemic dragonflies after removal of invasive alien trees', *Conservation Biology*, 24(1), 267–277. https://doi.org/10.1111/j.1523-1739.2009.01427.x

Schoeman, C.S. & Samways, M.J., 2011, 'Synergisms between alien trees and the Argentine ant on indigenous ant species in the Cape Floristic Region, South Africa', *African Entomology*, 19(1), 96–105. https://doi.org/10.4001/003.019.0117

Schoeman, C.S. & Samways, M.J., 2013, 'Temporal shifts in interactions between alien trees and the alien Argentine ant on native ants', *Journal of Insect Conservation*, 17(5), 911–919. https://doi.org/10.1007/s10841-013-9572-x

Schreuder, E. & Clusella-Trullas, S., 'Exotic trees modify the thermal landscape and food resources for lizard communities'. *Oecologia*. In press.

Steenkamp, H.E. & Chown, S.L., 1996, 'Influence of dense stands of an exotic tree, *Prosopis glandulosa* Benson, on a savanna dung beetle (Coleoptera: Scarabaeinae) assemblage in southern Africa', *Biological Conservation*, 78(3), 305–311. https://doi.org/10.1016/S0006-3207(96)00047-X

Stewart, D.A.B. & Samways, M.J., 1998, 'Conserving dragonfly (Odonata) assemblages relative to river dynamics in an African savanna game reserve', *Conservation Biology*, 12(3), 683–692. https://doi.org/10.1111/j.1523-1739.1998.96465.x

Trimble, M.J. & van Aarde, R.J., 2014, 'Amphibian and reptile communities and functional groups over a land-use gradient in a coastal tropical forest landscape of high richness and endemicity', *Animal Conservation*, 17(5), 441–453. https://doi.org/10.1111/acv.12111

van der Colff, D., Dreyer, L.L., Valentine, A., & Roets, F., 2015, 'Invasive plant species may serve as a biological corridor for the invertebrate fauna of naturally isolated hosts', *Journal of Insect Conservation*, 19(5), 863–875. https://doi.org/10.1007/s10841-015-9804-3

van der Merwe, M., Dippenaar-Schoeman, A.S. & Scholtz, C.H., 1996, 'Diversity of ground-living spiders at Ngome State Forest, Kwazulu/Natal: a comparative survey in indigenous forest and pine plantations', *African Journal of Ecology*, 34(4), 342–350. https://doi.org/10.1111/j.1365-2028.1996.tb00630.x

Yekwayo, I., Pryke, J.S., Roets, F. & Samways, M.J., 2016, 'Surrounding vegetation matters for arthropods of small, natural patches of indigenous forest', *Insect Conservation and Diversity*, 9(3), 224–235. https://doi.org/10.1111/icad.12160

Search 2: Effect mechanisms

Crous, C.J., Pryke, J.S. & Samways, M.J., 2015, 'Conserving a geographically isolated *Charaxes* butterfly in response to habitat fragmentation and invasive alien plants', *Koedoe*, 57(1), p.Art. #1297, 9 pages.

Gibson, M.R., Pauw, A. & Richardson, D.M., 2013, 'Decreased insect visitation to a native species caused by an invasive tree in the Cape Floristic Region', *Biological Conservation*, 157, 196–203. https://doi.org/10.1016/j.biocon.2012.07.011

Grass, I., Berens, D.G. & Farwig, N., 2014, 'Natural habitat loss and exotic plants reduce the functional diversity of flower visitors in a heterogeneous subtropical landscape', *Functional Ecology*, 28(5), 1117–1126. https://doi.org/10.1111/1365-2435.12285

Leslie, A.J. & Spotila, J.R., 2001, 'Alien plant threatens Nile crocodile (*Crocodylus niloticus*) breeding in Lake St. Lucia', South Africa, *Biological Conservation*, 98, 347–355. https://doi.org/10.1016/S0006-3207(00)00177-4

Remsburg, A.J., Olson, A.C. & Samways, M.J., 2008, 'Shade alone reduces adult dragonfly (Odonata: Libellulidae) abundance', *Journal of Insect Behavior*, 21(6), 460–468. https://doi.org/10.1007/s10905-008-9138-z

Wood, P.A. & Samways, M.J., 1991, 'Landscape element pattern and continuity of butterfly flight paths in an ecologically landscaped botanic garden, Natal, South Africa', *Biological Conservation*, 58(2), 149–166. https://doi.org/10.1016/0006-3207(91)90117-R

11

Grasses as invasive plants in South Africa revisited: Patterns, pathways and management

Authors:
Vernon Visser[1,2,3,4] ⓘ
John R.U. Wilson[3,4] ⓘ
Kim Canavan[5]
Susan Canavan[3]
Lyn Fish[6]
David Le Maitre[3,7] ⓘ
Ingrid Nänni[4]
Caroline Mashau[6]
Tim G. O'Connor[8] ⓘ
Philip Ivey[4]
Sabrina Kumschick[3,4] ⓘ
David M. Richardson[3] ⓘ

Affiliations:
[1]SEEC – Statistics in Ecology, the Environment and Conservation, Department of Statistical Sciences, University of Cape Town, South Africa

[2]African Climate and Development Initiative, University of Cape Town, South Africa

[3]Centre for Invasion Biology, Department of Botany and Zoology, Stellenbosch University, South Africa

[4]Invasive Species Programme, South African National Biodiversity Institute, Kirstenbosch Research Centre, South Africa

[5]Department of Zoology and Entomology, Rhodes University, South Africa

[6]National Herbarium, South African National Biodiversity Institute (SANBI), South Africa

[7]Natural Resources and the Environment, Stellenbosch, South Africa

[8]South African Environmental Observation Network (SAEON), Pretoria, South Africa

Background: In many countries around the world, the most damaging invasive plant species are grasses. However, the status of grass invasions in South Africa has not been documented recently.

Objectives: To update Sue Milton's 2004 review of grasses as invasive alien plants in South Africa, provide the first detailed species level inventory of alien grasses in South Africa and assess the invasion dynamics and management of the group.

Method: We compiled the most comprehensive inventory of alien grasses in South Africa to date using recorded occurrences of alien grasses in the country from various literature and database sources. Using historical literature, we reviewed past efforts to introduce alien grasses into South Africa. We sourced information on the origins, uses, distributions and minimum residence times to investigate pathways and patterns of spatial extent. We identified alien grasses in South Africa that are having environmental and economic impacts and determined whether management options have been identified, and legislation created, for these species.

Results: There are at least 256 alien grass species in the country, 37 of which have become invasive. Alien grass species richness increased most dramatically from the late 1800s to about 1940. Alien grass species that are not naturalised or invasive have much shorter residence times than those that have naturalised or become invasive. Most grasses were probably introduced for forage purposes, and a large number of alien grass species were trialled at pasture research stations. A large number of alien grass species in South Africa are of Eurasian origin, although more recent introductions include species from elsewhere in Africa and from Australasia. Alien grasses are most prevalent in the south-west of the country, and the Fynbos Biome has the most alien grasses and the most widespread species. We identified 11 species that have recorded environmental and economic impacts in the country. Few alien grasses have prescribed or researched management techniques. Moreover, current legislation neither adequately covers invasive species nor reflects the impacts and geographical extent of these species.

Conclusion: South Africa has few invasive grass species, but there is much uncertainty regarding the identity, numbers of species, distributions, abundances and impacts of alien grasses. Although introductions of alien grasses have declined in recent decades, South Africa has a potentially large invasion debt. This highlights the need for continued monitoring and much greater investment in alien grass management, research and legislation.

Introduction

In many parts of the world, grasses are among the most damaging and widespread alien plant species (D'Antonio, Stahlheber & Molinari 2011; D'Antonio and Vitousek 1992; Gaertner et al. 2014). In the Americas, Australia and on many tropical islands, grasses have transformed ecosystems, usually by altering the natural fire cycle (D'Antonio and Vitousek 1992; Gaertner et al. 2014). By contrast, South Africa has fewer invasive grasses, and alien plant control efforts are dedicated primarily to combating woody plant invasions. The only grass species that has been widely targeted for control operations by the Working for Water Programme is *Arundo donax*, although < 0.5% of the total budget for alien plant control was allocated to managing this species (Table 2 in van Wilgen et al. 2012). Grasses also do not feature in the National Strategy for Dealing with Biological Invasions in South Africa (Department of Environmental Affairs [DEA] 2014). The relative paucity of grass invasions in South Africa might be because of high fire frequencies in African grasslands and savannas excluding alien grasses (Visser et al. 2016). It would, therefore, seem that grass invasions in South Africa are generally neither common nor widespread and that they do not pose a major risk. However, there are several reasons to be concerned about undetected and possible future grass invasions.

The last review of alien grasses in South Africa was published more than a decade ago (Milton 2004). This review highlighted major gaps in our knowledge. We do not know how many alien grasses are in South Africa nor their identity or status on the introduction-naturalisation-invasion (INI) continuum (Blackburn et al. 2011; Richardson and Pyšek 2012). While it appears that South Africa experienced lower introduction effort of alien grasses relative to other regions of the world, it was one of the countries most actively engaged in trialling alien grasses at pasture research stations (Visser et al. 2016). However, there is no consolidated inventory of species cultivated in these pasture research trials. Without a comprehensive (or as near as possible) inventory of alien grasses, it is impossible to ascertain all the risks.

We also have inadequate knowledge of the introduction pathways of alien grasses, which is needed to determine 'introduction debt', the number of species that are likely to be introduced to South Africa in the future (Rouget et al. 2016). Introduced but not yet naturalised or invasive grasses might also represent an invasion debt as they might invade in the future, while current invasions might spread to new areas and cause increasing negative impacts (Rouget et al. 2016). However, we need information on the origins of alien grasses, their residence times (Wilson et al. 2007) and propagule pressure to be able to estimate establishment and spread debts (Rouget et al. 2016). We also have very poor knowledge of alien grass impacts in South Africa, with Milton's (2004) review relying mostly on published impacts of alien grasses in other parts of the world.

We should be particularly wary of changes in invasions in the group in the face of rising atmospheric CO_2 levels and concomitant global climatic change. Grass species using the C_4 photosynthetic pathway have higher nitrogen-use efficiency relative to those that use the C_3 photosynthetic pathway (Taylor et al. 2010). It has been suggested that the competitive dominance of C_4 grasses across much of South Africa is because of this comparative advantage, but that at higher CO_2 levels, this advantage disappears and alien C_3 grasses will be more likely to invade South African grasslands in the future (Milton 2004; Richardson et al. 2000). Grasses have also been shown to exhibit strong phylogenetic conservatism of climatic niches (Edwards and Smith 2010)

and climate change could have differential consequences for grasses of particular phylogenetic clades.

Since 2004, an increasing amount of information on alien grass distributions, origins, traits and impacts in South Africa has become available from online databases, surveys and research projects. The legal framework for alien grass management in South Africa has also changed substantially with the introduction of the Alien and Invasive Species (A&IS) regulations under the National Environmental Management: Biodiversity (NEM:BA) Act 10 of 2004. A number of alien grasses were listed as invasive in 2014 (NEM:BA Alien and Invasive Species List 2016; A&IS regulations) and now have specific management requirements or are prohibited from being imported. Moreover, the A&IS regulations require the publication of a national status report on listed invasive species every 3 years, the purpose of which is to monitor the status of listed species and the effectiveness of the regulations and associated control measures (Wilson et al. 2017).

Given the demonstrated potential for alien grasses to become problematic invaders elsewhere in the world, the increased availability of data in South Africa and new legislation regarding alien species, we aim to provide an updated review of the status of alien grasses in South Africa. To this end, we collate the first detailed species-level inventory of alien grasses in South Africa. We use this inventory, together with other literature, to address some of the information gaps identified above, including investigating (1) minimum residence times (MRTs) of alien grasses in South Africa, (2) pathways of introduction and spread, (3) areas that are potentially being most impacted by alien grasses, (4) impacting species and the nature of their impacts and (5) information on management of alien grasses. This information is needed to make recommendations for future management of alien grasses in South Africa.

Methods
Inventory

We produced an inventory of alien grasses in South Africa (Online Appendix) using additional data sources to contribute to an already extensive inventory that we used for a recent publication (Visser et al. 2016). The inventory is based on an extensive search of the scientific and grey literature and of distribution databases (Appendix 1). We first checked species names from all sources against The Plant List (www.theplantlist.org) and corrected them to accepted species. We changed infra-specific names to the species level and removed hybrid species, but kept unresolved names. In a final refinement of the list, we flagged all species for which there was only one data source for its occurrence in South Africa, or for which there were fewer than five distribution records (see below for more information on distribution data). These species were manually checked by inspecting the original data source (either a reference or herbarium specimen). As measures of confidence in the presence of an alien grass species in South Africa, we flagged all species (1) with only one data source for its occurrence in South Africa, (2) with no distribution data and (3) no herbarium records.

We also determined the status of each species on the INI continuum (introduced = species present outside of its native range either in cultivation or in the wild, but in the latter case, not yet reproducing. Hereafter, all references to species that are 'introduced' should be interpreted according to the aforementioned definition; naturalised = species that are reproducing in their alien range, but not spreading substantially; invasive = self-sustaining species that spread over large distances; Blackburn et al. 2011; Richardson and Pyšek 2012), relying on references to assign species' statuses (Appendix 1).

Minimum residence time

We obtained the earliest record of occurrence of a species in South Africa to calculate the MRT (Wilson et al. 2007). We checked GBIF (http://www.gbif.org), Plants of Southern Africa (POSA) and a number of historical and archaeobotanical studies for MRTs (Appendix 1), using the oldest date from all these databases for the MRT. Using these data, we created species accumulation curves for alien grasses in South Africa.

Pathways of introduction and spread

To investigate likely pathways of introduction and spread, we collected information on the uses of species (Quattrocchi 2006). Uses were assigned to one of the six categories (horticulture, animal food, food or beverage, raw material, soil stabilisation or none). We assumed that these uses will be correlated with both the initial reason for introduction and how and why species were spread by humans around South Africa (e.g. known pasture grasses were likely introduced as such and distributed to appropriate pasture lands). We investigated temporal patterns of alien grass introductions with respect to their primary uses, using the MRT for each species.

The origins of alien grasses can be useful for informing why species were introduced. We determined the native range of each species using the eMonocot database and manually assigned the native range of each species to one or more of six biogeographical realms: North America, South America, sub-Saharan Africa, temperate Eurasia, North Africa and Southeast Asia (Olson et al. 2001). We investigated temporal patterns of where species originated from using the same approach as for primary uses over time.

To test how the use and origin contribute to the progression of species across the INI continuum (from introduced to naturalised to invasive), we used ordinal logistic regression, with INI status as the response variable and use or origin as the predictor, using the R (R Core Team 2016) package ordinal (Christensen 2016).

To assess in more detail the role of the pasture industry in introducing alien grasses, we searched the literature for information on (1) the existence of pasture research stations in South Africa, (2) the duration that these stations operated for, and (3) the species that were trialled at these stations. These data were used to complement our inventory of alien grasses.

Prevalence

We collated species distribution data from online databases and scientific publications on alien grasses in South Africa (Appendix 1). We downscaled all coordinates to the centroid of the nearest quarter-degree-grid-cell (QDGC). We also recorded the year in which each grass occurrence was made. We calculated total alien species richness and numbers of records across South Africa. Observed species richness patterns likely suffer from sampling bias towards roads and urban centres, and rarefaction has been shown to be the best at reducing this bias, although it tends to underestimate richness (Engemann et al. 2015). We used the R package 'vegan' to estimate species richness, excluding QDGCs with less than 20 samples, because rarefaction is inaccurate for small sample sizes. We also investigated whether particular biomes have more alien grasses and whether protected areas have records of alien grasses, by overlaying observed alien grass species distributions (actual localities as well as overlap with QDGCs occupied) on a high-resolution map of South African biomes (Mucina et al. 2005) and of protected areas (DEA 2016). We then calculated numbers of alien grasses in each biome and the number of protected areas with alien grass records.

We calculated the area occupied by each species by summing the number of QDGCs occupied. We examined the area occupied by each species with respect to whether species have been recorded as having impacts (see below), INI status and legal status (see below).

Impacts

Empirical measures of impact (e.g. Hawkins et al. 2015) are unavailable for alien grasses in South Africa, and so we focus here on establishing a baseline of current understanding. We did this by assimilating information from the literature gathered for our inventory of alien grasses (Appendix 1). We selected all references that mention any changes caused to the native environment (mainly biodiversity) or harm caused on the socio-economy attributable to the alien species (cf. Jeschke et al. 2014). Studies mentioning the dominance or invasiveness of a species without referring to changes to the native environment were not considered (e.g. Musil, Milton & Davis 2005; Sharma et al. 2010). For the studies referring to environmental changes, we assigned the most likely Environmental Impact Classification of Alien Taxa (EICAT) score for South Africa noting the confidence level (Blackburn et al. 2014; Hawkins et al. 2015). Only references included in Appendix 1 were considered for the EICAT classification, and no standardised literature review was performed.

Management of alien grasses

To evaluate gaps in research on control measures for alien grasses, we searched the literature used for compiling our inventory (Appendix 1) and the Global Invasive Species Database (GISD; www.iucngisd.org/gisd/) for the 41 species that are either invasive or have impacts in South Africa. We grouped control measures into five categories: physical removal, fire, chemical, biocontrol and integrated control (based on the categories used in the GISD, with the addition of fire).

We used the prevalence and impact data collected in this study to evaluate the appropriateness of current NEM:BA categorisations (DEA 2016) of alien grass species [viz. prohibited species do not occur in the country and pose an unacceptable risk of invasion if they were to be introduced; category 1a species are those where eradication from the entire country is desirable and feasible (Wilson et al. 2013); category 1b species are those where ongoing control measures are required and all uses are prohibited; category 2 species can be used for commercial (or other) purposes provided that a permit is issued; otherwise, they are treated as category 1b species; finally, category 3 species are those where existing plantings can remain, but must be contained, no new plantings are allowed and a national management plan is required (DEA 2014)]. In brief, the differences between categories are because of whether an alien species is present in the country, whether eradication is feasible and some balance between benefits of plantings and risks of invasion. An accurate quantification for all species of both the feasibility of eradication and the net impact is beyond the scope of this study. Therefore, we used the spatial extent of each species as a proxy for eradication feasibility (Pluess et al. 2012a, 2012b; Rejmánek and Pitcairn 2002). Eradication feasibility is inversely proportional to the spatial extent (E), so we calculated a metric of relative invaded area (RIA) using the following formula:

$$RIA = 1 - \log(E)/\log(E_{max}), \quad\quad\quad\quad [Eqn 1]$$

where E_{max} is the spatial extent of the species with the highest number of QDGCs occupied. RIA is 0 for species that are not present and 1 for the most widespread invasive grass. To estimate the net impact of each species (negative, neutral or positive), we used the data collected on impacts (previous paragraph) and on the uses of each species. Species with recorded impacts (Table 4) or that were found to be invasive in South Africa, and that have only one use, were given a 'negative' relative benefit. Species with a 'neutral' benefit were defined as those with recorded impacts or that are invasive, but have more than one use, or non-invasive species that have no or just one use. Species with a 'positive' benefit were defined as those that are non-invasive, do not have an impact and have more than one use.

The future and providing a framework for assessing the status of grass invasions in South Africa

We highlighted two areas of concern for potential grass invasions in the future: invasion debt and global change. We do not attempt to calculate all aspects of invasion debt as

defined in Rouget et al. (2016), but instead focus on one of the key components of these calculations: identifying species that are not yet invasive in South Africa, but which are known to be invasive elsewhere (this has been shown to be a useful indicator of a species becoming invasive in novel regions, e.g. Kumschick and Richardson 2013; Panetta 1993). We used the weed status in the Global Compendium of Weeds (GCW; Randall 2012) to identify species that are invasive anywhere in the world (species with a GCW status of 'environmental weed', 'invasive' or 'noxious weed'), but that are non-invasive in South Africa (introduced or naturalised).

To provide an indication of future grass invasions because of global change, we investigated relative frequencies of photosynthetic pathway type (C_3, C_4 or intermediate C_3-C_4) and of taxonomic affiliation as these have been shown to influence grass biogeographical patterns in relation to climate. We used Osborne et al. (2014) to assign grass species' photosynthetic type. To assign taxonomic affiliation, we used grass tribe information from GrassBase (Clayton et al. 2006 onwards) together with a recent phylogeny of grasses (Soreng et al. 2015) to assign species to one of the nine grass subfamilies (Aristidoideae, Arundinoideae, Bambusoideae, Chloridoideae, Danthonioideae, Ehrhartoideae, Micrairoideae, Panicoideae and Pooideae).

To assist with the NEM:BA A&IS regulations requirement for a national status report (due October 2017, see Wilson et al. 2017) and to provide simple, useful indicators of the status of alien grasses in South Africa, we have proposed a framework that can be regularly updated and improved on over time (Table 1). This framework covers all aspects of grass invasions covered in this paper (species presence, pathways, prevalence, impacts and management). Where possible, we compare the current situation with that in 2004. We also make recommendations for improved indicators for 2020, when the next national status report is due to be published.

Results

Inventory

A total of 256 alien grass species were found to have been introduced to South Africa by human activity (Table 1; Online appendix). Of these species, 122 (48%) are considered naturalised, and 37 of these naturalised species have become invasive (representing ~14% of all introduced alien grass species and 30% of naturalised species; Table 1). For many species, there was little confidence regarding their presence in South Africa: 33% of species only had one record of occurrence in South Africa, 42% of species had no distribution data, 30% of species had only one reference and no distribution data, 30% had no herbarium record and 26% were lacking in all the aforementioned aspects (Table 1). A further 29 species are potentially native species, but we classified these as extra-limital because they are native to one of South Africa's neighbouring countries, but alien to South Africa based on our data sources (Online appendix). Some of these could potentially represent intra-African spread (Faulkner et al., 2017).

TABLE 1: Framework for monitoring alien grass invasions in South Africa.

Category	Indicator	Status in 2004	Status in 2016	Recommendation for monitoring and reporting status in 2020
Species	Lists	113 *naturalised* species	256 species in total 122 *naturalised* species 37 *invasive* species	Change in number that is invasive over time. Status as per Blackburn et al. (2011).
	Confidence in presence	Not recorded	Measures of confidence in presence: 83 species (33% of all alien grasses) have (A) only one reference for occurrence in South Africa. 106 species (42% of all species) have (B) no distribution data. 77 species (30% of all species) have (C) no herbarium records. 75 species (30% of all species) conform to both (A) and (B). 65 species (26% of all species) conform to (A), (B) and (C)	100% with at least one distribution record. 100% with herbarium records. Revisit sites with low confidence in presence.
Pathways	New introductions	Not recorded	Not recorded	Number of import permits issued, number of interceptions
	Total area occupied	330/462 QGDCs (71%) in South Africa with alien grasses	351/462 QGDCs (76%) in South Africa with alien grasses	Estimate of area of occupancy and extent of occurrence across spatial scales
	Occupancy of key sites (protected areas)	Not recorded	148 of 1097 protected areas with alien grass records	As above at protected areas
	Abundance.	Not recorded	Not recorded.	Percentage covers for A&IS listed species.
Impacts.	Impact scores for species.	Qualitative description of the impacts of ~12 species.	11 species with recorded impacts, 7 with impacts quantified.	Quantification of (global) impacts of all alien grasses that have recorded impacts elsewhere in the world.
Areas by impacts	Number of QDGCs for species with impacts	Not recorded	333/462 QDGCs (72%)	As previous
Efficacy of management and regulation.	Number of species listed.	CARA 2001 Listed: 9 species Prohibited: 33 taxa	NEM:BA A&IS regulations: 14 species listed; 38 species prohibited (3 of which are possibly already introduced).	Confirmation that all prohibited species are not in the country. Evidence for categorisation of listed species.
	Management spending.	Not recorded	0.33% of the total budget for Working for Water alien plant control between 1995 and 2008 spent on *Arundo donax* L. (van Wilgen et al. 2012).	Control costs for all alien grasses.
Research efforts	Number of theses on alien grasses.	Not recorded	16 theses up until July 2016 (out of ~254 on alien plants; 6%)	Amount of money invested in alien grass research.

Provided are various indicators covering all aspects ('Category') of grass invasions addressed in this paper. For each indicator we provide a status for 2004, 2016 (current) and recommendations for improvements in reporting for 2020. Results for 2004 are from our own results and from Milton (2004), when available.

Minimum residence time

The first alien grasses in South Africa were crops, such as *Eleusine coracana*, *Pennisetum glaucum* and *Sorghum bicolor*, which were brought to the region by Iron Age farmers early in the first millennium (Antonites and Antonites 2014). Maize (*Zea mays*) was introduced much later, sometime in the 17th or 18th centuries (Antonites and Antonites 2014). The oldest alien grass herbarium records were collected just over 200 years ago (in 1811) for *Arundo donax* near Tulbagh in the Western Cape and *Rostraria pumila* near Fraserburg in the Northern Cape (Online appendix). The number of alien grasses recorded in South Africa increased rapidly until about 1940 (90% of species were recorded before 1955); for many species, the first record of occurrence is 1938 because of introductions by pasture research stations (Figure 1, see discussion below). There are far fewer first records after 1940 (Figure 1). Species that have naturalised or become invasive have much longer residence times in South Africa than species that are not recorded as naturalised or invasive (mean MRT in years: introduced = 86, naturalised = 131, invasive = 123; one-way ANOVA, $F = 38.62$, $n = 234$, d.f. = 2, $P < 0.0001$; Figure 1).

Pathways of introduction and spread

Forage represents the most common use of alien grasses in South Africa (62.2%; Figure 2a). The other most common use categories are horticulture, soil stabilisation, food and beverages, raw materials and lastly those with no known use (Figure 2a). Species with no use ('none') were the most likely to be invasive, followed closely by species used for forage (Figure 2a; Appendix 2, Table 1-A2). Fewer species used for forage have been introduced into South Africa since about 1950 (as a proportion of all use categories; Figure 2b). Concomitantly, there has been an increase in species being introduced that have no use (Figure 2b).

Alien grasses in South Africa are native to (in decreasing order of contribution) Eurasia, Southeast Asia, sub-Saharan Africa, South America, North America, Australasia and the Pacific (Figure 3a). Species native to South America were proportionally the most likely to be invasive, followed by species native to Eurasia (Figure 3a; Appendix 2, Table 2-A2). Species native to North America were proportionately the least likely to be invasive (Figure 3a; Appendix, Table 2-A2). The proportion of species being introduced that are native to Eurasia has steadily declined over time, with a relative increase in the introduction of grasses native to Australasia (Figure 3b). More recently, since about the 1950s, an increasing proportion of introductions has been of species native to sub-Saharan Africa (Figure 3b, cf. Faulkner et al. 2017).

We found evidence of 14 different pasture research stations being active at some point in South Africa (Appendix 3, Table 1-A3), although some of these appear not to have existed for very long, or they cultivated very few alien grass species. Five stations (Prinshof, Athole, Leeuwkuil, Estcourt and Cedara) were responsible for introducing 95% of the 81 alien grass species cultivated at these stations, with Prinshof alone having cultivated 63 species (Appendix 3, Table 2-A3).

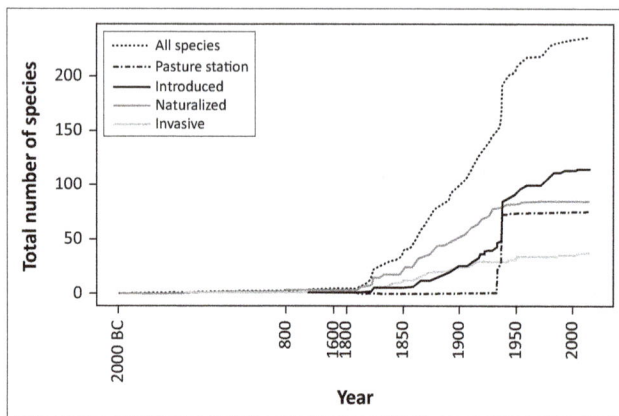

Note the scale on the x-axis is different before and after 1800. The total number of alien grass species is indicated by the dotted line. Species introduced by pasture research stations are indicated by the dot-dash line. The remaining solid lines represent species grouped by their status on the introduction-naturalisation-invasion continuum. The dark solid line represents species that are not yet naturalised or invasive, that is, 'introduced' on the introduction-naturalisation-invasion continuum. Naturalised, but not yet invasive species are represented by the slightly lighter solid line, and invasive species by the lightest solid line. New species were introduced at a relatively constant rate until 1938 when many species were introduced by pasture research stations. Most of the novel species introductions in 1938 did not result in new invasions as evidenced by the relatively constant total number of naturalised and invasive species after this date. After 1938 the rate of introductions has gradually slowed until the present day.

FIGURE 1: Cumulative number of alien grass species recorded in South Africa over time.

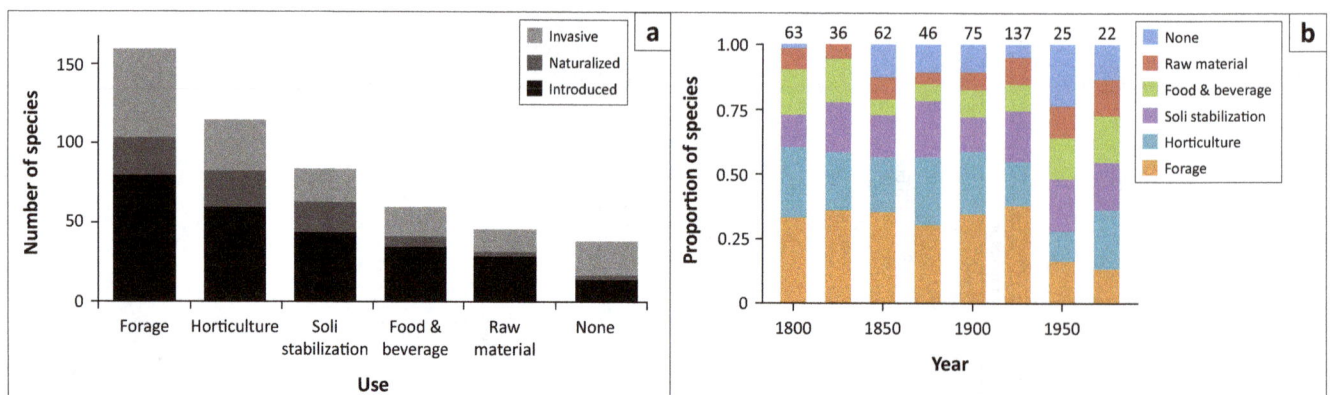

(b), The total number of species introduced in each interval is shown at the top of each bar.

FIGURE 2: (a) Uses of alien grasses in South Africa, with numbers of species per use category indicated by bar size. Each use category is further subdivided by species' statuses along the introduction-naturalisation-invasion continuum. (b) Trends over time (as deduced from earliest records of occurrence), per 25-year interval (apart from the first interval which represents the period 2000 BC to 1825), in the proportions of species being introduced to South Africa relative to use categories.

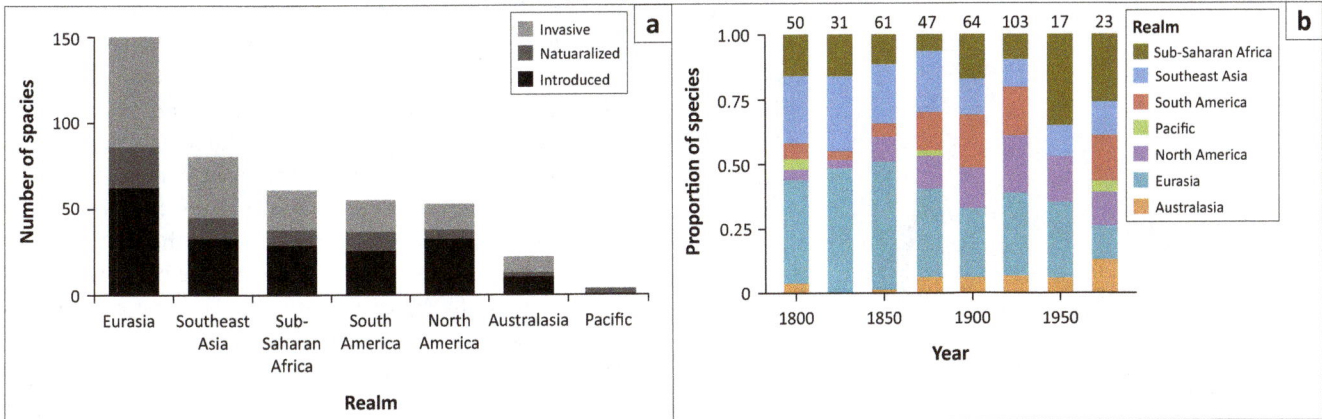

(b), The total number of species introduced in each interval is shown at the top of each bar.

FIGURE 3: (a) Origins of alien grasses in South Africa, with numbers of species per realm indicated by bar size. Each realm is further subdivided by species' statuses along the introduction-naturalisation-invasion continuum. (b) Trends over time (as deduced from earliest records of occurrence), per 25-year interval (apart from the first interval which represents the period 2000 BC to 1825), in the proportions of species being introduced to South Africa relative to their realms of origin.

Pasture research stations were responsible for introductions of 40 alien grass species in South Africa (~16% of all alien species), a conclusion reached because the starting date of trials at pasture stations involving these species preceded any other records of them in South Africa.

Prevalence

At present about 76% of QDGCs in South Africa have recorded occurrences of alien grasses, compared with 71% in 2004 (Table 1). The raw species occurrence data show that QDGCs with the most number of alien grass species are in the south-west of the country and that there are other notable pockets of high species richness around the major cities of Johannesburg and Port Elizabeth, and across much of the eastern escarpment (Figure 4a). A slightly different pattern emerges when we examine the *number of records* in QDGCs of these same species: the south-west of the country once again has the highest values, with the rest of the country generally having low numbers of records, apart from the areas around Johannesburg, Port Elizabeth and Durban (Figure 4b). However, as with most herbarium data, it is evident that there has been much more intensive collection of alien grasses along roads and around major urban centres (Figure 4a,b). After attempting to correct for sampling bias, it appears that alien grass species richness is still high in the Fynbos, but is possibly higher in the east of the country – in the grasslands of the Free State and in the southern Lowveld (Figure 4c). However, alien grass sampling was insufficient in many QDGCs, resulting in a much more restricted overview of alien grass species richness across the country compared with the raw data (Figure 4). When examining observed alien grass spatial patterns over time, we see that the high number of species and records in the south-west of the country is a fairly recent phenomenon, which has increased greatly in the last ~50 years, but is possibly the result of greater collection effort in this area during this time period (Appendix 4, Figure 1-A4).

In contrast to the high percentage of QDGCs occupied by alien grasses, only 148 of 1097 protected areas recorded the presence of alien grasses (Table 1). However, 195 protected areas occur in QDGCs where alien grasses were recorded. Alien grasses were recorded for the first time between 2004 and 2016 in an additional 24 protected areas.

In terms of the extent of individual species, we found that relatively few alien grasses occur across large areas of South Africa. Most alien grasses occupy relatively limited areas (Appendix 4, Figure 2-A4), although invasive grasses were much more widespread than other alien grasses (mean number of QDGCs ± SE: invasive = 88.08 ± 24.19, naturalised = 29.10 ± 7.67, introduced = 3.79 ± 2.44; Appendix 4, Figure 2-A4).

The results in terms of presence and abundance in the different biomes of South Africa are presented in detail in Appendix 4.

Impacts

We found recorded impacts for only 11 alien grass species in South Africa (Table 2). Of these, two species have major impacts (MR), two have moderate impact (MO), two minor impacts (MN) and five were data deficient (DD) according to our scoring of the EICAT due to a lack of environmental impact and the availability of only records of socio-economic impact (Table 2; Blackburn et al. 2014; Hawkins et al. 2015). Species with notable impacts were widespread, being recorded in 72% of QDGCs in South Africa (Table 1), although this is largely because of widespread species such as *Arundo donax* and *Pennisetum setaceum* – other species with notable impacts are not widespread (Appendix 4, Figure 2-A4). There was a great deal of uncertainty about ecological and socio-economic impacts caused by alien grasses in South Africa. Numerous studies have described alien grasses as dominant or invasive, without specifying the changes to the native environment (e.g. Musil et al. 2005; Rahlao et al. 2009; Sharma et al. 2010). However, few direct data exist: the EICAT score of only two species was with medium certainty and the rest with low certainty (Table 2).

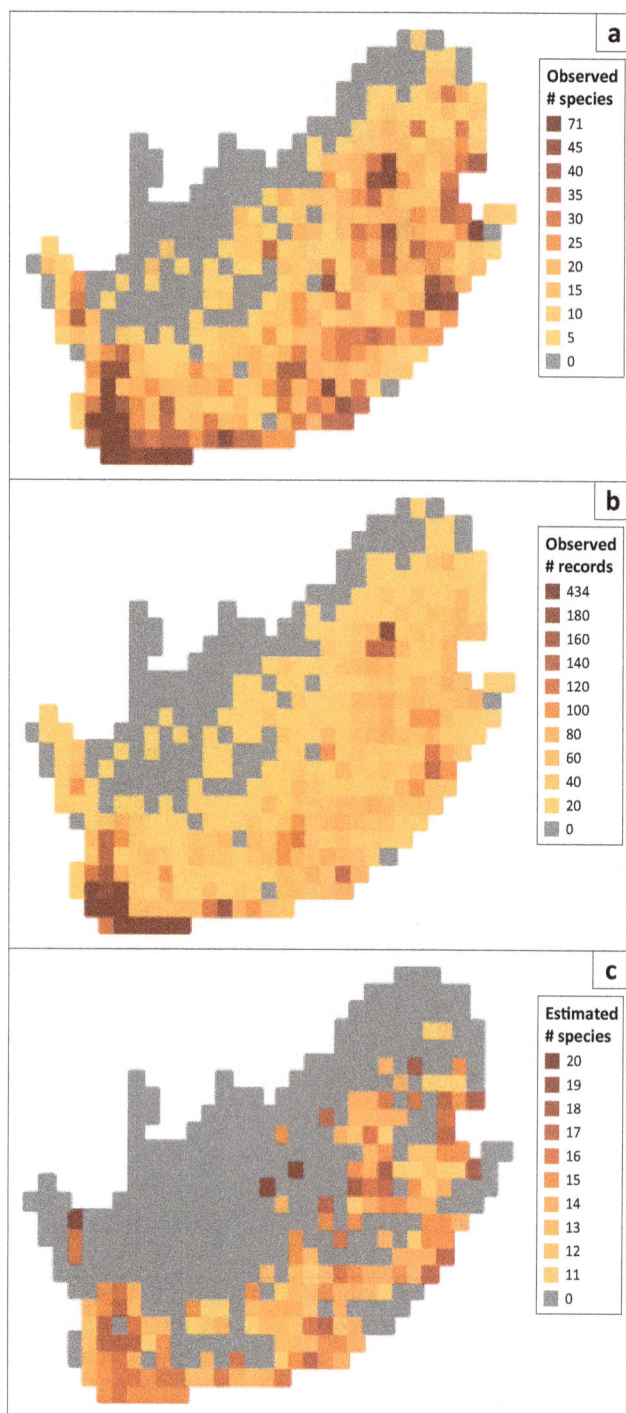

FIGURE 4: (a) Observed alien grass species richness, (b) observed numbers of records of alien grasses, and (c) estimated species richness in quarter-degree-grid-cells (QDGCs) across South Africa. Darker reds indicate higher species richness or numbers of records. Estimated species richness was calculated using rarefaction (see methods).

Management of alien grasses

A literature search revealed that of the 41 species identified as being invasive or having recorded impacts, management options have been described for only 11 species (27%) (Appendix 5). The most commonly suggested management strategy is physical removal (11 species), followed by chemical control (10 species), integrated control (8 species), using fire (6 species) and biological control (3 species) (Appendix 5).

Prior to 2004, only nine grass species were legislated for management and 33 taxa were prohibited from being introduced (Table 1; Appendix 6). Currently, under NEM:BA 14 species are legislated for management and 38 species are prohibited from being introduced (Table 1; Appendix 6). The current NEM:BA categorisation of species (NEM:BA Alien and Invasive Species List, 2016) is for the most part in accordance with our evaluation based on the spatial extent of a species' invaded area and the relative benefits of alien grasses in South Africa (Figure 5b). The one category 1a species (*Paspalum quadrifarium*), six of the eight category 1b species and the one category 3 species (*Ammophila arenaria*) were correctly categorised based on our scheme (Figure 5b). However, only 14 species, or 5.5% of all alien grass species in South Africa, are listed under the A&IS regulations (Figure 5b; Online appendix; Appendix 6). Based on our analysis, at least one other uncategorised species (*Bromus madritensis*) should be in category 1a; another 11 species in category 1b; 29 in category 2; 52 in category 3 and for 20, our analysis does not support their listing (Figure 5b). Some unlisted species are common agricultural grasses (e.g. wheat, *Triticum aestivum* etc.) and, therefore, do not require any regulation, but these only account for a small proportion of unlisted species. There is also one A&IS listed species where it is not clear whether it is present in South Africa [we could find no references for *Sasa ramosa* being present, but it is listed as category 3 (Appendix 6)]. We also found that 3 of 38 species on the NEM:BA A&IS prohibited list are already present in South Africa (*Panicum antidotale*, *Pennisetum polystachion* and *Themeda quadrivalvis*) (Appendix 6). However, they do not appear to have naturalised and are not yet widespread (Figure 5b; Online appendix).

The future and providing a framework for assessing the status of grass invasions in South Africa

To provide an indication of possible invasion debt, we investigated the number of non-invasive (introduced and naturalised) grasses in South Africa that are invasive elsewhere in the world. We found that 118 species (66% of 180 non-invasive species) are invasive elsewhere in the world (Online appendix). Of these species, 67 have naturalised in South Africa.

We also investigated the type of photosynthetic pathway used by alien grass species, and the taxonomic affinity of these species, as possible indicators of future invasion trends. Most alien grasses in South Africa use the C_3 photosynthetic pathway (61.4%; Appendix 7, Figure 1-A7), and these species are more common in the south-west of the country (Appendix 7, Figure 2-A7). However, most C_3 species belong to the subfamily Pooideae, with the next largest C_3 clade being represented by the clade Bambusoideae (Appendix 7, Figure 3-A7). There are only six C_3 alien grass species in South Africa in the largely C_4 Panicoideae clade (Appendix 7, Figure 3-A7).

TABLE 2: Alien grass species in South Africa with recorded impacts.

Species	Impacts	References	INI status	NEM:BA A&IS category 2016	Potential EICAT classification for South Africa	Certainty
Ammophila arenaria	Although indigenous species are able to co-exist with this species, these are a subset of the native dune flora (mostly smaller herbaceous plants) and the same species co-exist with *A. arenaria* across its entire distribution, unlike native dune communities which change in composition	Hertling and Lubke (1999)	Naturalised	3	MO	medium
Arundo donax	Dominates riparian areas, outcompeting native species and facilitating the establishment of other alien species. *A. donax* predicted to have greater impacts in drier, colder habitats where dry material can accumulate	Guthrie (2007)	Invasive	1b	MR	medium
	A. donax is thought to promote fires. Also causes accumulation of sediments and widening of river channels and locally excludes native species	Holmes et al. (2005)	-	-	MR	low
Avena barbata	Dominates old fields and disturbed habitats in lowland fynbos	Heelemann et al. (2013)	Invasive	None	MO	low
Avena fatua	Dominates old fields and disturbed habitats in lowland fynbos	Sharma et al. (2010)	Invasive	None	MN	low
Glyceria maxima	Outcompetes native wetland species	Mugwedi (2012)	Invasive	1b†	MR	low
Hordeum murinum	The long awns are especially a problem as they can cause injury to livestock by puncturing the mouth and throat and entering the skin and eyes as well as contaminating wool	Todd (2008)	Invasive	None	DD	-
Lolium multiflorum	Mentions highly competitive characteristics	Holmes (2008)	Invasive	None	MN	low
Nassella tenuissima	Poor grazing value so when dominant, decreases livestock productivity	Milton (2004)	Invasive	1b	DD	-
Nassella trichotoma	Poor grazing value so when dominant, decreases livestock productivity	Milton (2004)	Invasive	1b	DD	-
	The fine sharp seeds can contaminate wool	Esler, Milton and Dean (2006)	-	-	DD	-
Pennisetum setaceum	The leaves are barbed and unpalatable and the build-up of dead leaves increases the risk of veld fires	Esler, Milton and Dean (2006)	Invasive	1b‡	DD	-
Stipa capensis	Reduces grazing capacity and damages wool and hides	Steinschen, Gorne and Milton (1996)	Introduced	None	DD	-

Both, ecological and socio-economic impacts were considered as reported, but only negative changes are considered (cf. Jeschke et al. 2014). EICAT scores and certainty levels (according to Blackburn et al. 2014; Hawkins et al. 2015) are purely based on the studies performed on environmental impacts in South Africa (socio-economic impacts were not classified) and on references provided in this table rather than a standardised literature search. For full references see Appendix 1.

MR, major; MO, moderate; MN, minor; DD, data deficient.

†, 1b in protected areas and wetlands. Not listed elsewhere; ‡, Sterile cultivars or hybrids are not listed.

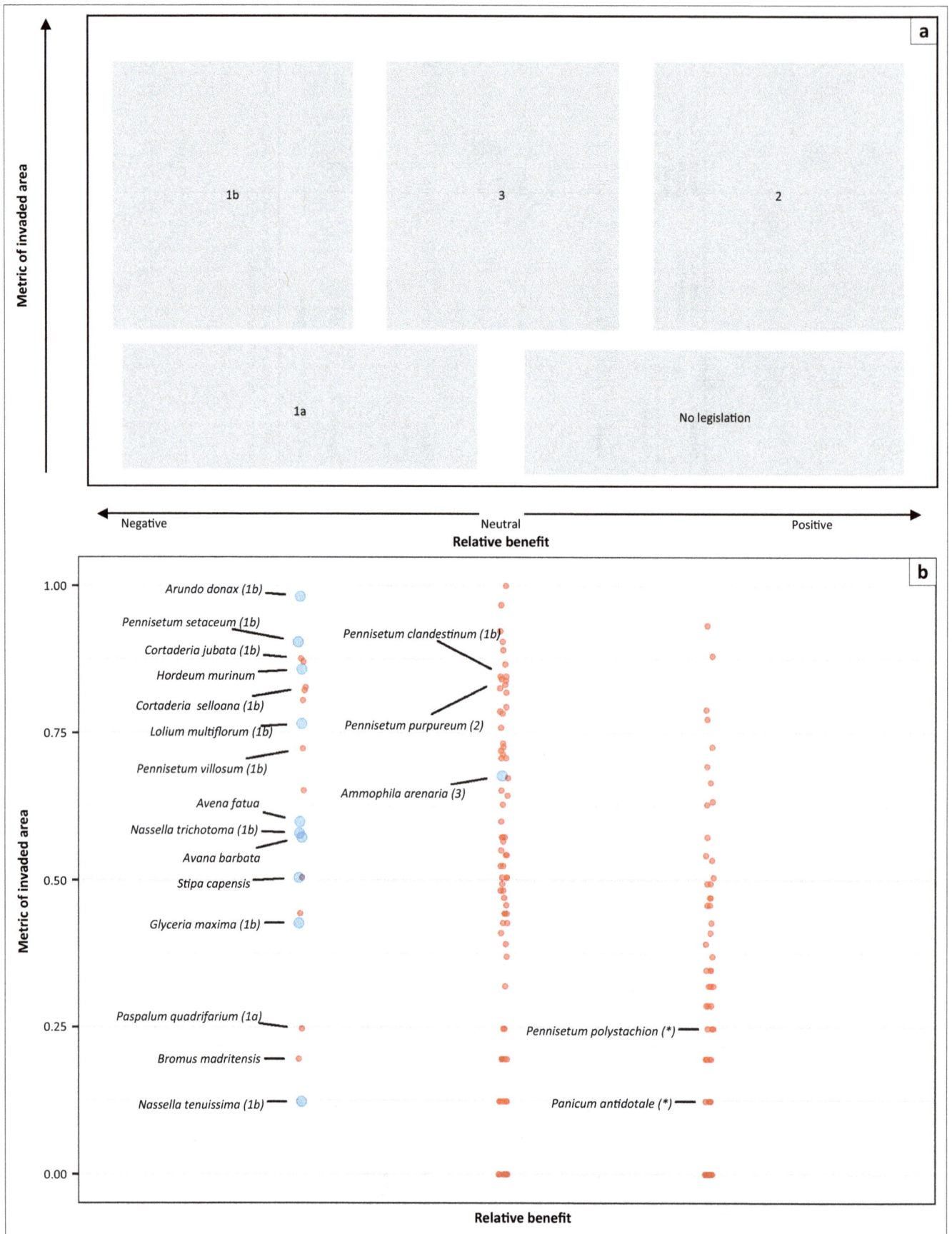

FIGURE 5: (a) Alien species in South Africa are regulated by the Alien and Invasive Species (A&IS) Regulations (2016) under the *National Environmental Management: Biodiversity* (NEM:BA) *Act* 10 of 2004. Guidelines for the categorisation of species (1a, 1b, 2 or 3) are primarily based on the spatial extent of the invaded area and the relative benefits of species. These two factors alone can possibly be used to provide an objective method of providing a NEM:BA category for alien species as shown in this figure. (b) Alien grass species plotted in relation to their and spatial extent and relative benefit (see methods for details). Also provided are species names for species categorised under NEM:BA and their associated categories (in parentheses). An asterisk represents species that are on the NEM:BA prohibited list, but were found to occur in South Africa. Species with recorded impacts are shown in blue.

Discussion

This study provides a much needed reassessment and improved clarity on the status of alien grass species in South Africa. Based on our inventory (Online appendix), at least 256 alien grass species have been introduced to the country. This list and its ancillary data can help inform alien grass research and management, as we shall discuss here.

In both absolute and relative terms, the number of invasive grasses in South Africa is lower than in many other countries (Visser et al. 2016). Around 14% of alien grasses introduced into South Africa have become invasive (Table 1), which is lower than Europe (19%) or the USA (34%) (Visser et al. 2016). Our knowledge of alien grasses in South Africa is generally very patchy; only 70% have herbarium records (Table 1). The identity of many alien grasses in South Africa is therefore uncertain. Moreover, the status of species on the INI continuum is often based solely on anecdotal published information. We might therefore be greatly underestimating the number of invasive grasses in the country. For a third of all alien grass species, we found one reference for their occurrence in South Africa; this, together with the large number of species with no herbarium records or distribution data (Table 1), makes it difficult to be certain that these species are indeed in the country. We suggest that future evaluations of alien grasses should make a concerted effort to address some of these shortcomings (Table 1).

Alien grasses are present throughout most of South Africa (Figure 4). However, this pattern is biased at least in part, by *ad hoc* botanical collections, with extensive sampling bias, for example, towards major urban centres (Figure 4c; see Engemann et al. 2015 and Richardson et al. 2005 for discussion on these collection biases and the implications for analyses). Nevertheless, it appears that the Fynbos Biome has among the highest number of alien grass species and that alien grasses in the fynbos tend to be more abundant and widespread than in other biomes (Figure 4; Appendix 4, Figure 4-A4). Of the 11 species with recorded impacts, six affect the Fynbos Biome. There is also considerable anecdotal evidence to suggest that alien grasses are having large impacts in the Fynbos Biome (Musil et al. 2005; Sharma et al. 2010; Vlok 1988), but these impacts have been poorly quantified. Our knowledge of alien grass impacts in South Africa in general is very poor (Table 2), and further research is needed. Interestingly, protected areas were mostly free of alien grasses (in terms of actual records of occurrence; Table 1; see also Foxcroft et al. 2017). One possible interpretation of this result is that protected areas are somehow more resistant to alien grass invasions, possibly because it is more difficult to invade undisturbed vegetation, or because fire and/or herbivores prevent the establishment of alien grasses (Mack and Thompson 1982; Visser et al. 2016). Another possibility is that sampling of alien grasses in protected areas has been poor. There is some justification to suspect the latter reason because of recent publications describing grass invasions in South African National Parks,

which are perhaps better monitored than other protected areas (Spear et al. 2011).

It is difficult to predict what the future holds with regard to grass invasions, but some trends are apparent. New introductions into South Africa have declined steadily over the last 70 years; although probably because of poor data, no new alien species have been recorded since 2004 (Figure 1). This suggests that the socio-economic factors that led to the introduction of many alien grasses in the past have changed and that the risk of this introduction pathway causing major problems in the future has been greatly reduced. It seems that South African ecosystems are inherently less open to invasion by alien grasses than those in many other parts of the world (Visser et al. 2016). However, it is likely that some of the species already present in South Africa will become naturalised or invasive in the future, representing a considerable invasion debt (~66% of non-invasive alien grass species in South Africa are known to be invasive elsewhere in the world). More than half (57%) of these species have naturalised and some are possibly on their way to becoming invasive. This is all the more likely because recent introductions were from regions with similar climates to South Africa, for example, Australia and sub-Saharan Africa (Figure 3b). Recent introductions were also relatively more likely to be of species with no known use (Figure 2b), those species that tended to be the mostly likely to become invasive. Pasture grasses were also more likely to be invasive than were species in other usage categories (Appendix 2, Table 1-A2). We found that the most common use for grasses in South Africa was for forage (Figure 2a), and that pasture research stations were responsible for introducing many novel alien grass species. Similar situations with regard to pasture grasses have been observed elsewhere in the world, for example, Australia and the USA (Cook and Dias 2006; Lonsdale 1994; Ryerson 1976; Visser et al. 2016). Australia is notable for the number of grass species that were introduced (~1600 species; Cook & Dias 2006; Visser et al. 2016) and also the subsequent number of problematic pasture grass invaders (Cook & Dias 2006). Of the 118 grass species that are not invasive in South Africa, but are invasive elsewhere, 74% are used for pasture, and 32% were trialled at pasture research stations. This suggests that pasture grasses present a considerable invasion debt for South Africa, as exemplified by the recent observation of a pasture species, *Glyceria maxima*, invading wetlands in KwaZulu-Natal (Mugwedi 2012). Another consideration for predicting future invasions is changing trends in the purposes for which grasses are used. We found few changes over the last two centuries in this regard (Figure 2b). However, recently there has been considerable interest in introducing grasses for novel uses such as biofuels (Blanchard et al. 2011) or for species that are thought to have potential for multiple purposes (e.g. bamboos; Canavan et al. 2016). Propagule pressure is likely to be high for these species, because they are likely to be cultivated in large-scale agricultural settings, and the chances of these species then naturalising in adjacent areas is all the more likely

(Simberloff 2009). Natural and human-mediated spread of introduced grasses between South Africa and Africa also represents a potentially increasing threat (Faulkner et al. 2017).

Another possible contributor to future grass invasions is global change (climate, atmospheric CO_2, N pollution, land-use change, etc.). One of the reasons that C_4 grasses are thought to dominate South African grasslands is their higher nitrogen-use efficiencies at pre-industrial CO_2 levels (Milton 2004; Richardson et al. 2000). Rising atmospheric CO_2 levels could, therefore, contribute to increased invasions by C_3 grasses (Milton 2004; Richardson et al. 2000). Moreover, altered nitrogen cycles because of nitrogen pollution or from nitrogen-fixing invasive species would similarly favour C_3 grass species. Weedy native and alien grasses have been documented to dominate nitrogen-enriched soils in fynbos and renosterveld (Sharma et al. 2010; Yelenik, Stock & Richardson 2004). However, C_4 grasses are also thought to be dominant in South African grasslands (and many other grasslands around the world) because of their higher water-use efficiencies under high light and low moisture and CO_2 conditions (Edwards et al. 2010; Edwards and Smith 2010; Osborne and Sack 2012). Therefore, reduced precipitation because of anthropogenic climate change would maintain the competitive advantage of native C_4 species, but perhaps allow for the establishment of alien C_4 species. Warmer temperatures would therefore also favour C_4 species. However, grasses have been shown to exhibit strong phylogenetic conservatism of climatic niches, principally in relation to temperature (Edwards and Smith 2010). The occurrence of C_3 grasses in cooler climes and C_4 grasses in warmer climes is now thought to be largely an artefact of the large number of species within the C_3 grass subfamily Pooideae and the large number of C_4 species in the subfamilies Panicoideae, Aristidoideae and Chloridoideae (Edwards & Smith 2010). Most grasses introduced into South Africa and most of the current invasive species belong to the subfamily Pooideae (Appendix 7). Given these species' affinity for cooler climates, it is unlikely that many new invaders will emerge from this clade and that these species will expand their distributions as the climate warms. Using a climate-envelope approach to predict the future distributions of grass invaders in South Africa, Parker-Allie, Musil & Thuiller (2009) provided support for such a notion. Overall, given the combination of the above factors (nitrogen-use efficiency, water-use efficiency and phylogenetic niche conservatism), we suggest that the alien grasses most likely to be favoured in South African by global change are C_3 Panicoideae species. Only six species in this group are known to have been introduced to South Africa (Appendix 7), which is possibly a contributing factor to the relative paucity of invasive grasses. Other aspects of global change such as land-use change and changing invasion pathways are also likely to affect possible future grass invasions. Transformation of natural environments can aid the establishment of invasive species, particularly many grass species (e.g. Rahlao et al. 2014; Veldman et al. 2009). Alien grasses are also commonly used for revegetation, particularly along roadsides (*pers. obs.*), and sometimes after mining operations have ended (Rahlao et al. 2014). We might therefore expect increasing land transformation to aid the spread and establishment of invasive grasses in South Africa. Novel invasion pathways, such as the introduction of grasses for biofuels or in carbon mitigation schemes, could also cause new grass invasions (Blanchard et al. 2011; Canavan et al. 2016).

Our results suggest numerous avenues for improved alien grass management in South Africa. Mechanical and chemical controls are the most commonly employed techniques for alien grass control (albeit for only a few species; Appendix 5), though biological control is used much less frequently when compared with other taxa (Hill and Coetzee 2017; Zachariades et al. 2017). These techniques are already widely employed by the Working for Water programme (van Wilgen et al. 2012). However, very few resources are being allocated to alien grass management (Table 1), and the current NEM:BA categorisations of grasses do not encourage much more investment, as only 14 species are listed (Figure 5). Our scheme to evaluate NEM:BA categorisations for species suggests that only about 8% of the 256 alien grasses in South Africa probably *do not* need to be on the NEM:BA A&IS 2016 list, but currently only 5.5% of species that should be listed *are* on this list. Further research in providing a simple, but objective and scientifically defensible method for categorising alien species is therefore urgently needed.

Conclusions

There are many alien but few invasive grass species in South Africa. Much uncertainty exists with respect to their identity, numbers of species, distributions, abundances and impacts. Given the potentially large grass invasion debt in South Africa, continued monitoring of alien grass distributions and abundances and much greater engagement with authorities is needed to limit future problems.

Acknowledgements

V.V. received funding from the South African National Department of Environmental Affairs through its funding of a South African National Biodiversity Institute's Invasive Species Programme post-doctoral fellowship, and a National Research Foundation Scarce Skills post-doctoral fellowship, and a research fellowship funded by the African Climate Development Initiative. D.L.M. was funded by the Natural Resource Management programmes, Department of Environmental Affairs, Cape Town. S.K. was funded by the South African National Biodiversity Institute's Invasive Species Programme and the DST-NRF Centre of Excellence for Invasion Biology.

Competing interests

The authors declare that they have no financial or personal relationship(s) that may have inappropriately influenced them in writing this article.

Authors' contributions

All authors conceived the idea and contributed to data collection. V.V. led the data analysis together with D.M.R. and J.R.U.W., with contributions from S.C. and S.K. V.V. led the writing of the article together with J.R.U.W. and D.M.R., with contributions from D.L.M., I.N., T.G.O'C. and S.K.

References

Antonites, A. & Antonites, A.R., 2014, 'The archaeobotany of farming communities in South Africa', in C.J. Stevens, S. Nixon, M.A. Murray & D.Q. Fuller (eds.), *Archaeology of African plant use*, pp. 225–232, UCL Institute of Archaeology Publications, Left Coast Press, Walnut Creek, CA.

Blackburn, T.M., Essl, F., Evans, T., Hulme, P.E., Jeschke, J.M., Kühn, I. et al., 2014, 'A unified classification of alien species based on the magnitude of their environmental impacts', *PLoS Biology* 12, e1001850. https://doi.org/10.1371/journal.pbio.1001850

Blackburn, T.M., Pyšek, P., Bacher, S., Carlton, J.T., Duncan, R.P., Jarošík, V. et al., 2011, 'A proposed unified framework for biological invasions', *Trends in Ecology and Evolution* 26, 333–339. https://doi.org/10.1016/j.tree.2011.03.023

Blanchard, R., Richardson, D.M., O'Farrell, P.J. & von Maltitz, P., G., 2011, 'Biofuels and biodiversity in South Africa', *South African Journal of Science* 107, 19–26. https://doi.org/10.4102/sajs.v107i5/6.186

Canavan, S., Richardson, D.M., Visser, V., Le Roux, J.J., Vorontsova, M.S. & Wilson, J.R.U., 2016, 'The global dissemination of bamboos (Poaceae: Bambusoideae): A review', *AoB Plants* 9, plw078. https://doi.org/10.1093/aobpla/plw078

Christensen, R.H.B., 2016, *Ordinal – Regression models for ordinal data*, R package, viewed 1 October 2016, from https://cran.r-project.org/

Cook, G.D. & Dias, L., 2006, 'It was no accident: Deliberate plant introductions by Australian government agencies during the 20th century', *Australian Journal of Botany* 54, 601–625. https://doi.org/10.1071/BT05157

D'Antonio, C.M., Stahlheber, K. & Molinari, N., 2011, 'Grasses and forbs', in D. Simberloff & M. Rejmánek (eds.), *Encyclopedia of biological invasions*, pp. 280–290, University of California Press, Berkeley and Los Angeles, California.

D'Antonio, C.M. & Vitousek, P.M., 1992, 'Biological invasions by exotic grasses, the grass/fire cycle, and global change', *Annual Review of Ecology and Systematics* 23, 63–87. https://doi.org/10.1146/annurev.es.23.110192.000431

Department of Environmental Affairs (DEA), 2016, *South Africa Protected Areas Database (SAPAD_OR_2016_Q1)*, viewed 12 April 2016, from http://egis.environment.gov.za/

Edwards, E.J., Osborne, C.P., Strömberg, C.A.E., Smith, S.A. & C_4 Grasses Consortium, 2010, 'The origins of C_4 grasslands: Integrating evolutionary and ecosystem science', *Science* 328, 587–591. https://doi.org/10.1126/science.1177216

Edwards, E.J. & Smith, S.A., 2010, 'Phylogenetic analyses reveal the shady history of C_4 grasses', *Proceedings of the National Academy of Sciences* 107, 2532–2537. https://doi.org/10.1073/pnas.0909672107

Engemann, K., Enquist, B.J., Sandel, B., Boyle, B., Jørgensen, P.M., Morueta-Holme, N. et al., 2015, 'Limited sampling hampers "big data" estimation of species richness in a tropical biodiversity hotspot', *Ecology and Evolution* 5, 807–820. https://doi.org/10.1002/ece3.1405

Faulkner, K.T., Hurley, B.P., Robertson, M.P., Rouget, M. & Wilson, J.R.U., 2017, 'The balance of trade in alien species between South Africa and the rest of Africa', *Bothalia* 47(2), a2157. https://doi.org/10.4102/abc.v47i2.2157

Foxcroft, L.C., van Wilgen, N., Baard, J. & Cole, N., 2017, 'Biological invasions in South African National Parks', *Bothalia* 47(2), a2158. https://doi.org/10.4102/abc.v47i2.2158

Gaertner, M., Biggs, R., Beest, M., Hui, C., Molofsky, J. & Richardson, D.M., 2014, 'Invasive plants as drivers of regime shifts: Identifying high-priority invaders that alter feedback relationships', *Diversity and Distributions* 20, 733–744. https://doi.org/10.1111/ddi.12182

Hawkins, C.L., Bacher, S., Essl, F., Hulme, P.E., Jeschke, J.M., Kühn, I. et al., 2015, 'Framework and guidelines for implementing the proposed IUCN Environmental Impact Classification for Alien Taxa (EICAT)', *Diversity and Distributions* 21, 1360–1363. https://doi.org/10.1111/ddi.12379

Henderson, L., 2007, 'Invasive, naturalized and casual alien plants in southern Africa: A summary based on the Southern African Plant Invaders Atlas (SAPIA)', *Bothalia* 37, 215–248. https://doi.org/10.4102/abc.v37i2.322

Hill, M.P. & Coetzee, J.A., 2017, 'The biological control of aquatic weeds in South Africa: Current status and future challenges', *Bothalia* 47(2), a2152. https://doi.org/10.4102/abc.v47i2.2152

Jeschke, J.M., Bacher, S., Blackburn, T.M., Dick, J.T.A., Essl, F., Evans, T. et al., 2014, Defining the impact of non-native species. *Conservation Biology* 28, 1188–1194. https://doi.org/10.1111/cobi.12299

Kumschick, S. & Richardson, D.M., 2013, Species-based risk assessments for biological invasions: Advances and challenges. *Diversity and Distributions* 19, 1095–1105. https://doi.org/10.1111/ddi.12110

Lonsdale, W.M., 1994, 'Inviting trouble: Introduced pasture species in northern Australia', *Australian Journal of Ecology* 19, 345–354. https://doi.org/10.1111/j.1442-9993.1994.tb00498.x

Mack, R.N. & Thompson, J.N., 1982, 'Evolution in steppe with few large, hooved mammals', *American Naturalist* 119, 757–773. https://doi.org/10.1086/283953

Milton, S.J., 2004, 'Grasses as invasive alien plants in South Africa', *South African Journal of Science* 100, 69–75.

Mucina, L., Rutherford, M.C., Powrie, L.W. & Rebelo, A.G., 2005, *Vegetation map of South Africa, Lesotho and Swaziland*, South African National Biodiversity Institute, Pretoria.

Mugwedi, L.F., 2012, 'Invasion ecology of *Glyceria maxima* in KZN rivers and wetlands', M.Sc. thesis, University of the Witwatersrand.

Musil, C.F., Milton, S.J. & Davis, G.W., 2005, 'The threat of alien invasive grasses to lowland Cape floral diversity: An empirical appraisal of the effectiveness of practical control strategies: Research in action', *South African Journal of Science* 101, 337–344.

Olson, D.M., Dinerstein, E., Wikramanayake, E.D., Burgess, N.D., Powell, G.V.N., Underwood, E.C. et al., 2001, 'Terrestrial ecoregions of the world: A new map of life on Earth. A new global map of terrestrial ecoregions provides an innovative tool for conserving biodiversity', *BioScience* 51, 933–938. https://doi.org/10.1641/0006-3568(2001)051[0933:TEOTWA]2.0.CO;2

Osborne, C.P. & Sack, L., 2012, 'Evolution of C_4 plants: A new hypothesis for an interaction of CO2 and water relations mediated by plant hydraulics', *Philosophical Transactions of the Royal Society B: Biological Sciences* 367, 583–600. https://doi.org/10.1098/rstb.2011.0261

Osborne, C.P., Salomaa, A., Kluyver, T.A., Visser, V., Kellogg, E.A., Morrone, O. et al., 2014, 'A global database of C_4 photosynthesis in grasses', *New Phytologist* 204, 441–446. https://doi.org/10.1111/nph.12942

Panetta, F., 1993, 'A system of assessing proposed plant introductions for weed potential', *Plant Protection Quarterly* 8, 10–14.

Parker-Allie, F., Musil, C.F. & Thuiller, W., 2009, 'Effects of climate warming on the distributions of invasive Eurasian annual grasses: A South African perspective', *Climatic Change* 94, 87–103. https://doi.org/10.1007/s10584-009-9549-7

Pluess, T., Cannon, R., Jarošík, V., Pergl, J., Pyšek, P. & Bacher, S., 2012a, 'When are eradication campaigns successful? A test of common assumptions', *Biological Invasions* 14, 1365–1378. https://doi.org/10.1007/s10530-011-0160-2

Pluess, T., Jarošík, V., Pyšek, P., Cannon, R., Pergl, J., Breukers, A. et al., 2012b, 'Which factors affect the success or failure of eradication campaigns against alien species?', *PLoS One* 7, e48157. https://doi.org/10.1371/journal.pone.0048157

Quattrocchi, U., 2006, *CRC world dictionary of grasses: Common names, scientific names, eponyms, synonyms, and etymology – 3 volume set*, CRC Press, Boca Raton, FL.

R Core Team, 2016, *R: A language and environment for statistical computing*, viewed 1 October 2016, from http://www.r-project.org/

Rahlao, S.J., Milton, S.J., Esler, K.J. & Barnard, P., 2014, 'Performance of invasive alien fountain grass (*Pennisetum setaceum*) along a climatic gradient through three South African biomes', *South African Journal of Botany* 91, 43–48. https://doi.org/10.1016/j.sajb.2013.11.013

Rahlao, S.J., Milton, S.J., Esler, S.J., van Wilgen, B.W. & Barnard, P., 2009, 'Effects of invasion of fire-free arid shrublands by a fire-promoting invasive alien grass species (*Pennisetum setaceum*) in South Africa', *Austral Ecology* 34, 920–928. https://doi.org/10.1111/j.1442-9993.2009.02000.x

Randall, R.P., 2012, *A global compendium of weeds, 2nd edition*, Department of Agriculture and Food, Perth, Western Australia.

Rejmánek, M. & Pitcairn, M., 2002, 'When is eradication of exotic pest plants a realistic goal?', in C.R. Veitch & M.N. Clout (eds.), *Turning the tide: The eradication of invasive species, Proceedings of the international conference on eradication of island invasives*, pp. 249–253, IUCN, Cambridge, UK.

Richardson, D.M., Bond, W.J., Dean, W.R.J., Higgins, S.I., Midgley, G., Milton, S.J. et al., 2000, 'Invasive alien species and global change: A South African perspective', in H.A. Mooney & R.J. Hobbs (eds.), *Invasive species in a changing world*, pp. 303–350, Island Press, Washington, DC.

Richardson, D.M. & Pyšek, P., 2012, 'Naturalization of introduced plants: Ecological drivers of biogeographical patterns', *New Phytologist* 196, 383–396. https://doi.org/10.1111/j.1469-8137.2012.04292.x

Richardson, D.M., Rouget, M., Ralston, S.J., Cowling, R.M., Van Rensburg, B.J. & Thuiller, W., 2005, 'Species richness of alien plants in South Africa: Environmental correlates and the relationship with indigenous plant species richness', *Ecoscience* 12, 391–402. https://doi.org/10.2980/i1195-6860-12-3-391.1

Rouget, M., Robertson, M.P., Wilson, J.R.U., Hui, C., Essl, F., Renteria, J.L. et al., 2016, 'Invasion debt – quantifying future biological invasions', *Diversity and Distributions* 22, 445–456. https://doi.org/10.1111/ddi.12408

Ryerson, K.A., 1976, 'Plant introductions', *Agricultural History* 50, 248–257.

Sharma, G.P., Muhl, S.A., Esler, K.J. & Milton, S.J., 2010, 'Competitive interactions between the alien invasive annual grass *Avena fatua* and indigenous herbaceous plants in South African Renosterveld: The role of nitrogen enrichment', *Biological Invasions* 12, 3371–3378. https://doi.org/10.1007/s10530-010-9730-y

Simberloff, D., 2009, 'The role of propagule pressure in biological invasions', *Annual Review of Ecology, Evolution and Systematics* 40, 81–102. https://doi.org/10.1146/annurev.ecolsys.110308.120304

Soreng, R.J., Peterson, P.M., Romaschenko, K., Davidse, G., Zuloaga, F.O., Judziewicz, E.J. et al., 2015, A worldwide phylogenetic classification of the Poaceae (Gramineae)', *Journal of Systematics and Evolution* 53, 117–137. https://doi.org/10.1111/jse.12150

Spear, D., McGeoch, M.A., Foxcroft, L.C. & Bezuidenhout, H., 2011, 'Alien species in South Africa's national parks', *Koedoe* 53, Art. #1032, 4 pages. https://doi.org/10.4102/koedoe.v53i1.1032

Taylor, S.H., Hulme, S.P., Rees, M., Ripley, B.S., Ian Woodward, F. & Osborne, C.P., 2010, 'Ecophysiological traits in C_3 and C_4 grasses: A phylogenetically controlled screening experiment', *New Phytologist* 185, 780–791. https://doi.org/10.1111/j.1469-8137.2009.03102.x

Van Wilgen, B., 2014, *A national strategy for dealing with biological invasions in South Africa*, DEA, Pretoria, South Africa.

van Wilgen, B.W., Forsyth, G.G., Le Maitre, D.C., Wannenburgh, A., Kotzé, J.D., van den Berg, E. et al., 2012, 'An assessment of the effectiveness of a large, national-scale invasive alien plant control strategy in South Africa', *Biological Conservation* 148, 28–38. https://doi.org/10.1016/j.biocon.2011.12.035

Veldman, J.W., Mostacedo, B., Peña-Clarosa, M. & Putz, F.E., 2009, 'Selective logging and fire as drivers of alien grass invasion in a Bolivian tropical dry forest', *Forest Ecology and Management* 258, 1643–1649. https://doi.org/10.1016/j.foreco.2009.07.024

Visser, V., Wilson, J.R.U., Fish, L., Brown, C., Cook, G.D. & Richardson, D.M., 2016, 'Much more give than take: South Africa as a major donor but infrequent recipient of invasive non-native grasses', *Global Ecology and Biogeography* 25, 679–692. https://doi.org/10.1111/geb.12445

Vlok, J., 1988, 'Alpha diversity of lowland fynbos herbs at various levels of infestation by alien annuals', *South African Journal of Botany* 54, 623–627. https://doi.org/10.1016/S0254-6299(16)31264-9

Wilson, J.R.U., Gaertner, M., Richardson, D.M. & van Wilgen, B.W., 2017, 'Contributions to the national status report on biological invasions in South Africa', *Bothalia* 47(2), a2207. https://doi.org/10.4102/abc.v47i2.2207

Wilson, J.R.U., Ivey, P., Manyama, P. & Nänni, I., 2013, 'A new national unit for invasive species detection, assessment and eradication planning', *South African Journal of Science* 109, 1–13.

Wilson, J.RU., Richardson, D.M., Rouget, M., Procheş, Ş., Amis, M.A., Henderson, L. et al., 2007, 'Residence time and potential range: Crucial considerations in modelling plant invasions', *Diversity and Distributions* 13, 11–22. https://doi.org/10.1111/j.1366-9516.2006.00302.x

Yelenik, S., Stock, W. & Richardson, D., 2004, 'Ecosystem-level impacts of invasive alien nitrogen-fixing plants. Ecosystem and community-level impacts of invasive alien *Acacia saligna* in the fynbos vegetation of South Africa', *Restoration Ecology* 12, 44–51. https://doi.org/10.1111/j.1061-2971.2004.00289.x

Zachariades, C., Paterson, I.D., Strathie, L.W., Hill, M.P. & van Wilgen, B.W., 2017, 'Assessing the status of biological control as a management tool for suppression of invasive alien plants in South Africa', *Bothalia* 47(2), a2142. https://doi.org/10.4102/abc.v47i2.2142

Appendix 1

TABLE 1-A1: Online and published sources used to compile a checklist of alien grass species in South Africa

Reference	Introduced	Naturalised	Invasive	Distribution	Date
Adams et al. (2012)	-	-	✓	-	✓
Baard and Kraaij (2014) INI status was assigned based on species' status in this reference †	-	✓	✓	-	-
Bromilow (2010)	-	✓	-	-	-
CABI (2016) INI status was assigned based on species' status in this reference †	✓	✓	✓	-	-
Clayton et al. (2006 onwards)	-	-	-	-	-
Cowan and Anderson (2014)	-	✓	-	✓	✓
Davies (1975)	-	-	-	-	✓
Department of Agriculture & Forestry (1940)	✓	-	-	-	✓
Foxcroft et al. (2008)	-	✓	✓	-	-
GBIF (2016)	✓	-	-	✓	✓
Glen (2002)	✓	-	-	-	-
GISD (2016) INI status was assigned based on species' status in this reference †	✓	✓	✓	-	-
Grootfontein Agricultural Development Institute (2016)	✓	-	-	✓	✓
Guthrie (2007)	-	-	✓	-	-
Harding (1982)	-	-	-	✓	✓
Heelemann et al. (2013)	-	✓	✓	✓	✓
Henderson (2007) Invasive if species recorded as being abundant or very abundant at more than one locality †	-	✓	✓	✓	✓
Hertling & Lubke (1999)	-	-	-	✓	✓
Holmes (2008)	-	-	✓	✓	✓
Joubert (1984)	-	-	✓	✓	✓
Lesoli et al. (2013)	-	✓	-	-	-
Maggs and Ward (1984)	-	-	-	-	✓
Manning and Goldblatt (2012)	✓	-	-	-	-
Masubelele et al. (2009)	-	✓	-	✓	✓
Milton et al. (1998)	-	✓	-	✓	✓
Mugwedi (2012)	-	-	✓	-	-
Musil et al. (2005)	-	-	✓	✓	✓
Plantsinstock (2016)	✓	-	-	-	-
Powrie (2012)	✓	-	-	✓	✓
Rahlao et al. (2010)	-	-	✓	-	-
Randall (2012) INI status was based on the statuses in this reference, as described in the relevant columns here †	Species occurring in SA with statuses in Randall (2012) that are not any of those listed for naturalised or invasive species	Species occurring in SA with statuses in Randall (2012) of: Cultivation Escape, Environmental Weed, Garden Escape, Invasive, Naturalised, Noxious Weed	Species occurring in SA with statuses in Randall (2012) of: Environmental Weed, Invasive, Noxious Weed	✓	-
Ratnasingham & Herbert (2007)	✓	-	-	-	-
SANBI (2009a) Naturalised, if described as such. †	✓	✓	-	-	-
SANBI (2009b)	✓	-	-	✓	✓
Scott (1982)	-	-	-	-	✓
Sharma et al. (2010)	-	-	✓	✓	✓
Shiponeni & Milton (2006)	-	-	-	✓	✓
Steinschen et al. (1996)	-	-	✓	✓	✓
Thunberg (1823)	-	-	-	-	✓
Todd (2008)	-	-	✓	✓	✓
USDA, ARS, National Genetic Resources Program (2016) Naturalised in South Africa, if described as such. †	✓	✓	-	-	-
Viljoen (1987)	-	-	-	✓	✓

Indicated are the references used to determine species' statuses along the introduction-naturalisation-invasion continuum (Blackburn et al. 2011). References with a ✓ indicate that the reference was used to assign that specific status to a species. References that were used as sources of distribution data are indicated with ticks in the 'Distribution' column, and those for obtaining dates of occurrences (and ultimately minimum residence times) in the 'Date' column.

To clarify the alien status of all species in South Africa we first compared our initial list of species against a list of native grasses obtained from the Plants of Southern Africa database (POSA; http://newposa.sanbi.org/). Thereafter, we further refined the list using distribution information on the native range of each species from the eMonocot database (http://emonocot.org) and the Germplasm Resources Information Network (http://ars-grin.gov/cgi-bin/npgs/html/taxgenform.pl?language=en). We flagged species that were possibly native to South Africa, but not recorded as such in POSA, and manually checked whether these are in fact native.

†, indicate special conditions.

Appendix 2

TABLE 1-A2: Statistical summary of differences among use categories in relation to the numbers of alien grass species of different statuses across the introduction-naturalisation-invasion continuum as determined by an ordinal logistic regression.

Use category	*β*	2.5%	97.5%	LRT	*P*
None	1.15	0.23	2.10	5.984	**< 0.05**
Forage	0.70	0.03	1.41	4.205	**< 0.05**
Horticulture	0.51	-0.05	1.08	3.125	0.07
Raw material	-0.53	-1.32	0.23	1.838	0.18
Soil stabilisation	0.27	-0.30	0.83	0.884	0.35
Food and beverage	-0.13	-0.83	0.56	0.135	0.71

A likelihood-ratio test (LRT) was used to test whether the association of each use category with introduction-naturalisation-invasion status was significantly different from zero. Significant *P*-values shown in bold. Also shown are correlation coefficients (*β*) and upper and lower 95% confidence intervals.

TABLE 2-A2: Statistical summary of differences among realms to which species are native in relation to the numbers of alien grass species of different statuses across the introduction-naturalisation-invasion continuum as determined by an ordinal logistic regression.

Realm	β	2.5%	97.5%	LRT	*P*
South America	2.23	1.26	3.27	21.61	**< 0.0001**
Eurasia	1.52	0.80	2.30	18.19	**< 0.0001**
North America	-1.29	-2.22	-0.42	8.73	**< 0.01**
Southeast Asia	0.25	-0.33	0.83	0.71	0.40
Australasia	0.34	-0.73	1.36	0.40	0.53
Pacific	0.55	-2.71	3.26	0.15	0.70

A likelihood-ratio test (LRT) was used to test whether the association of each use category with introduction-naturalisation-invasion status was significantly different from zero. Significant *P*-values shown in bold. Also shown are correlation coefficients (*β*) and upper and lower 95% confidence intervals.

Appendix 3

TABLE 1-A3: Pasture research stations established in South Africa.

Station	Establishment year	Closure year	Number species cultivated	Reference
Skinners Court, Pretoria	1903	1912	?	Smith and Rhind (1984)
Groenkloof, Pretoria	1912	1923?	2	Smith and Rhind (1984)
Burttholm, Vereeniging	1913	1919	3	Gunn and Codd (1981)
Prinshof, Pretoria	1923	1940?	63	DAF (1940); Smith and Rhind (1984)
Athole Research Station, Amsterdam, Mpumulanga	1934	?	26	Donaldson (1984)
Towoomba, Bela-Bela, Limpopo	1934	?	2	Donaldson (1984)
Albany Museum Herbarium, Grahamstown	1934	1937	1	Smith and Rhind (1984)
Rietondale, Pretoria	1934	1972	?	Smith and Rhind (1984)
Leeuwkuil, Vereeniging	1934	1938?	24	Donaldson (1984); Story (1938)
Potchefstroom	1946	< 1956?	2	Smith and Rhind (1984)
Cedara, Pietermaritzburg	1951	Still operational	13	Smith and Rhind (1984)
Rietvlei, Pretoria	1954	< 1956?	?	Smith and Rhind (1984)
Koolbank Grass Seed Station, Pretoria	1954	< 1956?	?	Smith and Rhind (1984)
Estcourt	1936	?	18	Scott (1966)

Source: Department of Agriculture & Forestry (DAF), 1940, *Pasture research in South Africa: Progress report no. 2. Division of soil and veld conservation*, Government Printer of the Union of South Africa, Pretoria

Provided for each station are the establishment year, closure year (when available) and supporting references.

TABLE 2-A3: Introduced grass species cultivated at pasture research stations in South Africa.

Station	Species	Year	Novel introduction	Reference
Prinshof	*Aegilops cylindrica* Host	1938	Yes	DAF (1940)
Estcourt	*Agropyron cristatum* (L.) Gaertn	1937	Yes	DAF (1940)
Leeuwkuil	*Agropyron cristatum* (L.) Gaertn	1938	-	DAF (1940)
Prinshof	*Agropyron cristatum* (L.) Gaertn	1938	-	DAF (1940)
Prinshof	*Agropyron desertorum* (Fisch. ex Link) Schult	1938	Yes	DAF (1940)
Prinshof	*Agropyron fragile* (Roth) P.Candargy	1938	Yes	DAF (1940)
Cedara	*Agropyron* spp.	1951	-	Smith and Rhind (1984)
Athole	*Agrostis capillaris* L.	1934	Yes	DAF (1940)
Estcourt	*Agrostis capillaris* L.	1937	-	DAF (1940)
Leeuwkuil	*Agrostis capillaris* L.	1938	-	DAF (1940)
Prinshof	*Agrostis capillaris* L.	1938	-	DAF (1940)
Cedara	*Agrostis* spp.	1951	-	Smith and Rhind (1984)
Athole	*Agrostis stolonifera* L.	1934	-	DAF (1940)
Estcourt	*Agrostis stolonifera* L.	1937	-	DAF (1940)
Leeuwkuil	*Agrostis stolonifera* L.	1938	-	DAF (1940)
Prinshof	*Andropogon gerardii* Vitman	1938	Yes	DAF (1940)
Prinshof	*Andropogon hallii* Hack	1938	Yes	DAF (1940)
Athole	*Arrhenatherum elatius* (L.) P.Beauv. ex J.Presl & C.Presl.	1934	-	DAF (1940)
Leeuwkuil	*Arrhenatherum elatius* (L.) P.Beauv. ex J.Presl & C.Presl.	1938	-	DAF (1940)
Prinshof	*Arrhenatherum elatius* (L.) P.Beauv. ex J.Presl & C.Presl.	1938	-	DAF (1940)
Prinshof	*Astrebla* spp.	1938	-	DAF (1940)
Prinshof	*Avena nuda* L.	1938	Yes	DAF (1940)
Athole	*Axonopus compressus* (Sw.) P.Beauv.	1934	-	DAF (1940)
Prinshof	*Bothriochloa saccharoides* (Sw.) Rydb.	1938	Yes	DAF (1940)
Prinshof	*Bouteloua chondrosioides* (Kunth) Benth. ex S.Watson	1938	Yes	DAF (1940)
Prinshof	*Bouteloua curtipendula* (Michx.) Torr.	1938	-	DAF (1940)
Prinshof	*Brachiaria mutica* (Forssk.) Stapf	1938	Yes	DAF (1940)
Athole	*Bromus catharticus* Vahl	1934	-	DAF (1940)
Cedara	*Bromus catharticus* Vahl	1951	-	Smith and Rhind (1984)
Estcourt	*Bromus catharticus* Vahl	1937	-	DAF (1940)
Leeuwkuil	*Bromus catharticus* Vahl	1938	-	DAF (1940)
Prinshof	*Bromus catharticus* Vahl	1938	-	DAF (1940)
Athole	*Bromus inermis* Leyss.	1934	-	DAF (1940)
Cedara	*Bromus inermis* Leyss.	1951	-	Smith and Rhind (1984)
Estcourt	*Bromus inermis* Leyss.	1937	-	DAF (1940)
Leeuwkuil	*Bromus inermis* Leyss.	1938	-	DAF (1940)
Prinshof	*Bromus inermis* Leyss.	1938	-	DAF (1940)
Prinshof	*Buchloe dactyloides* (Nutt.) Engelm.	1938	-	DAF (1940)
Prinshof	*Chondrosum eriopodum* Torr.	1938	Yes	DAF (1940)
Prinshof	*Chondrosum gracile* Kunth	1938	-	DAF (1940)
Prinshof	*Chondrosum hirsutum* (Lag.) Sweet	1938	Yes	DAF (1940)
Prinshof	*Coix lacryma-jobi* L.	1938	-	DAF (1940)
Athole	*Cynosurus cristatus* L.	1934	Yes	DAF (1940)
Prinshof	*Cynosurus cristatus* L.	1938	-	DAF (1940)
Athole	*Dactylis glomerata* L.	1934	-	DAF (1940)
Cedara	*Dactylis glomerata* L.	1951	-	Smith and Rhind (1984)
Estcourt	*Dactylis glomerata* L.	1937	-	DAF (1940)
Leeuwkuil	*Dactylis glomerata* L.	1938	-	DAF (1940)
Prinshof	*Dactylis glomerata* L.	1938	-	DAF (1940)
Leeuwkuil	*Echinochloa esculenta* (A.Braun) H.Scholz	1938	Yes	DAF (1940)
Athole	*Elymus caninus* (L.) L.	1934	-	DAF (1940)
Leeuwkuil	*Elymus caninus* (L.) L.	1938	-	DAF (1940)
Prinshof	*Elymus elongatus* (Host) Runemark	1938	Yes	DAF (1940)
Prinshof	*Elymus repens* (L.) Gould	1938	-	DAF (1940)
Prinshof	*Elymus smithii* (Rydb.) Gould	1938	Yes	DAF (1940)
Estcourt	*Elymus trachycaulus* (Link) Gould ex Shinners	1937	Yes	DAF (1940)
Prinshof	*Elymus trachycaulus* (Link) Gould ex Shinners	1938	-	DAF (1940)
Leeuwkuil	*Eragrostis tef* (Zucc.) Trotter	1938	-	DAF (1940)
Athole	*Festuca arundinacea* Schreb.	1934	-	DAF (1940)
Burttholm	*Festuca arundinacea* Schreb.	1914	-	Burtt-Davy (1920)
Cedara	*Festuca arundinacea* Schreb.	1951	-	Smith and Rhind (1984)

Appendix table continued on the next page →

TABLE 2-A3 (Continues...): Introduced grass species cultivated at pasture research stations in South Africa.

Station	Species	Year	Novel introduction	Reference
Estcourt	*Festuca arundinacea* Schreb.	1937	-	DAF (1940)
Groenkloof	*Festuca arundinacea* Schreb.	1912	-	Smith and Rhind (1984)
Leeuwkuil	*Festuca arundinacea* Schreb.	1938	-	DAF (1940)
Prinshof	*Festuca arundinacea* Schreb.	1938	-	DAF (1940)
Athole	*Festuca ovina* L.	1934	Yes	DAF (1940)
Leeuwkuil	*Festuca ovina* L.	1938	-	DAF (1940)
Leeuwkuil	*Festuca pratensis* Huds.	1938	Yes	DAF (1940)
Athole	*Festuca rubra* L.	1934	Yes	DAF (1940)
Estcourt	*Festuca rubra* L.	1937	-	DAF (1940)
Leeuwkuil	*Festuca rubra* L.	1938	-	DAF (1940)
Prinshof	*Festuca rubra* L.	1938	-	DAF (1940)
Athole	*Holcus lanatus* L.	1934	-	DAF (1940)
Cedara	*Holcus lanatus* L.	1951	-	Smith and Rhind (1984)
Leeuwkuil	*Holcus lanatus* L.	1938	-	DAF (1940)
Prinshof	*Holcus lanatus* L.	1938	-	DAF (1940)
Prinshof	*Leymus angustus* (Trin.) Pilg.	1938	Yes	DAF (1940)
Prinshof	*Leymus chinensis* (Trin.) Tzvelev	1938	Yes	DAF (1940)
Prinshof	*Leymus ramosus* (C.Richt.) Tzvelev	1938	Yes	DAF (1940)
Athole	*Lolium multiflorum* Lam.	1934	-	DAF (1940)
Cedara	*Lolium multiflorum* Lam.	1951	-	Smith and Rhind (1984)
Cedara	*Lolium multiflorum* Lam.	1951	-	Smith and Rhind (1984)
Estcourt	*Lolium multiflorum* Lam.	1937	-	DAF (1940)
Leeuwkuil	*Lolium multiflorum* Lam.	1938	-	DAF (1940)
Prinshof	*Lolium multiflorum* Lam.	1938	-	DAF (1940)
Athole	*Lolium perenne* L.	1934	-	DAF (1940)
Cedara	*Lolium perenne* L.	1951	-	Smith and Rhind (1984)
Estcourt	*Lolium perenne* L.	1937	-	DAF (1940)
Leeuwkuil	*Lolium perenne* L.	1938	-	DAF (1940)
Prinshof	*Lolium perenne* L.	1938	-	DAF (1940)
Athole	*Lolium rigidum* Gaudin	1934	-	DAF (1940)
Prinshof	*Oryzopsis hymenoides* (Roem. and Schult.) Ricker ex Piper	1938	Yes	DAF (1940)
Prinshof	*Panicum acrotrichum* Hook.f.	1938	Yes	DAF (1940)
Prinshof	*Panicum miliaceum* L.	1938	-	DAF (1940)
Prinshof	*Panicum obtusum* Kunth	1938	Yes	DAF (1940)
Athole	*Panicum phragmitoides* Stapf	1934	Yes	DAF (1940)
Towoomba	*Panicum phragmitoides* Stapf	1934	-	DAF (1940)
Prinshof	*Panicum plenum* Hitchc. and Chase	1938	Yes	DAF (1940)
Prinshof	*Panicum prolutum* F.Muell.	1938	Yes	DAF (1940)
Prinshof	*Panicum virgatum* L.	1938	Yes	DAF (1940)
Prinshof	*Paspalidium flavidum* (Retz.) A.Camus	1938	Yes	DAF (1940)
Athole	*Paspalum dilatatum* Poir.	1934	-	DAF (1940)
Cedara	*Paspalum dilatatum* Poir.	1951	-	Smith and Rhind (1984)
Estcourt	*Paspalum dilatatum* Poir.	1937	-	DAF (1940
Leeuwkuil	*Paspalum dilatatum* Poir.	1938	-	DAF (1940)
Prinshof	*Paspalum dilatatum* Poir.	1938	-	DAF (1940)
Prinshof	*Paspalum notatum* Flüggé	1938	-	DAF (1940)
Cedara	*Paspalum urvillei* Steud.	1951	-	Smith and Rhind (1984)
Athole	*Paspalum virgatum* L.	1934	Yes	DAF (1940)
Leeuwkuil	*Paspalum virgatum* L.	1938	-	DAF (1940)
Prinshof	*Paspalum virgatum* L.	1938	-	DAF (1940)
Athole	*Pennisetum clandestinum* Hochst. ex Chiov.	1934	-	DAF (1940)
Burttholm	*Pennisetum clandestinum* Hochst. ex Chiov.	1915	-	Burtt-Davy (1915a)
Groenkloof	*Pennisetum clandestinum* Hochst. ex Chiov.	1912	-	Smith and Rhind (1984)
Prinshof	*Pennisetum clandestinum* Hochst. ex Chiov.	1938	-	DAF (1940)
Leeuwkuil	*Pennisetum glaucum* (L.) R.Br.	1938	-	DAF (1940)
Potchefstroom	*Pennisetum glaucum* (L.) R.Br.		-	Smith and Rhind (1984)
Prinshof	*Pennisetum polystachion* Schult.	1938	-	DAF (1940)
Athole	*Pennisetum purpureum* Schumach.	1934	-	DAF (1940)
Burttholm	*Pennisetum purpureum* Schumach.	1915	-	Burtt-Davy (1915)
Cedara	*Pennisetum purpureum* Schumach.	1951	-	Smith and Rhind (1984)
Potchefstroom	*Pennisetum purpureum* Schumach.		-	Smith and Rhind (1984)

TABLE 2-A3 (Continues...): Introduced grass species cultivated at pasture research stations in South Africa.

Station	Species	Year	Novel introduction	Reference
Prinshof	*Pennisetum purpureum* Schumach.	1938	-	DAF (1940)
Prinshof	*Pennisetum setaceum* (Forssk.) Chiov.	1938	-	DAF (1940)
Athole	*Phalaris aquatica* L.	1934	-	DAF (1940)
Estcourt	*Phalaris aquatica* L.	1937	-	DAF (1940)
Leeuwkuil	*Phalaris aquatica* L.	1938	-	DAF (1940)
Prinshof	*Phalaris aquatica* L.	1938	-	DAF (1940)
Estcourt	*Phalaris arundinacea* L.	1937	-	DAF (1940)
Prinshof	*Phalaris arundinacea* L.	1938	-	DAF (1940)
Prinshof	*Phalaris coerulescens* Desf.	1938	Yes	DAF (1940)
Athole	*Phleum pratense* L.	1934	-	DAF (1940)
Estcourt	*Phleum pratense* L.	1937	-	DAF (1940)
Leeuwkuil	*Phleum pratense* L.	1938	-	DAF (1940)
Prinshof	*Phleum* sp.	1938	-	DAF (1940)
Prinshof	*Poa compressa* L.	1938	Yes	DAF (1940)
Athole	*Poa pratensis* L.	1934	-	DAF (1940)
Estcourt	*Poa pratensis* L.	1937	-	DAF (1940
Leeuwkuil	*Poa pratensis* L.	1938	-	DAF (1940)
Prinshof	*Poa pratensis* L.	1938	-	DAF (1940)
Athole	*Poa trivialis* L.	1934	-	DAF (1940)
Prinshof	*Poa trivialis* L.	1938	-	DAF (1940)
Estcourt	*Psathyrostachys juncea* (Fisch.) Nevski	1937	Yes	DAF (1940)
Prinshof	*Rytidosperma pilosum* (R.Br.) Connor and Edgar	1938	Yes	DAF (1940)
Athole	*Rytidosperma semiannulare* (Labill.) Connor and Edgar	1934	Yes	DAF (1940)
Prinshof	*Rytidosperma semiannulare* (Labill.) Connor and Edgar	1938	-	DAF (1940)
Estcourt	*Schedonnardus paniculatus* (Nutt.) Trel.	1937	Yes	DAF (1940)
Prinshof	*Schizachyrium scoparium* (Michx.) Nash	1938	Yes	DAF (1940)
Albany Museum	*Setaria* spp.		-	Smith and Rhind (1984)
Prinshof	*Sorghastrum nutans* (L.) Nash	1938	-	DAF (1940)
Leeuwkuil	*Sorghum × drummondii* (Nees ex Steud.) Millsp. and Chase	1938	-	DAF (1940)
Prinshof	*Sorghum × drummondii* (Nees ex Steud.) Millsp. and Chase	1938	-	DAF (1940)
Prinshof	*Sorghum halepense* (L.) Pers.	1938	-	DAF (1940)
Towoomba	*Sorghum halepense* (L.) Pers.	1934	-	DAF (1940)

Provided for each species are the station at which it was trialled, the year the species was trialled (or the pasture station establishment year if trial date unavailable), whether the species was previously not in South Africa (novel introduction), and supporting references.

Appendix 4

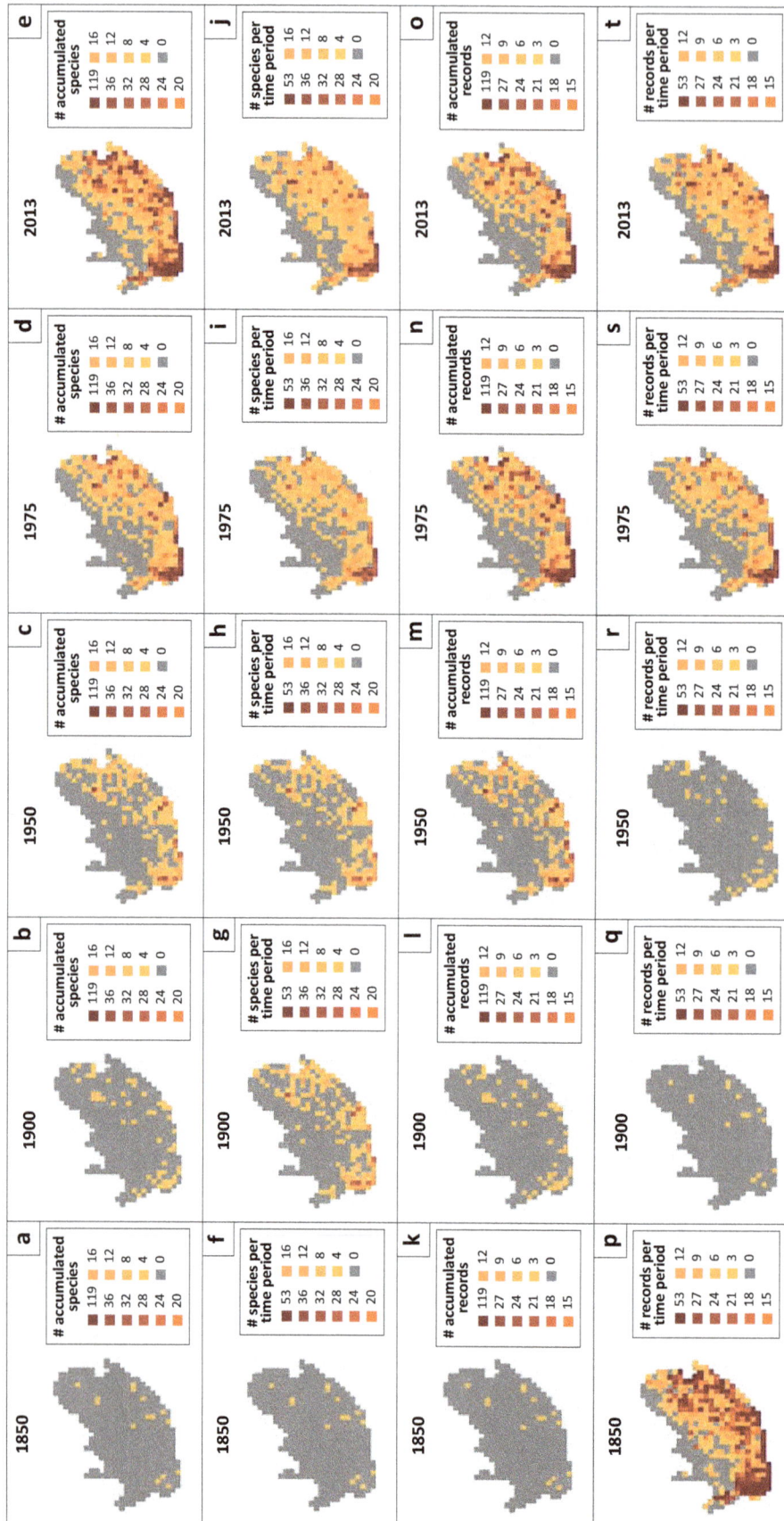

(a–e) accumulated alien grass species richness, (f–j) alien grass species richness per time interval (1811 to 1900, 1901 to 1950, 1951 to 1975, and 1976 to 2013), (k–o) accumulated number of alien grass records and (p–t) numbers of records of alien grasses per time interval. Species richness and numbers of records are aggregated to quarter-degree-grid-cells (QDGCs). Darker reds indicate higher species richness or numbers of records.

FIGURE 1-A4: Patterns over time.

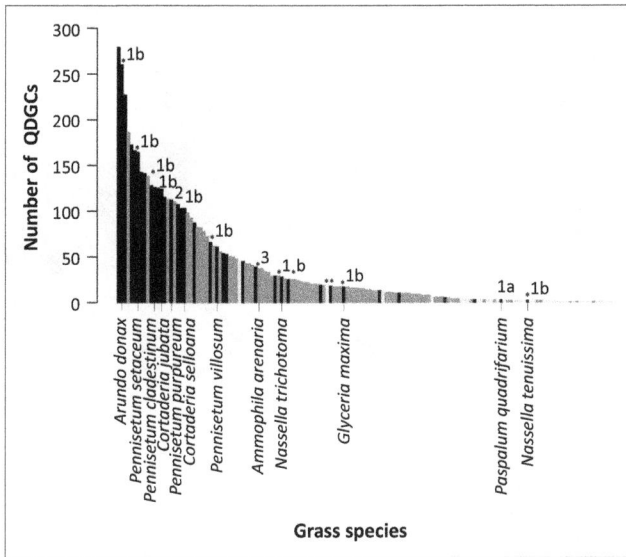

FIGURE 2-A4: Area occupied by alien grass species (measured in number of quarter-degree-grid-cells, QDGCs, occupied), plotted in descending order. The status of species on the introduction-naturalisation-invasion continuum is indicated by bar colour (light grey = introduced, dark grey = naturalised and black = invasive). Species with recorded impacts are indicated by an asterisk above the bars (see Table 2). Species that currently have legal requirements for their management are named and their legal status is indicated above the bars (see Appendix 5 for further details).

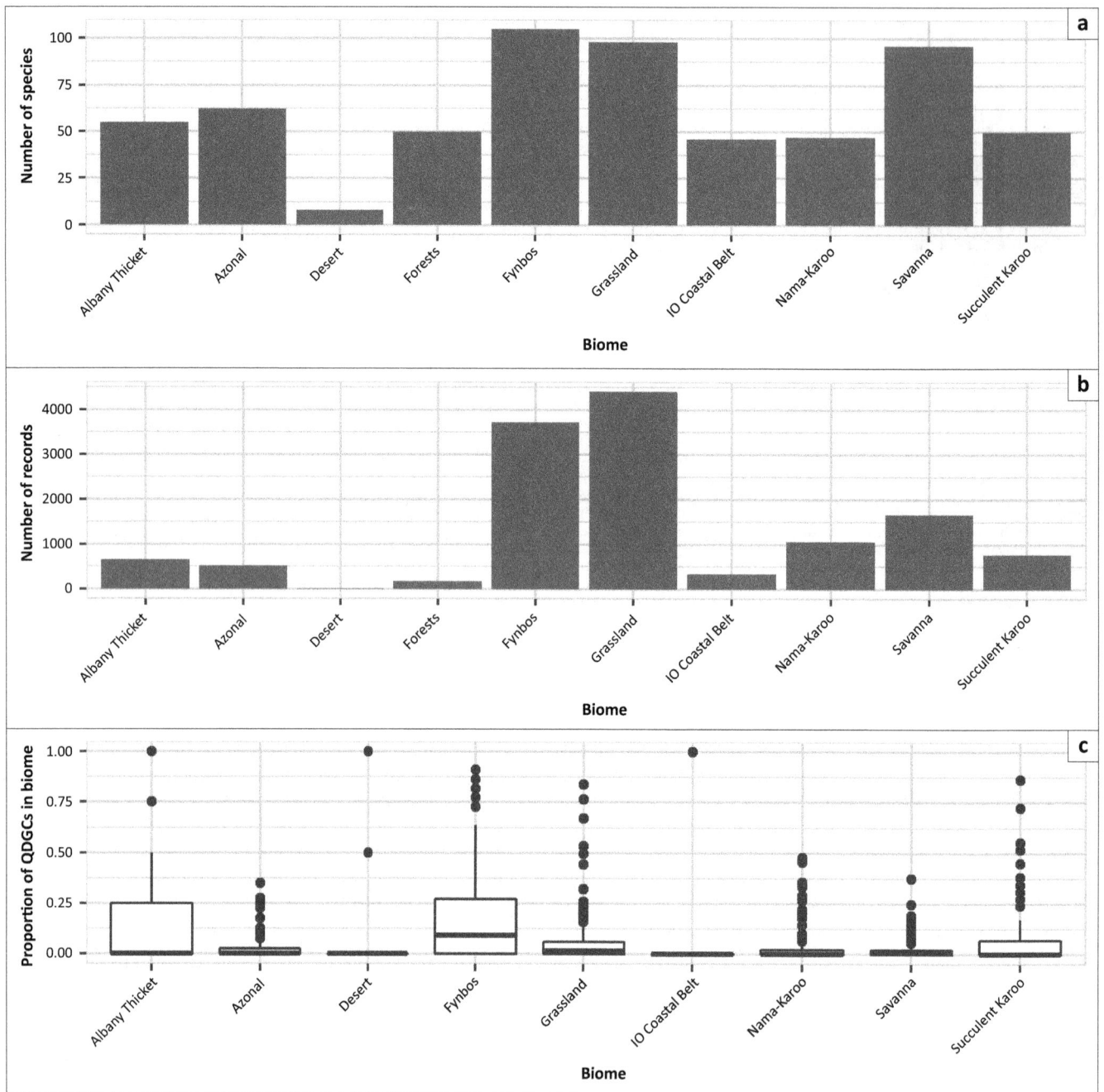

FIGURE 3-A4: (a) The number of alien grass species in each biome of South Africa. (b) The number of occurrence records of alien grass species in each biome of South Africa. (c) Area occupied by alien grass species in each biome of South Africa [measured as the proportion occupied by each species of quarter-degree-grid-cells (QDGCs) available in each biome]. The Fynbos, Grassland and Savanna biomes have the most alien grass species. The Grassland and Fynbos biomes also have the most occurrence records.

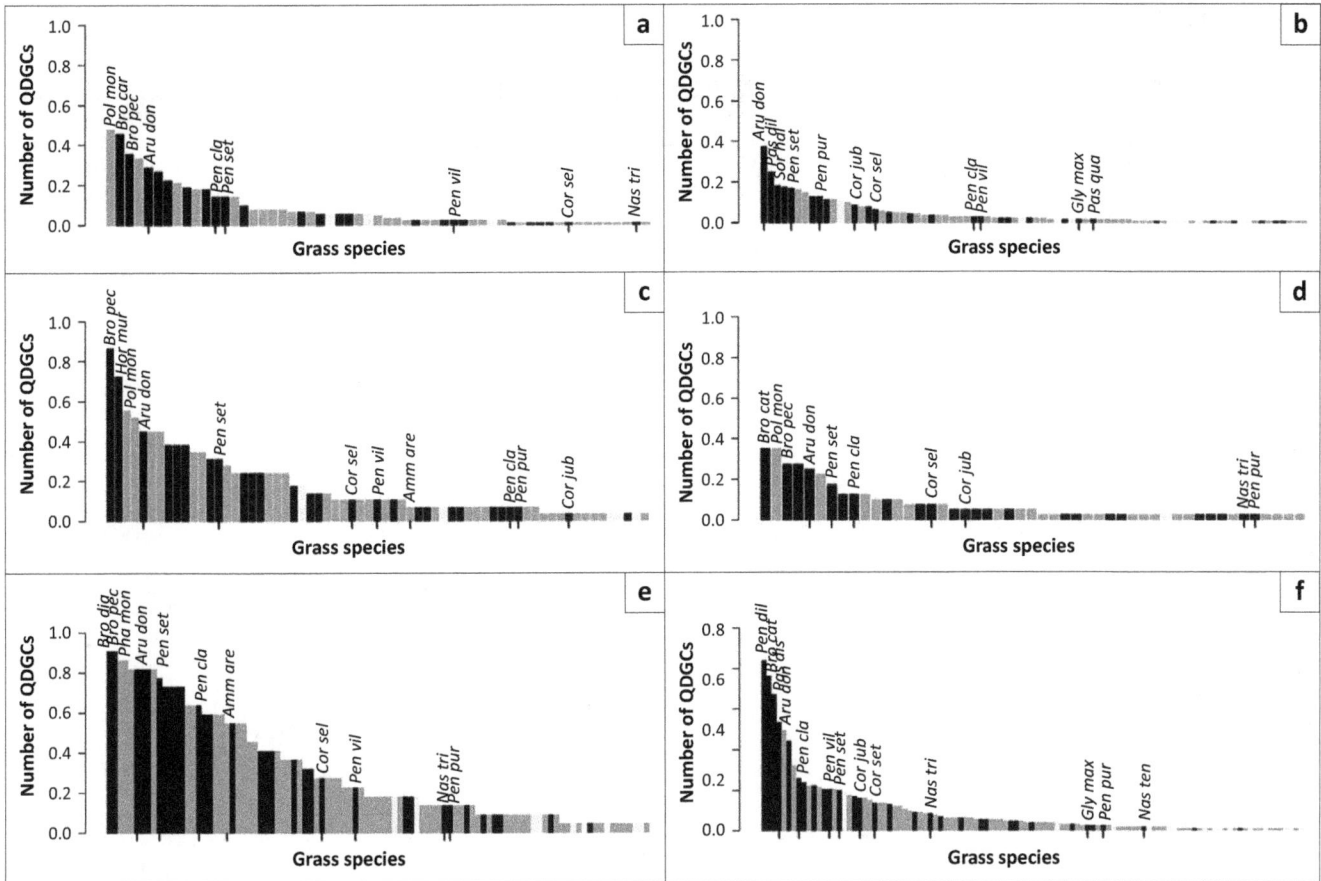

The status of species on the introduction-naturalisation-invasion continuum is indicated by bar colour (light grey = introduced, dark grey = naturalised and black = invasive). Species that currently have legal requirements for their management are named and indicated by a rug mark below the relevant bar (see Appendix 5 for further details). Also named are the three species with the greatest proportion of quarter-degree-grid-cells occupied in each biome. The Grassland and Succulent Karoo biomes also have some species that occur across almost all of these biomes, but these were far fewer in number than in the Fynbos Biome. The Fynbos Biome was exceptional in having a much higher number of species that occur across large areas of the biome.

FIGURE 4-A4: Area occupied by alien grass species as a proportion of quarter-degree-grid-cells in each biome (total number of quarter-degree-grid-cells in each biome provided in parentheses after biome name). (a) Nama-Karoo (90); (b) Savanna (142); (c) Succulent Karoo (29); (d) Azonal (40); (e) Fynbos (22); (f) Grassland (131).

Appendix 5

TABLE 1-A5: Management options for the 41 species that are either invasive or have impacts in South Africa.

Species	Physical removal	Fire	Chemical	Integrated	Biocontrol	Number of control measures	References
Ammophila arenaria (L.) Link	✓	✓	✓	✓	-	4	GISD (2016)
Arundo donax L.	✓	-	✓	✓	✓	4	GISD (2016); Guthrie (2007)
Avena barbata Pott ex Link	-	-	-	-	-	0	
Avena fatua L.	✓	✓	✓	✓	-	4	Musil et al. (2005); Todd (2008)
Bambusa balcooa Roxb.	-	-	-	-	-	0	
Bambusa vulgaris Schrad.	-	-	-	-	-	0	
Brachypodium distachyon (L.) P. Beauv.	-	-	-	-	-	0	
Briza maxima L.	-	-	-	-	-	0	
Briza minor L.	-	-	-	-	-	0	
Bromus catharticus Vahl	-	-	-	-	-	0	
Bromus diandrus Roth	✓	✓	✓	✓	-	4	Holmes (2008); Musil et al. (2005)
Bromus madritensis L.	-	-	-	-	-	0	
Bromus pectinatus Thunb.	✓	✓	✓	✓	-	4	Holmes (2008); Musil et al. (2005)
Bromus rigidus Roth	-	-	-	-	-	0	
Bromus rubens L.	-	-	-	-	-	0	
Bromus tectorum L.	-	-	-	-	-	0	
Cortaderia jubata (Lemoine ex Carrière) Stapf	-	-	-	-	-	0	
Cortaderia selloana (Schult. & Schult.f.) Asch. & Graebn.	-	-	-	-	-	0	
Glyceria maxima (Hartm.) Holmb.	-	✓	✓	-	-	2	GISD (2016); Mugwedi (2012)
Hordeum murinum L.	✓	-	-	-	-	1	Musil et al. (2005)
Lagurus ovatus L.	-	-	-	-	-	0	
Lolium multiflorum Lam.	✓	✓	✓	✓	-	4	Holmes (2008)
Lolium perenne L.	-	-	-	-	-	0	
Lolium rigidum Gaudin	-	-	-	-	-	0	
Nassella tenuissima (Trin.) Barkworth	-	-	-	-	-	0	Viljoen (1987)
Nassella trichotoma (Nees) Hack. & Arechav.	✓	✓	✓	✓	✓	4	Viljoen (1987)
Paspalum dilatatum Poir.	-	-	-	-	-	0	
Paspalum distichum L.	-	-	-	-	-	0	
Paspalum quadrifarium Lam.	-	-	-	-	-	0	
Paspalum urvillei Steud.	-	-	-	-	-	0	
Paspalum vaginatum Sw.	-	-	-	-	-	0	
Pennisetum clandestinum Hochst. ex Chiov.	✓	-	✓	✓	✓	4	GISD (2016)
Pennisetum purpureum Schumach.	-	-	-	-	-	0	
Pennisetum setaceum (Forssk.) Chiov.	✓	-	-	-	-	1†	Rahlao et al. (2014)
Pennisetum villosum Fresen.	-	-	-	-	-	0	
Phalaris aquatica L.	-	-	-	-	-	0	
Poa annua L.	✓	-	✓	-	-	2	GISD (2016)
Poa pratensis L.	-	-	-	-	-	0	
Sorghum halepense (L.) Pers.	-	-	-	-	-	0	
Stipa capensis Thunb.	-	-	-	-	-	0	
Vulpia myuros (L.) C.C.Gmel.	-	-	-	-	-	0	
Total number of species against which each control measure has been used	**11**	**6**	**11**	**8**	**3**		

†, Suggested control is to reduce seed production and establishment of *P. setaceum* through revegetation with native species.

All references as in Appendix 1.

Appendix 6

TABLE 1-A6: Grasses listed under the NEM:BA Alien and Invasive Species (A&IS) 2016 Regulations (excluding Marion and Prince Edward islands).

Name in NEM:BA A&IS 2016 regulations	The Plant List accepted name	NEM:BA A&IS status
Ammophila arenaria (L.) Link	Ammophila arenaria (L.) Link	3
Arundo donax L.	Arundo donax L.	1b
Cortaderia jubata (Lemoine ex Carriere) Stapf	Cortaderia jubata (Lemoine ex Carrière) Stapf	1b
Cortaderia selloana (Schult.) Asch. & Graebn.	Cortaderia selloana (Schult. & Schult.f.) Asch. & Graebn.	1b. Sterile cultivars or hybrids are not listed.
Glyceria maxima (Hartm.) Holmb.	Glyceria maxima (Hartm.) Holmb.	1b in protected areas and wetlands.
Nassella tenuissima (Trin.) Barkworth	Nassella tenuissima (Trin.) Barkworth	1b
Nassella trichotoma (Nees) Hack. ex Arechav.	Nassella trichotoma (Nees) Hack. & Arechav.	1b
Paspalum quadrifarium Lam.	Paspalum quadrifarium Lam.	1a
Pennisetum clandestinum Hochst. ex Chiov.	Pennisetum clandestinum Hochst. ex Chiov.	1b in Protected Areas and wetlands in which it does not already occur.
Pennisetum purpureum Schumach.	Pennisetum purpureum Schumach.	2
Pennisetum setaceum (Forssk.) Chiov.	Pennisetum setaceum (Forssk.) Chiov.	1b. Sterile cultivars or hybrids are not listed.
Pennisetum villosum R.Br. ex Fresen.	Pennisetum villosum Fresen.	1b
Sasa ramosa (Makino) Makino & Shibata	Sasa ramosa (Makino) Makino & Shibata	3
Spartina alterniflora Loisel.	Spartina alterniflora Loisel.	1a
Achnatherum brachychaetum (Godr.) Barkworth	Stipa brachychaeta Godr.	Prohibited.
Achnatherum caudatum (Trin.) S.W.L.Jacobs & J.Everett	Stipa caudata Trin.	Prohibited.
Aegilops cylindrica Host	Aegilops cylindrica Host	Prohibited.
Aegilops geniculata Roth	Aegilops geniculata Roth	Prohibited.
Aegilops species	Aegilops species	Prohibited.
Aegilops triuncialis L.	Aegilops triuncialis L.	Prohibited.
Andropogon bicornis L.	Andropogon bicornis L.	Prohibited.
Andropogon virginicus L.	Andropogon virginicus L.	Prohibited.
Arundinaria Michx. species	Arundinaria Michx.; Pleioblastus Nakai; Sasa Makino & Shibata; Pseudosasa Makino ex Nakai	Prohibited.
Cenchrus echinatus L.	Cenchrus echinatus L.	Prohibited.
Cenchrus longispinus (Hack.) Fernald	Cenchrus longispinus (Hack.) Fernald	Prohibited.
Chrysopogon aciculatus (Retz.) Trin.	Chrysopogon aciculatus (Retz.) Trin.	Prohibited.
Cortaderia richardii (Endl.) Zotov	Cortaderia richardii (Endl.) Zotov	Prohibited.
Cymbopogon refractus (R.Br.) A.Camus	Cymbopogon refractus (R.Br.) A.Camus	Prohibited.
Hymenachne amplexicaulis (Rudge) Nees	Hymenachne amplexicaulis (Rudge) Nees	Prohibited.
Imperata brasiliensis Trin.	Imperata brasiliensis Trin.	Prohibited.
Imperata brevifolia Vasey	Imperata brevifolia Vasey	Prohibited.
Ischaemum rugosum Salisb.	Ischaemum rugosum Salisb.	Prohibited.
Miscanthus floridulus (Labill.) Warb. ex K.Schum. & Lauterb.	Miscanthus floridulus (Labill.) Warb. ex K.Schum. & Lauterb.	Prohibited.
Muhlenbergia schreberi J.F.Gmel.	Muhlenbergia schreberi J.F.Gmel.	Prohibited.
Nassella charruana (Arechay.) Barkworth	Nassella charruana (Arechay.) Barkworth	Prohibited.
Nassella hyalina (Nees) Barkworth	Nassella hyalina (Nees) Barkworth	Prohibited.
Nassella leucotricha (Trin. & Rupr.) R.W.Pohl	Nassella leucotricha (Trin. & Rupr.) R.W.Pohl	Prohibited.
Neyraudia reynaudiana (Kunth) Keng ex Hitchc.	Neyraudia reynaudiana (Kunth) Keng ex Hitchc.	Prohibited.
Oryza rufipogon Griff	Oryza rufipogon Griff.	Prohibited.
Panicum antidotale Retz.	Panicum antidotale Retz.	Prohibited.
Pennisetum alopecuroides (L.) Spreng.	Pennisetum alopecuroides (L.) Spreng.	Prohibited.
Pennisetum pedicellatum Trin.	Pennisetum pedicellatum Trin.	Prohibited.
Pennisetum polystachion (L.) Schult.	Pennisetum polystachion (L.) Schult.	Prohibited.
Saccharum spontaneum L.	Saccharum spontaneum L.	Prohibited.
Setaria faberi R.A.W.Herrm.	Setaria faberi R.A.W.Herrm.	Prohibited.
Setaria palmifolia (J.Konig) Stapf	Setaria palmifolia (J.Koenig) Stapf	Prohibited.
Sorghum hybrid 'Silk'	Sorghum hybrid 'Silk'	Prohibited.
Sorghum X almum Parodi	Sorghum x almum Parodi	Prohibited.
Sporobolus indicus (L.) R.Br. var. major (Buse) Baaijens	Sporobolus fertilis (Steud.) Clayton	Prohibited.
Themeda quadrivalvis (L.) Kuntze	Themeda quadrivalvis (L.) Kuntze	Prohibited.
Themeda villosa (Pair.) A.Camus	Themeda villosa (Pair.) A.Camus	Prohibited.
Zizania latifolia (Griseb.) Turcz. ex Stapf	Zizania latifolia (Griseb.) Turcz. ex Stapf	Prohibited.

Provided are the species names, the type of management suggested (physical removal, use of fire, chemical control, integrated control – a combination of other management measures or biological control), the total number of control measures suggested and the references for these.

Appendix 7

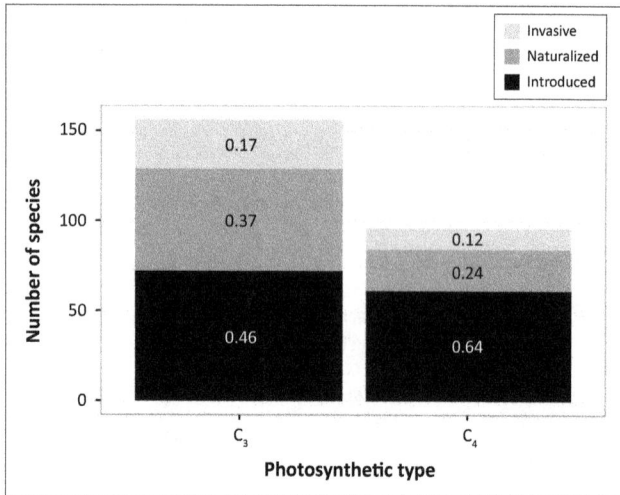

Within each photosynthetic pathway, the numbers of species of each status across the introduction-naturalisation-invasion continuum are shown (as well as the proportion, indicated by the numbers in each bar).

FIGURE 1-A7: Numbers of alien grass species using the C_3 or C_4 photosynthetic pathway.

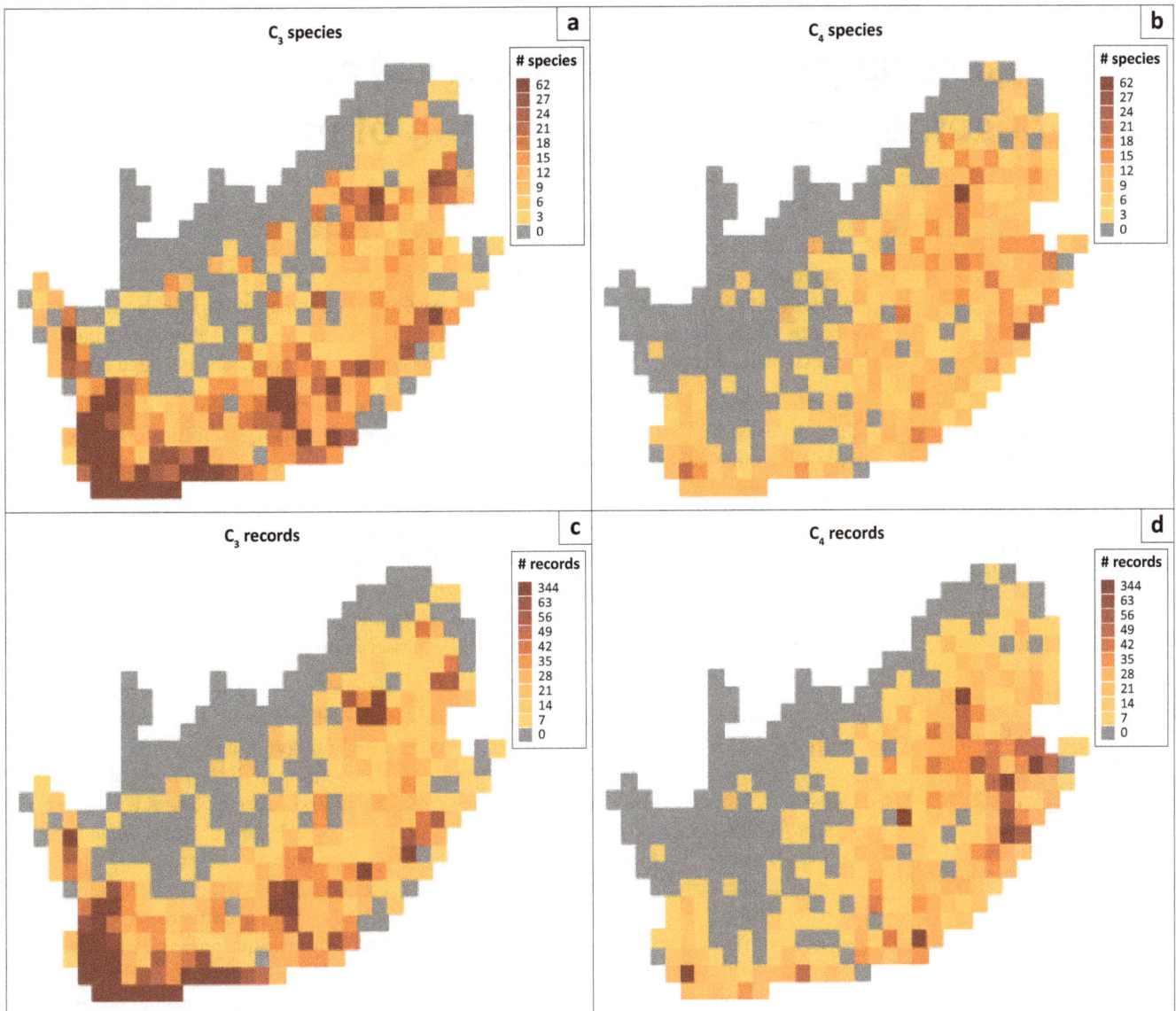

The geographical distribution of (a, b) species richness or (c, d) numbers of records of alien grasses using the (a, c) C_3 photosynthetic pathway or (b, d) the C_4 photosynthetic pathway. C_3 species richness and numbers of records are higher in the south-west of the country; C_4 species richness and numbers of records are higher in the eastern part of the country.

FIGURE 2-A7: Geographical distribution of grass species.

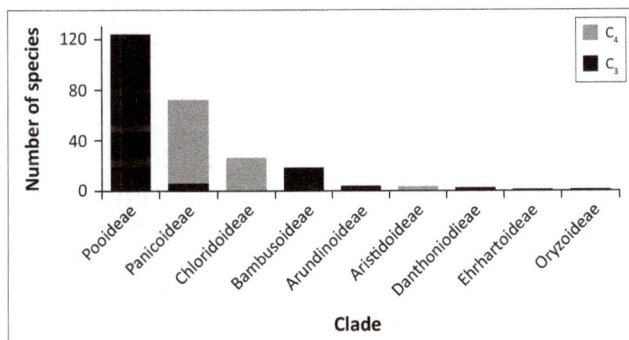

This figure clearly shows the strong phylogenetic signal of the type of photosynthetic pathway (C_3 or C_4). Most alien grasses in South Africa belong to the clade Pooideae, which only has species using the C_3 photosynthetic pathway and contributes the largest number of C_3 species to South Africa. The next most speciose clade of alien grasses in South Africa is Panicoideae, which mostly has species using the C_4 photosynthetic pathway, with only six species using the C_3 photosynthetic pathway in this clade. The clades Chloridoideae and Aristidoideae are completely C_4 dominated, and the remaining clades are completely C_3 dominated.

FIGURE 3-A7: Phylogenetic signals of the type of photosynthetic pathways.

12

Assessing the status of biological control as a management tool for suppression of invasive alien plants in South Africa

Authors:
Costas Zachariades[1,2] ®
Iain D. Paterson[3]
Lorraine W. Strathie[1]
Martin P. Hill[3] ®
Brian W. van Wilgen[4] ®

Affiliations:
[1]Plant Protection Research Institute, Agricultural Research Council, South Africa

[2]School of Life Sciences, University of KwaZulu-Natal, South Africa

[3]Department of Zoology and Entomology, Rhodes University, South Africa

[4]Centre for Invasion Biology, Department of Botany and Zoology, Stellenbosch University, South Africa

Corresponding author:
Costas Zachariades,
zachariadesc@arc.agric.za

Background: Biological control of invasive alien plants (IAPs) using introduced natural enemies contributes significantly to sustained, cost-effective management of natural resources in South Africa. The status of, and prospects for, biological control is therefore integral to National Status Reports (NSRs) on Biological Invasions, the first of which is due in 2017.

Objectives: Our aim was to evaluate the status of, and prospects for, biological control of IAPs in South Africa. We discuss expansion of biological control and suggest indicators to be used in the upcoming NSR to assess sufficient growth.

Method: We used published literature, unpublished work and personal communication to assess the status of biological control of IAPs. We propose indicators based on the targets for biological control that were proposed in the 2014 'National Strategy for dealing with biological invasions in South Africa'. To prioritise targets for future efforts, we used published lists of damaging IAPs and assessed the prospects for their biological control. Recommendations for using biological control as a management tool were made after discussion among the authors and with colleagues.

Results: Significant control of several Cactaceae, Australian *Acacia* species and floating aquatic plants, and many other IAPs has been achieved in South Africa since 1913. Recently, biological control has benefited from improved international collaboration, a streamlined application process for the release of new biological control agents (resulting in the approval of 19 agents against 13 IAP species since 2013), and increased funding and capacity. There is still a need to improve implementation and to better integrate biological control with other control methods. In order to maximise benefits from biological control, increased investment is required, particularly in implementation and post-release evaluation, and in targeting new IAPs. Proposed targets for growth between 2017 and 2020 include an increase in financial investment in research by 29%, implementation by 28% and mass-rearing by 68%. Research capacity should increase by 29%, implementation capacity by 63% and mass-rearing capacity by 61%. New research projects should be initiated on 12 new IAP targets, while post-release monitoring efforts should be expanded to another 31 IAPs.

Conclusion: Biological control of IAPs has contributed substantially to their management in South Africa, and continues to do so. Further investment in targeted aspects of IAP biological control will increase this contribution.

Introduction

Biological control of invasive alien plant (IAP) species is the use of introduced, highly selective natural enemies (usually herbivorous arthropods or pathogens) to control plants. It has been used in 130 countries as a valuable tool for the control of IAP species, with a total of over 550 biological control agents having been released (Winston et al. 2014). The benefits of biological control to natural ecosystems are significant (Van Driesch et al. 2010), with some specific examples of threatened indigenous species being protected by the action of biological control agents (Barton et al. 2007; Meyer, Fourdrigniez & Taputuarai 2011). Detailed analyses of programmes on biological control of IAPs have also clearly indicated that the risks of non-target effects from biological control agents are minimal (Fowler, Syrett & Hill 2000; Funasaki et al. 1988; Moran & Hoffmann 2015; Paynter et al. 2004; Pemberton 2000; Suckling & Sforza 2014). Less than 1% of all the agents released have a negative impact on non-target plant populations, and those that do could have been predicted to do so, and would not be released today (Suckling & Sforza 2014).

Biological control of IAPs is therefore an environment-friendly and safe control method that can result in effective and permanent control of IAPs, leading to the alleviation of their economic and ecological impacts.

In 2013, South Africa marked its 100-year anniversary in biological control of IAPs (Moran, Hoffmann & Zimmermann 2013). During the first few decades (1913 to the 1930s), invasive Cactaceae were the targets of biological control introductions (Klein 2011). These were all 'transfer projects', using insects that had already been successfully introduced as biological control agents in other countries. In the 1950s and 1960s, transfer projects were initiated on *Lantana camara* L. (Verbenaceae) and *Hypericum perforatum* L. (Clusiaceae), but the 1960s also saw the first novel South African projects, on *Hakea* species (Proteaceae) and *Leptospermum laevigatum* (Gaertn.) (Myrtaceae), which had predominantly invaded natural rather than agricultural ecosystems. During the 1970s and 1980s, more groups of IAPs, in particular floating aquatic plants and Australian *Acacia* species (Mimosaceae), became the targets of biological control efforts (Klein 2011). This was facilitated by the formation of a dynamic group of young researchers by Dr David Annecke (Plant Protection Research Institute, then under the national Department of Agriculture) in the early 1970s. Challenges during these decades included a paucity of funding and political restrictions on collecting candidate agents in many of the regions of origin. These challenges have since been resolved. The advent of democracy in South Africa in 1994 led to normalised international relations, and the inclusion of biological control in the funding model of the Working for Water programme (WfW) [now one of the Natural Resource Management Programmes (NRMP) within the national Department of Environmental Affairs (DEA)] in 1997 has led to substantial increases in the funding available for biological control of IAPs over the past 20 years. To date, about 93 species of insects, mites and plant pathogens have been established on 59 IAP species in the country (41 as primary targets and 18 as alternative hosts). An additional 25 species of plants have been worked on, or are currently being investigated, but do not have biological control agents established on them (Klein 2011, updated 2016: http://www.arc.agric.za/arc-ppri/Documents/Target weed species in South Africa.pdf). For several decades, South Africa has been recognised as one of the top five countries globally with regard to research on the biological control of IAPs (Cock et al. 2010; Moran & Hoffmann 2015).

Activities on biological control of IAPs in South Africa have been well documented through the publication of three sets of reviews of the local programmes (Hoffmann 1991; Olckers & Hill 1999; Moran, Hoffmann & Hill 2011). These reviews have contributed towards the success of IAP biological control in South Africa as they have encouraged the biological control community to reflect on successes and challenges, and provided an account of progress for local and international benefit (Moran et al. 2013).

For almost two decades prior to 2013, one of the main constraints to effective IAP biological control was an inefficient system by which agents were cleared for release through the relevant governmental departments. However, a far more effective system has been implemented (see section 'A streamlined biological control agent release process') and approval for a total of 19 biological control agents for 13 IAP species (or species complexes) has been granted since 2013. Of these, two agents were pathogens and the remainder were insect agents. Most have already been released, and in some cases establishment has been confirmed (Table 1).

In this article, we review the status of IAP biological control in South Africa and the contribution of biological control towards the management of the most problematic and damaging IAPs. We also discuss successes and constraints, and prospects both for the control of the most damaging of South Africa's IAPs and for the pro-active management of IAPs that are a threat but are not yet damaging to natural resources. Indicators to assess the effectiveness of a substantial investment in biological control, with targets for future national status reports, are proposed.

Measures of success of IAP biological control in South Africa

Control levels

In South Africa, biological control contributes significantly to the control of 34 of the 59 IAP species on which biological control agents are established. Fourteen of these target species are considered to be under complete control, with no need for any other control intervention (Klein 2011, updated 2016; see Box 1). Many of the most obvious successes have been against Australian *Acacia* species, cacti and floating aquatics (Klein 2011), although successes have certainly not been limited to these groups (e.g. Gordon & Kluge 1991; Hoffmann & Moran 1991).

The first-ever release of a biological control agent in South Africa was the cochineal insect *Dactylopius ceylonicus* (Green) (Hemiptera: Dactylopiidae) for the control of drooping prickly pear, *Opuntia monacantha* Haw. (Cactaceae), in 1913 (Moran et al. 2013). This agent was highly effective, resulting in a permanent reduction of the once extensive infestations of this plant to the point that it is no longer considered problematic (Moran & Zimmermann 1991; Moran et al. 2013). Since this early success, biological control agents have been released on another twenty cactus species, eight of which are considered to be under substantial control and seven under complete control (Klein 2011, updated 2016; Table 2). These include major IAP species, such as the sweet prickly pear, *Opuntia ficus-indica* (L.) Miller, which was reduced from an area of about 900 000 ha to less than 100 000 ha, and the Australian pest pear, *Opuntia stricta* (Haw.) Haw., which has declined in biomass by 90% in the Kruger National Park as a result of the impact of biological control (Paterson et al. 2011).

Other notable successes involved the use of seed attacking agents for the control of ten species of Australian *Acacia* using five seed-feeding weevils and two flower-galling

TABLE 1: Weed biological control agents for which permission to release in South Africa has been granted since the 2011 reviews were published.

Target weed	Natural enemy	Feeding guild	Agent status
ASTERACEAE			
Campuloclinium macrocephalum	*Liothrips tractabilis* (Thysanoptera: Phlaeothripidae)	Stem- and leaf-deformer	Released 2013, established
	Cochylis campuloclinium (Lepidoptera: Tortricidae)	Flower feeder	Release permit issued
Chromolaena odorata	*Dichrorampha odorata* (Lepidoptera: Tortricidae)	Stem-tip galler	Released 2013, establishment unconfirmed
	Recchia parvula (Coleoptera: Cerambycidae)	Stem- and root crown-borer	Released 2016, establishment unconfirmed
Parthenium hysterophorus	*Listronotus setosipennis* (Coleoptera: Curculionidae)	Stem borer	Released 2013, established
	Smicronyx lutulentus (Curculionidae)	Seed feeder	Released 2015, established
	Zygogramma bicolorata (Coleoptera: Chrysomelidae)	Leaf feeder	Released 2013, established
Tithonia rotundifolia	*Zygogramma piceicollis* (Coleoptera: Chrysomelidae)	Leaf feeder	Released 2015, establishment unconfirmed
	Zygogramma signatipennis (Coleoptera: Chrysomelidae)	Leaf feeder	Released 2015, establishment unconfirmed
BASELLACEAE			
Anredera cordifolia	*Plectonycha correntina* (Coleoptera: Chrysomelidae)	Leaf feeder	Release permit issued
BIGNONIACEAE			
Tecoma stans (L.) var. *stans*	*Mada polluta* (Coleoptera: Coccinellidae)	Leaf feeder	Released 2013, established
	Pseudonapomyza sp. (Diptera: Agromyzidae)	Leaf miner	Released 2014, establishment unconfirmed
CACTACEAE			
Pereskia aculeata	*Catorhintha schaffneri* (Hemiptera: Coreidae)	Stem wilter	Released 2014, established
HYDROCHARITACEAE			
Hydrilla verticillata	*Hydrellia purcelli* (Diptera: Ephydridae)	Leaf miner	Release permit issued
MIMOSACEAE			
Acacia baileyana, A. decurrens [*A. dealbata, A. podalyriifolia*]	*Dasineura pilifera* (Diptera: Cecidomyiidae)	Flower galler	Release permit issued
Paraserianthes lophantha	*Uromycladium* sp. (Pucciniales: Pileolariaceae)	Gall former	Release permit issued
PONTEDERIACEAE			
Eichhornia crassipes	*Megamelus scutellaris* (Hemiptera: Delphacidae)	Leaf sucker	Released 2013, established
SAPINDACEAE			
Cardiospermum grandiflorum	*Cissoanthonomus tuberculipennis* (Coleoptera: Curculionidae)	Seed feeder	Released 2013, established
VERBENACEAE			
Lantana camara	*Puccinia lantanae* (Pucciniales: Pucciniaceae)	Leaf- and stem-rust pathogen	Release permit issued

Source: Modified from Klein 2011

BOX 1: Evaluating success in biological control of IAPs in South Africa.

> Hoffmann (1995) first proposed the current system for evaluating success of biological control programmes. This system is now widely accepted and has been adapted and improved (Klein 2011). For each biological control programme, the level of success is estimated based on the reduction of alternative control methods owing to the impacts of biological control (Klein 2011). Each programme is assigned to one of the following categories:
>
> *Complete*: No other control measures are needed to reduce the IAP to acceptable levels, at least in areas where the agents are established.
>
> *Substantial*: Other methods are needed to reduce the IAP to acceptable levels, but less effort is required (e.g. less frequent herbicide applications or less herbicide needed per unit area).
>
> *Negligible*: In spite of damage inflicted by the agents, control of the IAP remains entirely reliant on the implementation of other control measures.
>
> *Not determined*: Either the release of the agents has been too recent for meaningful evaluation or the programme has not been evaluated.
>
> These categories are useful for estimating success across a wide variety of biological control programmes, many of which do not have quantitative post-release evaluation data available. To quantify success more accurately, long-term studies which measure the changes in IAP density and the associated reduction in the negative impact of the IAP should be conducted (e.g. changes to native biodiversity, water resources, fire regimes or agricultural productivity). It is essential that measurable pre-defined goals are set in order to evaluate success because complete eradication is not an appropriate goal for biological control (Morin et al. 2009; Paterson et al. 2011). There are very few studies where a reduction in the negative impacts of the IAP is quantified after control (e.g. Barton et al. 2007), but these studies will hopefully become more common if post-release evaluations are given a higher priority in biological control programmes.

midges (Impson et al. 2011). These *Acacia* species are considered among the most problematic and damaging invasive alien species in South Africa (Henderson 2001, 2007; van Wilgen et al. 2008), and biological control has resulted in extensive damage to six of the species, with considerable or moderate damage to a further three species (Impson et al. 2011). As a result, six Australian *Acacia* species are under substantial control (Klein 2011, updated 2016; Table 2). Although the impact of seed-attacking agents is less obvious and visible than agents that damage the structural tissue or leaves, the reduction in reproductive output caused by these agents is significant, even for IAPs with long-lived seedbanks, and has made control using manual and chemical methods economically viable (van Wilgen & Wannenburgh 2016). The successful long-term impact of seed-attacking agents on

populations of some woody invasive species has resulted in a call to stop manual removal of species that have effective agents in order to focus efforts on those without biological control (van Wilgen et al. 2016). The use of seed-attacking agents has led to a reduction in the reproductive output of a number of woody tree species while still allowing the species to be exploited commercially, thus avoiding conflicts of interest (Impson et al. 2011).

Five floating aquatic species have had agents released against them in South Africa. Four of these [*Pistia stratiotes* L. (Araceae) (water lettuce), *Salvinia molesta* D.S. Mitch. (Salviniaceae) (salvinia), *Myriophyllum aquaticum* (Vell. Conc.) Verd. (Halogaraceae) (parrot's feather) and *Azolla filiculoides* Lam. (Azollaceae) (red water fern)] have each had a single

TABLE 2: Invasive alien plants in South Africa which are under complete or substantial biological control, indicating whether deliberate, human-mediated biological control actions are required, and whether a change in the legal status of the IAP is desirable.

Species	Continue deliberate biological control actions? †	Type of intervention‡	Suggested change to legislation
Complete control§			
Ageratina riparia	No	Integrated control	-
Azolla filiculoides	No	n/a	Remove from NEM:BA
Cereus jamacaru and *C. hildmannianus*	Yes	Biological control only	-
Cylindropuntia fulgida var. *fulgida*	Yes	Biological control only	-
Cylindropuntia fulgida var. *mamillata*	Yes	Biological control only	-
Cylindropuntia leptocaulis	Yes	Biological control only	-
Harrisia martini	Yes	Biological control only	-
Hypericum perforatum	Yes	Biological control only	-
Myriophyllum aquaticum	Yes	Biological control only	-
Opuntia humifusa	Yes	Biological control only	-
Opuntia monacantha	Yes	Biological control only	-
Pistia stratiotes	Yes	Biological control only	-
Salvinia molesta	Yes	Biological control only	-
Sesbania punicea	Yes	Biological control only	-
Substantial control§			
Acacia cyclops	Yes	Integrated control	-
Acacia longifolia	No	Integrated control	-
Acacia mearnsii	Yes	Integrated control	-
Acacia melanoxylon	No	Integrated control	-
Acacia pycnantha	No	Integrated control	-
Acacia saligna	Yes	Integrated control	-
Cylindropuntia imbricata	Yes	Biological control only	-
Eichhornia crassipes	Yes	Integrated control	-
Hakea sericea	Yes	Integrated control	-
Harrisia balansae	Yes	Integrated control	-
Harrisia pomanensis	Yes	Integrated control	-
Harrisia tortuosa	Yes	Integrated control	-
Lantana camara (some varieties)	Yes	Integrated control	-
Opuntia aurantiaca	Yes	Biological control only	-
Opuntia ficus-indica	No	n/a	Remove from NEM:BA
Opuntia salmiana	Yes	Integrated control	-
Opuntia stricta	Yes	Biological control only	-
Paraserianthes lophantha	No	Integrated control	-
Solanum elaeagnifolium	Yes	Biological control only	-
Solanum sisymbriifolium	Yes	Biological control only	-

†, Where 'yes', efforts may range from minor, for weeds with small, isolated populations, to major, for weeds with more widespread populations.

‡, Integrated control here incorporates biological control in all cases (whether human-mediated or self-sustained), with other control methods (chemical, mechanical or fire). Restoration may also be included.

§, From Klein (2011), updated 2016. See also Henderson (2016 – this volume).

agent released and are considered to be under complete control (Klein 2011), whereby, if the agents are implemented correctly these IAPs no longer pose a threat to aquatic ecosystems (Coetzee et al. 2011; Hill & Coetzee 2017). Indeed, the programme against red water fern using the weevil *Stenopelmus rufinasus* Gyllenhal (Coleoptera: Curculionidae) has been so successful that it has led to widespread extirpation of this plant from most of the infested sites (McConnachie, Hill & Byrne 2004) and it should be removed from the *National Environmental Management: Biodiversity Act* (2004) (NEM:BA) list of invasive species in South Africa to allow utilisation (see section 'Integration of biological control in IAP management programmes'; Table 2). The biological control programme against the fifth species, *Eichhornia crassipes* (Mart.) Solms-Laub. (Pontederiaceae) (water hyacinth), was initiated in 1974 with the release of the first agent, *Neochetina eichhorniae* Warner (Coleoptera: Curculionidae) (Cilliers 1991). Since 1974, a further seven agents have been released against the species, of which six

have been confirmed as established (Coetzee et al. 2011). Biological control contributes substantially to the reduction in the invasiveness of water hyacinth (Klein 2011), but there are some areas of South Africa where additional measures, mostly herbicide application, are required to bring the plant under complete control. These are usually areas where the system is polluted with nitrates and phosphates (Coetzee & Hill 2012) and where winter temperatures result in a number of frost days (Hill & Olckers 2001). In these areas, biological control has to be correctly integrated with other control options (Hill & Coetzee 2008).

Fungal pathogens have played an important role in the success of biological control in South Africa. Fungal pathogens are often highly host specific and under suitable environmental conditions (temperature and humidity) can be extremely damaging to host plants. Introduced pathogens have achieved significant levels of biological control on several invasive plants, for example, the rust fungus

Uromycladium tepperianum (Sacc.) McAlpine (Pucciniales: Pileolariaceae) on *Acacia saligna* (Labill.) H.L. Wendl. (Mimosaceae), which after almost 30 years in the field has reduced densities in monitored stands by between 70% and 98% between fires (Wood & Morris 2007; A.R. Wood [ARC-PPRI], pers. comm., 15 August 2016) (although because of the high soil-stored seed bank, plants currently still regenerate at high densities following fires). Some pathogens require deliberate inoculation into stands of IAP populations but many, once established in a small area, spread rapidly under suitable conditions without, or with limited, further need for human intervention [e.g. *Puccinia eupatorii* Dietel (Pucciniales: Pucciniaceae) on *Campuloclinium macrocephalum* (Less.) DC. (Asteraceae) and *P. xanthii* Schwein. var. *parthenii-hysterophorae* Seier, H.C. Evans and Á. Romero on *Parthenium hysterophorus* L. (Asteraceae)]. Some indigenous fungi have been formulated into mycoherbicides, for example, *Colletotrichum acutatum* J.H. Simmonds (Sordariomycetidae: Glomerellaceae) (trade name Hakatak®) causing gummosis on *Hakea sericea* Schrad. and J.C. Wendl. (Proteaceae), and the saprophytic *Cylindrobasidium laeve* (Pers.) Chamuris (Agaricales: Physalacriaceae) (trade name Stumpout®) as a cut-stump treatment of *A. mearnsii* De Wild. (Fabaceae: Mimosoideae) (Morris et al. 1999) which can be applied in the field by IAP management teams with little biological control expertise.

For many biological control programmes, including those utilising seed-attacking agents, the impacts of the agents on the IAP populations may happen over a number of years or decades and are not always obvious without detailed post-release evaluations. For example, insecticide exclusion experiments have shown that the reproductive output and growth parameters of *L. camara* have been significantly reduced owing to the action of the suite of biological control agents released against it, although control of *L. camara* is not considered a complete success (Urban et al. 2011). Although partial success in biological control may be perceived by some as incomplete or failure to control, it is important to consider to what level an IAP population may have expanded if biological control had not been implemented (van Wilgen et al. 2004). A biological control agent that has not resulted in an obvious, visible decrease in the IAP population may still be making a considerable contribution through reducing the density and rates of increase, growth and spread of the IAP, through reduced photosynthetic ability, altered plant physiology or reduced ability to produce propagules. This highlights the need for detailed, quantitative post-release evaluations, an aspect of biological control that has been somewhat neglected, primarily because of the lengthy and sometimes difficult process. However, the assessment of agent impact and level of control attained is essential so that success can be quantified (Morin et al. 2009), so that supplementary biological or other control measures can be implemented if required, and so that control methods can be integrated. Such evaluations are also necessary to determine, for example, the most susceptible target IAPs and optimal selection of effective agents in order to improve the efficacy, success rate and cost-effectiveness of biological control programmes in future.

Economic costs and benefits of IAP biological control in South Africa

Compared with other options available to control IAPs (i.e. manual or chemical control), biological control can be extremely efficient in economic terms. The development of herbicides requires a great deal of expensive research (Rüegg, Quadranti & Zoschke 2007) and repeated herbicide applications are usually required, frequently in conjunction with manual clearing, to achieve any measurable IAP population reductions, while manual control is labour-intensive and therefore also expensive. Restoration after clearing is also expensive (Holmes & Richardson 1999), especially if it has to be applied over a large area, as is often the case. Biological control, by comparison, is effective because the research produces solutions that are cheap to apply, and that are usually self-sustaining. Biological control can also be integrated with other control methods in order to reduce the costs of control (van Wilgen & Wannenburgh 2016). In some cases, such integration of various control methods is essential for optimal control of the IAP.

The initial stages of IAP biological control programmes are the most costly because they require studies to confirm the origin of the IAP (which may include the need for molecular and taxonomic studies), exploratory surveys for candidate biological control agents in the region of origin and assessment of the suitability of candidate agents in both the region of origin and in quarantine in South Africa. In cases in which the initial phases of the biological control programme have already been conducted in other countries, significant savings in time and money can be made. Mass-rearing and releases of the agents to establish permanent, self-sustaining field populations (essential for the success of such a programme) may also be relatively costly. Several biological control research programmes that face complexities such as unknown origins or hybridisation of target plants, or conflicts of interest owing to the utilisation and benefits of the target plant, have required several decades of research, some having been initiated in the 1960s (Moran et al. 2013). Importantly, however, once biological control agents have established self-sustaining populations in the field, there are few further input costs, while the benefits continue to accrue indefinitely. Using South African case studies (six terrestrial IAP species from several functional groups), van Wilgen et al. (2004) calculated the total historical cost for the research programme to be ZAR41.1 million, expressed in values in 2000 currently, 14 ZAR ~ 1 US$). The total cost of biological control research on four 'functional groups' of terrestrial IAPs in five South African biomes was ZAR102 million, expressed in 2008 values (de Lange & van Wilgen 2010); the cost for individual groups covered an approximate five-fold range, from ZAR10 million for fire-adapted trees to ZAR50 million for subtropical shrubs.

Several assessments have been made of the benefits accruing from IAP biological control in South Africa, both at the time of the study and projected into the future. Versfeld, Le Maitre and Chapman (1998) estimated that existing biological control programmes had already reduced management costs by 20% (ZAR1.38 billion), with the potential to further reduce costs by more than 40% (ZAR2.89 billion). For the six terrestrial IAPs considered by van Wilgen et al. (2004), estimated current benefits of biological control varied from ZAR22 million for *Sesbania punicea* (Cav.) Benth. (Fabaceae) to ZAR6.1 billion for *Opuntia aurantiaca* Lindl. (Cactaceae), and the benefit to cost ratios from 8:1 for *S. punicea* to 709:1 for *O. aurantiaca*. Further, they estimated the future value of biological control for these species by calculating the final realised invasive range, compared with the benefits accrued with biological control. Here the benefits ranged from ZAR152 million for *S. punicea* to ZAR10.3 billion for *A. aurantiaca*. De Lange and van Wilgen (2010) estimated that the annual flow of benefits from the biological control of the four 'functional groups' of IAPs totalled ZAR11.6 billion, but that this varied widely between the groups, with fire-adapted trees (e.g. *Hakea* species) at ZAR67 million and invasive Australian trees at ZAR8.3 billion (subtropical shrubs and invasive succulents were intermediate, at ZAR205 million and ZAR3.0 billion, respectively). They estimated net present values of ecosystem service benefits which were attributable to biological control, for these four groups in the five biomes, to total almost ZAR145 billion, again with wide variation. McConnachie et al. (2003) conducted a cost–benefit analysis on the *A. filiculoides* biological control project in South Africa and showed that in 2000 the benefit to cost ratio was 2.5:1, but that this increased over time to 13:1 by 2005 and 15:1 by 2010 as less input was required, but with benefits still accruing. Economic evaluations therefore clearly indicate that biological control is a high value-for-money investment that should be given priority within control programmes (van Wilgen & de Lange 2011; van Wilgen et al. 2012).

Recent enablers and constraints
International cooperation

In the past, many surveys for biological control agents undertaken in countries of origin of the target plant by South African research organisations were carried out on a multi-species, opportunistic basis with little formal engagement with institutions in those countries. This was the result of restrictions on international field work as a result of a lack of cooperative international legislation governing the transfer of biodiversity or management of invasive species, lack of material transfer agreements between countries, political and bureaucratic difficulties and a lack of funding. While numerous successes were achieved in this manner, such a *modus operandus* is no longer tenable.

The international Convention on Biological Diversity (CBD) (https://www.cbd.int/convention/text/default.shtml), to which South Africa is party, was developed to provide a platform for a more structured approach to the transfer of biological diversity, such as potential IAP biological control agents, between countries. It has three primary objectives, namely to promote sustainable development through the conservation of biological diversity, the sustainable use of its components and sharing of the benefits arising from the

utilisation of genetic resources. A supplementary agreement to the CBD came into force in 2014 (https://www.cbd.int/abs/). It was developed to advance the CBD objective on the fair and equitable sharing of benefits arising out of the utilisation of genetic resources, by providing a transparent legal framework. However, in reality, IAP biological control practice, which must be recognised as non-commercial and of benefit to all recipients, has been hindered by the implementation of additional complex regulations and bureaucratic procedures for collection and exportation of potential biocontrol agents. These processes have made the practice of biological control more difficult, time-consuming and costly. Cock et al. (2010) described how biological control does not fit well into an access and benefit sharing regime, and advocated that national regulations should build on the existing multi-lateral practice of free exchange of biocontrol agents, with fair and equitable sharing of the benefits globally. Access and benefit sharing based on non-financial benefit sharing, such as cooperative research programmes, capacity building and technology transfer, were recommended (Cock et al. 2010).

International cooperation is key to the success of every IAP biological control programme, as many countries require any foreign parties wishing to access their biodiversity to operate through a local institution. Consequently, South African institutions undertaking research on IAP biological control have developed formal links with similar institutions in, *inter alia*, Argentina, Australia, Brazil, Jamaica, Mexico, New Zealand, the USA and Venezuela. These links have taken the form of Memoranda of Agreement or purpose-specific contracted research or tasks. Institutions in these countries undertake contracted preliminary research on prospective agents, and facilitate the acquisition of appropriate collection and export permits, as well as assisting with logistics and the collection and export of candidate biological control agents. Exploratory surveys and export of collected organisms would not be possible or advisable nowadays without such formal agreements. The funding of studies by postgraduate students at these institutions, and joint publication of findings, is encouraged, and it constitutes an example of 'south–south' collaboration. Comprehensive, focused exploratory surveys on several IAP species in countries of origin have yielded an array of prospective agents, allowing for prioritisation of those with the greatest potential (e.g. Paterson et al. 2014). Intensive surveys and prioritisation processes should be further developed in future, for additional species of IAP.

Transfer projects continue to be important to the biological control of IAPs in South Africa, and also require cooperation with other countries. Recent examples include those on *P. hysterophorus*, where agents were imported from their introduced range in Australia, *Tradescantia fluminensis* Vell. (Commelinaceae) (spiderwort) (from New Zealand) and *Arundo donax* L. (Poaceae) (Spanish reed) (from the USA). Selection of the most suitable agents for South African conditions, from those that are already well-established on these IAPs in their primary country of introduction, has resulted in considerable savings. South African research programmes on novel IAP species have also benefited, or could potentially benefit, other countries in which these species are invasive. These include *Cardiospermum grandiflorum* Swartz (Sapindaceae); *Dolichandra unguis-cati* (L.) Lohman (Bignoniaceae); *L. camara*; *Solanum mauritianum* Scop. and *S. elaeagnifolium* (Cav.) (Solanaceae); *Tecoma stans* (L.) Juss ex Kunth (Bignoniaceae); and *Tithonia diversifolia* (Hemsl.) A. Gray (Asteraceae) (see Moran et al. 2011).

South Africa is a leader in IAP biological control on the African continent, and has collaborated with other countries on both aquatic (e.g. *E. crassipes*, *P. stratiotes*) and terrestrial [e.g. *O. ficus-indica*, *Chromolaena odorata* (L.) R.M. King and H. Rob. (Asteraceae), *P. hysterophorus*] IAPs [see Winston et al. (2014) for agents that have been released in other countries via South Africa]. The natural spread of some IAP biological control agents, for example, the leaf-mining fly *Ophiomyia camarae* Spencer (Diptera: Agromyzidae) on *L. camara* or *Pareuchaetes insulata* (Walker) (Lepidoptera: Erebidae) on *C. odorata*, which have dispersed from South Africa to neighbouring countries, or even further north on the continent and adjacent islands, must have benefited the management of these plant species, although such benefits have rarely been measured. This ability of some biological control agents to disperse over large distances also illustrates that the potential for non-target attack on plant species present in other countries in sub-Saharan Africa must be considered when evaluating the suitability of agents for release (Faulkner et al. 2017). For example, *Guizotia abyssinica* (L.f.) Cass. (Asteraceae), a crop native to Ethiopia, was, under laboratory conditions, susceptible to damage by *Epiblema strenuana* (Walker) (Lepidoptera: Tortricidae), a potential biological control agent for *P. hysterophorus* (McConnachie 2015). This, together with the known strong dispersal ability of the moth and its use of multiple host plant species in the field in Australia, raised concerns over its possible spread to Ethiopia if it were to be released in South Africa. Further research is needed to determine whether this insect poses a risk to *G. abyssinica*, before it can be considered for release.

A streamlined biological control agent release process

Klein et al. (2011) reviewed the legislation associated with importing and releasing IAP biological control agents in South Africa. Until 2010, DEA legislation had proved the main stumbling block to releasing agents for over a decade, with the other agency, the Directorate: Plant Health within the national Department of Agriculture, Forestry and Fisheries (DAFF-DPH), having generally approved release applications timeously, using an evaluation panel. After 2010, when biological control was removed as an activity requiring a Basic Assessment to be conducted under DEA legislation, this obstacle fell away (albeit temporarily). However, as indicated by Klein et al. (2011), by this stage the DAFF process had also stalled, not through the presence of any adverse legislation, but rather through the legislative

confusion created by the *National Environmental Management: Biodiversity Act* (2004) (NEM:BA) (administered by DEA) and its potential overlap with the *Conservation of Agricultural Resources Act* (CARA), administered by DAFF. By late 2012 a substantial backlog (for 11 agents on 8 IAPs) of release applications had developed, dating back to 2009. The negative consequences of delayed processing of applications included (1) the inability to release biological control agents, while IAP populations continued to expand; (2) the utilisation of quarantine resources and capacity for the holding of agents, or the culling of cultures resulting in the need for recollection at a later stage, incurring additional costs, and (3) occupation of the limited quarantine space, preventing the importation of new potential biological control agents for assessment.

In 2012 and early 2013, the IAP biological control community, together with representatives from DEA: NRMP who were funding the research, thus engaged with DAFF-DPH to find a way forward. This resulted in the formation of the National Biological Control Release Application Review Committee (NBCRARC), chaired by the South African National Biodiversity Institute (SANBI), and with representatives from DAFF (both agriculture and forestry), DEA, private consultants (experts in IAP and insect biological control), and the Forestry and Agricultural Biotechnology Institute. Protocols were drawn up, including a standard application format and guidelines (which request the applicant to consider, *inter alia*, the need for biological control for the IAP; the identity, safety and potential efficacy of the candidate agent; a proposed release strategy: possible ecological effects its release would have; and plans for post-release mitigation, if needed) as well as a standard review format and guidelines. A list of potential expert external reviewers, at both national and international levels, was compiled (including IAP and insect biological control experts, entomologists, pathologists, botanists and others). The current process is as follows: applicants submit the application to DAFF-DPH which forwards it to the chair. After consultation with the committee, the chair distributes the application to the agreed reviewers. For each application, the chair solicits voluntary reviews from three experts (usually two national and one international reviewer for each application), with a timeline of four weeks. The reports from these experts are then read and discussed by the committee members, either over e-mail or more commonly at a meeting, and a recommendation is passed from the committee to DAFF-DPH. DAFF-DPH then either issues a release permit or passes on the committee's recommendation for further information, which may require either a desktop study/response, or further host range testing or other work. This process has enabled the release of a number of new agents in the past three years (Table 1).

Environmental legislation relevant to the importation of candidate biological control agents and their release falls under the ambit of NEM:BA. However, the Alien and Invasive Species Regulations for this Act were only published and put into effect in 2014 (Department of Environmental Affairs 2014a, 2014b;

http://www.invasives.org.za/legislation/invasive-species-legislation.html), at which time they supplanted the CARA legislation. Up to the present, this has not affected the process of import permits for biological control agents or candidate agents, which are still issued only by DAFF, or release permits, but this will certainly change in future, when Pest Risk Assessments will be evaluated by DEA. It is hoped, though, that the NBCRARC or a similar single, central channel ('one-stop shop') will continue to function. It is critical that a similar streamlined procedure is adopted and that this process is embedded in legislation soon.

The research versus implementation debate

Until the mid-1990s, South African researchers conducted or oversaw most work on IAP biological control, including the mass-rearing, release and post-release monitoring of agents. This often worked well with a relatively high rate of establishment of agents, but for some agents [e.g. *Pareuchaetes* species on *C. odorata* (Zachariades et al. 2011)] establishment could only be achieved by large scale mass-rearing which was beyond the capacity of research organisations. Furthermore, with an increase in the amount of work being conducted on control of IAPs owing to the launch of WfW in 1995, the demand for agents increased substantially, in order to achieve more widespread and faster biological control. An 'implementation' programme, embedded within WfW, was thus set up in the late 1990s and early 2000s (Gillespie, Klein & Hill 2004), with the aim of mass-rearing, field collection for redistribution, releases and basic monitoring of the establishment and spread of agents. Several mass-rearing centres were set up around the country, the existing insect-rearing facilities at the South African Sugarcane Research Institute (SASRI) were contracted and provincially based implementation officers employed. Interaction between researchers and implementers was encouraged, and facilitated by the annual 'Weed Biocontrol Workshops' that have been held since the 1970s (Wilson et al. 2017). Although this programme has facilitated thousands of releases of biological control agents throughout the country, there have been some limiting factors, such as several mass-rearing centres failing owing to funding issues; a lack of biological control expertise at the mass-rearing centres; implementation officers being co-opted into non-biological control activities; and a lack of structured cooperation and feedback loops between researchers and implementers (e.g. on which agents to mass-rear, numbers to be released or under what circumstances to make use of biological control). Often, inadequate distinction was made between agents which were still at an experimental phase (i.e. their establishment or efficacy was not yet proven) and agents which had already been shown to be effective but needed further redistribution as a management tool (see section 'Integration of biological control in IAP management programmes'). This also raised the question of where the dividing line between research and implementation lay, owing to the grey area along the continuum between these activities. Nevertheless, the implementation programme has substantially increased the number of biological control releases made in the country and the number of plants with active biological control implementation programmes in operation, increased capacity to some degree and has undoubtedly improved the level of control for many IAP species. Recently, quarterly meetings between researchers and implementers, and increased field interactions have closed the perceived gap between research and implementation further. Currently, there is a move back towards mass-rearing, field collection and releases being included within research programmes, but these will still need to be coordinated at a national level and should be structured so as not to distract researchers from their core functions.

Increases in funding and capacity

The growth in IAP biological control in South Africa in recent years can be largely attributed to the significant investment in the development of research and implementation by DEA: NRMP and its predecessor, WfW (which was based in the Department of Water Affairs), since 1997. Following economic analyses and recommendations by van Wilgen and de Lange (2011) and van Wilgen et al. (2012), a three-fold increase for 2014–2017 (3.4 years) from the previous funding contract has enabled the scaling up of research and implementation activities on terrestrial and aquatic IAPs. During 2014–2017, an investment of ZAR116.7 million was allocated to the development of IAP biological control research and implementation on 50 plant species, ZAR23.8 million in mass-rearing of selected biological control agents, and a further ZAR50.8 million in field implementation by DEA (officers whose mandate is the control of invasive alien species, including biological control of IAPs, in each province, with resources for biological control agent releases, redistribution and monitoring), a total of almost ZAR200 million (Table 3).

Capacity within South Africa has been built through various channels, including (1) expanded expertise through appointments at universities and national research organisations, (2) mass-rearing initiatives, (3) bursaries that have attracted increased numbers of postgraduate students, (4) internships, (5) an IAP biological control short course and (6) implementation officers stationed in each province to facilitate field activities. A Capacity Building Programme at the University of KwaZulu-Natal exposes university undergraduate students to experiential training periods of two months per annum at research institutions engaged in research and implementation activities on IAP biological control, stimulating knowledge of, and interest in, the practice (Downs 2010). A number of these students have gone on to work in this or related fields. Currently, IAP biological control in South Africa employs approximately 77 personnel in the field of research, at research organisations and universities, many of whom are also involved in mass-rearing of agents for releases (Table 3). An additional 32 personnel are involved in mass-rearing activities specifically, of whom seven have physical disabilities. Another 16 'biodiversity officers' (whose remit includes the implementation of IAP biological control) and managers, as well as 17 people in two field implementation

TABLE 3: Baseline indicators (between 2014 and 2017) and proposed targets (between 2017 and 2020) for a range of indicators that can be used to assess progress towards national strategic goals, and reported on at three-yearly intervals in the national status of biological invasions report.

Variable	Indicator	Baseline (2014–2017)	Proposed target (2017–2020) †
Funding (ZAR millions over 3 years)	Research‡	116.7	150 (29%)
	Implementation§	50.8	65 (28%)
	Mass-rearing§	23.8	40 (68%)
Human resources	Research capacity‡	39 researchers including consultants; 38 technical support personnel; 55 students (third year to postdoctorate)¶	50 researchers (28%); 50 technical support (32%); 70 students (mostly postgraduate) (27%)
	Implementation capacity§	16 implementation officers¶	26 implementation officers (63%)
	Mass-rearing capacity§	32 mass-rearing technicians¶	52 mass-rearing technicians (61%)
No. weed species targeted for biological control	Pre-release research (major species) ‡,††	34	41 (20%)
	Pre-release research (pro-active species) ‡,††	5	10 (100%)
	Post-release evaluations§	31	62 (100%)
	Mass-rearing§	30	37 (23%)

Monetary values are given as 2016 values.

†, Percentage increase over baseline indicated in brackets.

‡, Targets for 2017–2020 are based on the 100% growth over 10 years (2017–2027) for research recommended by the National Strategy for dealing with biological invasions in South Africa. This assumes that the growth over 10 years will be linear (i.e. 30% growth over 3 years). However, because, for example, infrastructure needs building, this may not be true.

§, Targets for 2017–2020 are based on the 100% growth over 5 years (2017–2022) for implementation recommended by the National Strategy. This assumes that the growth over 5 years will be linear (i.e. 60% growth over 3 years). However, because, for example, infrastructure needs building, this may not be true.

¶, Numbers for 2016.

††, Note that 'pre-release research' is distinct from 'pre-release monitoring' mentioned in the National Strategy. We include the former under research and the latter under implementation.

teams, are employed within the ranks of DEA and SA National Parks.

Increased investment has allowed for the expansion of infrastructure for research, through the development of new or extended quarantine facilities at national research and university institutions. There are currently six IAP biological control quarantine facilities around the country, four of which are run by ARC-PPRI [one in Pretoria, two in Stellenbosch (for insects and pathogens, respectively) and one in Cedara (near Pietermaritzburg)], with the others at Rhodes University in Grahamstown and the University of the Witwatersrand in Johannesburg. Such developments have facilitated research on new IAPs and additional agents on existing projects, and improved successes on existing projects.

Increased investment into mass-rearing of agents that have been approved for release has also been made by DEA, and is currently being undertaken at facilities in the provinces of KwaZulu-Natal (SASRI and ARC-PPRI), the Eastern Cape (Rhodes University, at both its Waainek and Uitenhage facilities), the Western Cape (City of Cape Town and ARC-PPRI), Gauteng (ARC-PPRI) and Limpopo (DEA: NRMP in collaboration with ARC-PPRI). Optimal mass-rearing can often not be conducted solely at research institutions because of a shortage of space, resources, capacity and in some cases, sub-optimal climatic conditions; mass-rearing techniques often differ substantially from methods used to rear small numbers of agents for research. In cases where mass-rearing has been conducted successfully at research institutions, it was because a separate mass-rearing unit was set up. Some of the 'mass-rearing' (a rather loose term) being carried out is currently on a small scale, for experimental releases of agents in order to try to establish them at sites with different climates or habitats, or to determine optimal release techniques, rather than large scale releases for distribution of established agents with known efficacy. Not all agents require intensive mass-rearing efforts but many benefit from such efforts, allowing

more successful establishment and a greater degree of IAP control to be achieved more rapidly. For some situations, repeated augmentative releases are required to achieve ongoing control of an IAP, although this is less so for terrestrial than for aquatic species, for which flooding events and water eutrophication create conditions in which classical biological control is insufficient (Hill & Coetzee 2017). Mass-rearing and field implementation of developed biological control agents has not yet been used to its full potential in South Africa as the optimal model for such endeavours is still evolving.

Currently, pre-release studies on 34 species of IAP are being undertaken, with post-release research on 31 species (Table 3). Of the 140 cases where 111 biological control agents have been released, occur locally or for which release permits have been issued, human-mediated redistribution is desirable in 76 cases on 48 IAP species (Figure 1). Of these 76 cases, mass-rearing alone is needed for 33, field collection alone for 24 and both of these methods are appropriate for 19. Currently, mass-rearing is being undertaken for 47 of the 52 cases for which it is desirable, for 30 of the 33 IAP species. However, for more than half of these, the mass-rearing is on a relatively small scale, experimental basis, and there are 35 cases (22 IAP species) for which an increase in mass-rearing effort is desirable (Figure 1).

Integration of biological control in IAP management programmes

The integration of IAP biological control into a wider control programme is routinely advocated (e.g. Watson & Wymore 1989), but in practice, in South Africa at least, this is often not achieved. In some cases, it is not easy to incorporate biological control, for example, when the aim of a control programme is to dramatically reduce the density of an IAP species in a defined area over a short period of time. For most IAP species, the biological control agent cannot be 'applied' like a

Agents and weeds

121 cases† in which agents are released and established; 5 have release permits but are not yet released; 5 released but establishment unconfirmed; 9 other‡ (total: 140)

of which

on

62 weed species in 18 families (including 21 Cactaceae and 11 Mimosaceae)

of which

Agent redistribution

64 cases† where redistribution is not needed at present [includes: 33 cases of strong dispersers or range fully occupied, 21 considered poor agents (not worth redistributing) and 10 'other']

76 cases† where species need redistribution (includes: 33 for mass-rearing only, 24 for field collection only and 19 for both)

Mass-rearing is desirable for 33 weed species; field collection is desirable for 32; no action required for 14

Mass-rearing

Ease of mass-rearing (52 cases†): 26 easy, 20 moderate and 6 difficult

Current mass-rearing: (47 cases†): 26 on small scale (experimental) and 21 on large scale (implementation)

35 cases where more mass-rearing is desirable or possibly desirable

Mass-rearing is currently being undertaken for agents on 30 of the 33 weed species, at either small scale or large scale

Upscaling of mass-rearing is desirable or possibly desirable for agents on 22 of the 33 weed species; in some cases, this is already planned

Field collection

Optimal season (43 cases†): 14 in spring, 25 in summer, 3 in autumn and 1 in winter

Ease of field collection (43 cases†): 9 easy, 33 moderate, 1 difficult

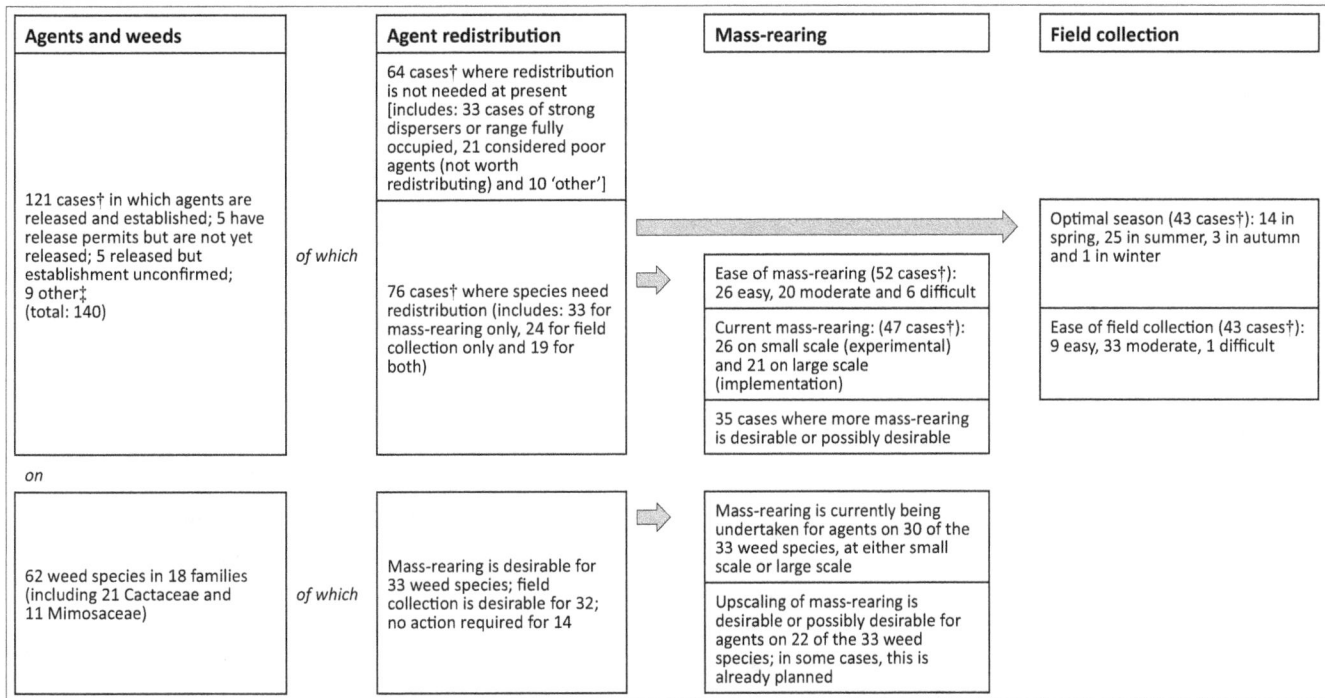

Source: Klein (2001), updated 2016, and the IAP biological control research and implementation community in South Africa

†, The term 'case' indicates a unique combination of a species of biological control agent on a species of weed. In some cases, the same species of agent is used on multiple weed species. The number of agent species established is 93, and total number released and established, with permits to release, or occurring in field naturally, is 111.

‡, Most of these occurred locally, and some were used as mycoherbicides.

FIGURE 1: Current and recommended mass-rearing and field collection of available weed biological control agents in South Africa.

herbicide, and although the agent may already be widely present and a good disperser, it might not reduce IAP density to the desired levels within the desired time frame. The agent may also be difficult to establish on the desired infestation, which could be a function of its biology or the local climatic conditions, or its abundance and performance may not be uniform throughout the invaded range, which is often the case owing to the variety of habitats that an IAP species invades.

Two scenarios are applicable in this context. (1) For a few groups of IAPs, biological control can be used within a designated area as part of a clearing programme, with other areas subject to herbicide or mechanical control measures. For many of the Cactaceae, cochineal (*Dactylopius* species) and the mealybug *Hypogeococcus festerianus* (Lizer y Trelles) (Hemiptera: Pseudococcidae) can be introduced onto individual plants or infestations and severely damage or kill most of the plants in the area. In the case of *H. festerianus*, fruiting is almost completely eliminated. This is a clear case where current control methods (manual and chemical) could in many instances be replaced by the correct use of biological control, thereby reducing costs (Table 2). Chemical and manual control may in fact impact negatively on the overall level of control of these species (Zimmermann 1979), so integration needs to be carefully considered. Some invasive alien aquatic species, such as *S. molesta*, *P. stratiotes* and *M. aquaticum*, can be completely controlled with augmentative releases of biological control agents, but control is disrupted by the use of herbicide applications (Coetzee et al. 2011). (2) There are also IAPs which are under such effective biological control at a regional or national level

(e.g. *A. filiculoides*, *O. ficus-indica*) that no human intervention is required, either with respect to manual or chemical clearing, or the redistribution of biological control agents. Under NEM:BA, it is permissible to use biological control alone to manage a category 1b species, and therefore it would not be necessary to change the categorisation of species described under (1) above (Table 2). However, it should become policy that only biological control is used for managing these species as no other control method is required. On the contrary, we recommend that *A. filiculoides* and *O. ficus-indica* [described under (2) above] be removed from the A&IS regulatory list so that they can be used as crops (both species have beneficial attributes), and so that resources are not wasted on their control (Table 2). Utilisation of *A. filiculoides* and *O. ficus-indica* could result in dispersal to new sites but the agents will disperse, or alternatively could be actively released to prevent the IAPs from becoming problematic. Biological control of *A. filiculoides* and *O. ficus-indica* has reduced the negative impacts of the plants to the point that the benefits of the plant now outweigh the negative impact to the environment and society. Species such as *S. punicea* and *H. perforatum* may require occasional, very limited redistribution of agents, and *Ageratina riparia* (Regel) R.M. King & H. Rob. (Asteraceae) may require manual or chemical control outside of areas in which its biological control agent [*Entyloma ageratinae* R.W. Barreto & H.C. Evans (Entylomatales: Entylomataceae)] provides complete control.

Calls by DEA: NRMP and other agencies for fuller integration of biological control into South African clearing programmes have been made for several years (e.g. A. Khan 2011, presentation to 39th Weed Biocontrol Workshop). DEA: NRMP has invested substantially in biological control of IAPs in South Africa, with the aim not only of developing new agents but also of more effective use of existing ones, in order to improve the control of IAPs and reduce the unnecessary use of resources. Integration of biological control into clearing operations [for appropriate agents, as described in (2) above] has occurred to some extent, but there is considerable room for improvement. Improved research-implementation feedback loops would facilitate management efforts. Management guidelines could be compiled for each IAP species, particularly those under complete or substantial control, with an indication of the areas in which biological control is likely to be effective and the areas in which the IAP should be prioritised for clearing using other methods. For example, *L. camara* is under more effective biological control along the South Coast of KwaZulu-Natal province than in higher-altitude regions and regions with different varieties of the plant. The coastal region should therefore be prioritised for biological control, whereas manual removal and chemical control could be used more intensively in high-altitude areas. Definitive guidelines may however not be practical to compile in all cases because of the multitude of environments and situations which a plant invades, interacting with variable performance of biological control agents under different environmental conditions. Such guidelines would be more easily compiled in later stage programmes once agents have had several years to be distributed and establish widely, and once the agents' field requirements and limitations are understood.

There are also other issues, including the inadvertent destruction of important biological control sites (e.g. release sites and well-established sites from which agents could be redistributed) by DEA: NRMP clearing teams and other agencies. A live, updatable centralised national database containing such data, which is available to all organisations involved in clearing operations and that could be integrated into their GIS-based IAP management planning systems, is recommended. Currently, although some release information is stored within the national DEA: NRMP database, it is not readily available to other organisations or to the general public and is not always used to full effect within DEA: NRMP's own clearing operations. Biological control needs to be better embedded as a management tool within the operational planning of DEA: NRMP and other organisations involved in clearing programmes.

The status of IAP biological control in South Africa, in the context of the National Strategy and National Status Reports

A 'National Strategy for dealing with biological invasions in South Africa' was commissioned by DEA in 2014 (https://sites.google.com/site/wfwplanning/strategy). This document was designed to provide a baseline for invasions in South Africa and to set targets for the future.

The success of the strategy, and progress in combatting IAPs, will be assessed every three years starting in 2017 in the form of National Status Reports on Biological Invasions in South Africa (Wilson et al. 2017).

The National Strategy recommends the following with regard to IAP biological control:

• Double the biological control research capacity in South Africa over the next decade.
• Expand the biological control programme to include species for which eradication from South Africa has been ruled out but for which widespread impacts are predicted but have not yet been demonstrated.
• Expand the capacity of the biological control implementation programme (100% growth over the next 5 years), with explicit targets and ring-fenced funding for pre- and post-release monitoring; mass-rearing; and agent distribution.

From the National Strategy targets, it is clear that significant growth in biological control would be required in both research and implementation for these goals to be met. While research capacity is recommended to increase by 100% in 10 years, implementation is required to increase by the same degree in half of this time. Implicit in the recommendations is that new targets for biological control should be worked on and that these targets should include species that are not yet widespread or damaging (i.e. pro-active targets; see Text Box 2). If the recommendations of the National Strategy are to be achieved, it is essential that (1) the most appropriate IAP species be targeted for biological control and (2) that development of capacity is facilitated. Capacity growth would need to incorporate expanded numbers of personnel at all relevant levels, as well as infrastructure.

In the following sections, we examine whether the species that have been targeted for biological control are aligned with priorities by determining how many of South Africa's worst IAPs are either under biological control or are targets for biological control. We report on prospects for control of recently targeted species and provide recommendations on future programmes, targeting both plants which are currently considered the worst of South Africa's invasive species as well as pro-active targets (see Box 2) which could become problematic in future. Finally, we discuss how growth in biological control capacity could be facilitated and provide specific 'indicators' that can be used to determine whether biological control is growing at a suitable rate to achieve the recommendations in the National Strategy (Table 3).

Prioritisation of targets for biological control

The recently revised Alien and Invasive Species Regulations under NEM:BA list 379 plant species that are considered problematic IAPs or plants that could become problematic in the country in future (Department of Environmental Affairs 2016). This large number of species highlights the

BOX 2: Prioritisation of targets for biological control.

A number of factors should be taken into consideration when selecting new targets for biological control programmes. A combination of these factors must be taken into account and not all factors should be evenly weighted. Although IAPs that are damaging to the natural resources of the country should be selected as targets, the threat of IAPs that are not yet damaging should also be taken into account. IAPs that are targeted using biological control before they have become very widespread and damaging are referred to as pro-active targets.

Factors that suggest that biological control is appropriate and likely to succeed:

1. Transfer programmes, where progress towards the development of a biological control programme of the target IAP made in other countries can reduce the costs of the biological control programme. Cases where biological control of the target IAP elsewhere in the world has already succeeded are best.
2. Successful control of congeners (or close relatives) of the target IAP.
3. Other control methods (e.g. herbicides; manual clearing) are ineffective, making biological control the most feasible option.
4. IAPs are very hard to reach or in inaccessible areas.

Factors that suggest that biological control may be inappropriate and unlikely to succeed, or that are likely to make a programme more difficult or lengthy:

1. Failed attempts at using biological control for the target IAP elsewhere in the world.
2. A high diversity of indigenous and commercially important plant species that are closely related to the target weed are present in the region where biological control is desired. This increases the number of plant species that must be included in host-specificity testing, and thus the costs of the biological control programme, and may prolong the process of finding a suitably specific biological control agent.
3. The origin of the target IAP is unknown or disputed, or the taxonomy of the plant and its close relatives is unresolved. This makes sourcing suitable agents difficult and increases the complexity of host-specificity testing, resulting in the programme becoming more expensive and reducing the chances of rapid success.
4. The target IAP is a beneficial plant that is utilised commercially or non-commercially, causing a conflict of interest (Zengeya et al. 2017). Biocontrol options become more limited and restricted to certain plant parts.

scale of the IAP problem in South Africa and raises the question of how to select and prioritise species for biological control. There are too many IAPs for all of them to be targeted for biological control, some species are better suited for biological control than others, and others cannot be targeted because of conflicts of interest (Box 2); therefore, a prioritisation process is essential. In the past, selection of candidate target species has been done by the biological control community on an *ad hoc* basis in collaboration with funding bodies such as DEA: NRMP, and through discussions and consensus at the annual Weed Biocontrol Workshops. Although selecting targets in this way has been largely successful because of the considerable cumulative expertise of the group, a more formal system that could rank the IAPs listed in the NEM:BA regulations in order of appropriateness as targets for biological control would be beneficial. It would reduce subjectivity and therefore provide stronger evidence to funders for the need to work on the most appropriate targets. Plants that are not appropriate for biological control could then be prioritised for manual removal and/or chemical control.

A number of systems have been developed for the prioritisation of target plants for biological control (Hansen & Bloem 2006; McClay 1989; Paynter et al. 2009; Peschken & McClay 1995; Syrett 2002; van Klinken, Morin & Sheppard 2016), which could be modified to form a system to prioritise species for biological control in South Africa. The most recent prioritisation framework was developed for IAPs impacting the livestock industry in Australia and drew on the expert opinions of biological control researchers during a two-day workshop (van Klinken et al. 2016). A similar method could be adopted by the South African biological control community in order to better structure and justify prioritisation decisions. Although van Klinken et al. (2016) warned against the use of quantitative scoring systems to the exclusion of expert knowledge, they suggest that the more qualitative methods used in their study could be combined with a quantitative scoring system. The development of a prioritisation system for South Africa, which could rank species according to their suitability as candidate targets for biological control based on both quantitative assessments and expert opinion, should be considered.

Current and future targets for biological control of IAPs in South Africa

Van Wilgen et al. (2008, 2012) present the most recent and well supported lists of the most important IAPs in the country, and we used them in the absence of any other objective list, in full recognition that not all of the species should be considered as transformers. The list presented by van Wilgen et al. (2012) was compiled based on previous studies which estimated the area that each species occupied (Kotzé et al. 2010; Le Maitre, Versfeld & Chapman 2000) and the prominence value as defined by Henderson (2007). The list included 18 taxa of the worst IAPs in the country, combining all *Eucalyptus* spp., *Pinus* spp., *Populus* spp., *Prosopis* spp. and, importantly, all Cactaceae, into a single taxon (van Wilgen et al. 2012). Van Wilgen et al. (2008) list the worst IAPs in terms of the negative impact to ecosystem services. These species were selected based on the impact to surface water, grazing potential and biodiversity. Fifty-six taxa were included in that list, all as individual species rather than taxonomically related groups of species.

Here we categorise each of the worst IAPs from the work of van Wilgen et al. (2008) in terms of the contribution that biological control has, or will likely have, towards the management of the species. Each species was allocated to one of the following categories in terms of the current effectiveness of biological control: (1) under complete control, (2) under substantial control, (3) control effectiveness is currently negligible and (4) not determined (see Box 1). These four categories are widely used for evaluating success of biological control since they were first used in the 1999 biological control reviews (Hoffmann 1995; Klein 2011). We include two additional categories: (5) active research programme (for programmes with no agents released but where research to develop new agents is being undertaken) and (6) no research programme (for taxa that have no research programme at present) (Table 4). Information for each biological control programme was sourced from Klein (2011) as well as the biological control community for more recent developments.

For species that are not considered to be under complete control, we evaluated the prospects for improving control by reporting on any new potential agents, recently released agents or other ways that control could be improved, such as better integration with other control methods and increased implementation efforts (Table 4). For those species that have no active biological control research programme in the country, we first report whether the programme would be a novel programme in South Africa or whether it would be a repeat programme from elsewhere in the world. Biological control programmes require a substantial initial investment of funds (although these costs are miniscule compared to either the damage that the IAPs inflict or the cost of alternative control methods); therefore, transfer programmes are significantly less expensive and good value-for-money (Paynter et al. 2009). We then suggest which species are poor, possible, good or excellent candidates for biological control. Factors that were considered in allocating each species to the various categories are also given (Table 4). Poor candidates are those that should not be considered for biological control. Possible candidates include species that could be appropriate for biological control, but there is little evidence to support whether the programme would be successful or not. Good candidates are those with some

evidence to support that biological control is possible and may be successful, and excellent candidates are those where potential agents are readily available and the chances of success are high (Table 4).

Thirteen (72%) of the 18 worst taxa from the work of van Wilgen et al. (2012) have biological control agents released on them already, and seven (39%) of these are under either complete or substantial biological control. When considering the more extensive list of species presented in van Wilgen et al. (2008), a list of 56 most damaging species to ecosystem functioning, 19 (34%) have active biological control and 13% are considered to be under substantial control. The higher percentage of taxa with biological control in van Wilgen et al. (2012) is probably because of the fact that the worst invaders have been targeted for biological control and only the worst 18 species were included in van Wilgen et al. (2012) while many other less damaging species were included in van Wilgen et al. (2008). Furthermore, of the 60 species in South Africa on which biological control is established, 21 are Cactaceae, which appear in van Wilgen et al. (2012) but not in van Wilgen et al. (2008). In our opinion, of the 37 species

TABLE 4: The status of and prospects for biological control of the 56 most damaging invasive alien plant species in South Africa. The list of target weed species is from van Wilgen et al. (2008).

Species	Degree of control achieved by biological control	Biological control has reached full potential	Prospects for biological control					Comments/reference
			New agent	Excellent candidate	Good candidate	Possible candidate	Poor candidate	
Acacia baileyana	Negligible	-	X	X	-	-	-	New agent released but too soon to evaluate impact; [+] successful biological control of other Australian *Acacia* species
Acacia cyclops	Substantial	X	-	-	-	-	-	Impson et al. (2011)
Acacia dealbata	Negligible	-	X	X	-	-	-	New agent released but too soon to evaluate impact; [+] successful biological control of other Australian *Acacia* species
Acacia decurrens	Negligible	-	X	X	-	-	-	New agent released but too soon to evaluate impact; [+] successful biological control of other Australian *Acacia* species
Acacia longifolia	Substantial	X	-	-	-	-	-	Impson et al. (2011)
Acacia mearnsii	Substantial	-	X	X	-	-	-	New agent spreading and damaging (J.H. Hoffmann [UCT] pers. comm., 2 September 2015); [+] successful biological control of other Australian *Acacia* species
Acacia melanoxylon	Substantial	X	-	-	-	-	-	Redistribution could improve control further (Impson et al. 2011)
Acacia saligna	Substantial	X	-	-	-	-	-	Impson et al. (2011)
Achyranthes aspera	No research programme	-	-	-	-	X	-	Novel programme; [-] uncertainties regarding native distribution of the target weed reduce the potential for biological control
Agave americana	No research programme	-	-	-	-	X	-	Novel programme; [-] possible conflict of interest as close relatives are grown commercially
Arundo donax	Active research programme, no agents released yet	-	-	X	-	-	-	Transfer programme; [+] agents have been released in USA
Atriplex lindleyi	No research programme	-	-	-	-	X	-	Novel programme; [-] plant has native congeners which would limit the potential for finding a host specific agent
Atriplex nummularia	No research programme	-	-	-	-	X	-	Novel programme; [-] plant has native congeners which would limit the potential for finding a host specific agent
Cortaderia selloana	No research programme	-	-	-	-	X	-	[+] Preliminary investigations into natural enemies have been conducted; [-] manual and herbicidal control is effective on a congener (DiTomaso, Drewitz & Kyser 2008)
Caesalpinia decapetala	Negligible	-	-	-	X	-	-	[+] Natural enemies known (Byrne, Witkowski & Kalibbala 2011)
Cestrum laevigatum	Research programme shelved	-	-	-	X	-	-	[+] Natural enemies known (D.O. Simelane [ARC-PPRI] pers. comm., 26 April 2016)

Table 4 continues on the next page →

TABLE 4 (continues...): The status of and prospects for biological control of the 56 most damaging invasive alien plant species in South Africa. The list of target weed species is from van Wilgen et al. (2008).

Species	Degree of control achieved by biological control	Biological control has reached full potential	Prospects for biological control					Comments/reference
			New agent	Excellent candidate	Good candidate	Possible candidate	Poor candidate	
Chromolaena odorata	Not determined	-	X	-	X	-	-	Two new agents released and more under consideration; [+] biological control has been successful elsewhere
Cuscuta campestris	No research programme	-	-	-	-	X	-	[+] Natural enemies known; [-] failed attempts at biological control elsewhere; [-] plant has native congeners which would limit the potential for finding a host specific agent
Datura stramonium	No research programme	-	-	-	X	-	-	[+] Fungal biological control agents known
Dolichandra (Macfadyena) unguis-cati	Negligible	-	-	-	X	-	-	Agents have been released too recently to determine impact; [+] other control methods are ineffective
Echinopsis spachiana	No research programme	-	-	-	X	-	-	[+] Success on other Cactaceae
Eucalyptus camaldulensis	No research programme	-	-	-	-	-	X	Novel programme; [-] conflict of interest with bee-keeping industry; [-] biological control investigated and no suitable candidates are available (all possible agents stop flowering and therefore conflict with bee-keeping industry). Seed feeder already present (Klein et al. 2015)
Eucalyptus grandis	No research programme	-	-	-	-	-	X	Novel programme; [-] conflict of interest with commercial forestry;
Eucalyptus conferruminata	No research programme	-	-	-	-	-	X	Novel programme; [-] conflict of interest with bee-keeping industry; [-] biological control investigated and no suitable candidates are available (all possible agents stop flowering and therefore conflict with bee-keeping industry)
Hakea drupacea	Research programme shelved	-	-	-	X	-	-	[+] Natural enemies known; [+] successful biological control of congeners
Hakea gibbosa	Negligible	X But see comments	-	-	-	-	-	Better integration with other control methods could improve success, agents have been released too recently to determine impact (Gordon & Fourie 2011)
Hakea sericea	Substantial	X But see comments	-	-	-	-	-	Better integration with other control methods could improve success, agents have been released too recently to determine impact (Gordon & Fourie 2011)
Ipomoea indica	No research programme	-	-	-	-	-	X	Novel programme; [-] uncertainties regarding native distribution of the target weed reduce the potential for biological control, indigenous and commercially valuable relatives (including congeners) which would limit the potential for finding a host specific agent; [+] other control methods ineffective
Jacaranda mimosifolia	No research programme	-	-	-	-	X	-	Novel programme; potential conflict of interest but seeds could be targeted; [-] seed attacking surveys were conducted and no suitable agents were discovered
Lantana camara	Negligible - substantial	-	X	-	-	X	-	One new agent and others have also been released too recently to determine impact (Urban et al. 2011)
Leptospermum laevigatum	Negligible	-	-	-	-	-	X	[-] Known natural enemies are considered unpromising (Gordon 2011)
Melia azedarach	Research programme shelved	-	-	-	-	X	-	Novel programme; [-] uncertainties regarding native distribution; [-] preliminary surveys failed to find promising candidates
Nicotiana glauca	No research programme	-	-	-	X	-	-	Transfer programme; [+] biological control successful elsewhere; [-] potential conflict with tobacco growers
Paraserianthes lophantha	Substantial	-	X	-	X	-	-	New agent released but too soon to evaluate impact; [+] successful biological control with close relatives (Australian *Acacia*)
Pennisetum clandestinum	No research programme	-	-	-	-	X	-	Novel programme; [-] little past success with grasses; [-] conflict of interest as plant is grown commercially
Pinus elliottii	No research programme	-	-	-	-	-	X	Novel programme; [-] biological control of pines investigated and no suitably specific candidates were available; [-] conflict of interest
Pinus halepensis	No research programme	-	-	-	-	-	X	Novel programme; [-] biological control of pines investigated and no suitably specific candidates were available; [-] conflict of interest

Table 4 continues on the next page →

TABLE 4 (continues...): The status of and prospects for biological control of the 56 most damaging invasive alien plant species in South Africa. The list of target weed species is from van Wilgen et al. (2008).

Species	Degree of control achieved by biological control	Biological control has reached full potential	Prospects for biological control					Comments/reference
			New agent	Excellent candidate	Good candidate	Possible candidate	Poor candidate	
Pinus patula	No research programme	-	-	-	-	-	X	Novel programme; [-] biological control of pines investigated and no suitably specific candidates were available; [-] conflict of interest
Pinus pinaster	Research programme shelved	-	-	-	-	-	X	Novel programme; [-] biological control of pines investigated and no suitably specific candidates were available; [-] conflict of interest
Pinus radiata	No research programme	-	-	-	-	-	X	Novel programme; [-] biological control of pines investigated and no suitably specific candidates were available; [-] conflict of interest
Populus alba	No research programme	-	-	-	-	X	-	Novel programme
Populus canescens	No research programme	-	-	-	-	X	-	Novel programme
Prosopis glandulosa	Negligible	-	X	-	-	X	-	Conditional release permit for new agent granted
Prunus persica	No research programme	-	-	-	-	-	X	Novel programme; [-] conflict of interest with commercial growers
Psidium guajava	No research programme	-	-	-	-	-	X	Novel programme; [-] conflict of interest with commercial growers
Pyracantha angustifolia	No research programme	-	-	-	X	-	-	Novel programme; [+] no close relatives in South Africa; [+] known native distribution
Robinia pseudoacacia	No research programme	-	-	X	-	-	-	Novel programme; [+] extensive knowledge of effective agents
Rubus cuneifolius	Research programme shelved	-	-	-	-	-	X	Novel programme; [-] closely related plants are valuable; [-] taxonomic uncertainties
Rubus fruticosus	Research programme shelved	-	-	-	-	-	X	Novel programme; [-] closely related plants are valuable; [-] taxonomic uncertainties; [-] conflict of interest
Salix babylonica	No research programme	-	-	-	-	X	-	Novel programme
Senna didymobotrya	No research programme	-	-	-	-	X	-	Novel programme; [-] manual and herbicidal control effective
Senna occidentalis	No research programme	-	-	-	-	X	-	Novel programme; [-] manual and herbicidal control effective
Solanum mauritianum	Negligible	-	-	-	X	-	-	Impact of recently released agents not determined; [+] success with two congeners (Hoffmann, Moran & Impson 1998; Klein 2011); [-] congeners of economic importance
Solanum seaforthianum	No research programme	-	-	-	X	-	-	Novel programme; [+] success with two congeners (Hoffmann et al. 1998; Klein 2011); [-] congeners of economic importance
Solanum sisymbriifolium	Substantial	X	-	-	-	-	-	
Xanthium strumarium	No research programme	-	-	-	X	-	-	Repeat programme; [+] successful agent available
TOTAL		7	8	6	13	16	14	-

The degree of control for each species was taken from Klein (2011) updated in 2016 (http://www.arc.agric.za/arc-ppri/Documents/Target weed species in South Africa.pdf). Species where biological control has not reached full potential were classified as either poor, possible, good or excellent candidates for biological control (see section 'Current and future targets for biological control of IAPs in South Africa' for definitions). The factors that were considered in allocating each species to one of these categories are included under comments. In the comments column, a minus sign indicates that the factor suggests that biological control is inappropriate or unlikely to succeed. A plus sign indicates that the factor suggests that biological control is appropriate and likely to succeed. A novel programme refers to one that has not been developed in any other country before while an active research programme is one that is active in South Africa specifically.

in the list of worst invaders from van Wilgen et al. (2008) that do not have biological control agents released, there are 2 excellent candidates [*A. donax* and *Robinia pseudoacacia* L. (Fabaceae)] for biological control, 8 good candidates, 14 possible candidates and 13 poor candidates (Table 4). If the species that have new agents or biological control programmes that have not yet reached full potential are included, then there are 6 excellent candidates for biological control, 13 good candidates, 16 possible candidates and 14 poor candidates (Table 4). Of the species that are considered under negligible control, there is one good candidate for future biological control research, one poor candidate and six species with recently released agents (Table 4). We recommend that biological control should be considered for all but the poor candidates. Research into biological control of excellent and good candidates should be supported and possible candidates should be considered depending on the availability of capacity and funds.

While biological control has contributed significantly towards the control of several major IAPs in South Africa, there are many more species that have also been targeted for biological control in the past. Excluding the species with active biological control programmes that coincide with van Wilgen et al. (2008) leaves 37 species with biological control agents established in South Africa. Some of these (such as *O. ficus-indica* and *E. crassipes*) should certainly be considered major IAPs, but many are currently rather minor. This does not necessarily indicate that these target species for biological control were poorly selected because pro-actively targeting

IAPs for biological control before they become widespread invaders can be even more beneficial than targeting well-established species. The contribution of biological control to pro-active management of IAPs in South Africa should not be overlooked (Olckers 2004) but is more difficult to quantify than impacts to abundant species, as it is difficult to predict how much of a problem they would have become if biological control had been implemented against them only at a later date. It is probable that some species (e.g. *H. perforatum* and *A. riparia*) would be far more problematic today if they had not been the targets of earlier biological control programmes. Transfer projects are ideal for IAP species that have not yet become problematic in the country but are likely to cause problems in future, because the costs of these projects are dramatically reduced compared to novel targets, and the project can be implemented rapidly before the IAP increases extensively in abundance and distribution. Biological control should however only be considered an option once it is clear that eradication of the IAP is no longer possible, as eradication is often more cost-effective (see Wilson et al. 2013 for a review of eradication attempts in South Africa to date).

The initiation of biological control programmes on new target species should be carefully considered beforehand, as each constitutes a potentially major commitment of research time and funding. Most IAP biological control programmes require a minimum intensive commitment of about 10 years, with an average of approximately 4 years to develop each agent (Moran, Hoffmann & Zimmermann 2005). Some transfer projects may have shorter timeframes [an average of 2.8 years per agent, as opposed to novel projects which have an average of 5.0 years (Moran et al. 2005)]. Several of the older, successful programmes have taken decades to realise their full potential, generally owing to varying complexities particular to each target species, as well as owing to the inherent slow process of biological control and natural systems. In some of the previous contracts with DEA: NRMP (e.g. the early to mid-2000s), the biological control community may have been overambitious in their delivery targets. A better approach for certain IAPs may be to obtain funding for feasibility studies first, and thorough molecular studies in some cases [e.g. *Melia azedarach* L. (Meliaceae)] to determine origin or genetic diversity of populations of the target species, as well as thorough systematic surveys for potential agents, with preliminary screening of host range. It should be emphasised to the funder from the outset that by funding a project (particularly on a novel IAP species), they should consider that a long-term funding commitment may be needed to deliver good results (see Scott, Yeoh & Michael 2016). Furthermore, as many habitats are susceptible to secondary invasion following the biological control of the primary invader (e.g. cacti in arid areas, invasive aquatics and disturbed habitats), in some instances habitats (i.e. groups of IAPs) rather than individual IAPs should be prioritised. It should be understood that the initial development phase of every IAP biological control programme offers low returns on value for money, but that benefits increase incrementally and indefinitely once the initial stages of the implementation phase are passed.

It should also be understood that not all programmes result in the release of biological control agents, nor are agent releases guaranteed to succeed.

Finally, we have recommended target numbers of IAPs to work on for the second status report due in 2020 (Table 3). The average increase for pre-release studies across both major and pro-active targets is 28%, which is in line with the level of growth suggested in the National Strategy. A higher percentage of growth to pro-active targets is indicated because many of these are small studies and because few pro-active species have been targeted in the past. Similarly, the suggested overall increase in the number of species for post-release evaluation and mass-rearing is 64% (Table 3). Sixty-two IAP species were recommended for post-release evaluation because this is the total number with biological control agents released on them.

Building IAP biological control capacity and creating job opportunities

Three main aspects need to be addressed for the expansion of capacity. Firstly, long-term training through expanded growth at universities, in both undergraduate and particularly postgraduate programmes, will be required to produce numerous qualified candidates who could be employed as researchers and skilled technicians in IAP biological control research programmes at universities and research organisations. Qualified expertise will also be required for mass-rearing facilities and implementation. As many opportunities for employment will exist in such structures, capacity growth aligns with the job creation priorities of national government and the Expanded Public Works Programme. People with disabilities can also be trained and employed within the field of biological control.

Secondly, short-term training in aspects of IAP biological control is required. At present, several such initiatives are in operation, which include an annual five-day short course, a short course on Cactaceae, eight-week undergraduate experiential training from one university, 12-months in-service training of final year University of Technology Diploma students at a research institution, public exhibitions and Farmers' Day events. Considerably greater scope is possible. New short courses at several levels are required to train research and technical support personnel, managers, implementers (mass-rearing, releases and field teams) and unskilled workers. A national IAP biological control internship programme after graduation from universities, within research and mass-rearing institutions, would be beneficial. National and international exchange programmes as well as exposure to top international researchers would broaden expertise. The retention of skilled expertise is essential to maintain consistency and development of the practice because of the long-term nature of IAP biological control.

Thirdly, expansion of infrastructure is required to build IAP biological control capacity. While this is a costly exercise, if the desired levels of IAP biological control are to be achieved, additional facilities will be required. Existing facilities are constrained in terms of space or operational mandates. Expansion of existing infrastructures, as well as erection of new, appropriate facilities (quarantine facilities, greenhouses, laboratories, offices, etc.), would be required. To achieve a 100% increase in implementation activities over the 5-year period indicated by the National Strategy, a significant investment (several million ZAR) would be required for infrastructure expansion within a short time frame. Involvement of organs of state, private companies and nongovernmental organisations (NGOs) operating in the environmental and agricultural sectors may be beneficial in mass-rearing operations. However, qualified, skilled and experienced entomologists and pathologists are critical to achieve the desired levels of success as well as to foster a strong link with research programmes. Regardless of the parties involved, a long-term funding commitment through multiple 3-year funding cycles will be necessary.

Our recommendations for targets in increasing capacity by 2020, in line with the National Strategy, are provided in Table 3. While the recommended increase in funding for research is in line with the National Strategy, that for implementation funding is lower (40% instead of 60% overall, for implementation and mass-rearing combined). In our opinion, a 60% increase over 3 years is too great, given that considerable additional facilities for mass-rearing will have to be developed. The higher relative budget for mass-rearing is designed with the development of infrastructure in mind. With regard to human resources, our recommended increases both in research and implementation capacity (Table 3) are consistent with the National Strategy. Most of the 10 additional implementation officers would be employed under existing research institutes. Most of the 20 additional mass-rearing technicians envisaged would only be employed once infrastructure is in place.

Currently, most provinces have one or two 'biodiversity officers' employed by DEA for field implementation of IAP biological control. However, additional duties within their portfolio include managing aquatic clearing teams and dealing with invasive fauna. During peak season in particular, there is a need for greater capacity on IAP biological control field implementation, including site selection, releases and monitoring of release sites. Furthermore, it is important to break down provincialism in the context of the implementation of IAP biological control, as virtually all IAPs with biological control agents released invade more than one province; therefore a national, coordinated outlook is required (e.g. through a national mass-rearing strategy which assigns each biological control agent to one or a few mass-rearing centres, for national or regional distribution).

The nature of funding (Expanded Public Works Programme) prescribes that the maximum number of people be employed in manual and chemical clearing operations, a principle that is often thought to be contradictory to the aims of biological control. However, despite the large investment in manual and chemical clearing programmes since 1995, the scale of plant invasions in South Africa remains so great that large scale employment, using these clearing methods, continues to be guaranteed for the foreseeable future. Furthermore, mass-rearing and redistribution of biological control agents have the potential to employ a substantial number of people, sometimes in higher-skilled, higher-paid positions than manual and chemical IAP clearing operations. The new NEMBA legislation requires the development of national management programmes for every IAP species that is categorised as 1b, which may also assist in promoting biological control to an appropriate status within clearing programmes.

A potential pitfall of the current *status quo* of IAP biological control in South Africa is the largely single funding source, which relies on the current priorities of government. Diversification of funding sources would protect against future radical priority shifts to retain capacity and infrastructure investments in IAP biological control, but may not be feasible. Another issue is the short-term allocation of funding because of national objectives for 3-year funding cycles. Longer-term allocation of funds would be preferable, particularly in cases where infrastructure and capacity expertise for IAP biological control research and implementation activities need to be expanded.

Conclusions

South Africa has a long and successful history of IAP biological control and has made considerable progress in cost-effectively protecting the country's natural resources from IAPs. We are currently in a very favourable position, both in terms of funding and policy. South Africa has already increased capacity for IAP biological control and can do so to a greater extent, in line with the National Strategy recommendations. A relatively large proportion of this growth should be invested in implementation and post-release evaluations so that existing biological control agents are utilised to their full potential and the contribution of biological control can be better quantified. Research on selected new pro-active targets should also be undertaken to address the large number of NEM:BA-listed species. This will also provide a substantial number of higher-skilled, higher-paid employment opportunities in line with government policies. South Africa has already benefited substantially from previous investments into biological control of IAPs, but the full potential has most certainly not been reached. Continued and increased investment into biological control will result in greater protection of the country's natural resources and benefit the people who depend on them.

Acknowledgements

We appreciate inputs by the biological control community towards the mass-rearing and field collection data as well as prioritisation of target IAP species. Helmuth Zimmermann is thanked for comments on Table 2, while Hildegard Klein,

Candice-Lee Lyons and Lesley Henderson (ARC-PPRI) are thanked for comments on Table 4. Alana den Breeyen and Alan Wood provided valuable comments on pathogens. Philip Ivey (SANBI) is thanked for inputs on the release approval process.

The considerable and sustained financial support for most IAP biological control programmes since 1997 to date by the WfW programme (now one of the NRMP within the Department of Environmental Affairs) is gratefully acknowledged as crucial to the successes achieved. Part of the funding for work on this paper was provided by the South African Research Chairs Initiative of the Department of Science and Technology and the National Research Foundation (NRF) of South Africa. Any opinion, finding, conclusion or recommendation expressed in this material is that of the authors and the NRF does not accept any liability in this regard.

Competing interests

The authors declare that they have no financial or personal relationship(s) that may have inappropriately influenced them in writing this article.

Authors' contributions

C.Z. initiated the review paper and solicited co-authors, as well as wrote and edited sections of the paper, and obtained data on mass-rearing and field-distribution. I.D.P. suggested the structure of the paper, analysed data on prospects for new biological control programmes, wrote sections of the paper and contributed to the entire paper. L.W.S. obtained data on personnel numbers, wrote sections of the paper and contributed to the entire paper. M.P.H. wrote the sections on aquatic plants and commented on the paper as a whole. B.W.V.W. provided valuable insight into the National Strategy and status report, and commented on the paper as a whole.

References

Barton, J., Fowler, S.V., Gianotti, A.F., Winks, C.J., de Beurs, M., Arnold, G.C. et al., 2007, 'Successful biological control of mist flower (*Ageratina riparia*) in New Zealand: Agent establishment, impact and benefits to the native flora', *Biological Control* 40, 370–385. https://doi.org/10.1016/j.biocontrol.2006.09.010

Byrne, M.J., Witkowski, E.T.F. & Kalibbala, F.N., 2011, 'A review of recent efforts at biological control of *Caesalpinia decapetala* (Roth) Alston (Fabaceae) in South Africa', *African Entomology* 19, 247–257. https://doi.org/10.4001/003.019.0201

Cilliers, C.J., 1991, 'Biological control of water hyacinth, *Eichhornia crassipes* (Pontederiaceae), in South Africa', *Agriculture, Ecosystems and Environment* 37, 207–217. https://doi.org/10.1016/0167-8809(91)90149-R

Cock, M.J.W., van Lenteren, J.C., Brodeur, J., Barratt, B.I.P., Bigler, F., Bolckmans, K. et al., 2010, 'Do new access and benefit sharing procedures under the convention on biological diversity threaten the future of biological control?', *BioControl* 55, 199–218. https://doi.org/10.1007/s10526-009-9234-9

Coetzee, J.A. & Hill, M.P., 2012, 'The role of eutrophication in the biological control of water hyacinth, *Eichhornia crassipes*, in South Africa', *BioControl* 57, 247–261. https://doi.org/10.1007/s10526-011-9426-y

Coetzee, J.A., Hill, M.P., Byrne, M.J. & Bownes, A., 2011, 'A review of the biological control programmes on *Eichhornia crassipes* (C. Mart.) Solms (Pontederiaceae), *Salvinia molesta* D.S. Mitch. (Salviniaceae), *Pistia stratiotes* L. (Araceae), *Myriophyllum aquaticum* (Vell.) Verdc. (Haloragaceae) and *Azolla filiculoides* Lam. (Azollaceae) in South Africa', *African Entomology* 19, 451–468. https://doi.org/10.4001/003.019.0202

De Lange, W.J. & van Wilgen, B.W., 2010, 'An economic assessment of the contribution of biological control to the management of invasive alien plants and to the protection of ecosystem services in South Africa', *Biological Invasions* 12, 4113–4124. https://doi.org/10.1007/s10530-010-9811-y

Department of Environmental Affairs, 2014a, *Notice No. R. 598. National Environmental Management: Biodiversity Act (10/2004): Alien and Invasive Species Regulations, 2014*, Government Gazette 37885, Pretoria, 1 August 2014. (Regulation Gazette No. 10244).

Department of Environmental Affairs, 2014b, *Notice No. 599. National Environmental Management: Biodiversity Act (10/2004): Alien and Invasive Species List, 2014*, Government Gazette 37886, Pretoria, 1 August 2014.

Department of Environmental Affairs, 2016, *Notice No. 864. National Environmental Management: Biodiversity Act, 2004 (Act No. 10 Of 2004). Alien and invasive species lists, 2016*, Government Gazette 40166, pp. 31–104.

DiTomaso, J.M., Drewitz, J.J. & Kyser, G.B., 2008, 'Jubata grass (*Cordateria jubata*) control using chemical and mechanical methods', *Invasive Plant Science and Management* 1, 82–90. https://doi.org/10.1614/IPSM-07-028

Downs, C.T., 2010, 'Is vacation apprenticeship of undergraduate life science students a model for human capacity development in the life sciences?', *International Journal of Science Education* 32, 687–704. https://doi.org/10.1080/0950069 0903075004

Faulkner, K.T., Hurley, B.P., Robertson, M.P., Rouget, M. & Wilson, J.R.U., 2017, 'The balance of trade in alien species between South Africa and the rest of Africa', *Bothalia* 47(2), a2157. https://doi.org/10.4102/abc.v47i2.2157

Fowler, S.V., Syrett, P. & Hill, R.L., 2000, 'Success and safety in the biological control of environmental weeds in New Zealand', *Austral Ecology* 25, 553–562. https://doi.org/10.1046/j.1442-9993.2000.01075.x

Funasaki, G.Y., Lai, P.Y., Nakahara, L.M., Beardsley, J.W. & Ota, A.K., 1988, 'A review of biological control introductions in Hawaii: 1980 to 1985', *Proceedings of the Hawaii Entomological Society* 28, 105–160.

Gillespie, P., Klein, H. & Hill, M.P., 2004, 'Establishment of a weed biocontrol implementation programme in South Africa', in J.M. Cullen, D.T. Briese, D.J. Kriticos, W.M. Lonsdale, L. Morin & J.K. Scott (eds.), *Proceedings of the XIth International Symposium on Biological control of Weeds*, CSIRO Entomology, Canberra, Australia, pp. 400–406.

Gordon, A.J., 2011, 'Biological control endeavours against Australian myrtle, *Leptospermum laevigatum* (Gaertn.) F. Muell. (Myrtaceae), in South Africa', *African Entomology* 19, 349–355. https://doi.org/10.4001/003.019.0206

Gordon, A.J. & Fourie, A., 2011, 'Biological control of *Hakea sericea* Schrad. & J.C. Wendl. and *Hakea gibbosa* (Sm.) Cav. (Proteaceae) in South Africa', *African Entomology* 19, 303–314. https://doi.org/10.4001/003.019.0205

Gordon, A.J. & Kluge, R.L., 1991, 'Biological control of St. John's Wort, *Hypericum perforatum* (Clusiaceae), in South Africa', *Agriculture, Ecosystems and Environment* 37, 77–90. https://doi.org/10.1016/0167-8809(91)90140-S

Hansen, R. & Bloem, K., 2006, *USDA-APHIS_PPQ Biological targets for pest canvassing and evaluation 2005–2006, Final Report*, viewed 6 May 2016, from www.aphis.usda.gov/plant_health/plant_pest_info/biocontrol/download/pest-canvassing.pdf

Henderson, L., 2001, *Alien weeds and invasive plants*. Plant Protection Research Institute Handbook No. 12, Agricultural Research Council, Pretoria, South Africa.

Henderson, L., 2007, 'Invasive, naturalised and casual alien plants in southern Africa: A summary based on the Southern African Plant Invaders Atlas (SAPIA)', *Bothalia* 37, 215–248. https://doi.org/10.4102/abc.v37i2.322

Hill, M.P. & Coetzee, J.A., 2008, 'Integrated control of water hyacinth (*Eichhornia crassipes*) in Africa', *EPPO Bulletin/Bulletin OEPP* 38, 452–457. https://doi.org/10.1111/j.1365-2338.2008.01263.x

Hill, M.P. & Olckers, T., 2001, 'Biological control initiatives against water hyacinth in South Africa: Constraining factors, success and new courses of action', in M.H. Julien, M.P. Hill, T.D. Center & D. Jianqing (eds.), *Biological and integrated control of water hyacinth, Eichhornia crassipes. Proceedings of the Second Global Working Group Meeting for the Biological and Integrated Control of Water Hyacinth*, ACIAR Proceedings No. 102, Beijing, China, October 9–12, 2000, pp. 33–38.

Hill, M.P. & Coetzee, J.A., 2017, 'The biological control of aquatic weeds in South Africa: Current status and future challenges', *Bothalia* 47(2), a2152. https://doi.org/10.4102/abc.v47i2.2152

Hoffmann, J.H. (ed.), 1991, 'Biological control of weeds in South Africa', *Agriculture, Ecosystems and Environment* 37, 1–255.

Hoffmann, J.H., 1995, 'Biological control of weeds: The way forward, a South African perspective', *BCPC Symposium Proceedings* 64, 77–89.

Hoffmann, J.H. & Moran, V.C., 1991, 'Biological control of *Sesbania punicea* (Fabaceae) in South Africa', *Agriculture, Ecosystems and Environment* 37, 157–173. https://doi.org/10.1016/0167-8809(91)90144-M

Hoffmann, J.H., Moran, V.C. & Impson, F.A.C., 1998, 'Promising results from the first biological control programme against a solanaceous weed (*Solanum elaeagnifolium*)', *Agriculture, Ecosystems and Environment* 70, 145–150. https://doi.org/10.1016/S0167-8809(98)00120-0

Holmes, P.M. & Richardson, D.M., 1999, 'Protocols for restoration based on recruitment dynamics, community structure, and ecosystem function: Perspectives from South African fynbos', *Restoration Ecology* 7, 215-230. https://doi.org/10.1046/j.1526-100X.1999.72015.x

Impson, F.A.C., Kleinjan, C.A., Hoffmann, J.H., Post, J.A. & Wood, A.R., 2011, 'Biological control of Australian *Acacia* species and *Paraserianthes lophantha* (Willd.) Nielsen (Mimosaceae) in South Africa', *African Entomology* 19, 186–207. https://doi.org/10.4001/003.019.0210

Klein, H., 2011, 'A catalogue of the insects, mites and pathogens that have been used or rejected, or are under consideration, for the biological control of invasive alien plants in South Africa', *African Entomology* 19, 515–549. https://doi.org/10.4001/003.019.0214

Klein, H., Hill, M.P., Zachariades, C. & Zimmermann, H.G., 2011, 'Regulation and risk assessment for importations and releases of biological control agents against invasive alien plants in South Africa', *African Entomology* 19, 488–497. https://doi.org/10.4001/003.019.0215

Klein, H., Hoffmann, J.H., Neser, S. & Dittrich-Schröder, G., 2015, 'Evidence that *Quadrastichodella nova* (Hymenoptera: Eulophidae) is the only gall inducer among four hymenopteran species associated with seed capsules of *Eucalyptus camaldulensis* (Myrtaceae) in South Africa', *African Entomology* 23, 207–223. https://doi.org/10.4001/003.023.0117

Kotzé, J.D.F., Beukes, B.H., Van Den Berg, E.C. & Newby, T.S., 2010, *National invasive alien plant survey*, Report No. GW/A/2010/21, Agricultural Research Council: Institute for Soil, Climate and Water, Pretoria, South Africa.

Le Maitre, D.C., Versfeld, D.B. & Chapman, R.A., 2000, 'The impact of invading alien plants on surface water resources in South Africa: A preliminary assessment', *Water SA* 26, 397–408.

McClay, A.S., 1989, *Selection of suitable targets of weeds for classical biological control in Alberta*, AECV89-RI, Alberta Environmental Central, Vegreville, AB, Canada.

McConnachie, A.J., 2015, 'Host range tests cast doubt on the suitability of *Epiblema strenuana* as a biological control agent for *Parthenium hysterophorus* in Africa', *BioControl* 60, 1–9. https://doi.org/10.1007/s10526-015-9675-2

McConnachie, A.J., de Wit, M.P., Hill, M.P. & Byrne, M.J., 2003, 'Economic evaluation of the successful biological control of *Azolla filiculoides* in South Africa', *Biological Control* 28, 25–32. https://doi.org/10.1016/S1049-9644(03)00056-2

McConnachie, A.J., Hill, M.P. & Byrne, M.J., 2004, 'Field assessment of a frond-feeding weevil, a successful biological control agent of red water fern, *Azolla filiculoides*, in southern Africa', *Biological Control* 29, 326–331. https://doi.org/10.1016/j.biocontrol.2003.08.010

Meyer, J.Y., Fourdrigniez, M. & Taputuarai, R., 2011, 'Restoring habitat for native and endemic plants through the introduction of a fungal pathogen to control the alien invasive tree *Miconia calvescens* in the island of Tahiti', *Biological Control* 57, 191–198.

Moran, V.C. & Hoffmann, J.H., 2015, 'The Fourteenth International Symposium on Biological Control of Weeds, 1969–2014: Delegates, demographics and inferences from the debate on non-target effects', *Biological Control* 87, 23–31. https://doi.org/10.1016/j.biocontrol.2015.04.008

Moran, V.C., Hoffmann, J.H. & Hill, M.P. (eds.), 2011, 'Biological control of invasive alien plants in South Africa (1999–2010)', *African Entomology* 19, 177–549. https://doi.org/10.4001/003.019.0218

Moran, V.C., Hoffmann, J.H. & Zimmermann, H.G., 2005, 'Biological control of invasive alien plants in South Africa: Necessity, circumspection, and success', *Frontiers in Ecology and the Environment* 3, 77–83. https://doi.org/10.2307/3868513

Moran, V.C., Hoffmann, J.H. & Zimmermann, H.G., 2013, '100 years of biological control of invasive alien plants in South Africa: History, practice and achievements', *South African Journal of Science* 109(9/10), Art. #a0022, 1–6. https://doi.org/10.1590/sajs.2013/a0022

Moran, V.C. & Zimmermann, H.G., 1991, 'Biological control of cactus weeds of minor importance in South Africa', *African Entomology* 19, 37–55. https://doi.org/10.1016/0167-8809(91)90138-n

Morin, L., Reid, A.M., Sims-Chilton, N.M., Buckley, Y.M., Dhileepan, K., Hastwell, G.T. et al., 2009, 'Review of approaches to evaluate the effectiveness of weed biological control agents', *Biological Control* 51, 1–15. https://doi.org/10.1016/j.biocontrol.2009.05.017

Morris, M.J., Wood, A.R. & den Breeÿen, A, 1999, 'Plant pathogens and biological control of weeds in South Africa: A review of projects and progress during the last decade', *African Entomology Memoir* 1, 129–137.

Olckers, T., 2004, 'Targeting emerging weeds for biological control in South Africa: The benefits of halting the spread of alien plants at an early stage of their invasion', *South African Journal of Science* 100, 64–68.

Olckers, T. & Hill, M.P. (eds.), 1999, 'Biological control of weeds in South Africa (1990–1998)', *African Entomology Memoir* 1, 1–182.

Paterson, I.D., Hoffmann, J.H., Klein, H., Mathenge, C.W., Neser, S. & Zimmermann, H.G., 2011, 'Biological control of Cactaceae in South Africa', *African Entomology* 19, 230–246. https://doi.org/10.4001/003.019.0221

Paterson, I.D., Vitorino, M.D., de Cristo, S.C., Martin, G.D. & Hill, M.P., 2014, 'Prioritisation of potential agents for the biological control of the invasive alien weed, *Pereskia aculeata* (Cactaceae), in South Africa', *Biocontrol Science and Technology* 24, 407–425. https://doi.org/10.1080/09583157.2013.864382

Paynter, Q.E., Fowler, S.V., Gourlay, A.H., Haines, M.L., Harman, H.M., Hona, S.R. et al., 2004, 'Safety in New Zealand weed biocontrol: A nationwide survey for impacts on non-target plants', *NZ Plant Protection* 57, 102–107.

Paynter, Q., Hill, R., Bellgard, S. & Dawson, M., 2009, *Improving targeting of weed biological control projects in Australia*, Land and Water Australia, Canberra.

Pemberton, R.W., 2000, 'Predictable risk to native plants in weed biological control', *Oecologia* 125, 489–494. https://doi.org/10.1007/s004420000477

Peschken, D.P. & McClay, A.S., 1995, 'Picking the target – A revision of McClay's scoring system to determine the suitability of weeds for classical biological control', in E.S. Delfosse & R.R. Scott (eds.), *Proceedings of the Eighth International Symposium on Biological Control of Weeds*, Canterbury, New Zealand, February 2–7, 1992, CSIRO, Melbourne, pp. 137–143.

Rüegg, W.T., Quadranti, M. & Zoschke, A., 2007, 'Herbicide research and development: Challenges and opportunities', *Weed Research* 47, 271–275. https://doi.org/10.1111/j.1365-3180.2007.00572.x

Scott, J.K., Yeoh, P.B. & Michael, P.J., 2016, 'Methods to select areas to survey for biological control agents: An example based on growth in relation to temperature and distribution of the weed *Conyza bonariensis*', *Biological Control* 97, 21–30. https://doi.org/10.1016/j.biocontrol.2016.02.014

Suckling, D.M. & Sforza, R.F.H., 2014, 'What magnitude are observed non-target impacts from weed biocontrol?', *PLoS One* 9, e84847. https://doi.org/10.1371/journal.pone.0084847

Syrett, P., 2002, *Biological control of weeds on conservation land: Priorities for the Department of Conservation*, Internal series no. 82, Department of Conservation, Wellington, New Zealand.

Urban, A.J., Simelane, D.O., Retief, E., Heystek, F., Williams, H.E. & Madire, L.G., 2011, 'The invasive 'Lantana camara L.' hybrid complex (Verbenaceae): A review of research into its identity and biological control in South Africa', *African Entomology* 19, 315–348. https://doi.org/10.4001/003.019.0225

Van Driesche, R.G., Carruthers, R.I., Center, T., Hoddle, M.S., Hough-Goldstein, J., Morin, L. et al., 2010, 'Classical biological control for the protection of natural ecosystems', *Biological Control* 54, S2–S33. https://doi.org/10.1016/j.biocontrol.2010.03.003

Van Klinken, R.D., Morin, L. & Sheppard, A., 2016, 'Experts know more than just facts: Eliciting functional understanding to help prioritise weed biological control targets', *Biological Invasions*. https://doi.org/10.1007/s10530-016-1175-5

Van Wilgen, B.W. & de Lange, W.J., 2011, 'The costs and benefits of biological control of invasive alien plants in South Africa', *African Entomology* 19, 504–514. https://doi.org/10.4001/003.019.0228

Van Wilgen, B.W., De Wit, M.P., Anderson, H.J., Le Maitre, D.C., Kotze, I.M., Ndala, S., Brown, B. & Rapholo, M.B., 2004, 'Costs and benefits of biological control of invasive alien plants: Case studies from South Africa', *South African Journal of Science* 100, 113–122.

Van Wilgen, B.W., Fill, J.M., Baard, J., Cheney, C., Forsyth, A.T. & Kraaij, T., 2016, 'Historical costs and projected future scenarios for the management of invasive alien plants in protected areas in the Cape Floristic Region', *Biological Conservation* 200, 168–177. https://doi.org/10.1016/j.biocon.2016.06.008

Van Wilgen, B.W., Forsyth, G.G., Le Maitre, D.C., Wannenburgh, A., Kotzé. J.D.F., van den Berg, E. et al., 2012, 'An assessment of the effectiveness of a large, national-scale invasive alien plant control strategy in South Africa', *Biological Conservation* 148, 28–38. https://doi.org/10.1016/j.biocon.2011.12.035

Van Wilgen, B.W., Reyers, B., Le Maitre, D.C., Richardson, D.M. & Schonegevel, L., 2008, 'A biome-scale assessment of the impact of invasive alien plants on ecosystem services in South Africa', *Journal of Environmental Management* 89, 336–349. https://doi.org/10.1016/j.jenvman.2007.06.015

Van Wilgen, B.W. & Wannenburgh, A., 2016, 'Co-facilitating invasive species control, water conservation and poverty relief: Achievements and challenges in South Africa's *Working for Water* programme', *Current Opinion in Environmental Sustainability* 19, 7–17. https://doi.org/10.1016/j.cosust.2015.08.012

Versfeld, D.B., Le Maitre, D.C. & Chapman, R.A., 1998, *Alien invading plants and water resources in South African: A preliminary assessment*, WRC Report No. TT 99/98, Water Research Commission, Pretoria, South Africa.

Watson, A.K. & Wymore, L.A., 1989, 'Biological control, a component of integrated weed management', in E.S. Delfosse (ed.), *Proceedings of the VII International Symposium for Biological Control of Weeds*, Istituto Sperimentale per la Patologia Vegetale, Ministero dell'Agricoltura e delle Foreste, Rome, Italy, March 6-11, 1988, pp. 101–106.

Wilson, J.R.U., Gaertner, M., Richardson, D.M. & van Wilgen, B.W., 2017, 'Contributions to the National Status Report on Biological Invasions in South Africa', *Bothalia* 47(2), a2207. https://doi.org/10.4102/abc.v47i2.2207

Wilson, J.R.U., Ivey, P., Manyama, P. & Nänni, I., 2013, 'A new national unit for invasive species detection, assessment and eradication planning', *South African Journal of Science* 109(5/6), Art. #0111. https://doi.org/10.1590/sajs.2013/20120111

Winston, R.L., Schwarzländer, M., Hinz, H.L., Day, M.D., Cock, M.J.W. & Julien, M.H. (eds.), 2014, *Biological control of weeds: A world catalogue of agents and their target weeds*, 5th edition, FHTET-2014-04, USDA Forest Service, Forest Health Technology Enterprise Team, Morgantown, West Virginia.

Wood, A.R. & Morris, M.J., 2007, 'Impact of the gall-forming rust fungus *Uromycladium tepperianum* on the invasive tree *Acacia saligna* in South Africa: 15 years of monitoring', *Biological Control* 41, 68–77. https://doi.org/10.1016/j.biocontrol.2006.12.018

Zachariades, C., Strathie, L.W., Retief, E. & Dube, N., 2011, 'Progress towards the biological control of *Chromolaena odorata* (L.) R.M. King & H. Rob (Asteraceae) in South Africa', *African Entomology* 19, 282–302. https://doi.org/10.4001/003.019.0229

Zengeya, T., Ivey, P., Woodford, D.J., Weyl, O., Novoa, A., Shackleton, R. et al., 2017, 'Managing conflict-generating invasive species in South Africa: Challenges and trade-offs', *Bothalia* 47(2), a2160. https://doi.org/10.4102/abc.v47i2.2160

Zimmermann, H.G., 1979, 'Herbicidal control in relation to distribution of *Opuntia aurantiaca* Lindley and effect on cochineal populations', *Weed Research* 19, 89–93. https://doi.org/10.1111/j.1365-3180.1979.tb01580.x

13

Promise and challenges of risk assessment as an approach for preventing the arrival of harmful alien species

Authors:
Reuben P. Keller[1]
Sabrina Kumschick[2,3]

Affiliations:
[1]Institute of Environmental Sustainability, Loyola University, United States

[2]Centre for Invasion Biology, Department of Botany & Zoology, Stellenbosch University, South Africa

[3]Invasive Species Programme, South African National Biodiversity Institute, South Africa

Corresponding author:
Reuben Keller,
rkeller1@luc.edu

Background: Harmful alien species impose a growing environmental, economic and human well-being burden around the globe. A promising way to reduce the arrival of new species that may become harmful is to utilise pre-border risk assessment (RA) tools that relate the traits of introduced species to whether those species have become established and harmful. These tools can be applied to species proposed for intentional introduction so that informed decisions can be made about whether each species poses an acceptable risk and should be allowed for import.

Objectives: A range of approaches to RA tool development have emerged, each relying on different assumptions about the relationships between traits and species impacts, and each requiring different levels and types of data. We set out to compare the qualities of each approach and make recommendations for their application in South Africa, a high biodiversity developing country that already has many invasive species.

Method: We reviewed five approaches to pre-border RA and assessed the benefits and drawbacks of each. We focused on how pre-border RA could be applied in South Africa.

Results: Recent legislation presents a framework for RA to evaluate species introductions to South Africa, but we find that this framework assumes an approach to RA that is relatively slow and costly and that does not leverage recent advances in RA tool development.

Conclusion: There is potential for proven RA approaches to be applied in South Africa that would be less costly and that could more rapidly assess the suite of species currently being introduced.

Introduction

Harmful alien species continue to be a major driver of biodiversity change across the globe, as well as causing enormous economic costs and impacts to human health and livelihoods (Keller et al. 2009; Lodge et al. 2006; Pimentel 2011; Shackleton et al. 2007). The problems related to alien and invasive species have intensified as globalisation has produced a range of vectors that intentionally and unintentionally transport live organisms across borders. The diversity of pathways and organisms involved presents a challenge for national policies that aim to reduce the harm from alien species (Essl et al. 2015; Faulkner et al. 2017; Hulme et al. 2008; Lodge et al. 2006).

Established alien species are usually difficult and costly to manage, and they are rarely eradicated, which means that their costs should be seen as perpetual (van Wilgen et al. 2016). Preventing the arrival of harmful alien species, as opposed to managing them once they become established, is thus seen as a wise approach. However, preventing the arrival of all new alien species is not desirable because many of the trades that move these species provide large benefits to society. At the other extreme, allowing all alien species for import is not without costs because of the harms that arise from the subset that become harmful invaders (Keller & Springborn 2014). A more desirable goal is to prevent the introduction of species that are likely to become harmful while allowing the import of all others. For most taxonomic groups, the proportion of species that become invasive and have negative impact is relatively small (Kumschick et al. 2015a) so that most species could remain available for trade.

A range of pre-border risk assessment (RA) approaches have emerged for predicting the likely harm that alien species will cause. Making such predictions is challenging because the risks of these species need to be accurately identified using only information that is known before they are introduced. This paper focuses on RA approaches that are or could be applied to species intentionally introduced through trade when the identity of the species is known. We do not address RA approaches that have been designed for addressing *pathways* (e.g. ballast water, pests of nursery plants) of unintentional introduction.

Pre-border RA tools ideally have at least five qualities. Firstly, they should be reasonably accurate so that small proportions of species are misclassified. Secondly, they should be transparent so that the rationale for the result is clear. Thirdly, they should be rapid so that decisions are quickly available and trade interruptions are minimised. Fourthly, they should produce consistent results so that different people performing assessments arrive at the same conclusions. Finally, implementation of these RA tools must require a realistic level of resources in terms of finances and skilled practitioners so that the many thousands of species in trade can be assessed.

RA tools are already implemented in some countries, and more countries are actively developing them (Kumschick & Richardson 2013). Australia and New Zealand have each implemented RA programmes for animal and plant introductions for over a decade and these programmes are some of the best developed globally (Keller & Drake 2009). Analyses have shown that national programmes – such as those in Australia and New Zealand – that identify and keep out harmful invaders can reap large economic benefits for the importing nation (e.g. Keller, Lodge & Finnoff 2007b; Keller & Springborn 2014; Springborn et al. 2015; Springborn, Romagosa & Keller 2011). These economic benefits are in addition to the environmental benefits from prevention of further biodiversity impacts and the benefits to human health and livelihood. Importantly, international agreements and standard setting organisations, such as the International Plant Protection Convention and the World Organisation for Animal Health, explicitly allow restrictions on species in trade if those measures are scientifically justified in the form of RA (Keller & Perrings 2011; Perrings et al. 2005).

The goal of this paper is to evaluate five of the most prominent approaches that have emerged for developing pre-border invasive species RA tools. These tools can be applied to species that are intentionally introduced and for which the identity of the species is known. Following the review of approaches, we discuss some general considerations and South Africa's new *National Environmental Management: Biodiversity Act* (NEM:BA), *Alien and Invasive Species Regulations* (Department of Environmental Affairs 2014). We emphasise here that this paper reviews approaches to RA rather than actual tools, and we refer readers to other papers for more comprehensive coverage of the tools available (e.g. Hayes & Barry 2008; Keller & Drake 2009; Kolar & Lodge 2001; Kumschick & Richardson 2013; Leung et al. 2012; Pysĕk & Richardson 2007).

History of risk assessment for alien species

Herbert Baker (1965) carried out some of the first work to investigate the link between species traits and invasion potential. Recent work has built upon this foundation and expanded the goals to include making explicit predictions about the likely behaviour of species not yet introduced. To do this, researchers have usually dealt with smaller taxonomic units and geographic areas so that traits associated with invasion are more likely to be consistent (Kolar & Lodge 2001). For example, while Baker (1965) sought generalisations across all plant species and continents, more recent RAs have focused on, for example, predicting the likelihood of fish species spreading into the North American Great Lakes because of climate change (Mandrak 1989) and the likely impacts from woody plant invasions in the South African fynbos (Tucker & Richardson 1995).

Baker's (1965) approach was to relate the behaviour of a species within its ecosystem to its traits, and modern RA tool development does the same (but see *Detailed* and *Mechanistic* approaches below). Different approaches are applied to search for patterns in traits that can explain the observed invasion history, and if strong correlations between traits and invasion history are found, it is assumed that they will be useful predictors of future invasions. These correlations can then be formalised into specific tools for RA. Although this logic is common across much RA tool development, there are important differences in approaches that in turn represent different beliefs about the relationships between traits and invasiveness.

Trait-based risk assessment

Before a trait-based RA tool can be developed, the taxonomic unit and geographical area need to be set, along with the step(s) in the invasion sequence of interest (Blackburn et al. 2011). Limiting each should lead to traits that are more consistently related to passage through the invasion step, and may thus be more useful for prediction (Kolar & Lodge 2001). For example, the traits associated with fishes moving from introduced to established in South Africa are more likely to be consistent than the traits associated with all vertebrates making the same transition across all of Africa. Conversely, increasing geographic area, taxonomic breadth or the number of invasion sequence steps can be beneficial because there will have been more previous introductions from which data can be gathered for RA development. This trade-off is made more complex because RA tools are usually developed for use within political geographical areas (e.g. a nation or region), and the boundaries of these areas are rarely based on ecological factors. Existing legislation may also present constraints, for example, if an agency controls plants imported for agriculture but not plants imported for ornamental purposes.

These complexities and trade-offs ultimately need to be dealt with on a case-by-case basis. We note here that RA tools have been successfully developed at a large range of geographic areas and taxonomic units, and for all steps in the invasion sequence (Kumschick & Richardson 2013). Once the parameters for the RA are set, the process of developing the RA tool can proceed. Different approaches to this are reviewed in the following section.

Trait scoring

The *Trait Scoring* approach to RA is based on the belief that many traits can make a species more likely to pass through a step in the invasion sequence and, thus, that species possessing more of these traits are most likely to become established and/or cause harm. This is the most commonly applied approach for developing pre-border RA tools, with the resulting tools usually consisting of a list of questions about the presence or absence of traits (Leung et al. 2012). Presence of a trait is scored as a positive number (usually +1) and absence is scored as a 0 or negative number (usually –1). Once all questions have been answered, the scores for each question are summed to a final score, with higher final scores indicating greater likelihood that the species will become established or harmful. *Trait Scoring* RAs are thus conceptually simple and can be implemented in a basic spreadsheet. Examples of this approach are the Australian Weed Risk Assessment (WRA; Pheloung, Williams & Halloy 1999), which was developed for all plant introductions to Australia; the Fish Invasiveness Scoring Kit, which was developed in the United Kingdom and has now been applied to many regions (Centre for Environment, Fisheries and Aquaculture Science 2013; XX, 2016 [*this volume*]); and the New Zealand Aquatic Plants Risk Assessment (NZ AqWRA), which has recently been adapted for the United States (Gantz et al. 2015; Gordon et al. 2012).

The first step in developing these tools is for experts to develop a list of traits that they believe are associated with invasiveness. Scores are assigned to the presence/absence of each trait based on their perceived importance, and this produces a RA tool. Next, the tool is validated by collecting data about the traits of species that have previously been introduced to the region, assessing those species and comparing their scores to the known outcomes from these introductions (i.e. introduced vs. established, benign vs. harmful). If necessary, the RA tool can be tuned by modifying the traits used and the scores assigned to each. If, in the final RA, the group of species that successfully passed through the step in the invasion sequence consistently receive higher scores than the group that failed, a score threshold can be set to discriminate between these groups. This threshold can then be used for prediction when species that have not yet been introduced are assessed.

A drawback to the *Trait Scoring* approach is that it does not consider interactions among traits. It is likely that some traits influence invasiveness depending on the presence of other traits, and thus that while two traits alone may not be predictive, the combination may be a strong predictor. Although the *Trait Scoring* approach could

TABLE 1: Five representative trait questions (out of 49 total) from the Australian Weed Risk Assessment, an example of the *Trait Scoring* approach.

Trait	Score
Produces spines, thorns or burrs	Yes = 1, No = 0
Unpalatable to grazing animals	Yes = 1, No = –1
Causes allergies or is otherwise toxic to humans	Yes = 1, No = 0
Self-fertilisation	Yes = 1, No = –1
Propagules bird dispersed	Yes = 1, No = –1

Source: Adapted from Pheloung et al. 1999

in theory consider such interactions, in practice it would be cumbersome and we are not aware of a tool that does this. A second drawback is that correlations among traits may lead to double counting (Leung et al. 2012). For example, in the application of the NZ AqWRA to the Laurentian Great Lakes, plants are assessed on their tolerance for a range of habitats, with a maximum score for a species that can live from dry land to fully aquatic. Plants are separately assessed on their tolerance to periodic flooding and drying, with a maximum score for species with high tolerance (Gantz et al. 2015). Species scoring highly on the first trait are also likely to tolerate flooding, meaning that they effectively receive double points. Despite these logical drawbacks, *Trait Scoring* approaches have been extensively evaluated and found to be acceptable for policy (Keller & Drake 2009; Kumschick & Richardson 2013).

The Australian WRA (Table 1) is the most prominent example of the *Trait Scoring* approach and is designed to assess the potential that alien plants will become established and harmful. It consists of 49 questions about invasion history, biology, environmental tolerance, ecology and reproduction (Pheloung et al. 1999). These traits were selected by experts, assembled into an RA and then tested by assessing 370 species that had previously been introduced to Australia. This RA tool has been used to make decisions about plant imports to Australia since 1997 and has been adapted for testing and use in many other regions (Gordon et al. 2008; Kumschick & Richardson 2013).

Statistical approach

Many recent developments in RA have come through the *Statistical* approach. Development of *Statistical* RAs begins similarly to *Trait Scoring* with a list of traits that experts believe are associated with invasiveness. Next, the set of species from the taxonomic unit of interest that have previously been introduced to the geographic area of interest is determined, and a matrix is created that includes trait data about each species and the outcome of each introduction. This matrix is analysed with a statistical or machine learning algorithm to find patterns in traits that are correlated with outcomes. Algorithms used include logistic regression, discriminant analysis, categorical and regression trees and neural networks (Keller, Kocev & Džeroski 2011). Resulting models are most commonly validated with leave-one-out cross-validation to determine performance (Keller et al. 2011).

The logic of this approach differs from *Trait Scoring* in three important ways. Firstly, the *Statistical* approach holds that

TABLE 2: Risk assessment for alien fishes in the North American Great Lakes, an example of the *Statistical* approach using decision trees. Outcomes are in italics.

Attribute/Trait	If Yes...	If No...
1) Climate match greater than 71.7	Go to 2	*Fail to establish*
2) Includes fish in diet	*Establish, high impacts*	Go to 3
3) Fecundity (number of eggs) > 1 013 000	*Establish, high impacts*	*Establish, low impacts*

Source: Adapted from Howeth et al. 2016

TABLE 3: Basic framework for *Rapid Screening* approach to risk assessment for alien species. Outcomes are in italics.

Attribute	If Yes...	If No...
1) Strong climate match between native range and region of interest	Go to 2	*Not a harmful invader*
2) History of causing harm as an alien species	*Harmful invader*	*Not a harmful invader*

Source: See text for discussion of development of the Rapid Screening approach

that it is possible for just one or a few traits to explain invasiveness, and *Statistical* RA tools generally require data on about one to five traits to perform an assessment (Table 2). An RA using logistic regression for established alien molluscs in the Laurentian Great Lakes, for example, found that annual fecundity was sufficient to explain which species become harmful (Keller, Drake & Lodge 2007a). Similarly, a study of environmentally harmful Cactaceae found that the size of a species' native range is a strong predictor of spread and impacts in South Africa (Novoa et al. 2016).

Secondly, this approach holds that interactions among traits may be important and the algorithms used are designed to find such interactions. As discussed above, such interactions are rarely, if ever, included in *Trait Scoring* RAs and may even be masked by scoring a species separately on these traits.

Thirdly, the *Statistical* approach holds that the available data should inform the structure of the RA model. In the *Trait Scoring* approach experts determine which traits should be included and how they should be scored. In *Statistical* RA the practitioner determines the traits that will be available to the model, but how these traits are incorporated largely depends on the algorithm, which in turn relies on the historical data. This reduces any potential bias on the part of the RA developer and may lead to surprising and non-intuitive outcomes, providing new insight into the invasion process.

The *Statistical* approach also has a number of drawbacks. Firstly, the models created are often mathematically complex and based on algorithms that are not widely understood. This can reduce acceptance because managers and policy-makers may not be prepared to support methods that they do not fully understand. Secondly, the small number of steps of the tools may be problematic because it conflicts with beliefs that invasion is a highly complex process. Again, this may limit acceptance of the resulting models. Finally, lack of data can be a greater issue for these shorter RA tools because missing data about one trait may make it impossible to reach a conclusion. In comparison, most *Trait Scoring* tools are robust to some level of missing data. Although many *Statistical* RA tools have been developed and although they have been shown to have high accuracy and produce rapid results (Keller & Drake 2009; Lodge et al. 2016), we are not aware of any jurisdiction that currently implements them.

Rapid screening

The *Rapid Screening* approach has seen development in the last few years and shows a lot of promise as a stand-alone approach to RA, for the creation of watch lists, and as a way

to prioritise species for more detailed RA. This approach is usually based on just two species attributes. The first of these is climate match – the degree to which the climate in the alien range is similar to that in the species' native range. The second is whether the species has a history of causing harm elsewhere in its alien range. If a species has both strong climate match and a history of impacts, it is designated as likely to cause harm in its new range. If it lacks either, it is considered unlikely to cause harm. The logic of this approach comes from the observation that climate match and invasion history are the two attributes most often correlated to the likelihood that an alien species will become established and cause harm (see Hayes & Barry 2008; Table 3).

The *Rapid Screening* approach is simple, rapid to implement, appealingly intuitive and can generally be applied across all taxa and geographic areas. However, a major drawback is that it will not be useful for species that may become established for the first time because no records are available about their impacts elsewhere (Kumschick et al. 2015b). The United States Fish and Wildlife Service has developed a *Rapid Screening* RA tool that addresses this concern by treating assessments as 'Uncertain' if the species in question has not had alien established populations in at least one place for at least 10 years or has not been in trade for at least 10 years (Hoff 2014).

Faulkner et al. (2014) recently published a *Rapid Screening* RA tool that they applied to 394 alien species in South Africa. These species come from a range of taxa and habitats, and the RA tool showed reasonable performance. Faulkner et al. (2014) argue that the tool could be used for creating watch lists and that these watch lists could guide import decisions or be used to prioritise species for further RA. They also note that species can be quickly assessed using readily available data, making this approach particularly applicable to jurisdictions lacking the resources to conduct more involved RAs.

Other approaches to risk assessment

Neither of the following two approaches is explicitly based on species traits, although each requires extensive information about the species being assessed. Firstly, the *Mechanistic* approach is based on the logic that to become harmful an alien species must cross certain barriers to invasion (as outlined by Blackburn et al. 2011) and be transported, introduced, released, become established, spread and cause negative impacts. This approach treats these steps separately and considers, for example, that if a species is highly unlikely to be introduced, then it poses a low overall risk regardless of

TABLE 4: Five representative questions (out of 25 total) from the Harmonia⁺ Risk Assessment, an example of the *Mechanistic* approach.

Stage in Invasion Sequence	Question
Introduction	The probability for The Organism to be introduced into The Area's wild by natural means is [low/medium/high].
Establishment	The area provided [non-optimal/sub-optimal/optimal] climate for establishment of The Organism.
Spread	The Organism's capacity to disperse within The Area by natural means is [very low/low/medium/high/very high].
Impacts: environmental targets	The organism has an [inapplicable/low/medium/high] effect on native species through predation, parasitism or herbivory.
Impacts: human targets	The organism has a(n) [inapplicable/very low/low/medium/high/very high] effect on human health, through parasitism.

Source: Adapted from D'Hondt et al. 2015

its potential impacts. The Harmonia⁺ RA was recently developed for use in Belgium (D'hondt et al. 2015; Table 4) and could be readily adapted for use elsewhere. This tool requires users to estimate separately the likelihood that a species will pass through each step in the invasion sequence, with the questions addressing such outcomes rather than traits. Thus, the RA is more transferable between taxonomic units and geographic areas, but it relies strongly on users to make difficult judgements about the likelihoods of specific outcomes. The time taken to perform the RA will depend very much on decisions made by the user about how much detail to include and their expertise on the species.

In the *Detailed* approach, all available details and information about the species, the region into which it may be introduced and the circumstances of its introduction are included. This usually begins with extensive literature review, may include interviews with experts and proceeds from this information to scenarios of likely outcomes from allowing the species for import. It resembles more closely a risk analysis approach as it often also considers how risks can be managed and the potential benefits of a species (e.g. European Food Safety Authority 2012). Again, the time taken to complete a *Detailed* RA depends on decisions about how complex it should be and what data are relevant. However, we note that this approach often takes years to complete and is thus far more expensive than other approaches. The main benefit of this approach is that it can provide detailed predictions, for example, that a species will have different impacts in different areas. We believe that its use will only be justified for pre-border RA when the species in question has the potential for both benefits and adverse impacts that are considered significant. For example, it was used in Canada to assess the risk posed by five species of Asian carps (Mandrak & Cudmore 2004). These species were all in trade, and were all considered potentially very harmful if they became established.

Important qualities of risk assessment tools

An ideal RA tool is transparent, cheap and rapid to implement, accurate and consistent so that different people assessing the same species arrive at the same conclusion. Trade-offs among these qualities will need to be made because, for example, the fastest RA tool may not achieve required levels of accuracy. It is ultimately a policy question as to how these qualities should be balanced.

As previously mentioned, the most commonly applied approach to invasive species RA has been *Scored Questions*. In Australia and New Zealand, tools based on this approach have been in use for well over a decade with little controversy. Species assessments generally take 1–2 days and the tools have accuracies that are almost always > 80% (Lodge et al. 2016). This level of accuracy has been shown to produce economic benefits in addition to the environmental benefits from keeping out harmful alien species (Keller et al. 2007b). *Statistical* RA tools achieve similar accuracy and can be completed more quickly because they require fewer data (Lodge et al. 2016). A disadvantage of *Scored Questions* and *Detailed* approaches is that development of tools for a given taxa and region can take months to years. The *Rapid* approach to RA is the fastest and can usually be completed in less than an hour per species and possibly even more quickly if many species are being assessed using the same data source. The accuracy of this approach has not yet been rigorously tested, but preliminary results are encouraging (Lodge et al. 2016). In contrast, the *Detailed* approach usually requires extensive resources and time, and we are not aware of any attempt to determine its accuracy. Indeed, because this approach is usually different in every application, it is difficult to imagine how its accuracy could be assessed. Finally, the *Mechanistic* approach is an interesting addition to the RA toolbox, and similar to the *Rapid* and *Detailed* approaches, it can be used across taxonomic groups. It is too recent for us to assess its accuracy or time taken to apply it, although we believe it would take quite a lot longer than the first three approaches reviewed.

The large differences summarised above, in RA approaches inevitably make it somewhat confusing for new programmes to decide how to proceed, and existing legislation and administrative structures will need to be considered. We make two further observations that can assist with the development and use of RA tools. Firstly, improvements in the availability of trait and invasion history data can be leveraged. For example, *FishBase* (Froese & Pauly 2015) is a freely accessible online database with information about the biogeography, invasion history, physical traits and environmental tolerances of most fish species. Pantheria (Jones et al. 2009) provides trait data for mammals and TRY (try-db.org; Kattge et al. 2011) for plants. Furthermore, databases on invasive and alien species, like the Global Invasive Species Database (http://www.iucngisd.org/gisd/) and Global Register of Introduced and Invasive Species (http://www.griis.org/) as well as CABI's large collections of data (e.g. http://www.cabi.org/isc/; Randall 2012), are useful sources of data about invasion history. Developing RA tools that leverage such data can reduce the cost of tool development and use.

Secondly, cost savings may be possible by adopting, with appropriate modifications, RA tools developed for other regions. The Australian WRA has been shown to be effective in several regions around the world (Gordon et al. 2008; Kumschick & Richardson 2013), although work is usually required to calibrate the threshold between harmful and benign species (e.g. Nishida et al. 2009). The Australian WRA has also been modified to apply to several aquatic taxonomic groups (CEFAS 2013) and the resulting tools have been successfully used in a range of regions (e.g. Lawson et al. 2012) including being applied to fishes in South Africa (Marr et al. 2017). Likewise, an RA tool developed in New Zealand for assessing risks from alien aquatic plants is effective in multiple regions (Gantz et al. 2015; Gordon et al. 2012).

Recommendations for pre-border risk assessment in South Africa

Permits to import alien species to South Africa are given by the Department of Environmental Affairs (DEA) and the Department of Agriculture, Forestry and Fisheries. We focus here on the framework for new importations and permit applications recently produced by DEA (DEA 2014). This framework is part of the NEM:BA, Alien and Invasive Species Regulations published in August 2014 and guides RA for individual species that have been proposed for import and which require a permit (i.e. new imports and species listed as Category 2 under the NEM:BA regulations). Under the framework, a pre-border RA must consider the biology, ecology and invasion history of the species, the proposed use of the species in South Africa, characteristics of the environment that the species is likely to encounter, risks of hitch-hiker species or diseases arriving with the species and several other factors including the cost of control should the species escape. This list covers a broad range of the factors that are known to be important predictors of invasion, and many of the ways that harmful alien species can cause impacts. However, the comprehensiveness of the list means that conducting an RA that meets these standards would most resemble the *Detailed* approach. The disadvantages of such an approach are detailed above and include the extensive resources that would be required to assess a significant proportion of the species that may enter the country. Additionally, the consistency of such an approach may be low because there is no published guidance as to how different factors should be weighed or the extent and type of information required to adequately assess each factor. However, we note that such guidance could be produced in the future.

While the exact ways that the new NEM:BA regulations will manifest in RA for individual alien species are not yet known, for three reasons we believe that the challenges for alien species introduction to South Africa could be better addressed with other approaches. Firstly, South Africa has a great diversity of ecosystems and species, many of which are already severely impacted by harmful alien species and all of which are at risk from future invasions. Secondly, the resources available for pre-border RA are not sufficient to assess a large proportion of introduced alien species with the *Detailed* approach. This issue is compounded as the number of species in international trade increases, making it reasonable to expect that the number of species proposed for import to South Africa will likewise increase over coming years. Thirdly, the *Detailed* approach to RA is difficult to defend in terms of accuracy and consistency. The difficulties for assessing accuracy are described above, and the difficulty for consistency arises because the structure of *Detailed* RA and the data accessed and used will inevitably differ among users. It would thus be possible for different stakeholders to reach different conclusions while each being able to claim that they are using the process outlined in the NEM:BA regulations.

We suggest that a two-tiered RA system could better meet the needs of South Africa to prevent the arrival of harmful alien species while acknowledging resource limitations. Our suggested system is similar to that suggested by Faulkner et al. (2014). As a first tier, we suggest that species be assessed with a *Rapid* RA that could be based on tools already developed (e.g. Faulkner et al. 2014; Hoff 2014). The results of these assessments should be publicised online and in other relevant forums and should be initially used to determine which species are allowed and disallowed for import. The second tier would consist of either a *Scored Questions* or *Statistical* RA. This would only be used if there were a request for further assessment, which could come from either a person believing that a species banned from import presents low risks or that a species allowed for import poses unacceptably high risks. In either case, the person could petition the DEA to conduct a second tier RA, which would be final. All RA tools and results from assessments should be reviewed by independent experts prior to implementation, but given our suggestion that relatively simple approaches to RA be used this review could be rapid.

Such an approach, if designed to leverage readily available data, could be used to quickly assess a large number of species in trade. The *Rapid* RA approach is straightforward to conduct, and it is likely that personnel with a graduate degree in biology could perform the assessments. A main challenge to our suggested approach would be the development of second tier RA tools for all taxonomic groups. However, we note that a tool now exists for fishes (Marr et al. 2017) and that it may be possible to calibrate existing tools, such as the Australian WRA, for use in South Africa. These options would greatly shorten the time to having a full suite of Tier 2 tools available and would reduce costs for development. Alternatively, *Mechanistic* tools like Harmonia⁺ provide a trade-off between the time needed for assessments and the need to develop tools as they do not require separate tools for different taxa. Such *Mechanistic* tools could be used until others are available, or over longer periods if they are deemed appropriate and adequate resources are available.

Conclusions

Pre-border RA tools have advanced over recent decades and are now often applied to protect nations from the effects of harmful alien species. As well as the environmental case for implementing these tools, there is strong evidence that they protect the economy of the importing nation. Indeed, we are not aware of an economic analysis of an RA tool that has not shown support for its application. Despite the support for RA, there remain challenges to implementation, including deciding which approach will best meet the needs of the importing nation.

Application of pre-border RA in South Africa presents many challenges but could be extremely beneficial. South Africa contains several unique biomes where alien taxa already cause significant impacts (e.g. Richardson & van Wilgen 2004) and pre-border RA could aid in the protection of this exceptional biodiversity. Additionally, developing countries need to implement cost-effective solutions to potential risks posed to their economies (e.g. van Wilgen et al. 2001) and people's livelihoods (Shackleton et al. 2007). Implementing a robust pre-border RA programme would offer the opportunity to prevent the arrival of additional harmful species and thus reduce economic and social risks. A framework for such a programme has recently been suggested under the legal umbrella of NEM:BA, but it does not explicitly leverage recent advances in RA tools. In particular, it appears to require a *Detailed* assessment of all species and this likely makes it infeasible to assess and appropriately manage the total number of species that pose risks. We have suggested an alternative framework that builds upon recent advances in RA for alien species and that would make it possible to assess many more species in a much shorter amount of time. While our suggested framework is not without challenges, we believe that it could ultimately be a much more realistic and effective way for South Africa to increase its protection from invasive species.

Acknowledgements

We thank the three referees who each provided helpful comments on an earlier version of the manuscript. Additionally, we thank the many conference attendees who provided feedback on R.P.K.'s presentation.

S.K. was supported by the South African National Department of Environment Affairs through its funding of the South African National Biodiversity Institute Invasive Species Programme.

Competing interests

The authors declare that they have no financial or personal relationship(s) that may have inappropriately influenced them in writing this article.

Authors' contributions

R.P.K. and S.K. conceived and outlined the paper. R.P.K. wrote most of the first draft. R.P.K. and S.K. worked together to edit it into its current form.

References

Baker, H.G., 1965, 'Characteristics and modes of origin of weeds', in H.G. Baker & G.L. Stebbins (eds.), *The genetics of colonizing species*, pp. 147–168, Academic Press, New York.

Blackburn, T.M., Pyšek, P., Bacher, S., Carlton, J.T., Duncan, R.P., Jarosík, V. et al., 2011, 'A proposed unified framework for biological invasions', *Trends in Ecology and Evolution* 26, 333–339. https://doi.org/10.1016/j.tree.2011.03.023

Centre for Environment, Fisheries and Aquaculture Science, 2013, *Decision support tools: Invasive species identification kits*, CEFAS, Lowestoft.

Department of Environmental Affairs, 2014, *National Environmental Management: Biodiversity Act 2004 (Act no. 10 of 2004). Alien and invasive species regulations, 2014*, Vol. 590, No. 37885, Government Gazette, 1 August 2014.

D'hondt, B., Vanderhoeven, S., Roelandt, S., Mayer, F., Versteirt, V., Adriaens, T. et al., 2015, 'Harmonia⁺ and Pandora⁺: Risk screening tools for potentially invasive plants, animals, and their pathogens', *Biological Invasions* 17, 1869–1883. https://doi.org/10.1007/s10530-015-0843-1

Essl, F., Bacher, S., Blackburn, T.M., Booy, O., Brundu, G., Brunel, S. et al., 2015, 'Crossing frontiers in tackling pathways of biological invasions', *BioScience* 65, 769–782. https://doi.org/10.1093/biosci/biv082

European Food Safety Authority (EFSA) Scientific Committee, 2012, 'Scientific opinion on risk assessment terminology', *EFSA Journal* 10(5), 2664. https://doi.org/10.2903/j.efsa.2012.2664

Faulkner, K.T., Hurley, B.P., Robertson, M.P., Rouget, M. & Wilson, J.R.U., 2017, 'The balance of trade in alien species between South Africa and the rest of Africa', *Bothalia* 47(2), a2157. https://doi.org/10.4102/abc.v47i2.2157

Faulkner, K.T., Robertson, M.P., Rouget, M. & Wilson, J.R.U., 2014, 'A simple, rapid methodology for developing invasive species watchlists', *Biological Conservation* 179, 25–32. https://doi.org/10.1016/j.biocon.2014.08.014

Froese, R. & Pauly, D. (eds.), 2015, *FishBase*, viewed 15 November 2015, from www.fishbase.org

Gantz, C.A., Gordon, D.R., Jerde, C.L., Keller, R.P., Chadderton, W.L., Champion, P.D. et al., 2015, 'Managing the introduction and spread of non-native aquatic plants in the Great Lakes: A regional risk assessment approach', *Management of Biological Invasions* 6, 45–55. https://doi.org/10.3391/mbi.2015.6.1.04

Gordon, D.R., Gantz, C.A., Jerde, C.L., Chadderton, W.L., Keller, R.P. & Champion, P.D., 2012, 'Weed risk assessment for aquatic plants: Modification of a New Zealand system for the United States', *PLoS One* 7, e40031. https://doi.org/10.1371/journal.pone.0040031

Gordon, D.R., Onderdonk, D.A., Fox, A.M. & Stocker, R.K., 2008, 'Consistent accuracy of the Australian weed risk assessment system across varied geographies', *Diversity and Distributions* 14, 234–242. https://doi.org/10.1111/j.1472-4642.2007.00460.x

Hayes, K.R. & Barry, S.C., 2008, 'Are there any consistent predictors of invasion success?', *Biological Invasions* 10, 483–506. https://doi.org/10.1007/s10530-007-9146-5

Hoff, M.H., 2014, *Standard operating procedures: Rapid screening of species risk of establishment and impact in the U.S.*, United States Fish and Wildlife Service, viewed 15 November 2016, from http://www.fws.gov/injuriouswildlife/pdf_files/Standard_Operating_Procedures_01_08_14.pdf

Howeth, J.G., Gantz, C.A., Angermeier, P.L., Frimpong, E.A., Hoff, M.H., Keller, R.P. et al., 2016, 'Predicting invasiveness of species in trade: Climate match, trophic guild, and fecundity influence invasive success of nonnative freshwater fish', *Diversity and Distributions* 22, 148–160. https://doi.org/10.1111/ddi.12391

Hulme, P.E., Bacher, S., Kenis, M., Klotz, S., Kuhn, I., Minchin, D. et al., 2008, 'Grasping the routes of biological invasions: a framework for integrating pathways into policy' *Journal of Applied Ecology* 45, 403–414. https://doi.org/10.1111/j.1365-2664.2007.01442.x

Jones, K.E., Bielby, J., Cardillo, M., Fritz, S.A., O'Dell, J., Orme, C.D.L. et al., 2009, 'PanTHERIA: A species-level database of life history, ecology, and geography of extant and recently extinct mammals', *Ecology* 90, 2648. https://doi.org/10.1890/08-1494.1

Kattge, J., Diaz, S., Lavorel, S., Prentice, I.C., Leadley, P., Bönisch, G. et al., 2011, 'TRY – A global database of plant traits', *Global Change Biology* 17, 2905–2935. https://doi.org/10.1111/j.1365-2486.2011.02451.x

Keller, R.P. & Drake, J.M., 2009, 'Trait based risk assessment for invasive species', in R.P. Keller, D.M. Lodge, M.A. Lewis & J.F. Shogren (eds.), *Bioeconomics of invasive species: Integrating ecology, economics, policy and management*, pp. 44–62, Oxford University Press, New York.

Keller, R.P., Drake, J.M. & Lodge, D.M., 2007a, 'Fecundity as a basis for risk assessment of nonindigenous freshwater molluscs', *Conservation Biology* 21, 191–200. https://doi.org/10.1111/j.1523-1739.2006.00563.x

Keller, R.P., Kocev, D. & Džeroski, S., 2011, 'Trait-based risk assessment for invasive species: High performance across diverse taxonomic groups, geographic ranges and machine learning/statistical tools', *Diversity and Distributions* 17, 451–461. https://doi.org/10.1111/j.1472-4642.2011.00748.x

Keller, R.P., Lodge, D.M. & Finnoff, D.C., 2007b, 'Risk assessment for invasive species produces net bioeconomic benefits', *Proceedings of the National Academy of Sciences USA* 104, 203–207. https://doi.org/10.1073/pnas.0605787104

Keller, R.P., Lodge, D.M., Lewis, M.A. & Shogren, J.F. (eds.), 2009, *Bioeconomics of invasive species*, Oxford University Press, New York.

Keller, R.P. & Perrings, C., 2011, 'International policy options for reducing the environmental impacts of invasive species', *BioScience* 61, 1005–1012. https://doi.org/10.1525/bio.2011.61.12.10

Keller, R.P. & Springborn, M., 2014, 'Closing the screen door to new invasions', *Conservation Letters* 7, 285–292. https://doi.org/10.1111/conl.12071

Kolar, C. & Lodge, D.M., 2001, 'Progress in invasion biology: Predicting invaders', *Trends in Ecology and Evolution* 16, 199–204. https://doi.org/10.1016/S0169-5347(01)02101-2

Kumschick, S., Bacher, S., Marková, Z., Pergl, J., Pyšek, P., Vaes-Petignat, S. et al., 2015a, 'Comparing impacts of alien plants and animals using a standard scoring system', *Journal of Applied Ecology* 52, 552–561. https://doi.org/10.1111/1365-2664.12427

Kumschick, S., Gaertner, M., Vilà, M., Essl, F., Jeschke, J.M., Pyšek, P. et al., 2015b, 'Ecological impacts of alien species: quantification, scope, caveats and recommendations', *BioScience* 65, 55–63. https://doi.org/10.1093/biosci/biu193

Kumschick, S. & Richardson, D.M., 2013, 'Species-based risk assessments for biological invasions: Advances and challenges', *Diversity and Distributions* 19, 1095–1105. https://doi.org/10.1111/ddi.12110

Lawson, L.L., Vilizzi, L., Hill, J.E., Hardin, S. & Copp, G.H., 2012, 'Revisions of the Fish Invasiveness Scoring Kit (FISK) for its application in warmer climatic zones, with particular reference to peninsular Florida', *Risk Analysis* 33, 1414–1431. https://doi.org/10.1111/j.1539-6924.2012.01896.x

Leung, B., Roura-Pascual, N., Bacher, S., Heikkila, J., Brotons, L., Burgman, M.A. et al., 2012, 'TEASIng apart alien-species risk assessments: a framework for best practices', *Ecology Letters* 15, 1475–1493. https://doi.org/10.1111/ele.12003

Lodge, D.M., Simonin, P.W, Burgiel, S.W., Keller, R.P., Bossenbroek, J.M., Jerde, C.L. et al., 2016, 'Risk analysis and bioeconomics of invasive species to inform policy and management', *Annual Review of Environment and Resources* 41, 453–488. https://doi.org/10.1146/annurev-environ-110615-085532

Lodge, D.M., Williams, S., MacIsaac, H.J., Hayes, K.R., Leung, B., Reichard, S. et al., 2006, 'Biological invasions: Recommendations for policy and management', *Ecological Applications* 16, 2034–2054. https://doi.org/10.1890/1051-0761(2006)016[2035:BIRFUP]2.0.CO;2

Mandrak, N.E., 1989, 'Potential invasion of the Great Lakes by fish species associated with climate warming', *Journal of Great Lakes Research* 15, 306–316. https://doi.org/10.1016/S0380-1330(89)71484-2

Mandrak, N.E. & Cudmore, B., 2004, *Risk assessment for Asian carps in Canada*, Department of Fisheries and Oceans Canada, Burlington, Ontario.

Marr, S.M., Ellender, B.R., Woodford, D.J., Alexander, M.E., Wasserman, R.J., Ivey, P. et al., 2017, 'Evaluating invasion risk for freshwater fishes in South Africa', *Bothalia* 47(2), a2177. https://doi.org/10.4102/abc.v47i2.2177

Nishida, T., Yamashita, N., Asai, M., Kurokawa, S., Enomoto, T., Pheloung, P.C. et al., 2009, 'Developing a pre-entry weed risk assessment system for use in Japan', *Biological Invasions* 11, 1319–1333. https://doi.org/10.1007/s10530-008-9340-0

Novoa, A., Kumschick, S., Richardson, D.M., Rouget, M. & Wilson, J.R.U., 2016, 'Native range size and growth form in Cactaceae predict invasiveness and impact', *NeoBiota* 30, 75–90. https://doi.org/10.3897/neobiota.30.7253

Perrings, C., Dehnen-Schmutz, K., Touza, J. & Williamson, M., 2005, 'How to manage biological invasions under globalization', *Trends in Ecology & Evolution* 20, 212–215. https://doi.org/10.1016/j.tree.2005.02.011

Pheloung, P., Williams, P.A. & Halloy, S.R., 1999, 'A weed risk assessment model for use as a biosecurity tool evaluating plant introductions', *Journal of Environmental Management* 57, 239–251. https://doi.org/10.1006/jema.1999.0297

Pimentel, D., 2011, *Biological invasions: Economic and environmental costs of alien plant, animal, and microbe species*, CRC Press, Boca Raton, FL.

Pyšek, P. & Richardson, D.M., 2007, 'Traits associated with invasiveness in alien plants: Where do we stand?', in W. Nentwig (ed.), *Biological invasions*, pp. 97–126, Springer-Verlag, Berlin.

Randall, R.P., 2012, *A global compendium of weeds*, 2nd edn., Department of Agriculture and Food, Western Australia.

Richardson, D.M. & van Wilgen, B.W., 2004, 'Invasive alien plants in South Africa: How well do we understand the ecological impacts?', *South African Journal of Science* 100, 45–52.

Shackleton, C.M., McGarry, D., Fourie, S., Gambiza, J., Shackleton, S.E. & Fabricius, C., 2007, 'Assessing the effects of invasive alien species on rural livelihoods: Case examples and a framework from South Africa', *Human Ecology* 35, 113–127. https://doi.org/10.1007/s10745-006-9095-0

Springborn, M.R., Keller, R.P., Elwood, S., Romagosa, C.R., Zambrana-Torrelio, C. & Daszak, P., 2015, 'Integrating risk assessment for invasion and disease risk in live animal trade', *Diversity and Distributions* 21, 101–110. https://doi.org/10.1111/ddi.12281

Springborn, M.R., Romagosa, C.M. & Keller, R.P., 2011, 'The value of nonindigenous species risk assessment in international trade', *Ecological Economics* 70, 2145–2153. https://doi.org/10.1016/j.ecolecon.2011.06.016

Tucker, K.C. & Richardson, D.M., 1995, 'An expert system for screening potentially invasive alien plants in the South African fynbos', *Journal of Environmental Management* 44, 309–338. https://doi.org/10.1016/S0301-4797(95)90347-X

van Wilgen, B.W., Fill, J.M., Baard, J., Cheney, C., Forsyth, A.T. & Kraaij, T., 2016, 'Historical costs and projected future scenarios for the management of invasive alien plants in protected areas in the Cape Floristic Region', *Biological Conservation* 200, 168–177. https://doi.org/10.1016/j.biocon.2016.06.008

van Wilgen, B.W., Richardson, D.M., Le Maitre, D.C., Marais, C. & Magadlela, D., 2001, 'The economic consequences of alien plant invasions: Examples of impacts and approaches to sustainable management in South Africa', *Environmental Development and Sustainability* 3, 145–168. https://doi.org/10.1023/A:1011668417953

14

Fungi and invasions in South Africa

Author:
Alan R. Wood[1] (ORCID)

Affiliation:
[1]Weeds Division, ARC-Plant Protection Research Institute, South Africa

Corresponding author:
Alan R. Wood,
wooda@arc.agric.za

Background: Fungi are a major component of the functioning of all terrestrial ecosystems.

Objectives: To increase awareness of fungi as drivers of ecosystem processes, including invasion biology.

Method: Here, I reviewed the information available regarding fungal invasions of native ecosystems in South Africa in the context of the National Status Report on Biological Invasions.

Results: Only seven fungal species are regulated as invaders (all category 1b) under the National Environmental Management: Biodiversity Act (NEM:BA) A&IS regulations. Four of these species are not yet known to occur in South Africa. Similarly, under the NEM:BA A&IS regulations, two of the four species listed as prohibited (i.e. not present in the country but which would pose a threat if introduced) are already present in the country.

The actual number of alien fungi in South Africa is much greater. A preliminary listing of alien fungal species is made, with a total of 9 pathogenic species known to attack indigenous plants, 11 saprotrophic species, 1 fish pathogen, 23 host-specific pathogens of listed alien terrestrial plants, 61 ectomycorrhizal species and 7 host-specific pathogens deliberately introduced as biological control agents. The majority of fungal species were introduced to South Africa most likely via the introduction of crop plants as passengers, although there are as yet very little details available on pathways of introduction into South Africa.

Conclusion: For almost all aspects considered, it is concluded that there is simply not sufficient data to begin to understand the role and impact of fungal invasions in South Africa.

Introduction

The fungi are a highly diverse eukaryotic kingdom, with approximately 100 000 known species. Estimates of the likely total number of species vary widely, largely due to there being less information available for them in contrast to the relatively well known plant and animal kingdoms. An earlier widely accepted conservative estimate was 1.5 million species (Hawksworth 1991, 2001), although this was revised upwards with time to 5.1 million species (Blackwell 2011). The advent of high-throughput sequencing technology has allowed a more accurate estimation than previously, with an estimate based on soil fungi from 365 localities across the world suggesting fungal diversity would be within the range of 2–3.4 million species (Tedersoo et al. 2014). Thus, the number of described species represents between 2% and 6.7% of the estimated total fungal diversity. Although there are insufficient data available to give an accurate estimate for the number of fungi indigenous to South Africa, plant to fungi ratios generated by international studies would suggest there would be at least 171 500 indigenous species (Crous et al. 2006). In addition, there are a number of groups that were traditionally considered as fungi but are now accepted as belonging to the Amoebozoa (e.g. slime moulds) and Stramenopiles (e.g. water moulds, including downy mildews and *Phytophthora*).

Fungi are important components of every terrestrial ecosystem, as decomposers, food sources, pathogens or mutualists. Despite this importance, it is only recently that a growing body of published research is documenting how important fungi are as drivers of ecosystem function. Fungi, including organisms traditionally considered as fungi (hereafter included in the general term 'fungi'), are poorly known both in terms of their biodiversity and their ecology relative to what is known about plants and animals (Desprez-Loustau et al. 2007). Likewise, fungi have been little considered in invasion biology, despite numerous international invasions by fungal pathogens causing devastating economic and ecological damage (Fisher et al. 2012; Loo 2009). In addition to increased research on the threat posed to indigenous organisms (plants, animals and microbes) by invasive fungal pathogens, there is also a need for research on the ecological impact of invasive saprobic and mutualist fungi, and on the impact of other invasive organisms on indigenous fungi and the ecological services they provide (Desprez-Loustau et al. 2007;

Litchman 2010). Many fungi grow and reproduce rapidly, traits associated with invasion by a range of organisms. An extreme demonstration of their ability to produce large numbers of spores is the 2.175×10^9 spores day^{-1} cm^{-2} of exposed spore-producing surface recorded for the bracket fungus *Trametes pubescens* (Schumach.) Pilát (Rockett & Kramer 1974). Intercontinental natural dispersal of spores by wind is well known in fungi, which led to a widespread belief that fungi were largely cosmopolitan, their distribution only limited by the availability of suitable environmental conditions. However, there is now an appreciation that many fungi are endemic to particular areas (Desprez-Loustau et al. 2007; Litchman 2010), and that many have been transported around the world by humans and have invaded new geographic ranges (Desprez-Loustau et al. 2007; Fisher et al. 2012; Litchman 2010; Loo 2009; Roy et al. 2014; Wingfield et al. 2001).

A number of reviews from an international perspective have been published on fungal invasions (Desprez-Loustau 2007; Fisher et al. 2012; Litchman 2010; Loo 2009), mutualists and plant invasion (Nuñez & Dickie 2014; Richardson et al. 2000), population biology (Gladieux et al. 2015) and soil mycobiota and plant invasion (Dawson 2015; Dawson & Schrama 2016; Reinhart & Callaway 2006). This review focusses on what is currently known about fungi and invasion in South Africa. In particular, invasion as relevant to the National Environmental Management: Biodiversity Act (Act No. 10 of 2004) (NEM:BA). A simple matrix illustrates various possible interactions between alien and indigenous fungi with alien and indigenous plants (or ecological communities) (Figure 1). This review is organised in a manner reflecting this matrix.

In South Africa, pathogens of crop plants have been well documented (Crous, Phillips & Baxter 2000; Doidge & Bottomley 1931; Doidge et al. 1953; Gorter 1977); however, they have not received attention as invaders of natural ecosystems. Doidge (1950) noted that with the introduction of crop plants into South Africa, there have been many introductions of pathogens of these crop plants, and that many were widespread and occurred wherever their hosts were cultivated. Indications are that the rate of introductions is increasing (Wingfield et al. 2001). These pathogens may potentially infect indigenous plants and invade natural ecosystems, yet it is only in the last few years that this potential threat is beginning to be assessed (e.g. Adams, Roux &

Wingfield 2006; Cruywagen et al. 2016; Mehl et al. 2016). Many other non-pathogenic fungi have also been introduced, but no consideration has been given to these as invaders or to any potential negative impacts on indigenous ecosystems.

For the purpose of this review, the large literature on pathogens of crop and plantation plants in South Africa is not considered unless these fungi are known to affect indigenous biodiversity. This in no way implies that fungal pathogens of crop plants are not important, but control of these disease-causing organisms falls under the jurisdiction of the Plant Health and Animal Health directorates of the Department of Agriculture, Forestry and Fisheries (DAFF) and is regulated by the Agricultural Pests Act, (Act No. 36 of 1983). There is a need to investigate the spill over of introduced crop pathogens into natural ecosystems. Aquatic and marine fungi, and fungi introduced for food production, are not considered in this review. Despite an increasing awareness of invasive fungal diseases of vertebrate wildlife (Allender et al. 2015; Berger et al. 2016; Tomkins et al. 2015), only one report from South Africa was found, namely a disease-causing water mould, *Aphanomyces invadans* Willoughby, Roberts & Chinabut, causal agent of mycotic granulomatosis in fish. This has been found in widespread localities in southern Africa, including the Western Cape province (Huchzermeyer & Van der Waal 2012). Fungal pathogens of vertebrates (including humans) are also not further considered in this review, although there is a clear need for vigil against potential invaders and studies on species already present.

Diversity of indigenous fungi

Knowledge of indigenous biodiversity is fundamental to being able to evaluate whether an organism is indigenous or alien (Desprez-Loustau et al. 2007). In South Africa, there has been a long tradition of documenting indigenous fungi, starting with European plant collectors who explored South Africa such as Carl Peter Thunberg (1743–1828) and William John Burchell (1781–1863) (Doidge 1950; Rong & Baxter 2006). The National Collection of Fungi (PREM) houses less than 70 000 specimens, which includes all historical collections that were previously housed in other herbaria in the country, as well as specimens from other continents, although much of the early collections (including type specimens) are housed in various European herbaria (Rong & Baxter 2006). Doidge (1950) remains the only comprehensive

Ecosystem interactions between Indigenous fungi and Indigenous plants	Alien fungi and Indigenous plants (p. 8, Table 2)	Host Plant Indigenous
• Decomposers	• Naturalised non-host specific pathogens (p. 8)	
• Growth promotion (positive plant–soil microbiome feedback)	• Saprotrophic fungi (p. 9)	
• Mutualists (mycorrhizae, endophytes)		
• Pathogens		
Indigenous fungi and Alien plants (p. 10)	Alien fungi and Alien plants (p. 5, Tables 3 and 4)	Host Plant Alien
• Mutualists (VA mycorrhizae) facilitating invasion (p. 10)	• Ectomycorrhizae (p. 6, Table 4)	
• Accumulation of local pathogens (p. 5)	• Intentionally introduced biological control agents (p. 5, Table 3)	
• Changes of soil microbiome communities in favour of invader plants (p. 5)	• Naturalised host specific pathogens (p. 5, Table 3)	
• Biotic resistance (p. 5)		
• New association biological control agents (p. 10, Table 3)		
Fungus Indigenous	**Fungus Alien**	

FIGURE 1: Schematic of interactions between Indigenous and alien fungi and their host plants (or ecological community). These interactions can have positive (no shading), negative (dark grey shading) or unknown (light grey shading) impacts on indigenous biodiversity. The relevant section (and sub-sections) of this review are indicated.

listing of fungi recorded from southern Africa. Many of the listed fungi have so far only been recorded as pathogens of introduced crop plants. A total of 4748 species were listed (Doidge 1950), only a small fraction of the total expected (~171 500, Crous et al. 2006). Even with an increase in the rate of species described from South Africa since 1950, particularly since the advent of molecular phylogenetic techniques, the number of named indigenous species is only a small fraction (~3%) of the likely total diversity.

Indigenous fungi remain little documented, other than certain groups of plant pathogens, macrofungi and lichens. For example, arbuscular mycorrhizae (AM) are common mutualists of South African plants (Allsopp & Stock 1993; Gaur et al. 1999; Hawley & Dames 2004; Straker, Weiersbye & Witkowski 2007) and are considered essential to ecosystem functioning (van der Heijden et al. 2015). To date only five species have been named from indigenous plants in South Africa (Blaszkowski et al. 2010; Gaur et al. 1999), and a further four species have been recorded from cassava (Straker, Hilditch & Rey 2010). In contrast, a total of 43 species have been recorded from semi-arid regions in Namibia (Stutz et al. 2000; Uhlmann et al. 2004) and 147 virtual taxa were obtained by 454 pyrosequencing from Gorongosa National Park, Mozambique (Rodriguez-Echeverria et al. 2016), suggesting that most of the indigenous biodiversity of this group of fungi is yet to be documented in South Africa. Many such examples from a wide variety of major taxa could be cited.

It can be concluded that, despite efforts by the few mycologists who have operated in South Africa, baseline biodiversity information is too little to be of any use in determining the alien or indigenous status of fungi newly discovered in South Africa, except in some specific cases.

Alien fungi

The human-assisted distribution of fungi around the world has been, and remains, the most important pathway of introduction (Anderson et al. 2004; Desprez-Loustau et al. 2007; Fisher et al. 2012; Roy et al. 2014; Wingfield et al. 2001). The trade in live plants is an important route of introduction to, and further spread within, continents and countries (e.g. Carnegie & Cooper 2011; Moralejo et al. 2009; Roy et al. 2014). Although there is a general trend amongst pests and pathogens of crop plants introduced around the world that dispersed range increases with wider host range, introduced fungi have the narrowest host range but the widest introduced range (Bebber, Holmes & Gurr 2014), suggesting that many fungus species are well adapted for human-assisted dispersal and invasion. Propagule characteristics related to successful long-distance dispersal, allowing for rapid spread once introduced, were good predictors of invasion success by fungi (Philibert et al. 2011). As a result of several hundred years of introducing horticultural, crop and plantation plants, and plant products, it is likely that large numbers of fungi have been introduced to South Africa.

There is no literature available specifically on fungi and natural ecosystem invasions in South Africa. The first

attempt at listing invasive fungi was published in the NEM:BA A&IS regulations (Notice No. 599, 1 Aug. 2014), which lists a total of seven fungi (list 11 of the regulations), all placed in category 1b (Table 1). Of these, three are recorded only from species of *Eucalyptus*, and two are recorded only from *Pinus* species. In addition, four of these species have not yet been recorded in South Africa (Farr & Rossman undated). Thus, this list does not in any way reflect alien microbial species important in natural ecosystems in South Africa. Only one of the listed species is well known as a pathogen of indigenous plants (*Phytophthora cinnamomi* Rands; Lübbe & Mostert 1991; Nesamari, Coutinho & Roux 2016; Von Broembsen 1984a; Von Broembsen & Kruger 1985). Rather this reflects the reality that there is not a single researcher employed in South Africa with the specific mandate to undertake studies on alien fungi in natural ecosystems, nor has there been one. What is known is scattered piecemeal in the mycological and plant pathological literature. An attempt is made here to start to bring some of this literature together. Information provided here can be used as a springboard to compile future lists that take into account both invasive status and whether a National Management Programme could be developed and implemented for invasive species, as required by the legislation for each listed organism.

How can fungi be recognised as alien? In general, host-specific pathogenic or mutualistic fungi can be recognised as alien when their hosts are themselves alien. However, few fungal species that can be recognised as introduced have been assessed as to whether they attack indigenous plants or invade natural ecosystems. For example, the polypore *Laetiporus sulphureus* (Bull.) Murrill is well known as a pathogen of various Northern Hemisphere trees in their native and introduced ranges, but it has also been recorded on the native *Olea europaea* L. ssp. *africana* (Mill.) P.S. Green in South Africa (Doidge 1950). This old record suggests

TABLE 1: Fungi and fungus-like organisms listed in NEM:BA Alien and Invasive Species Lists, 2014.

Species	Category	Recorded hosts†	Present in South Africa†
National lists of invasive species in terms of section 70(1)			
Tetratosphaeria destructans (as *Kirramyces destructans*)	1b	*Eucalyptus* spp.	Yes‡
Teratosphaeria eucalypti (as *Kirramyces eucalypti*)	1b	*Eucalyptus* spp.	No
Phytophthora kernoviae	1b	Various	No
Phytophthora pinifolia	1b	*Pinus radiata*	No
Phytophthora cinnamomi	1b	Various	Yes
Teratosphaeria cryptica	1b	*Eucalyptus* spp.	No
Fusarium circinatum	1b	*Pinus* spp.	Yes
List of prohibited alien species in terms of section 67(1)			
Fusarium oxysporum f. sp. *cubense*	-	*Musa*	Yes§
Phytophthora ramorum	-	Various	No
Puccinia psidii	-	Myrtales	Yes¶
Nosema ceranae	-	Bees	No††

Source: Government Gazette No. 37886(3), Notice No. 599, 1 Aug. 2014
†, Farr & Rossman (undated) http://nt.ars-grin.gov/fungaldatabases/
‡, First recorded in 2015 (Greyling et al. 2016).
§, Recorded as present by the 1940s (Doidge 1950).
¶, First recorded in 2013 (Roux et al. 2013).
††, Strauss et al. (2013).

that it might also damage native trees, but this requires confirmation. Non-specific pathogens, saprotrophs, endophytes and mutualistic fungi are more difficult to assess as alien. Morphologically characteristic macrofungi could be identified as being alien if they were not recorded in the early South African literature. For example, the Australian *Aseroë rubra* Labill. was only confirmed as present in South Africa since the 1950s (Coetzee 2010). Likewise, the European *Coprinopsis picacea* (Bull.) Redhead, Vilgalys & Moncalvo, was not listed by Doidge (1950) or Pearson (1950) but has recently been reported on the i-spot website (http://www.ispotnature.org/communities/southern-africa). Detailed molecular analyses comparing genetic diversity within South African isolates, or comparing local genotypes to overseas ones, will be necessary to identify most alien species. The European *Armillaria mellea* (Vahl) P. Kumm. and *Armillaria gallica* Marxm. & Romagn. were identified as introduced by such detailed studies (Coetzee et al. 2001, 2003). Therefore, only a few fungi can be singled out which are definitely alien and have naturalised in South Africa (Table 2), and even fewer have been recognised as having a detrimental impact on our natural environments.

Fungi and theories of invasion biology

Many hypotheses have been put forward to explain why and how species become invasive after introduction to a region where they are not native (Catford, Jansson & Nilsson 2009). Fungi are considered to have positive or negative impacts on invasive plants in various of these hypotheses (Catford et al. 2009), although as yet there has been no consideration of which of the hypotheses are relevant directly to fungi as invaders.

Recently there has been an interest in the interaction of soil biota, including fungi other than mycorrhizas, and invasive plants. Soil microbiota may have a direct positive (Cui & He 2009; Gundale et al. 2014), negative (Lankau 2010; Vitullo et al. 2014) or no impact (Bennett & Strauss 2013; Birnbaum & Leishman 2013) on the growth and invasiveness of alien plants. However, these interactions are dependent on many variables including the ecologically functional groups of the microorganisms considered, as well as the surrounding vegetation and interactions with other organisms (Reinhart & Callaway 2006). Phylogenetically close plant relatives may produce a selective pressure resulting in similar soil microbiota on their roots (Wehner et al. 2014), which can facilitate invasion by alien close relatives more than the microbiota associated with non-relatives (Hill & Kotanen 2012; Reinhart & Callaway 2006). Indirect interactions also occur, a positive soil microbe–plant feedback has been shown to increase growth of *Impatiens glandulifera* Royle and increase foliar endophyte fungus levels that may provide enhanced protection from herbivores (Pattison et al. 2016). In general, native plants grow better in 'home' soil whereas invasive plants are more catholic and less affected by the soil source (i.e. derived from below native or invasive plants). Overcoming the native plants' 'home' soil advantage by invasive plants is dependent on exceeding a threshold of

TABLE 2: A preliminary list of fungi and fungus-like organisms that are naturalised in South Africa, including species recorded as pathogens of plants indigenous to South Africa or saprotrophs that are considered to be native to other continents.

Fungus	Order	Life style	Origin	References
Ascomycota				
Ceratocystis pirilliformis	Microascales	PP	Australia	Lee et al. 2016
Helvella crispa	Pezizales	S	Europe	Gryzenhout 2010 (as *Pseudocraterellus undulatus*)
Helvella lacunosa	Pezizales	S	Europe	Doidge 1950 (as *H. mitra*)
Morchella esculenta	Pezizales	S	Europe	Doidge 1950 (as *M. conica*)
Ophiostoma quercus	Ophiostomatales	PP	N. Hemisphere	Kamgan et al. 2008
Badidiomycota				
Armillaria mellea†	Agaricales	PP	Europe	Coetzee et al. 2001; Wingfield et al. 2010
Armillaria gallica†	Agaricales	PP	Europe	Coetzee et al. 2003; Wingfield et al. 2010
Aseroë rubra	Phallales	S	Australia	Coetzee 2010
Chlorophyllum rhacodes	Agaricales	S	N. Hemisphere	van der Westhuizen & Eicker 1994
Clathrus archeri	Phallales	S	Australia	Bottomley 1948; Coetzee 2010
Coprinopsis picacea	Agaricales	S	Europe	i-spot‡
Erythricium salmonicolor†	Corticales	PP	Asia?	Roux & Coetzee 2005
Hygrocybe nigrescens	Agaricales	S	Europe	Doidge 1950
Ileodicttyon gracile	Phallales	S	Australia	Bottomley 1948; Coetzee 2010
Lysurus cruciatus	Phallales	S	Australia?	Bottomley 1948; Coetzee 2010
Macrolepiota procera	Agaricales	S	N. Hemisphere	Doidge 1950
Puccinia lagenophorae†	Pucciniales	PP	Australia	Scholler et al. 2011
Puccinia psidii†	Pucciniales	PP	South America	Roux et al. 2013; Roux et al. 2016
Oomycota				
Aphanomyces invadans	Saprolegniales	Fish path.	Asia	Huchzermeyer & Van der Waal 2012
Phytophthora cinnamomi†	Peronosporales	PP	Asia?	Von Broembsen 1984a Von Broembsen & Kruger 1985
Pythium irregulare†	Peronosporales	PP	?	Bahramisharif et al. 2014

Ectomycorrhizal fungi and pathogens of introduced crop and plantation plants are excluded.
S, saprotroph; PP, plant pathogen.
†, Recorded as causing disease symptoms or death of indigenous plants.
‡, http://www.ispotnature.org/communities/southern-africa

abundance, external disturbances or having a competitive or dispersal advantage relative to the native plants (Suding et al. 2013).

Invasive plants alter the soil microbiota composition (Elgersma & Ehrenfeld 2011), altering abiotic soil properties such as pH and nutrient levels (Duchicela et al. 2012; Kourtev, Ehrenfeld & Häggblom 2003). In turn, these altered soil conditions affect indigenous plants negatively, although the level of impact is species specific (Scharfy et al. 2010).

Differences in above-ground and below-ground fungal pathogens occur between the native and invaded range of alien plants. The Enemy Release hypothesis suggests that plants succeed as invaders because they have been introduced without their herbivores and pathogens. The practice of classical biological control, the introduction of co-evolved host-specific pathogens (or other organisms), is based on this hypothesis. Release from soil pathogens (Reinhart et al. 2010; Van Grunsven et al. 2009), as well as from generalist foliar and seed pathogens (Halbritter et al. 2012), have also been reported as important factors in the invasion by introduced plants. However, enemy escape also occurs on a local scale in the native range of invasive plants (MacKay & Kotanen 2008) and is not correlated with invasiveness of many plants (van Kleunen & Fischer 2009). Also, escape from one functional guild of pathogens can be counteracted by non-escape from other guilds (Agrawal et al. 2005), and the accumulation of pathogens native to the invaded range which damage the invasive plant may counter benefits derived from the initial enemy release (Flory & Clay 2013; Flory, Kleczewski & Clay 2011; Vitullo et al. 2014). Native pathogens that limit or reduce initial invasion are considered to be a Biotic Resistance mechanism (Catford et al. 2009). Contrary to this last premise, the Accumulation of Local Pathogens hypothesis suggests that if the invasive plant is less susceptible than native plants, accumulating high levels of local generalist pathogens increases invasion success by suppressing native plant growth (Eppinga et al. 2006; Mangla, Inderjit & Callaway 2008). Dawson (2015) and Dawson and Schrama (2016) provide a theoretical framework incorporating these hypotheses and plant–soil feedbacks to understand how changes in soil mycobiota facilitate invasion by plants.

Plant–soil microbe interactions or invasion processes may apply inconsistently to individual plant species across their invasive range or change with time since invasion. In California (USA), Ammophila arenaria (L.) Link accumulated local pathogens suppressing native plants (Eppinga et al. 2006), whereas in South Africa this was not the case. Rather, Biotic Resistance associated with native grasses was an important determinant of where this plant was more or less invasive (Knevel et al. 2004). Release from the soil pathogens in its native range, South Africa – Enemy Release – has been demonstrated to be important to Carpobrotus edulis (L.) L. Bolus invasion of the Mediterranean (Van Grunsven et al. 2009) or not (de la Peña et al. 2010). In the latter study, Biotic Resistance was shown to be an important initial factor but

with time this was overcome and the plant modified the soil microbiota in its favour. Native fungi accumulated on roots of Vincetoxicum rossicum (Kleopow) Pobed. have variously been shown to increase its growth or not, and to have negative impacts on only some native plants (Day, Dunfield & Antunes 2015, 2016).

Modifying the soil biota to favour the invader plant and be deleterious to native plants may also produce phenotypic changes in the invader. Ageratina adenophora Spreng. modifies the soil microbiota in China, by reducing AM fungi and increasing the ratio of bacteria to fungi to the detriment of native plants (Niu et al. 2007; Xingjun et al. 2005). In addition, plants in the China had higher leaf nitrogen levels and surface area compared with plants from the native range (Feng et al. 2011). The accumulation of native Fusarium root pathogens benefitted Chromolaena odorata (L.) R.M. King and H. Rob. in India (Mangla et al. 2008) but not in South Africa (te Beest et al. 2009). However, in the latter study, plants increased stem biomass and height growth when grown in unsterilised soil from the invaded range (te Beest et al. 2009). The above few studies of fungal interactions with alien plants in South Africa need to be expanded, and other invasive plants included, to determine whether there are any generalities of invasion mechanisms peculiar to South Africa. Comparing mechanisms of invasion in South Africa with that occurring elsewhere in the world would provide useful information; for instance, does A. ageratina also benefit from modifying the soil microbiota here as it does in China? And why does soil biota affect the invasion of A. arenaria and C. odorata differently in South Africa compared with other parts of the world?

Alien fungi and alien plants
Host-specific pathogens

Amongst the numerous pathogenic fungi accidentally introduced along with their alien host plants (including crop, plantation, horticultural and accidentally introduced invasive species) are species that attack invasive plants. Some of these, primarily host-specific pathogens, may well help to suppress the populations of their hosts, providing a free biological control service (Table 3). The list provided is not complete, as the plants listed are primarily declared weeds (NEM:BA A&IS) and could be greatly expanded if other alien plants are included (e.g. Uromyces bidenticola Arthur and Entyloma bidentis Henn. are both common on Bidens pilosa L.).

The deliberate introduction of alien fungi as biological control agents of their co-evolved host plants began in South Africa with the release of Uromycladium tepperianum (Sacc.) McAlpine against Acacia saligna (Labill.) Wendl. in 1987. To date, a total of nine fungi have been approved for release, permission for release of two was obtained in late 2015 and efforts are currently underway to release these in the field. Of the other fungi released, five established well whereas two failed to establish. Two of the established fungi have a considerable impact on their host's population densities

TABLE 3: Fungi recorded on listed alien weeds that either are considered indigenous to South Africa (in bold) or were deliberately (as biological control agents) or probably accidentally introduced with their host.

Plant	Cat.†	Fungi recorded	References
Acacia cyclops	1b	*Calonectria scoparia*	Schoch et al. 1999
		Pseudolagarobasidium acaciicola	Wood & Ginns 2006
Acacia longifolia	1b	*Calonectria scoparia*	Hagemann & Rose 1988
Acacia mearnsii‡	2	**Ceratocystis albifundus**	Morris, Wingfield & De Beer 1993; Morris et al. 1999;
		Cylindrobasidium laeve§	Morris et al. 1999
		Uromycladium acaciae	McTaggart, Doungsa-ard & Wingfield 2015
Acacia saligna	1b	*Uromycladium tepperianum¶*	Zachariades et al. 2017
Ageratina adenophora	1b	*Passalora ageratinae¶*	Zachariades et al. 2017
Ageratina riparia	1b	*Entyloma ageratinae¶*	Zachariades et al. 2017
Ageratum conyzoides	1b	*Passalora perfoliati*	Morris & Crous 1994
Campuloclinium macrocephalum	1b	*Puccinia eupatorii¶*	Zachariades et al. 2017
Canna indica	1b	*Puccinia thaliae*	van Jaarsveld, Kriel & Minaar 2006
Convolvulus arvensis	1b	*Septoria* sp.	unpublished data ARC-PPRI
Dolichandra unguiscati	1b	**Cercosporella dolichandrae**	Crous et al. 2014
Echium plantagineum	1b	*Cerospora echii*	Morris & Crous 1994
Eichornia crassipes	1b	*Acremonium zonatum*	Morris et al. 1999
		Alternaria eichhorniae	Morris et al. 1999
		Cercospora piaropi	Morris 1990; Morris et al. 1999
		Cercospora rodmanii¶	Morris 1990; Morris et al. 1999
		Myrothecium roridum	unpublished data ARC-PPRI
Hakea sericea	1b	**Colletotrichum acutatum§**	Morris 1991; Morris et al. 1999
Lantana camara	1b	**Mycovellosiella lantaniphila**	Morris & Crous 1994
		Pseudocercospora formosana	Crous & Braun 1996
Malva arborea (as *M. dendromorpha*)	1b	*Puccinia malvacearum*	Doidge 1950
Paraserianthes lophantha	1b	*Uromycladium* sp. ¶	Zachariades et al. 2017
Parthenium hysterophorus	1b	*Puccinia abrupta* var. *partheniicola*	Wood & Scholler 2002
		Puccinia xanthii¶	Zachariades et al. 2017
Pennisetum clandestinum	(1b)	*Phakopsora apoda*	Adendorff & Rijkenberg 1995
Pistia stratiotes	1b	*Cercospora pistiae*	Morris & Crous 1994
Populus X canescens	2	*Melampsora* sp.	unpublished data ARC-PPRI
Prosopis hybrid	1b/3	**Coniothyrium prosopidis**	Crous, Groenewald & Wood 2013
		Peyronellaea prosopidis	Crous et al. 2013
Pueraria montana	1a	*Phakopsora pachyrhizi*	Pretorius, Visser & du Preez 2007
Ricinus communis	2	*Melampsora ricini*	Doidge 1950
Rubus cuneifolius	1b	**Kuehneola uredinis**	Morris et al. 1999
Rubus fruticosus	2	*Pseudocercospora rubi*	Crous & Braun 1996
Sesbania punicea	1b	*Erysiphe pisi*	Gorter 1993
Solanum mauritianum	1b	*Mycovellosiella brachycarpa*	Crous & Braun 1996
Xanthium strumarium	1b	*Puccinia xanthii*	Pretorius, van Wyk & Kriel 2000

Many pathogens have been recorded on *Eucalyptus* spp. and *Pinus* spp., including species listed as declared weeds in South Africa; these fungi are not listed here as the majority are known from commercial plantations or nurseries only.

Listing is restricted to those species that have been observed to be damaging at least some times or are at least widespread in occurrence.

†, Category according to NEM:BA A&IS.

‡, Many more pathogens have been recorded on Acacia mearnsii in South Africa from plantations; these are not listed here.

§, Exploited as biological control agents.

¶, Pathogens deliberately introduced to South Africa and which were successfully established as biological control agents.

(*U. tepperianum, Entyloma ageratinae* R.W. Barreto & H.C. Evans). Reviews of biological control of invasive alien plants using plant pathogens in South Africa have been published (Morris 1991; Morris, Wood & den Breeÿen 1999). See Zachariades et al. (2017) for a recent review of the effectiveness of biological control in general in South Africa.

Ectomycorrhizal fungi

The role of mycorrhizas in plant invasions has recently been reviewed (Nuñez & Dickie 2014; Richardson et al. 2000). The majority of invasive plants in South Africa are AM or ectomycorrhizal (EM); some invasive plants are however non-mycorrhizal (e.g. *Hakea* species) (Richardson et al. 2000).

Pinus and *Eucalyptus* are both obligatory EM plant genera, and a large number of EM fungi have been introduced into South Africa, either deliberately or accidentally (Vellinga, Wolfe & Pringle 2009), allowing the widespread cultivation of these plantation plants. Vellinga et al. (2009) listed 45 species from South Africa, although an additional 17 species have also been recorded (e.g. Hawley, Taylor & Dames 2008; Martin et al. 2002; van der Westhuizen & Eicker 1994) (Table 4), and therefore the total number of introduced EM fungi may be more than currently known.

Pines are recognised as being important invaders in mountain catchment areas in South Africa. This invasion is a dual invasion by the plants, which are obligately EM and their

TABLE 4: Introduced ectomycorrhizal fungi recorded from South Africa.

Fungus	Host tree genus	Natural distribution	References
Albatrellus ovinus	*Pinus*	N. Hemisphere	Hawley et al. 2008
Amanita excelsa	*Pinus*	Europe	Vellinga et al. 2009
Amanita marmorata	*Eucalyptus*	Australia	Vellinga et al. 2009
Amanita muscaria	*Pinus, Quercus*	N. Hemisphere	Vellinga et al. 2009
Amanita pantherina	*Pinus, Quercus*	N. Hemisphere	Vellinga et al. 2009
Amanita phalloides	*Quercus, Populus, Pinus*	Europe, Asia	Vellinga et al. 2009
Amanita rubescens	Various	Europe	Vellinga et al. 2009
Amanita spissa	*Pinus*	N. Hemisphere	Hawley et al. 2008
Astraeus hygrometricus	*Pinus*	N. Hemisphere	Vellinga et al. 2009
Boletus aestivalis	*Pinus*	Europe	van der Westhuizen & Eicker 1994
Boletus edulis	*Pinus*	N. Hemisphere	Vellinga et al. 2009
Buchwaldoboletus hemichrysus	*Pinus*	North America	Vellinga et al. 2009
Chalciporus piperatus	*Pinus*	N. Hemisphere	Vellinga et al. 2009
Clavulina cristata	*Pinus*	N. Hemisphere	Vellinga et al. 2009
Clitopilus prunulus	*Pinus*	N. Hemisphere	van der Westhuizen & Eicker 1994
Gyroporus castaneus	*Quercus*	N. Hemisphere	van der Westhuizen & Eicker 1994
Hebeloma crustuliniforme	*Pinus*	N. Hemisphere	Vellinga et al. 2009
Hebeloma cylindrosporum	*Pinus*	Europe	van der Westhuizen & Eicker 1994
Inocybe curvipes	*Pinus*	Europe	Vellinga et al. 2009
Inocybe euthelea	*Pinus*		van der Westhuizen & Eicker 1994
Itajahya galericulata	*Jacaranda*	S. America	van der Westhuizen & Eicker 1994
Laccaria fraterna	*Eucalyptus*	Australia	Vellinga et al. 2009
Laccaria laccata	*Pinus*	N. Hemisphere	Vellinga et al. 2009
Lactarius deliciosus	*Pinus*	N. Hemisphere	Vellinga et al. 2009
Lactarius hepaticus	*Pinus*	N. Hemisphere	van der Westhuizen & Eicker 1994
Leccinum duriusculum	*Populus*	Europe	Vellinga et al. 2009
Paxillus involutus	Various	Europe	Vellinga et al. 2009
Pisolithus microcarpus	*Eucalyptus*	Australia	Martin et al. 2002
Pisolithus arhizus	*Pinus*	Spain	Martin et al. 2002
Pisolithus tinctorius	*Pinus, Quercus*	N. Hemisphere	Martin et al. 2002
Rhizopogon luteolus	*Pinus*	Europe	Vellinga et al. 2009
Rhizopogon roseolus	*Pinus*	Europe	Vellinga et al. 2009
Rhizopogon rubescens	*Pinus*	North America	van der Westhuizen & Eicker 1994
Russula caerulea	*Pinus*	Europe	van der Westhuizen & Eicker 1994
Russula capensis	*Pinus*	unknown	Vellinga et al. 2009
Russula cyanoxantha	*Pinus*	N. Hemisphere	Vellinga et al. 2009
Russula fallax sensu Cooke	*Pinus*	Europe (?)	Vellinga et al. 2009
Russula grisea	*Pinus*	N. Hemisphere	Vellinga et al. 2009
Russula pectinata	*Pinus*	Europe, North America	Vellinga et al. 2009
Russula sardonia	*Pinus*	Europe	Vellinga et al. 2009
Russula sororia	*Quercus, Pinus*	Europe, North America	Vellinga et al. 2009
Russula xerampelina	*Pinus*	N. Hemisphere	Vellinga et al. 2009
Scleroderma cepa	*Pinus*	N. Hemisphere	Vellinga et al. 2009
Scleroderma citrinum	*Pinus*	N. Hemisphere	van der Westhuizen & Eicker 1994
Scleroderma verrucosum	*Pinus*	N. Hemisphere	Vellinga et al. 2009
Suillus bellinii	Various	Europe	Vellinga et al. 2009
Suillus bovinus	*Pinus*	Europe, Asia	Vellinga et al. 2009
Suillus granulatus	*Pinus*	N. Hemisphere	Vellinga et al. 2009
Suillus luteus	*Pinus*	N. Hemisphere	van der Westhuizen & Eicker 1994
Suillus salmonicolor	*Pinus*	North America, Asia	Vellinga et al. 2009
Thelephora intybacea	*Pinus*	Europe	Vellinga et al. 2009
Thelephora penicillata	*Pinus*	Europe	Vellinga et al. 2009
Thelephora terrestris	*Pinus*	N. Hemisphere	Vellinga et al. 2009
Tricholoma albobrunneum	Various	Europe	Vellinga et al. 2009
Tricholoma eucalypticum	*Eucalyptus*	Australia	Vellinga et al. 2009
Tricholoma meridianum	*Pinus*	unknown	Vellinga et al. 2009
Tricholoma saponaceum	Various	N. Hemisphere	Vellinga et al. 2009
Tricholoma ustale	Various	Europe	Vellinga et al. 2009
Tuber rapaeodorum	*Pinus*	Europe	Vellinga et al. 2009
Xerocomus badius	Various	Europe, Asia	Vellinga et al. 2009
Xerocomus chrysenteron	Various	N. Hemisphere	Vellinga et al. 2009

associated EM fungi (Collier & Bidartondo 2009; Dickie et al. 2010; Hynson et al. 2013). Initially, the fungal inoculum available is low, slowing the invasion of the plants. However, with time, the pines and their EM fungi proliferate and replace the indigenous vegetation and their mycorrhizal fungi (Collier & Bidartondo 2009). Currently, there have been no studies investigating which EM fungi are co-invading South Africa's natural ecosystems; this may be a small subset of the total diversity of introduced species (Dickie et al. 2010; Hynson et al. 2013).

It has been found that usually the EM fungal communities on alien plantation trees consist of introduced species normally associated with those plants in their native habitats, even when indigenous EM fungi are present in the surrounding natural ecosystems (Dickie et al. 2010; Jairus et al. 2011). Yet, several EM fungi have been recorded as invading natural ecosystems, including *Amanita muscaria* (L.) Lam. and *Amanita phalloides* (Vaill. ex Fr.) Link (Dunk, Lebel & Keane 2012; Pringle et al. 2009). The presence of indigenous EM fungi and host plants potentially allow for host shifts by introduced EM fungi and invasion of natural ecosystems, given time to adapt to local conditions (Jairus et al. 2011).

Approximately 400 species of indigenous putatively EM fungi have been reported from tropical Africa (Verbeeken & Buyck 2002). In South Africa, the only indigenous fungi known to be EM symbionts are the desert truffles (*Kalaharituber pfeilii* (Henn.) Trappe & Kagan-Zur, *Eremiomyces echinulatus* (Trappe & Marasas) Trappe & Kagan-Zur, *Mattirolomyces austroafricanus* (Marasas & Trappe) Kovács, Trappe & Claridge) (Trappe et al. 2008). Introduced EM fungi likely pose little direct threat to natural ecosystems in South Africa, although north of South Africa's borders there is a potential threat. Controlling invading pines, or other EM plants, would control any invasion by their associated EM fungi. However, the co-invasion of EM plants and associated EM fungi can reduce indigenous mycorrhizal fungi (Becklin, Pallo & Galen 2012). Once reduced or lost, it may take many years for the indigenous AM fungi to recover (Lankau et al. 2014), reducing the ability of AM plants to recover post control of the invaders. Many fynbos plants are associated with AM fungi (Allsopp & Stock 1993). In a study conducted in South Africa, a single cycle of invasion (by ectomycorrhizal *Pinus pinaster* Aiton and the non-mycorrhizal *Hakea sericea* Schrad. & J.C.Wendl.) did not reduce the development of AM fungi in indigenous plants following clearing, likely due to the persistence of AM plants at low densities within the invaded area during the invaded period (Allsopp & Holmes 2001). However, the impact of successive cycles of invasion has not been studied. The impact would likely be greatest on ericoid mycorrhizae as these are very slow to re-establish (Allsopp & Holmes 2001).

Alien fungi and indigenous plants
Non-host specific pathogens

Amongst the many fungi introduced to South Africa are those which are not host specific, but with a broader host range they may potentially attack indigenous plants, either closely related to their natural host or to a wide range of phylogenetically unrelated plants. Internationally, severe ecological impacts have been recorded by a wide range of fungal pathogens (Desprez-Loustau et al. 2007; Fisher et al. 2012). *P. cinnamomi* is one of the most serious invasive alien pathogens of natural ecosystems, as well as a pathogen of numerous crop species, around the world including in South Africa. It is the causal organism of Jarrah Dieback in Western Australia, killing a broad range of plants. Widespread in Australia, it kills many plant species in various ecosystems (Cahill et al. 2008) and is also associated with disease of naturally growing trees in other parts of the world (e.g. Balci et al. 2007; Camilo-Alves, Clara & Ribeiro 2013; Vettraino et al. 2005). In South Africa, it is the causal organism of a root and crown rot of *Leucadendron argenteum* (L.) R. Br. (van Wyk 1973), dieback of *Ocotea bullata* (Burch.) E. Meyer (Lübbe & Mostert 1991) and recently has been found to have caused some mortality of *Encephalartos transvenosus* Stapf & Burtt Davy (Nesamari et al. 2016). It has been recorded as causing mortality of a wide range of indigenous fynbos plants, especially members of the Proteaceae (Von Broembsen 1984a; Von Broembsen & Brits 1985; Von Broembsen & Kruger 1985), including in remote pristine mountain fynbos (Von Broembsen 1984b) and has been found in headwaters of rivers in the fynbos biome (Von Broembsen 1984b). Low genetic diversity in South African populations indicates that it is an introduced species (Linde, Drenth & Wingfield 1999).

Several other species of *Phytophthora* are likely invasive in South Africa. *Phytophthora multivora* P.M. Scott & T. Jung and *P. capensis* Bezuid, Denman, A. McLeod & S.A. Kirk have been isolated from several indigenous plants (Bezuidenhout et al. 2010). The origin of these species is not known, but they may well be aliens. *Phytophthora niederhauserii* Z.G. Abad & J.A. Abad is a recently described species known to infect grapevines in South Africa. Considering that it has been recorded from a wide variety of plant species worldwide (Abad et al. 2014), it is likely to also be invasive in South Africa. Other species known so far only as crop pathogens in South Africa may also be invasive. The related genus *Pythium* also contains some species which have been isolated from indigenous plants or associated soil (McLeod et al. 2009) including *P. irregulare* Buisman isolated from *Aspalathus liearis* (Burm.f.) R. Dahlgren in natural ecosystems (Bahramisharif et al. 2014). An unidentified *Pythium* species has been associated with dieback and mortality of fynbos shrubs in the Western Cape (Jacobsen et al. 2012).

Puccinia lagenophorae Cooke is a rust fungus originating in Australia and which has since been introduced around the world. It was present in South Africa by 1918, and has been recorded on 48 indigenous species in the Asteraceae, and is now distributed from the Cape Peninsula to the Richtersveld and Mpumalanga (Scholler et al. 2011). It has also been recorded as having formed a hybrid with an unknown, presumably indigenous, species in KwaZulu-Natal (Morin et al. 2009); hybridisation is an important means of speciation amongst fungi leading to the emergence of new invasive species (Brasier 2001; Stukenbrock 2016).

The myrtle rust *Puccinia psidii* G. Winter, originally from South America, has been introduced around the world in *Eucalyptus* plantations and has caused significant damage to indigenous Myrtales plants in Australia (Carnegie et al. 2016; Pegg et al. 2014). This fungus is listed as a prohibited species in the NEM:BA:A&IS regulations but has recently been recorded as present in South Africa (Roux et al. 2013). A risk assessment of selected indigenous species in the Myrtales found that this rust fungus does pose a threat to our indigenous flora (Roux et al. 2015). This fungus has now been found to be widespread in South Africa, ranging from southern KwaZulu-Natal to Limpopo provinces, and occurring on a range of introduced and indigenous plants in the Myrtales, but so far not on *Eucalyptus*. The genotype found in South Africa is distinct to that which is invasive elsewhere in the world (Roux et al. 2016).

A. mellea was introduced to the Company Garden, Cape Town, possibly as long as 300 years ago (Coetzee et al. 2001). It has, until recently, been limited to this small area by the development of the City of Cape Town isolating this locality, but has now been recorded from Kirstenbosch on indigenous Proteaceae (Coetzee et al. 2003; Wingfield et al. 2010). Likewise, *Armillaria gallica* has also been recorded from Kirstenbosch (Coetzee et al. 2003). Both these fungi are from the Northern Hemisphere and are highly destructive wood-rot pathogens of a wide range of tree species. They may well pose a significant threat to indigenous forests if they establish therein.

The endemic and endangered wild rye (*Secale strictum* C. Presl ssp. *africanum* (Stapf) K. Hammer) is susceptible to, and infected by, *Puccinia graminis* Pers., *P. striiformis* Westend. and *P. recondita* Dietel & Holw., all introduced and widespread pathogens of cereal crops in South Africa. Infection by these rust fungi may have contributed to the almost complete loss of this grass, now restricted to a single farm on the Roggeveld (Pretorius, Bender & Visser 2015).

Ceratocystis pirilliformis I. Barnes & M.J. Wingf., an introduced fungus which colonises wounds on *Eucalyptus* (Nkuekam et al. 2009), has recently been found on *Rapanea melanophloeos* (L.) Mez, and it is suggested that ongoing distribution of this pathogen is by anthropogenic activities (Lee et al. 2016).

There are likely many more alien fungi attacking indigenous plants. Non-specific pathogens introduced with alien host plants can be more damaging to native plants than to the original host, facilitating invasion by the alien plant (Li et al. 2014). However, recognition of these as invaders is difficult, and there have been few studies till recently looking for alien fungi in natural systems in South Africa. *Erythricium salmonicolor* (Berk. & Broome) Burds. has been recorded as causing disease and even death of indigenous trees such as *Podocarpus* spp. and *Dais cotonifolia* L. near to plantations of *Eucalyptus* and *Acacia mearnsii* De Wild.. However, it was not conclusively determined if this was an alien pathogen introduced in association with the plantation industry, or indigenous (Roux & Coetzee 2005). It is likely alien. There is a need to investigate many pathogens of commercial and plantation crops as potential invaders of natural habitats in South Africa. The Botryosphaeriaceae illustrate this point; they are a highly diverse family of woody plant inhabiting fungi that may be endophytes or latent pathogens, the latter only causing disease under conditions of environmental stress on the host plant (Slippers & Wingfield 2007). Many species have been recorded from introduced crop plants and indigenous plants such as members of the Proteaceae (Marincowitz et al. 2008), *Syzygium cordatum* Hochst. ex Krauss (Pavlic et al. 2007), *Adansonia digitata* L. (Cruywagen et al. 2016) and *Sclerocarya birrea* subsp. *caffra* (Sond.) Kokwaro (Mehl et al. 2016) in South Africa. Species recorded from indigenous plants include wide spread species recorded from many host plants, whereas there is also a high diversity of indigenous members of this family. There is a need to determine which are alien and which may cause disease in indigenous plants of sufficient severity to impact these plants' population dynamics, if any.

Saprotrophic fungi

Pearson (1950:277) noted that the identification of species of mushrooms collected in and around Cape Town proved to be easy as 'Most of the agarics and all the boleti collected are the same as found in Europe'. This was ascribed to most trees being of foreign origin, and few mushrooms occurring in natural forest and fynbos (Pearson 1950). It is therefore likely that many saprophytic macro- and microfungi are introduced and naturalised. A high proportion of the macrofungi named to species level in South Africa have been recognised because they are widespread European or Northern Hemisphere species (e.g. Gryzenhout 2010; van der Westhuizen & Eicker 1994). An example is the morel *Morchella esculenta* (L.) Pers. (Doidge 1950; Gryzenhout 2010); this genus is recognised as indigenous to the Northern Hemisphere (O'Donnell et al. 2011). Because macrofungi are relatively well known throughout much of the developed world, the majority of saprophytic fungi that are recognised as naturalised are these rather than microfungi. In France, 26% (59 of 227 species) of fungi recorded as alien were saprotrophs; all were macrofungi, and all but one species belonged to the Agaricales, Polyporales and Phallales (Basidiomycota) (Desprez-Loustau et al. 2010).

Many naturalised alien saprotrophs are, however, likely to simply occupy ecological space rather than damaging natural ecosystem functions (Vacher et al. 2010). *Favolaschia calocera* R. Heim has been recorded as having invaded many countries around the world but is not considered damaging to the environment anywhere (Vizzini, Zotti & Mello 2009). However, this requires testing as these fungi have not been investigated as to whether they potentially disrupt or change ecosystem processes, or replace indigenous species.

Indigenous fungi and alien plants

A number of fungi regarded as indigenous to South Africa cause diseases of plantation trees which are also invasive alien plants, including *Ceratocystis albifundus* M.J. Wingf., De Beer & M.J. Morris (wilt disease of *Acacia mearnsii*) (Roux et al. 2007) and *Chrysoporthe austroafricana* Gryzenh. & M.J. Wingf. (canker disease of *Eucalyptus*) (Heath et al. 2006). Several related fungi described from indigenous trees have proven to be pathogenic to plantation trees in inoculation experiments (Kamgan et al. 2008; Nakabonge et al. 2006; Pavlic et al. 2007). Various other fungi likely to be indigenous to South Africa have been recorded on naturalised weeds or invasive plants (Table 3).

These new associations allow the deliberate use of these pathogens as biological control agents. Two fungi have been developed as biocontrol agents of alien plants in South Africa: *Colletotrichum acutatum* J.H. Simmonds which causes gummosis disease of *H. sericea* (Morris 1983, 1989) and the wood-rotting *Cylindrobasidium laeve* (Pers.) Chamuris which is used to prevent coppicing of felled *Acacia mearnsii* trees (Morris et al. 1999). Another is under development, *Pseudolagarobasidium acaciicola* Ginns, which causes a dieback disease of *Acacia cyclops* G. Don (Kotze, Wood & Lennox 2015; Wood & Ginns 2006). More could be developed.

The high biomass or seed production of invasive plants may lead to high levels of associated generalist local pathogens, not sufficiently damaging to the invader's population dynamics to exert control, but which can spill over onto indigenous plants and can reduce the native populations (Beckstead et al. 2010). Accumulation of local pathogens can also provide a mechanism promoting invasion (see above). Accumulation of local pathogens can persist even after the removal of the alien plants, leaving a legacy of increased damage to recovering indigenous vegetation (Maoela et al. 2016).

Arbuscular mycorrhizae

AM fungi are widespread in terrestrial ecosystems and are considered to be non-specific, and therefore available to invasive plants to form symbiotic associations (Redecker & Raab 2006; Richardson et al. 2000). Although able to infect most plants, AM fungi can show host preferences, thus the interaction between invasive plants and the available AM fungi may result in a positive feedback which changes the composition of the AM community to the benefit of invasive plants and to the detriment of indigenous plants (Zhang et al. 2010). Various impacts of AM have been recorded facilitating invasion by alien plants. *A. saligna* (Labill.) Wendl. has a larger root system than indigenous Fabaceae of the fynbos biome, and therefore has more extensive AM fungal colonisation providing more efficient nutrient uptake which promotes invasion (Hoffman & Mitchell 1986). Disturbance can reduce the amount of AM fungus inoculum available, providing a window of opportunity for non-dependent invasive plants to establish dominance over AM-dependent native plants (Carvalho et al. 2010; Owen et al. 2013). Allelopathic invasive plants can also reduce AM fungi, resulting in reduced growth of native plants dependent on these fungi (Hale, Lapointe & Kalisz 2016; Koch et al. 2011), although this effect can be offset by soil microbiota, possibly by the degradation of the allelopathic chemicals (Lankau 2010).

Fungi and arthropods

Little work has been done on fungi associated with insects and other arthropods in South Africa. Rong & Grobbelaar (1998) listed records of fungi (excluding microsporidia) found associated with arthropods in South Africa. Amongst these are fungi that have potential to be developed as biological control agents of arthropod pests (Rong & Grobbelaar 1998). Several microorganisms have been developed internationally as commercial biopesticides, such as the fungi *Beauveria bassiana* (Bals.-Criv.) Vuill. and *Metarhizium anisopliae* (Metschn.) Sorokin, and the bacterium *Bacillus thuringiensis* Berliner. Several of these products are marketed in South Africa; however, no ecological risk assessment of using alien strains of these microorganisms, isolated and commercially produced on other continents, has been undertaken in South Africa. The development of indigenous isolates, adapted to the local environment, requires investigation. There are a few local companies developing these. Naturally existing entomopathogenic fungi can also be managed as part of an integrated pest management programme of introduced pests of crops (Hatting et al. 1999; Hatting, Poprawski & Miller 2000) and invasive alien insects.

Control measures

The author is not aware of any attempts in South Africa to control an invasive fungus in a natural ecosystem. Preventing the introduction of potentially ecologically damaging fungi is the best method of mitigating impacts (Dickie et al. 2016; Hansen 2008; Wingfield et al. 2001). Part of the quarantine process is to prohibit the introduction of named species considered to pose a risk to the country. The current list of prohibited species in the NEM:BA:A&IS regulations include four fungus species (Table 1), two of which already occur in the country. *Fusarium oxysporum* Schltdl. f. sp. *cubense* W.C. Snyder & H.N. Hansen, cause of Panama disease of bananas, was recorded as present in South Africa by the 1940s (Doidge 1950). Various pathogens can be recognised as likely posing a risk to native ecosystems, and need to be added to the list of prohibited species. An example is *Armillaria luteobubalina* Watling & Kile, an Australian species pathogenic to many plant species (e.g. Shearer et al. 1997, Shearer & Tippert 1988). This list must be compiled in cooperation with DAFF who administer their own list, and administer inspection and quarantine services. *Phytophthora kernoviae* Brasier, Beales & S.A. Kirk is better placed in the list of prohibited species rather than in the invasive species lists. A shortfall of such lists is that undescribed but little known but potentially devastating pathogens will not be listed.

If a damaging fungus is introduced and has become widespread, control measures that can be implemented are focused on (1) stopping further spread of the fungus, (2) reducing inoculum levels within the area affected and (3) restoring damaged vegetation (Hansen 2008). Early control methods relied on stopping further spread by restricting entry to infected areas and strict hygiene imposed on any person or vehicle entering infected areas, for *Phytophthora* species (Hansen et al. 2000; Shearer & Tippert 1988). These remain as important control measures, and may be the only practical measures that can be applied (Dickie et al. 2016). Identification of high hazard areas, where the fungus is most likely to invade and cause damage, focusses these control methods where most needed (Cahill et al. 2008; Shearer & Tippert 1988). Fungicides can be applied for control (Cahill et al. 2008; Dickie et al. 2016; Dunstan et al. 2010; Hardy, Barrett & Shearer 2001; Hill, Tippert & Shearer 1995); however, these can have negative impacts on native plants (phytotoxicity) (Hardy et al. 2001) or fungi (Dickie et al. 2016).

Eradication of pathogens that have a limited host range can be successful, following early detection of outbreaks in geographically limited areas, by eradication of host plants (Ganley & Bulman 2016; Sosnowski et al. 2009). This has been unsuccessful when applied in natural ecosystems to pathogens that have efficient long-distance dispersal mechanisms (e.g. *P. psidii*, Carnegie & Cooper 2011) or broad host ranges (e.g. *P. cinnamomi*, Hill et al. 1995). However, the combination of eradicating host plants, soil fumigation and installation of a physical root barrier proved to be successful in eradicating *P. cinnamomi* in a small area adjacent to a high-value conservation area. The topography at this site limited the spread of this pathogen by water-borne spores (Dunstan et al. 2010).

Conclusions and recommendations

The purpose of this review has been to provide a very brief summary of some of the information that has become available on fungi and invasions in the scientific literature, to provide a springboard for future investigations of fungal invasion in South Africa. The broad phylogenetic diversity and numerous ecological roles of fungi result in multiple interactions with other organisms. The best known ecological impacts by invasive fungi are of introduced tree pathogens that have caused significant reductions to their new-association host's abundance (Cahill et al. 2008; Fisher et al. 2012; Loo 2009). In South Africa, the most widespread and damaging is *P. cinnamomi*, although the impact locally in natural ecosystems has not yet been quantified. But, not all naturalised species cause ecological damage to natural ecosystem functioning. Many fungi, including pathogens, are assimilated into existing plant–fungus ecological networks (Vacher et al. 2010). Internationally, it is only relatively recently that an interest in the fungal dimension in invasion science has developed, with inventory, impact and invasion process studies beginning to be published. Apart from terrestrial plant pathogens, very little is still

known about naturalised saprotrophs (Desprez-Loustau et al. 2007; Litchman 2010), plant mutualists, animal pathogens and aquatic fungi.

One consequence of invasions by fungi is the increased potential for hybridisation leading to either novel genotypes or even new species that may be more aggressive, or attack different hosts, than the parents (Brasier 2001; Callaghan & Guest 2015; Stukenbrock 2016). Fungi can rapidly adapt to new environments and novel hosts, and then these novel genotypes can rapidly disperse, facilitating invasion (Gladieux et al. 2015). Traits that have been associated with fungal invasions include (1) characteristics associated with long-distance dispersal; (2) sexual reproduction; (3) spore shape, size and number of cells; (4) pre-adaption to environmental conditions of introduced range; and (5) parasitic specialisation (Philibert et al. 2011). Fungi have been implicated in a number of hypotheses explaining invasion by plants but have not yet been investigated as to which are relevant to their invasions. In addition to direct impacts of invasion, fungi (and other microbes) interact in many different ways with other organisms within ecological networks, so that there are also many indirect impacts of fungi (both indigenous and alien) on invasions (Dawson & Schrama 2016; Desprez-Loustau et al. 2007).

In terms of the National Status Report as set out in the NEM:BA A&IS regulations (Government Notice No. R598, 1 August 2014), it can be concluded that the fungi listed as category 1b are mainly not damaging to natural ecosystems (five of the seven listed are plantation crop pathogens, four are not yet present in the country) and therefore irrelevant to the regulations, that no risk assessment of any fungus species as a potential invader has been undertaken and no control measures have been implemented by the Department of Environmental Affairs (DEA) against any listed species.

This situation exists because there is simply no information available to evaluate the invasive status of most fungi. Interventions required include the development of capacity to document the fungi indigenous to South Africa. This will also provide capacity to document invasive fungi, and record new arrivals. Capacity to investigate ecological processes driven by or influenced by fungi also needs to be developed, so that the impacts of invasive species can be determined, including by invasive mutualistic and saprotrophic fungi. This capacity would also be relevant to determine possible means of reducing impacts by invasive fungi, although there is limited scope for direct control. Fungi also have potential as biological control organisms, and efforts to develop agents against invasive alien plants should continue. There is under-exploited potential for the development of locally occurring agents against invasive or economically important pathogens (e.g. *Trichoderma* spp.), arthropods (e.g. *B. bassiana*) and other animal groups such as nematodes.

If an alien fungus is recorded in South Africa, what control measures would be appropriate? Which government agency should be responsible for devising and carrying out a control or containment programme? For example, the myrtle rust pathogen, *P. psidii*, was recorded for the first time in 2013 (Roux et al. 2013); this is a listed prohibited species in the NEM:BA A&IS regulations. The author is unaware of efforts led by any government agency to control this pathogen. Protocols or plans need to be developed and formulated by a government agency that has specifically been made responsible for control efforts. It may be noted here that despite a rapid response by Australian organs of state, the spread of this fungus was not halted in Australia (Carnegie et al. 2016; Pegg et al. 2014).

There is little information available on pathways of introduction, despite preventing introduction being in most cases the only practical means of control. It is recognised that most introductions have been human mediated (Fisher et al. 2012; Moralejo et al. 2009; Roy et al. 2014; Wingfield et al. 2001), but there is a need to identify potential routes and substrates of introductions. One of the most likely important means of introduction in the past was the transport of living crop plants growing in soil to our shores. The transport of live plants in the forestry and horticultural industries continues this trend (Moralejo et al. 2009; Roy et al. 2014). The nursery trade is also an important means of spreading alien fungi within countries. Bulk timber consignments, and timber products, are also important introduction pathways (Roy et al. 2014; Wingfield et al. 2001).

Currently, the importation of all organisms is controlled by DAFF, who administer applications for import by issuing permits, and through inspection and quarantine services. It is therefore important for cooperation between DEA and DAFF to ensure that risks to the environment are part of any risk assessment undertaken when considering an application for import. This should include applications for the import of microbial remedies such as *Trichoderma* species and *B. bassiana*, or mutualists such as AM fungi. In such cases, a precautionary principle may be appropriate, that developing commercial products using locally adapted indigenous strains of the species of interest would be preferable to importing alien strains. However, it needs to be recognised that in many cases we know so little about what fungi are native to South Africa, and their ecological importance, that a risk assessment may be impossible to compile. Standards for packing material and any medium in which plants, plant parts or animals are transported may also need modification.

There is a clear need to revise the current list of declared alien microbial species, as well as the list of prohibited species. It will be important at the same time to consider if any control options are possible against any listed species, as the NEM:BA act requires control to be implemented against these. The decision to include species should not only be based solely on the fact that they are recorded as alien but also on that a plan of action to at least ameliorate negative ecological impacts can be implemented.

Acknowledgements

The Department of Environmental Affairs: Natural Resources Management Programme (DEA:NRMP) is thanked for providing funding, and three unknown reviewers for their valuable comments which improved the text.

Competing interests

The author declares that he has no financial or personal relationship(s) that may have inappropriately influenced him in writing this article.

Authors' contributions

A.R.W. conceptualised this review, researched literature and wrote the manuscript.

References

Abad, Z.G., Abad, J.A., Cacciola, S.O., Pane, A., Faedda, R., Moralejo, E., et al., 2014, '*Phytophthora niederhauserii* sp. nov., a polyphagous species associated with ornamentals, fruit trees and native plants in 13 countries', *Mycologia* 106, 431–447. https://doi.org/10.3852/12-119

Adams, G.C., Rpux, J. & Wingfield, M.J., 2006, '*Cytospora* species (Ascomycota, Diapothales, Valsaceae): introduced and native pathogens of trees in South Africa', *Australasian Plant Pathology* 35, 521–548. https://doi.org/10.1071/AP06058

Adendorf, R., & Rijkenberg, F.H.J., 1995, 'New report on rust on Kikuyu grass in South Africa caused by *Phakopsora apoda*', *Plant Disease* 79, 1187. https://doi.org/10.1094/PD-79-1187B

Agrawal, A.A., Kotanenen, P.M., Mitchell, C.E., Power, A.G., Godsoe, W. & Klironomos, J., 2005, 'Enemy release? An experiment with congeneric plant pairs and diverse above- and belowground enemies', *Ecology* 86, 2979–2980. https://doi.org/10.1890/05-0219

Allender, M.C., Raudabaugh, D.B., Gleason, F.H. & Miller, A.N., 2015, 'The natural history, ecology, and epidemiology of *Ophidiomyces ophidiicola* and its potential impact on free-ranging snake populations', *Fungal Ecology* 17, 187–196. https://doi.org/10.1016/j.funeco.2015.05.003

Allsopp, N. & Holmes, P.M., 2001, 'The impact of alien plant invasion on mycorrhizas in mountain fynbos vegetation', *South African Journal of Botany* 67, 150–156. https://doi.org/10.1016/S0254-6299(15)31113-3

Allsopp, N. & Stock, W.D., 1993, 'Mycorrhizal status of plants growing in the Cape Floristic Region, South Africa', *Bothalia* 23, 91–104.

Anderson, P.K., Cunningham, A.A., Patel, N.G., Morales, F.J., Epstein, P.R. & Daszak, P., 2004, 'Emerging infectious diseases of plants: Pathogen pollution, climate change and agrotechnology drivers', *Trends in Ecology and Evolution* 19, 535–544. https://doi.org/10.1016/j.tree.2004.07.021

Bahramisharif, A., Lamprecht, S.C., Spies, C.F.J., Botha, W.J., Calitz, F.J. & McLeod, A., 2014, '*Pythium* spp. associated with rooibos seedlings, and their pathogenicity towards rooibos, lupin and oat', *Plant Disease* 98, 223–232. https://doi.org/10.1094/PDIS-05-13-0467-RE

Balci, Y., Balci, S., Eggers, J., MacDonald, W.L., Juzwik, J., Long, R.P., et al., 2007, '*Phytophthora* spp. associated with forest soils in eastern and north-central U.S. Oak ecosystems', *Plant Disease* 91, 705–710. https://doi.org/10.1094/PDIS-91-6-0705

Bebber, D.P., Holmes, T. & Gurr, S.J., 2014, 'The global spread of crop pests and pathogens', *Global Ecology and Biogeography* 23, 1398–1407. https://doi.org/10.1111/geb.12214

Becklin, K.M., Pallo, M. & Galen, C., 2012, 'Willows indirectly reduce arbuscular mycorrhizal fungal colonization in understorey communities', *Journal of Ecology* 100, 343–351. https://doi.org/10.1111/j.1365-2745.2011.01903.x

Beckstead, J., Meyer, S.E., Connolly, B.M., Huck, M.B. & Street, L.E., 2010, 'Cheatgrass facilitates spillover of a seed bank pathogen onto native grass species', *Journal of Ecology* 98, 168–177. https://doi.org/10.1111/j.1365-2745.2009.01599.x

Bennett, A.E. & Strauss, S.Y., 2013, 'Response to soil biota by native, introduced non-pest, and pest grass species: Is responsiveness a mechanism for invasion?' *Biological Invasions* 15, 1343–1353. https://doi.org/10.1007/s10530-012-0371-1

Berger, L., Roberts, A.A., Voyles, J., Longcore, J.E., Murray, K.A. & Skerratt, L.F., 2016, 'History and recent progress on chytridiomycosis in amphibians', *Fungal Ecology* 19, 89–99. https://doi.org/10.1016/j.funeco.2015.09.007

Bezuidenhout, C.M., Denman, S., Kirk, S.A., Botha, W.J., Mostert, L. & McLeod, A., 2010, '*Phytophthora* taxa associated with cultivated *Agathosma*, with emphasis on the *P. citricola* complex and *P. capensis* sp. nov.', *Persoonia* 25, 32–49. https://doi.org/10.3767/003158510X538371

Birnbaum, C. & Leishman, M.R., 2013, 'Plant-soil feedbacks do not explain invasion success of *Acacia* species in introduced range populations in Australia'. *Biological Invasions* 15, 2609–2625. https://doi.org/10.1007/s10530-013-0478-z

Blackwell, M., 2011, 'The fungi; 1, 2, 3, ... 5.1 million species?' *American Journal of Botany* 98, 426–438. https://doi.org/10.3732/ajb.1000298

Blaszkowski, J., Kovács, G.M., Balázs, T.K., Orlowska, E., Sadravi, M., Wubet, T., et al., 2010, '*Glomus africanum* and *G. iranicum*, two new species of arbuscular mycorrhizal fungi (Glomeromycota)', *Mycologia* 102, 1450–1462. https://doi.org/10.3852/09-302

Brasier, C.M., 2001, 'Rapid evolution of introduced plant pathogens via interspecific hybridization', *BioScience* 51, 123–133. https://doi.org/10.1641/0006-3568 (2001)051[0123:REOIPP]2.0.CO;2

Bottomley, A.M., 1948, 'Gasteromycetes of South Africa', *Bothalia* 4, 473–810. https://doi.org/10.4102/abc.v4i3.1859

Cahill, D.M., Rookes, J.E., Wilson, B.A., Gibson, L. & McDougall, K.L., 2008, '*Phytophthora cinnamomi* and Australia's biodiversity: Impacts, predictions and progress towards control', *Australasian Journal of Botany* 56, 279–310. https://doi.org/10.1071/BT07159

Callaghan, S. & Guest, D., 2015, 'Globalisation, the founder effect, hybrid *Phytophthora* species and rapid evolution: New headaches for biosecurity', *Australasian Plant Pathology* 44, 255–262. https://doi.org/10.1007/s13313-015-0348-5

Camilo-Alves, C.S.P., Clara, M.I.E. & Ribeiro, N.M.C.A., 2013, 'Decline of Mediterranean oak trees and its association with *Phytophthora cinnamomi*: A review', *European Journal of Forest Research* 132, 411–432. https://doi.org/10.1007/s10342-013-0688-z

Carnegie, A.J. & Cooper, K., 2011, 'Emergency response to the incursion of an exotic myrtaceous rust in Australia', *Australasian Plant Pathology* 40, 346–359. https://doi.org/10.1007/s13313-011-0066-6

Carnegie, A.J., Kathuria, A., Peg, G.S., Entwistle, P., Nagel, M. & Giblin, F.R., 2016, 'Impact of the invasive rust *Puccinia psidii* (myrtle rust) on native Myrtaceae in natural ecosystems in Australia', *Biological Invasions* 18, 127–144. https://doi.org/10.1007/s10530-015-0996-y

Carvalho, L.M., Antunes, P.M., Martins-Loução, M.A. & Klironomous, J.N., 2010, 'Disturbance influences the outcome of plant-soil biota interactions in the invasive *Acacia longifolia* and in native species', *Oikos* 119, 1172–1180. https://doi.org/10.1111/j.1600-0706.2009.18148.x

Catford, J.A., Jansson, R. & Nilsson, C., 2009, 'Reducing redundancy in invasion ecology by integrating hypotheses into a single theoretical framework', *Diversity and Distributions* 15, 22–40. https://doi.org/10.1111/j.1472-4642.2008.00521.x

Coetzee, M.P.A., Wingfield, B.D., Harrington, T.C., Steimel, J., Coutinho, T.A. & Wingfield, M.J., 2001, 'The root rot fungus *Armillaria mellea* introduced into South Africa by early Dutch settlers', *Molecular Ecology* 10, 387–396. https://doi.org/10.1046/j.1365-294x.2001.01187.x

Coetzee, M.P.A., Wingfield, B.D., Roux, J., Crous, P.W., Denman, S. & Wingfield, M.J., 2003, 'Discovery of two northern hemisphere *Armillaria* species on Proteaceae in South Africa', *Plant Pathology* 52, 604–612. https://doi.org/10.1046/j.1365-3059.2003.00879.x

Coetzee, J.C., 2010, 'Taxonomic notes on the Clathraceae (Phallales: Phallomycetidae) *sensu* Bottomley and a new key to the species in southern Africa', *Bothalia* 40, 155–159. https://doi.org/10.4102/abc.v40i2.205

Collier, F.A. & Bidartondo, M.I., 2009, 'Waiting for fungi: The ectomycorrhizal invasion of lowland heathlands', *Journal of Ecology* 97, 950–963. https://doi.org/10.1111/j.1365-2745.2009.01544.x

Crous, P.W. & Braun, U., 1996, 'Cercosporoid fungi from South Africa', *Mycotaxon* 57, 233–321.

Crous, P.W., Groenewald, J.Z., den Breeÿen, A. & King, A., 2014, '*Cercosporella dolichandrae*, Fungal Planet 243', *Persoonia* 32, 232–233.

Crous, P.W., Groenewald, J.Z. & Wood, A.R., 2013, '*Coniothyrium prosopidis* & *Peyronellaea prosopidis*, Fungal Planet 165 & 166', *Persoonia* 31, 206–207.

Crous, P.W., Phillips, A.J.L. & Baxter, A.P., 2000, *Phytopathogenic fungi from South Africa*, University of Stellenbosch, Department of Plant Pathology Press, Stellenbosch. ISBN 0-7972-0777-5.

Crous, P.W., Rong, I.H., Wood, A.R., Lee, S., Glen, H., Botha, W., et al., 2006. 'How many species of fungi are there at the tip of Africa?' *Studies in Mycology* 55, 13–33. https://doi.org/10.3114/sim.55.1.13

Cruywagen, E.M., Slippers, B., Roux, J. & Wingfield, M.J., 2016, 'Phylogenetic species recognition and hybridization in *Lasiodiplodia*: A case study on species from baobabs', *Fungal Biology* https://doi.org/10.1016/j.funbio.2016.07.014

Cui, Q.-G. & He, W.-M., 2009, 'Soil Biota, but not soil nutrients, facilitate the invasion of *Bidens pilosa* relative to a native species *Saussurea deltoidea*', *Weed Research* 49, 201–206. https://doi.org/10.1111/j.1365-3180.2008.00679.x

Dawson, W., 2015, 'Release from belowground enemies and shifts in root traits as interrelated drivers of alien plant invasion success: A hypothesis', *Ecology and Evolution* 5, 4505–4516. https://doi.org/10.1002/ece3.1725

Dawson, W. & Schrama, M., 2016, 'Identifying the role of soil microbes in plant invasions', *Journal of Ecology* 104, 1211–1218. https://doi.org/10.1111/1365-2745.12619

Day, N.J., Dunfield, K.E. & Antunes, P.M., 2015, 'Temporal dynamics of plant-soil feedback and root associated fungal communities over 100 years of invasion by a non-native plant', *Journal of Ecology* 103, 1557–1569. https://doi.org/10.1111/1365-2745.12459

Day, N.J., Dunfield, K.E. & Antunes, P.M., 2016, 'Fungi from a non-native invasive plant increase its growth but have different growth effects on native plants', *Biological Invasions* 18, 231–243. https://doi.org/10.1007/s10530-015-1004-2

de la Peña, E., de Clerq, N., Bonte, D., Roiloa, S., Rodríguez-Echeverría, S. & Freitas, H., 2010, 'Plant-soil feedback as a mechanism of invasion by *Carpobrotus edulis*', *Biological Invasions* 12, 3637–3648. https://doi.org/10.1007/s10530-010-9756-1

Desprez-Loustau, M.-L., Courtescuisse, R., Robin, C., Husson, C., Moreau, P.-A., Blancard, D., et al., 2010, 'Species diversity and drivers of spread of alien fungi *(sensu lato)* in Europe with particular focus on France', *Biological Invasions* 12, 157–172. https://doi.org/10.1007/s10530-009-9439-y

Desprez-Loustau, M.-L., Robin, C., Buée, M., Courtecuisse, R., Garbaye, J., Suffert, F., et al., 2007, 'The fungal dimension in biological invasions', *TRENDS in Ecology and Evolution* 22, 472–480. https://doi.org/10.1016/j.tree.2007.04.005

Dickie, I.A., Bolstridge, N., Cooper, J.A. & Peltzer, D.A., 2010, 'Co-invasion by *Pinus* and its mycorrhizal fungi', *New Phytologist* 187, 475–484.

Dickie, I.A., Nuñez, M.A., Pringle, A., Lebel, T., Tourtellot, S.G. & Johnston, P.R., 2016, 'Towards management of invasive ectomycorrhizal fungi', *Biological Invasions* 18, 3383–3395. https://doi.org/10.1007/s10530-016-1243-x

Doidge, E.M., 1950, 'The South African fungi and lichens to the end of 1945', *Bothalia* 5, 1–1094.

Doidge, E.M. & Bottomley, A.M., 1931, 'A revised list of plant diseases occurring in South Africa', *Botanical Survey of South Africa* 11, 1–78.

Doidge, E.M., Bottomley, A.M., van der Plank, J.E. & Pauer, G.D., 1953, 'A revised list of plant diseases in South Africa', *Science Bulletin* 346, 1–122.

Duchicela, J., Vogelsang, K.M., Schultz, P.A., Kaonongbua, W., Middleton, E.L. & Bever, J.D., 2012, 'Non-native plants and soil microbes: Potential contributions to the consistent reduction in soil aggregate stability caused by the disturbance of North American grasslands', *New Phytologist* 196, 212–222. https://doi.org/10.111/j.1469-8137.2012.04233.x

Dunk, C.W., Lebel, T. & Keane, P.J., 2012, 'Characterization of ectomycorrhizal formation by the exotic fungus *Amanita muscaria* with *Nothofagus cunninghamii* in Victoria, Australia', *Mycorrhiza* 22, 135–147. https://doi.org/10.1007/s00572-011-0288-9

Dunstan, W.A., Rudman, T., Shearer, B.L., Moore, N.A., Paap, T., Calver, M.C., et al., 2010, 'Containment and spot eradication of a highly destructive, invasive plant pathogen (*Phytophthora cinnamomi*) in natural ecosystems', *Biological Invasions* 12, 913–925. https://doi.org/10.1007/s10530-009-9512-6

Elgersma, K.J. & Ehrenfeld, J.G., 2011, 'Linear and non-linear impacts of a non-native plant invasion on soil microbial community structure and function', *Biological Invasions* 13, 757–768. https://doi.org/10.1007/s10530-010-9866-9

Eppinga, M.B., Rietkerk, M., Dekker, S.C., De Ruiter, P.C. & Van der Putten, W.H., 2006, 'Accumulation of local pathogens: A new hypothesis to explain exotic plant invasions', *Oikos* 114, 168–176. https://doi.org/10.1111/j.2006.0030-1299.14625.x

Farr, D.F. & Rossman, A.Y., undated, *Fungal databases, Systematic Mycology and Microbiology Laboratory, ARS, USDA*, viewed March – November 2016, from http://nt.ars-grin.gov/fungaldatabases/

Feng, Y.-L., Li, Y.-P., Wang, R.-F., Callaway, R.M., Valiente-Banuet, A. & Inderjit, 2011, 'A quicker return energy-use strategy by populations of a subtropical invader in the non-native range: A potential mechanism for the evolution of increased competitive ability', *Journal of Ecology* 99, 1116–1123. https://doi.org/10.1111/j.1365-2745.2011.01843.x

Fisher, M.C., Henk, D.A., Briggs, C.J., Brownstein, J.S., Madoff, L.C., McCraw, S.L. et al., 2012, 'Emerging fungal threats to animal, plant and ecosystem health', *Nature* 484, 186–194. https://doi.org/10.1038/nature10947

Flory, S.L. & Clay, K., 2013, 'Pathogen accumulation and long-term dynamics of plant invasions', *Journal of Ecology* 101, 607–613. https://doi.org/10.1111/1365-2745.12078

Flory, S.L., Kleczewski, N. & Clay, K., 2011, 'Ecological consequences of pathogen accumulation on an invasive grass', *Ecosphere* 2, 120. https://doi.org/10.1890/ES11-00191.1

Gaur, A., van Greuning, J.V., Sinclair, R.C. & Eicker, A., 1999, 'Arbuscular mycorrhizas of *Vangueria infausta* Burch. subsp. *infausta* (Rubiaceae) from South Africa', *South African Journal of Botany* 65, 434–436. https://doi.org/10.1016/S0254-6299(15)31036-X

Ganley, R.J. & Bulman, L.S., 2016, 'Dutch elm disease in New Zealand: Impacts from eradication and management programmes', *Plant Pathology* 65, 1047–1055. https://doi.org/10.1111/ppa.12527

Gladieux, P., Feurtey, A., Hood, M.E., Snirc, A., Clavel, J., Dutech, C., et al., 2015, 'The population biology of fungal invasions', *Molecular Ecology* 24, 1969–1986. https://doi.org/10.1111/mec.13028

Gorter, G.J.M.A., 1977, 'Index of plant pathogens and the diseases they cause in cultivated plants in South Africa', *Science Bulletin* 392, 1–177.

Gorter, G.J.M.A., 1993, 'A revised list of South African Erysiphaceae (powdery mildews) and their host plants', *South African Journal of Botany* 59, 566–568. https://doi.org/10.1016/S0254-6299(16)30671-8

Greyling, I., Wingfield, M.J., Coetzee, M.P.A., Marincowitz, S. & Roux, J. 2016, 'The *Eucalyptus* shoot and leaf pathogen *Teratosphaeria destructans* recorded in South Africa', *Southern Forests* 2016, 1–7. https://doi.org/10.2989/20702620.2015.113 6504

Gryzenhout, M., 2010, *Mushrooms of South Africa*, Struik, Cape Town, South Africa.

Gundale, M.J., Kardol, P., Nilsson, M.-C., Nilsson, U., Lucas, R.W. & Wardle, D.A., 2014, 'Interactions with soil biota shift from negative to positive when a tree species is moved outside its native range', *New Phytologist* 202, 415–421. https://doi.org/10.1111/nph.12699

Hagemann, G.D. & Rose, P.D., 1988, 'Leaf spot and blight on *Acacia longifolia* caused by *Cylindrocladium scoparium*: A new host record', *Phytophlactica* 20, 311–316.

Halbritter, A.H., Carroll, G.C., Güsewell, S. & Roy, B.A., 2012, 'Testing assumptions of the enemy release hypothesis: Generalist versus specialist enemies of the grass *Brachypodium sylvaticum*', *Mycologia* 104, 34–44. https://doi.org/10.3852/11-071

Hale, A.N., Lapointe, L. & Kalisz, S., 2016, 'Invader disruption of belowground plant mutualisms reduces carbon acquisition and alters allocation patterns in a native forest herb', New Phytologist 209, 542–549. https://doi.org/10.1111/nph.13709

Hansen, E.M., 2008, 'Alien forest pathogens: Phytophthora species are changing world forests', Boreal Environment Research 13, 33–41.

Hansen, E.M., Goheen, D.J., Jules, E.S. & Ullian, B., 2000, 'Managing Port-Orford-Cedar and the introduced pathogen Phytophthora lateralis', Plant Disease 84, 4–10. https://doi.org/10.1094/PDIS.2000.84.1.4

Hardy, G.E.St.J., Barrett, S. & Shearer, B.L., 2001, 'The future of phosphite as a fungicide to control the soilborne plant pathogen Phytophthora cinnamomi in natural ecosystems', Australasian Plant Pathology 30, 133–139. https://doi.org/10.1071/AP01012

Hatting, J.L., Humber, R.A., Poprawski, T.J. & Miller, R.M., 1999, 'A survey of fungal pathogens of aphids from South Africa, with special reference to cereal aphids', Biological Control 16, 1–12. https://doi.org/10.1006/bcon.1999.0731

Hatting, J.L., Poprawski, T.J. & Miller, R.M., 2000, 'Prevalences of fungal pathogens and other natural enemies of cereal aphids (Homoptera: Aphididae) in wheat under dryland and irrigated conditions in South Africa', BioControl 45, 179–199. https://doi.org/10.1023/A:1009981718582

Hawksworth, D., 1991, 'The fungal dimension of biodiversity: Magnitude, significance, and conservation', Mycological Research 95, 641–655. https://doi.org/10.1016/S0953-7562(09)80810-1

Hawksworth, D., 2001, 'The magnitude of fungal diversity: The 1.5 million species estimate revisited', Mycological Research 105, 1422–1432. https://doi.org/10.1017/S0953756201004725

Hawley, G.L. & Dames, J.F., 2004, 'Mycorrhizal status of indigenous tree species in a forest biome of the Eastern Cape, South Africa', South African Journal of Science 100, 633–637.

Hawley, G.L., Taylor, A.F.S. & Dames, J.F., 2008, 'Ectomycorrhizas in association with Pinus patula in Sabie, South Africa', South African Journal of Science 104, 273–283.

Heath, R.N., Gryzenhout, M., Roux, J. & Wingfield, M.J., 2006, 'Discovery of the canker pathogen Chrysopothe austroafricana on native Syzygium spp. in South Africa', Plant Disease 90, 433–438. https://doi.org/10.1094/PD-90-0433

Hill, S.B. & Kotanen, P.M., 2012, 'Biotic interactions experienced by a new invader: Effects of its close relatives at the community scale', Botany 90, 35–42. https://doi.org/10.1139/B11-084

Hill, T.C.J., Tippert, J.T. & Shearer, B.L., 1995, 'Evaluation of three treatments for eradication of Phytophthora cinnamomi from deep, leached sands in southwest Australia', Plant Disease 79, 122–127. https://doi.org/10.1094/PD-79-0122

Hoffman, M.T. & Mitchell, D.T., 1986, 'The root morphology of some legume spp. in the south-western Cape and the relationship of vesicular-arbuscular mycorrhizas with dry mass and phosphorus content of Acacia saligna seedlings', South African Journal of Botany 52, 316–320. https://doi.org/10.1016/S0254-6299(16)31527-7

Huchzermeyer, K.D.A. & Van der Waal, B.C.W., 2012, 'Epizootic ulcerative syndrome: Exotic fish disease threatens Africa's aquatic ecosystems', Journal of the South African Veterinary Association 83, Art.#204, 1–6. https://doi.org/10.4102/jsava.v83i1.204

Hynson, N.A., Merckx, V.S.F.T., Perry, B.A. & Treseder, K.K., 2013, 'Identities and distributions of the co-invading ectomycorrhizal fungal symbionts of exotic pines in the Hawaiian Islands', Biological Invasions 15, 2373–2385. https://doi.org/10.1007/s10530-013-1458-3

Jacobsen, A.L., Roets, F., Jacobs, S.M., Esler, K.J. & Pratt, R.B., 2012, 'Dieback and mortality of South African fynbos shrubs is likely driven by a novel pathogen and pathogen-induced hydraulic failure', Austral Ecology 37, 227–235. https://doi.org/10.1111/j.1442-9993.2011.02268.x

Jairus, T., Mpumba, R., Chinoya, S. & Tedersoo, L., 2011, 'Invasion potential and host shifts of Australian and African ectomycorrhizal fungi in mixed eucalypt plantations', New Phytologist 192, 179–187. https://doi.org/10.1111/j.1469-8137.2011.03775.x

Kamgan, N.G., Jacobs, K., de Beer, Z.W., Wingfield, M.J. & Roux J., 2008, 'Ceratocystis and Ophiostoma species, including three new taxa, associated with wounds on native South African trees', Fungal Diversity 29, 37–59.

Knevel, I.C., Lans, T., Menting, F.B.J., Hertling, U.M. & van der Putten, W., 2004, 'Release from native root herbivores and biotic resistance by soil pathogens in a new habitat both affect the alien Ammophila arenaria in South Africa', Oecologia 141, 502–510. https://doi.org/10.1007/s00442-004-1662-8

Koch, A.M., Antunes, P.M., Barto, E.K., Cipollini, D., Mummey, D.L. & Klironomos, J.N., 2011, 'The effects of arbuscular mycorrhizal (AM) fungal and garlic mustard introductions on native AM fungal diversity', Biological Invasions 13, 1627–1639. https://doi.org/10.1007/s10530-010-9920-7

Kotzé, L.J.D., Wood, A.R. & Lennox, C.L., 2015, 'Risk assessment of the Acacia cyclops dieback pathogen, Pseudolagarobasidium acaciicola, as a mycoherbicide in the South African strandveld and limestone fynbos', Biological Control 82, 52–60. https://doi.org/10.1016/j.biocontrol.2014.12.011

Kourtev, P.S., Ehrenfeld, J.G. & Häggblom, M., 2003, 'Experimental analysis of the effect of exotic and native plant species on the structure and function of soil microbial communities', Soil Biology & Biochemistry 35, 895–905. https://doi.org/10.1016/S0038-071(03)00120-2

Lankau, R., 2010, 'Soil microbial communities after allelopathic competition between Alliaria petiolate and a native species', Biological Invasions 12, 2059–2068. https://doi.org/10.1007/s10530-009-9608-z

Lankau, R.A., Bauer, J.T., Anderson, M.R. & Anderson, R.C., 2014, 'Long-term legacies and partial recovery of mycorrhizal communities after invasive plant removal', Biological Invasions 16, 1979–1990. https://doi.org/10.1007/s10530-014-0642-0

Lee, D.H., Roux, J., Wingfield, B.D., Barnes, I. & Wingfield, M.J., 2016, 'New host range and distribution of Ceratocystis pirilliformis in South Africa', European Journal of Plant Pathology 146, 483–496. https://doi.org/10.1007/s10658-016-0933-7

Li, H., Zhang, X., Zheng, R., Li, X., Elmer, W.H., Wolfe, L.M., et al., 2014, 'Indirect effects of non-native Spartina alterniflora and its fungal pathogen (Fusarium palustre) on native slatmarsh plants in China', Journal of Ecology 102, 1112–1119. https://doi.org/10.1111/1365-2745.12285

Linde, C., Drenth, A. & Wingfield, M.J., 1999, 'Gene and genotypic diversity of Phytophthora cinnamomi in South Africa and Australia revealed by DNA polymorphisms', European Journal of Plant Pathology 105, 667–680. https://doi.org/10.1023/A:1008755532135

Litchman, E., 2010, 'Invisible invaders: Non-pathogenic invasive microbes in aquatic and terrestrial ecosystems', Ecology Letters 13, 1560–1572. https://doi.org/10.1111/j.1461-0248.2010.01544.x

Loo, J.A., 2009, 'Ecological impacts of non-indigenous invasive fungi as forest pathogens', Biological Invasions 11, 81–96. https://doi.org/10.1007/s10530-008-9321-3

Lübbe, W.A. & Mostert, G.P., 1991, 'Rate of Ocotea bullata decline in association with Phytophthora cinnamomi at three study sites in the southern Cape indigenous forests', South African Forestry Journal 159, 17–24. https://doi.org/10.1080/00382167.1991.9630390

MacKay, J. & Kotanen, P.M., 2008, 'Local escape of an invasive plant, common ragweed (Ambrosia artemisiifolia L.), from above-ground and below-ground enemies in its native area', Journal of Ecology 96, 1152–1161. https://doi.org/10.1111/j.1365-2745.2008.01426.x

Mangla, S., Inderjit & Callaway, R.M., 2008, 'Exotic invasive plant accumulates native soil pathogens which inhibit native plants', Journal of Ecology 96, 58–67. https://doi.org/10.1111/j.1365-2745.2007.01312.x

Maoela, M.A., Jacobs, S.M., Roets, F. & Esler, K.J., 2016, 'Invasion, alien control and restoration: Legacy effects linked to folivorous insects and phylopathogenic fungi', Austral Ecology https://doi.org/10.1111/aec.12383

Marincowitz, S., Groenewald, J.Z., Wingfield, M.J. & Crous, P.W., 2008, 'Species of Botryosphaeriaceae occurring on Proteaceae', Persoonia 21, 111–118. https://doi.org/10.3767/003158508X372387

Martin, F., Díez, J., Dell, B. & Delaruelle, C., 2002, 'Phylogeography of the ectomycorrhizal Pisolithus species as inferred from nuclear ribosomal DNA ITS sequences', New Phytologist 153, 345–357. https://doi.org/10.1046/j.0028-646X.2001.00313.x

McLeod, A., Botha, W.J., Meitz, J.C., Spies, C.F.J., Tewoldemedhin, Y.T. & Mostert, L., 2009, 'Morphological and phylogenetic analysis of Pythium species in South Africa', Mycological Research 113, 933–951. https://doi.org/10.1016/j.mycres.2009.04.009

McTaggart, A.R., Doungsa-ard, C. & Wingfield, M.J., 2015, 'Uromycladium acacia, the cause of a sudden, severe disease epidemic on Acacia mearnsii in South Africa', Australasian Plant Pathology 44, 637–645. https://doi.org/10.1007/s13313-015-0381-4

Mehl, J.W.M., Slippers, B., Roux, J. & Wingfield, M.J., 2016, 'Overlap of latent pathogens in the Botryosphaeriaceae on a native and agricultural host', Fungal Biology https://doi.org/10.1016/j.funbio.2016.07.015

Moralejo, E., Pérez-Sierra, A.M., Álvarez, L.A., Belbahri, L., Lefort, F. & Descals, E., 2009, 'Multiple alien Phytophthora taxa discovered on disease ornamental plants in Spain', Plant Pathology 58, 100–110. https://doi.org/10.1111/j.1365-3059.2008.01930.x

Morin, L., van der Merwe, M., Hartley, D. & Müller, P., 2009, 'Putative natural hybrid between Puccinia lagenophorae and an unknown rust fungus on Senecio madagascariensis in KwaZulu-Natal, South Africa', Mycological Research 113, 725–736. https://doi.org/10.1016/j.mycres.2009.02.008

Morris, M.J., 1983, 'Evaluation of field trials with Colletotrichum gloeosporiodes for the biological control of Hakea sericea', Phytophylactica 15, 13–16.

Morris, M.J., 1989. 'A method for controlling Hakea sericea Schad. seedlings using the fungus Colletotrichum gloeosporiodes (Penz.) Sacc.', Weed Research 29, 449–454. https://doi.org/10.1111/j.1365-3180.1989.tb01317.x

Morris, M.J., 1990, 'Cercospora priaropi recorded on the aquatic weed, Eichhornia crassipes, in South Africa', Phytophylactica 22, 255–256.

Morris, M.J., 1991, 'The use of plant pathogens for biological weed control in South Africa', Agriculture, Ecosystems and Environment 37, 239–255. https://doi.org/10.1016/0167-8809(91)90153-O

Morris, M.J. & Crous, P.W., 1994, 'New and interesting records of South African fungi. XIV. Cercosporoid fungi from weeds', South African Journal of Botany 60, 325–332. https://doi.org/10.1016/S0254-6299(16)30587-7

Morris, M.J., Wingfield, M.J. & De Beer, C., 1993, 'Gummosis and wilt of Acacia mearnsii in South Africa caused by Ceratocystis fimbriata', Plant Pathology 42, 814–817. https://doi.org/10.1111/j.1365-3059.1993.tb01570.x

Morris, M.J., Wood, A.R. & den Breeÿen, A., 1999, 'Plant pathogens and biological control of weeds in South Africa: A review of projects and progress during the last decade', African Entomology memoir 1, 129–137

Nakabonge, G., Gryzenhout, M., Roux, J., Wingfield, B.D. & Wingfield, M.J., 2006, 'Celoporthe dispersa gen. et sp. nov. from native Myrtales in South Africa', Studies in Mycology 55, 255–267. https://doi.org/10.3114/sim.55.1.255

Nesamari, R., Coutinho, T.A. & Roux, J., 2016, 'Investigations into Encephalartos insect pests and diseases in South Africa and identification of Phytophthora cinnamomi as a pathogen of the Mojadji cycad', Plant Pathology. https://doi.org/10.1111/ppa.12619

Niu, H.B., Liu, W.X., Wan, F.H. & Liu, B., 2007, 'An invasive aster (*Ageratina adenophora*) invades and dominates forest understories in China: Altered soil microbial communities facilitate the invader and inhibit natives', *Plant Soil* 294, 73–85. https://doi.org/.1007/s11104-007-9230-8

Nkuekam, G.K., Barnes, I., Wingfield, M.J. & Roux, J., 2009, 'Distribution and population diversity of *Ceratocystis pirilliformis* in South Africa', *Mycologia* 101, 17–25. https://doi.org/10.3852/07/171

Nuñez, M.A. & Dickie, I.A., 2014, 'Invasive belowground mutualists of woody plants', *Biological Invasions* 16, 645–661. https://doi.org/10.1007/s10530-013-0612-y

O'Donnell, K., Rooney, A.P., Mills, G.L., Kuo, M. & Weber, N.S., 2011, 'Phylogeny and historical biogeography of true morels (*Morchella*) reveals an early Cretaceous origin and high continental endemism and provincialism in the Holarctic', *Fungal Genetics and Biology* 48, 252–265. https://doi.org/10.1016/j.fgb.2010.09.006

Owen, S.M., Sieg, C.H., Johnson, N.C. & Gehring, C.A., 2013, 'Exotic cheatgrass and loss of soil biota decrease the performance of a native grass', *Biological Invasions* 15, 2503–2517. https://doi.org/10.1007/s10530-013-0469-0

Pattison, Z., Rumble, H., Tanner, R.A., Jin, I. & Gange, A.C., 2016, 'Positive plant-soil feedbacks of the invasive *Impatiens glandulifera* and their effects on above-ground microbial communities', *Weed Research* 56, 198–207. https://doi.org/10.1111/wre.12200

Pavlic, D., Slippers, B., Coutinho, T.A. & Wingfield, M.J., 2007, 'Botrysphaeriaceae occurring on native *Syzygium cordatum* in South Africa and their potential threat to *Eucalyptus*', *Plant Pathology* 56, 624–636. https://doi.org/10.1111/j.1365-3059.2007.01608.x

Pearson, A.A., 1950, 'Cape agarics and boleti', *Transactions of the British Mycological Society* 33, 276–316. https://doi.org/10.1016/S0007-1536(50)80080-3

Pegg, G.S., Giblin, F.R., McTaggart, A.R., Guymer, G.P., Taylor, H., Ireland, K.B., et al., 2014, '*Puccinia psidii* in Queensland, Australia: Disease symptoms, distribution and impact', *Plant Pathology* 63, 1005–1021. https://doi.org/10.1111/ppa.12173

Philibert, A., Desprez-Loustau, M.-L., Fabre, B., Frey, P., Halkett, F., Husson, C., et al., 2011, 'Predicting invasion success of forest pathogenic fungi from species traits', *Journal of Applied Ecology* 48, 1381–1390. https://doi.org/10.1111/j.1365-2664.2011.02039.x

Pretorius, Z.A., Bender, C.M. & Visser, B., 2015, 'The rusts of wild rye in South Africa', *South African Journal of Botany* 96, 94–98. https://doi.org/10.1016/j.sajb.2014.10.005

Pretorius, Z.A., van Wyk, P.S. & Kriel, W.M., 2000, 'Occurrence of *Puccinia xanthii* on sunflower in South Africa', *Plant Disease* 84, 924. https://doi.org/10.1094/PDIS.2000.84.8.924A

Pretorius, Z.A., Visser, B. & du Preez, P.J., 2007, 'First report of Asian Soybean Rust caused by *Phakopsora pachyrhizi* on Kudzu in South Africa', *Plant Disease* 91, 1364. https://doi.org/10.1094/PDIS-91-10-1364C

Pringle, A., Adams, R.I., Cross, H.B. & Bruns, T.D., 2009, 'The ectomycorrhizal fungus *Amanita phalloides* was introduced and is expanding its range on the west coast of North America', *Molecular Ecology* 18, 817–833. https://doi.org/10.1111/j.1365-294X.2008.04030.x

Redecker, D. & Raab, P., 2006, 'Phylogeny of the Glomeromycota (arbuscular mycorrhizal fungi): Recent developments and new gene markers', *Mycologia* 98, 885–895. https://doi.org/10.3852/mycologia.98.6.885

Reinhart, K.O. & Callaway, R.M., 2006, 'Soil biota and invasive plants', *New Phytologist* 170, 445–457. https://doi.org/10.1111/j.1469-8137.2006.01715.x

Reinhart, K.O., Tytgat, T., Van der Putten, W.H. & Clay, K., 2010, 'Virulence of soil-borne pathogens and invasion by *Prunus serotina*', *New Phytologist* 186, 484–495. https://doi.org/10.1111/j.1469-8137.2009.03159.x

Richardson, D.M., Allsopp, N., D'Antonio, C.M., Milton, S.J. & Rejmánek, M., 2000, 'Plant invasions – The role of mutualisms', *Biological Review* 75, 65–93. https://doi.org/10.1017/S0006323199005435

Rockett, T.R. & Kramer, C.L., 1974, 'Periodicity and total spore production by lignicolous basidiomycetes', *Mycologia* 66, 817–829. https://doi.org/10.2307/3758202

Rodriguez-Echeverria, S., Teixeira, H., Correia, M., Timóteo, S., Heleno, R., Öpik, M. et al., 2016, 'Arbuscular mycorrhizal fungal communities from tropical Africa reveal strong ecological structure', *New Phytologist* 213, 380–390. https://doi.org/10.1111/nph.14122

Rong, I.H. & Baxter, A.P., 2006, 'The South African National Collection of Fungi: Celebrating a centenary 1905–2005', *Studies in Mycology* 55, 1–12. https://doi.org/10.3114/sim.55.1.1

Rong, I.H. & Grobbelaar, E., 1998, 'South African records of associations between fungi and arthropods', *African Plant Protection* 4, 43–63.

Roux, J. & Coetzee, M.P.A., 2005, 'First report of Pink Disease on native trees in South Africa and phylogenetic placement of *Erythricium salmonicolor* in the Homobasidiomycetes', *Plant Disease* 89, 1158–1163. https://doi.org/10.1094/PD-89-1158

Roux, J., Germishuizen, I., Nadel, R., Lee, D.J., Wingfield, M.J. & Pegg, G.S., 2015, 'Risk assessment for *Puccinia psidii* becoming established in South Africa', *Plant Pathology* 64, 1326–1335. https://doi.org/10.1111/ppa.12380

Roux, J., Granados, G.M., Shuey, L., Barnes, I., Wingfield, M.J. & McTaggart, A.R., 2016, 'A unique genotype of the rust pathogen, *Puccinia psidii*, on Myrtaceae in South Africa', *Australasian Plant Pathology* 45, 645–652. https://doi.org/10.1007/s13313-016-0447-y

Roux, J., Greyling, I., Coutinho, T.A., Verleur, M. & Wingfield, M.J., 2013, 'The myrtle rust pathogen, *Puccinia psidii*, discovered in Africa', *IMA Fungus* 4, 155–159. https://doi.org/10.5598/imafungus.2013.04.01.14

Roux, J., Heath, R.N., Labuschagne, L., Nkuekam, G.K. & Wingfield, M.J., 2007, 'Occurrence of the wattle wilt pathogen, *Ceratocystis albifundus* on native South African trees', *Forest Pathology* 37, 292–302. https://doi.org/10.1111/j.1439-0329.2007.00507.x

Roy, B.A., Alexander, H.M., Davidson, J., Campbell, F.T., Burdon, J.J., Sniezko, R. et al., 2014, 'Increasing forest loss worldwide from invasive pests require new trade regulations', *Frontiers in Ecology and Environment* 12, 457–465. https://doi.org/10.1890/130240

Scharfy, D., Güsewell, S., Gessner, M.O. & Venterink, H.O., 2010, 'Invasion of *Solidago gigantean* in contrasting experimental plant communities: Effects on soil microbes, nutrients and plant-soil feedbacks', *Journal of Ecology* 98, 1379–1388. https://doi.org/10.1111/j.1365-2745.2010.01722.x

Schoch, C.L., Crous, P.W., Wingfield, B.D. & Wingfield, M.J., 1999, 'The *Cylindrocladium candelabrum* species complex includes four distinct mating populations', *Mycologia* 91, 286–298. https://doi.org/10.2307/3761374.

Scholler, M., Lutz, M., Wood, A.R., Hagendorn, G. & Mennicken, M., 2011, 'Taxonomy and phylogeny of *Puccinia lagenophorae*: A study using rDNA sequence data, morphological and host range features', *Mycological Progress* 10, 175–187. https://doi.org/10.1007/s11557-010-0687-0

Shearer, B.L., Byrne, A., Dillon, M. & Buehrig, R., 1997, 'Distribution of *Armillaria luteobubalina* and its impact on community diversity and structure in *Eucalyptus wandoo* woodland in southern Western Australia', *Australian Journal of Botany* 45, 151–165. https://doi.org/10.1071/BT95083

Shearer, B.L. & Tippert, J.T., 1988, 'Distribution and impact of *Armillaria luteobubalina* in the *Eucalyptus marginata* forest in South-western Australia', *Australian Journal of Botany* 36, 433–445. https://doi.org/10.1071/BT9880433

Shearer, B.L. & Tippert, J.T., 1989, *Jarrah dieback: The dynamics and management of Phytophthora cinnamomi in the Jarrah* (Eucalyptus marginata) *forests of South-western Australia*, Research Bulletin No. 3. Department of Conservation and Land Management, Como, Australia. ISSN 1032-8106.

Slippers, B. & Wingfield, M.J., 2007, 'Botryosphaeriaceae as endophytes and latent pathogens of woody plants: Diversity, ecology and impact', *Fungal Biology Reviews* 21, 90–106. https://doi.org/10.1016/j.fbr.2007.06.002

Sosnowski, M.R., Fletcher, J.D., Daly, A.M., Rodoni, B.C. & Viljanen-Rollinson, S.L.H., 2009, 'Techniques for the treatment, removal and disposal of host material during programmes for plant pathogen eradication', *Plant Pathology* 58, 621–635. https://doi.org/10.1111/j.1365-3059.2009.02042.x

Straker, C.J., Hilditch, A.J. & Rey, M.E.C., 2010, 'Arbuscular myccorhizal fungi associated with cassava (*Manihot esculenta* Crantz) in South Africa', *South African Journal of Botany* 76, 102–111. https://doi.org/10.1016/j.sajb.2009.09.005

Straker, C.J., Weiersbye, I.M. & Witkowski, E.T.F., 2007, 'Arbuscular mycorrhiza status of gold and uranium tailings and surrounding soils of South Africa's deep level gold mines: I. Root colonization and spore levels', *South African Journal of Botany* 73, 218–225. https://doi.org/10.1016/j.sajb.2006.12.006

Strauss, U., Human, H., Gauthier, L., Crewe, R.M., Dietemann, V. & Pirk, C.W.W., 2013, 'Seasonal prevalence of pathogens and parasites in the savannah honeybee (*Apis mellifera scutellata*)', *Journal of Invertebrate Pathology* 114, 45–52. https://doi.org/10.1016/j.jip.2013.05.003

Stukenbrock, E.H., 2016, 'The role of hybridization in the evolution and emergence of new fungal plant pathogens', *Phytopathology* 106, 104–112. https://doi.org/10.1094/PHYTO-08-15-0184-RVW

Stutz, J.C., Copeman, R., Martin, C.A, & Morton, J.B., 2000, 'Patterns of species composition and distribution of arbuscular mycorrhizal fungi in arid regions of southwestern North America and Namibia, Africa', *Canadian Journal of Botany* 78, 237–245. https://doi.org/10.1139/b99-183

Suding, K.N., Harpole, W.S., Fukami, T., Kulmatiski, A., MacDougall, A.S., Stein, C., et al. 2013, 'Consequences of plant-soil feedbacks in invasion', *Journal of Ecology* 101, 298–308. https://doi.org/10.1111/1365-2745.12057

te Beest, M., Stevens, N., Olff, H. & van der Putten, W.H., 2009, 'Plant-soil feedback induces shift in biomass allocation in the invasive plant *Chromolaena odorata*', *Journal of Ecology* 97, 1281–1290. https://doi.org/10.1111/j.1365-2745.2009.01574.x

Tedersoo, L., Bahram, M., Põlme, S., Kõljalj, U., Yourou, N.S., Wijesundera, R., et al., 2014, 'Global diversity and geography of soil fungi', *Science* 346, 1256688. https://doi.org/10.1126/science.1256688

Tomkins, D.M., Carver, S., Jones, M.E., Krkošek, M. & Skerratt, L.F., 2015, 'Emerging infectious diseases of wildlife: A critical perspective', *Trends in Parasitology* 31, 149–159. https://doi.org/10.1016/j.pt.2015.01.007

Trappe, J.M., Claridge, A.W., Arora, D. & Smit, W.A., 2008, 'Desert truffles of the African Kalahari: Ecology, ethnomycology, and taxonomy', *Economic Botany* 62, 521–529.

Uhlmann, E., Görke, C., Petersen, A. & Oberwinkler. F., 2004, 'Arbuscular mycorrhizae from semiarid regions of Namibia', *Canadian Journal of Botany* 82, 645–653. https://doi.org/10.1139/B04-039

Vacher, C., Daudin, J.-J., Piou, D. & Desprez-Loustau, M.-L., 2010, 'Ecological integration of alien species into a tree-parasitic fungus network', *Biological Invasions* 12, 3249–3259. https://doi.org/10.1007/s10530-010-9719-6

van der Heijden, M.G.A., Martin, F.M., Selosse, M.-A. & Sanders, I.R., 2015, 'Mycorrhizal ecology and evolution: The past, the present, and the future', *New Phytologist* 205, 1406–1423. https://doi.org/10.1111/nph.13288

van der Westhuizen, G.C.A. & Eicker, A., 1994, *Field guide to Mushrooms of southern Africa*, Struik, Cape Town, South Africa.

Van Grunsven, R.H.A., Bos, F., Ripley, B.S., Suehs, C.M. & Veenendaal, E.M., 2009, 'Release from soil pathogens plays an important role in the success of invasive *Carpobrotus* in the Mediterranean', *South African Journal of Botany* 75, 172–175. https://doi.org/10.1016/j.sajb.2008.09.003

van Jaarsveld, L.C., Kriel, W.M. & Minaar, A., 2006, 'First report of *Puccinia thaliae* on Canna Lilly in South Africa', *Plant Disease* 90, 113. https://doi.org/10.1094/PD-90-0113C

van Kleunen, M. & Fischer, M., 2009, 'Release from foliar and floral fungal pathogen species does not explain the geographic spread of naturalized North American plants in Europe', *Journal of Ecology* 97, 385–392. https://doi.org/10.1111/j.1365-2745.2009.01483.x

van Wyk, P.S., 1973, 'Root and crown rot of silver trees', *Journal of South African Botany* 39, 255–260.

Vellinga, E.C., Wolfe, B.E. & Pringle, A., 2009, 'Global patterns of ectomycorrhizal introductions', *New Phytologist* 181, 960–973. https://doi.org/10.1111/j.1469-8137.2008.02728.x

Verbeeken, A. & Buyck, B., 2002, 'Diversity and ecology of tropical ectomycorrhizal fungi in Africa', in R. Watling, J.C. Frankland, A.M. Ainsworth, S. Isaac, & C.H. Robinson (eds.), *Tropical mycology Vol. 1, Macromycetes,* pp.11–24, CAB International, Wallingford, UK.

Vettraino, A.M., Morel, O., Perlerou, C., Robin, C., Diamandis, S. & Vannini, A., 2005, 'Occurrence and distribution of *Phytophthora* species in European chestnut stands, and their association with ink disease and crown decline', *European Journal of Plant Pathology* 111, 169–180. https://doi.org/10.1007/s10658-004-1882-0

Vitullo, D., De Curtis, F., Palmieri, D. & Lima, G., 2014, 'Milkwort (*Polygala myrtifolia* L.) decline is caused by *Fusarium oxysporum* and *F. solani* in Southern Italy', *European Journal of Plant Pathology* 140, 883–886. https://doi.org/10.1007/s10658-014-0514-6

Vizzini, A., Zotti, M. & Mello, A., 2009, 'Alien fungal species distribution: The study case of *Favolaschia calocera*', *Biological Invasions* 11, 417–429. https://doi.org/10.1007/s10530-008-9259-5

Von Broembsen, S.L., 1984a, 'Occurrence of *Phytophthora cinnamomi* on indigenous and exotic hosts in South Africa, with special reference to the south-western Cape Province', *Phytophylactica* 16, 221–225.

Von Broembsen, S.L., 1984b, 'Distribution of *Phytophthora cinnamomi* in rivers of the south-western Cape Province', *Phytophylactica* 16, 227–229.

Von Broembsen, S.L. & Brits, G.J., 1985, 'Phytophthora root rot of commercially cultivated Proteas in South Africa', *Plant Disease* 69, 211–213. https://doi.org/10.1094/PD-69-211

Von Broembsen, S.L. & Kruger, F.J., 1985, '*Phytophthora cinnamomi* associated with mortality of native vegetation in South Africa', *Plant Disease* 69, 715–717. https://doi.org/10.1094/PD-69-715

Wehner, J., Powell, J.R., Muller, L.A.H., Caruso, T., Veresoglou, S.D., Hempel, S. et al., 2014, 'Determinants of root-associated fungal communities within Asteraceae in a semi-arid- grassland', *Journal of Ecology* 102, 425–436. https://doi.org/10.1111/1365-2745.12197

Wingfield, M.J., Coetzee, M.P.A., Crous, P.W., Six, D. & Wingfield, B.D., 2010, 'Fungal phoenix rising from the ashes?' *IMA Fungus* 1, 149–153. https://doi.org/10.5598/imafungus.2010.01.02.06

Wingfield, M.J., Slippers, B., Roux, J. & Wingfield, B.D., 2001, 'Worldwide movement of exotic forest fungi, especially in the tropics and the southern hemisphere', *BioScience* 51, 134–140. https://doi.org/10.1641/0006-3568(2001)051[0134:WMOEFF]2.0.CO;2

Wood, A.R. & Ginns, J., 2006, 'A new dieback disease of *Acacia cyclops* in South Africa caused by *Psuedolagarobasidium acaciicola* sp. nov.', *Canadian Journal of Botany* 84, 750–758. https://doi.org/10.1039/B06-032

Wood, A.R. & Scholler, M., 2002, '*Puccinia abrupta* var. *partheniicola* on *Parthenium hysterophorus* in Southern Africa', *Plant Disease* 86, 327. https://doi.org/10.1094/PDIS.2002.86.3.327A

Xingjun, Y., Dan, Y., Zhijun, L. & Keping, M., 2005, 'A new mechanism of invader success: Exotic plant inhibits natural vegetation restoration by changing soil microbe community', *Chinese Science Bulletin* 50, 1105–1112. https://doi.org/10.1360/04WC0280

Zachariades, C., Paterson, I.D., Strathie, L.W., Hill, M.P. & Wilgen, B.W.V., 2017, 'Assessing the status of biological control as a management tool for suppression of invasive alien plants in South Africa', *Bothalia* 47(2), a2142. https://doi.org/10.4102/abc.v47i2.2142

Zhang, Q., Yang, R., Tang, J., Yang, H., Hu, S. & Chen, X., 2010, 'Positive feedback between mycorrhizal fungi and plants influences plant invasion success and resistance to invasion', *PLoS One* 5, e12380. https://doi.org/10.1371/journal.pone.0012380

15

Railway side mapping of alien plant distributions in Mpumalanga, South Africa

Authors:
Ndifelani Mararakanye[1] ⓘ
Modau N. Magoro[2]
Nomakhazi N. Matshaya[3]
Matome C. Rabothata[2]
Sthembele R. Ncobeni[3]

Affiliations:
[1]Directorate: Information Services, Department of Agriculture, Rural Development, Land and Environmental Affairs, South Africa

[2]Directorate: Veld, Pasture Management and Nutrition, Department of Agriculture, Rural Development, Land and Environmental Affairs, South Africa

[3]Rail Network Division, Transnet Freight Rail, South Africa

Corresponding author:
Ndifelani Mararakanye, nmararak@gmail.com

Background: Alien plant invasions are among the major threats to natural and semi-natural ecosystems in South Africa on approximately 18 million hectares of land. Much of the available data are not suitable for planning of local scale management because it is presented at a quarter degree grid square scale, which makes accurate location and estimates of invaded areas difficult.

Objectives: The aim was to identify the dominant alien plant species and quantify their areal extent along a 479 km railway corridor in the Mpumalanga province.

Method: The extent of the invaded area was obtained by manual digitising of alien plant distribution and density from Satellite Pour l'Observation de la Terre 5 imagery and by further applying an Iterative Self-Organising Data Analysis technique of the unsupervised classification method. Species' occurrences were located and identified in the field using a Global Positioning System.

Results: The most dominant invaders in terms of the number of individual polygons and the infested area were *Eucalyptus* spp., *Acacia* spp., *Populus alba* L., *Pinus patula* Schltdl & Cham., *Salix babylonica* L. and *Caesalpinia decapetala* (Roth) Alston. These species have also been previously classified as major invaders, although the *Conservation of Agricultural Resources Act* regulations permit their planting provided spreading to adjacent areas is avoided except for *C. decapetala*, which must be cleared under all circumstances.

Conclusion: Knowledge of the species' occurrence and their extent will assist landowners and relevant authorities to control the spread of alien plants, which impact rail safety, agricultural production, water availability and biodiversity.

Introduction

Invasive alien plants are a major threat to biodiversity because of their effects on the population dynamics of native species, effects on community dynamics (e.g. species richness, diversity and trophic structure) and disruption of ecosystem processes and functioning (Van Wilgen et al. 2001). Invasive alien plants threaten agricultural productivity, forestry, human health and biodiversity (e.g. Le Maitre, Versfeld & Chapman 2000; Richardson & Van Wilgen 2004). Richardson et al. (2000) define alien species as taxa whose presence in a landscape is because of human introduction either accidentally or intentionally. Alien plants were introduced to South Africa for a range of purposes such as crops and garden ornamentals, for timber and firewood, for stabilising sand dunes and as hedge plants. Some introduced species became naturalised, reproducing consistently and have become well established, spreading rapidly (Van Wilgen et al. 2001). Certain alien plant species (particularly when growing in dense uncontrolled infestations) consume more water than indigenous plants and may contribute to declining underground water resources, surface water run-off and stream flow reduction (Le Maitre et al. 2000).

It was previously estimated that approximately 10 000 000 ha (8%) of land in South Africa was invaded by various taxa of alien species with the majority in the Western Cape followed by the Mpumalanga, KwaZulu-Natal and Limpopo provinces (Le Maitre et al. 2000). Recent estimates show that the figure has almost doubled and now alien plants infest more than 18 000 000 ha of South Africa (Kotzé et al. 2010). Plant invasions occur in both natural ecosystems and man-made environments or disturbed areas such as arable land, road and railway sides (Dar, Reshi & Shah 2015; Pysek et al. 2012; Rutkovska et al. 2013). Roads are often regarded as a contributing factor to the spatial spread because they act as pathways for alien plant species (Barbosa et al. 2010; Dar et al. 2015). Unlike roadside alien invasions which have received considerable attention in the literature (e.g. Barbosa et al. 2010; Christen & Matlack 2006; Dar et al. 2015; Milton & Dean 1998), invasions along railway lines have not been widely investigated, particularly in

South Africa. The Southern African Plant Invaders Atlas (SAPIA) database (2016) provides information on railway side invasion in certain areas of South Africa, including parts of the Mpumalanga province. Woody alien plants along railway sides are a concern to railway infrastructure and rail operations because dense infestations often conceal railway crossings, endangering human life (Rutkovska et al. 2013). Infestations along railways may also act as a starting point for the spread of species to the neighbouring agricultural or natural landscape.

An inventory of alien invasive species, an understanding of invasion processes and the management history of a region would constitute the baseline data necessary for effective management (Masubelele, Foxcroft & Milton 2009). Richardson et al. (2005) described three data sources of alien plants for South Africa namely: the National Herbarium's Pretoria Computerised Information System (PRECIS, now called the Botanical Database of Southern Africa), the Catalogue of Problem Plants (Wells et al. 1986) and SAPIA. PRECIS contains locality records of 1 300 000 herbarium specimens of indigenous and naturalised plants at a quarter degree square level (Williams & Crouch 2017), with records of some 1460 taxa of alien plants, including cultivated species. The Catalogue of Problem Plants (Wells et al. 1986) includes records of 711 alien species (Richardson et al. 2005). SAPIA is the most comprehensive database on the spatial distribution of alien plants in South Africa containing 601 naturalised and casual alien plant species (Henderson 2007) with information on their distribution, abundance, habitat preferences and date of introduction summarised for the country (Henderson 1999, 2007; Richardson et al. 2005). SAPIA was developed in early 1994 by collating roadside surveys (Henderson 1999) and later incorporated data from other habitats including railway sides (Henderson 2007). SAPIA gives an acceptable idea of the invaded area and extent at regional scale, but because the capturing of information on alien plants was done at a quarter degree scale (Henderson 1999), it cannot be easily converted to estimate the location and extent of the invaded area at local scale (see Van Wilgen et al. 2001). Determining the extent of the invaded area is required for effective planning and implementation of control measures.

In Mpumalanga province, our initial field observations showed that woody alien plants along railway sides are increasing and are spreading to and/or from the neighbouring land because of the lack of appropriate control. This is a major concern to landowners who are bound by conservation laws and regulations to manage these populations (Department of Agriculture 2001; Department of Environmental Affairs 2014a). Effective management of alien invasive plants should be based on knowledge of species' distributions, modes and rates of spread, potential and known effects, and control methods (Crimmins, Mauzy & Studd 2008).

In this context, we aimed to identify the dominant alien plant species and quantify their distribution and density along 479 km of railway corridors through endangered ecosystems and other adjacent areas in the Mpumalanga province,

South Africa. To achieve this aim, the study had the following objectives: (1) to map all visible woody plant species from satellite imagery; (2) to survey and identify different alien plant species within each mapping unit including other visible herbaceous, shrub, grass and succulent alien species; and (3) to compare the results with other databases available for the study area. The data will be the first step towards developing proper control measures for implementation by landowners in order to manage and reduce the spread and impact of alien vegetation, which is a problem for railway managers and also for ecosystem functioning.

Research method and design
Materials

In this study, Satellite Pour l'Observation de la Terre (SPOT) 5 satellite imagery acquired in 2014 and 2015 was used. This imagery has four spectral bands (green, red, near infrared and shortwave infrared) and was accessed from the Mpumalanga provincial Department of Agriculture, Rural Development, Land and Environmental Affairs. It was preferred over Landsat (http://landsat.usgs.gov/) and Moderate Resolution Imaging Spectroradiometer (MODIS) (https://earthdata. nasa.gov/earth-observation-data) satellite imageries that are freely available on the Internet. Although Landsat and MODIS have more spectral bands (11 for Landsat 8 and 36 for MODIS) than SPOT 5 (4 multispectral bands), their spatial resolution (15 m to 30 m and 250 m to 1 km, respectively) is poor compared to the 10 m of SPOT 5 imagery, which enables the detection of a single large tree. The imagery was obtained preprocessed of the geometric and radiometric distortions by the South African National Space Agency and is available for the whole of South Africa. Aerial photographs with a higher spatial resolution of 0.5 m were available but could not be used because of their poor temporal resolution and the latest photographs in the study area were from 2010.

Study site

The study area is within the grassland biome of the Mpumalanga province, South Africa (Figure 1), and comprises multiple railway line sections including Ermelo to Machadodorp (119 km), Buhrmanskop to Lothair (49 km), Trichardt to Davel (64 km), Ermelo to Davel (37 km), Maviristad to Ermelo (24 km), Davel to Gelukplaas (47 km), Gelukplaas to Wonderfontein (56 km), Ogies to Gelukplaas (59 km) and Ogies to Blackhill (24 km). The area surveyed was up to 200 m on both sides of the railway line. Vegetation types are dominated by Eastern Highveld, Soweto Highveld, Kangwane Montane grasslands and Lydenburg Thornveld (Mucina & Rutherford 2006). According to Ferrar and Lotter (2007), the ecosystems in this region are highly endangered, and they have undergone degradation and are at risk of further transformation because of coal mining and associated infrastructure development as well as agriculture. The study area forms part of the Highveld region, inland of the great escarpment with elevation ranging from 1500 to 1800 m above sea level. Most of the landscape in the study area is underlain by rocks belonging to the Vryheid Formation of the broader Karoo supergroup,

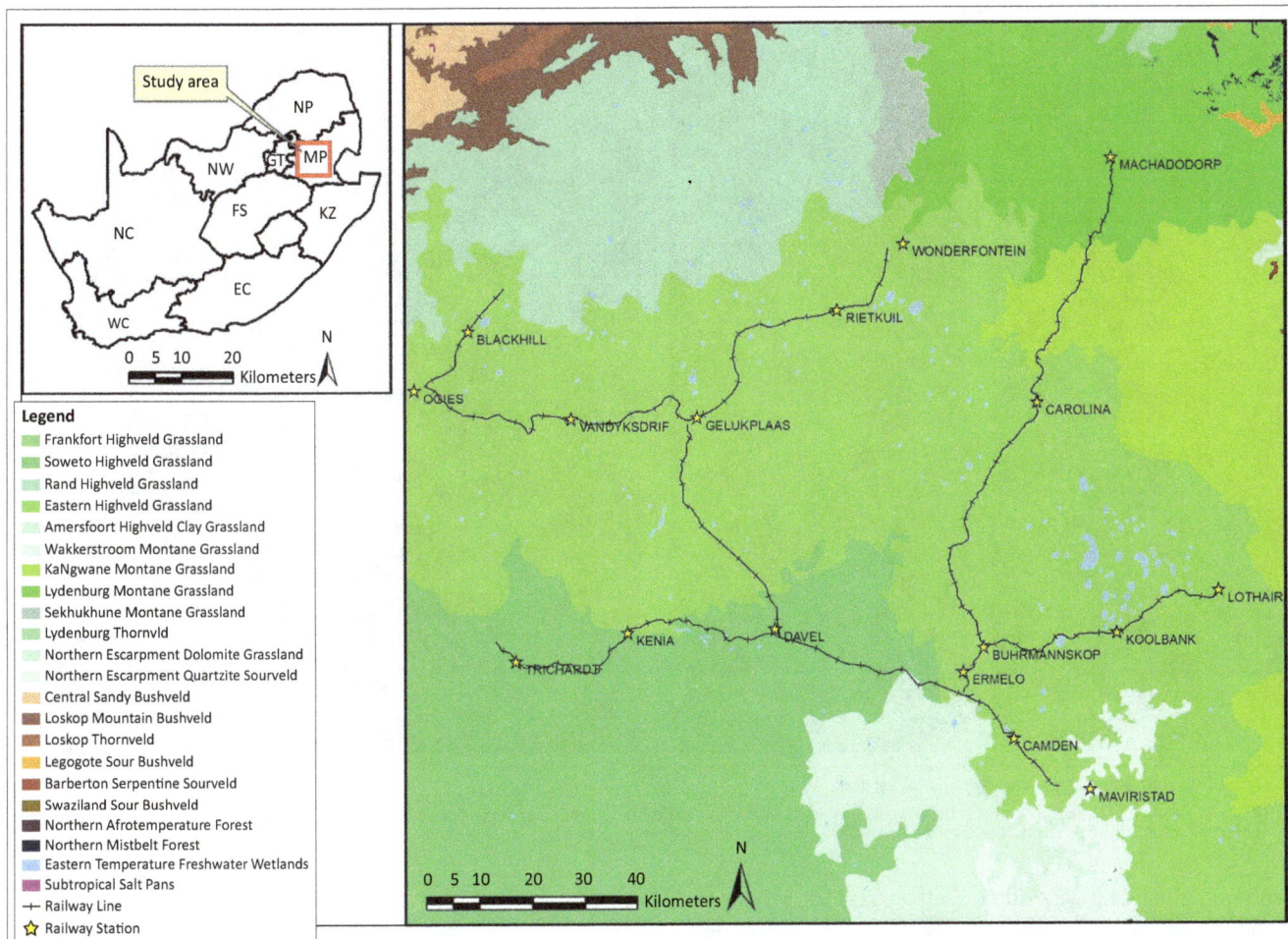

Inset map notes: WC, Western Cape; NC, Northern Cape; EC, Eastern Cape; KZ, KwaZulu-Natal; FS, Free State; NW, North West; GT, Gauteng; MP, Mpumalanga; LP, Limpopo provinces.

FIGURE 1: Location of the selected railway line in the study area and vegetation type boundaries from a database edited by Mucina and Rutherford (2006).

which is intruded by dolerite rocks (Geological Survey 1986a, 1986b, 1986c, 1986d).

The Vryheid Formation is characterised by grey micaceous shale, course-grained sandstone and subordinate grit and coal beds found at the basin margin (Johnson 1976; Turner 2000). It hosts some of the major coalfields in South Africa including Witbank, Highveld and Ermelo (Jeffrey 2005), which has resulted in extensive usage of the railway networks. Soils developed from this Vryheid Formation favour arable land use and are characterised by red and yellow apedal soils (Land types of South Africa and soil inventory databases 1984). The area is a summer rainfall region with a mean annual rainfall ranging from 600 to 800 mm. Earlier field observations indicated that alien vegetation was widespread along the railway lines in this province, leading us to choose the study site in Figure 1.

Mapping of alien plant using remote sensing

The first step of the mapping phase was to separate the woody vegetation from herbaceous and grass vegetation. This was based on visual interpretation of the SPOT 5 satellite imagery and manual digitising of the background woody vegetation boundary in ArcMap 10.3.1 (ArcGIS ESRI 2015) along a 200 m buffer of railway line. Visual interpretation

was made easier because of the sharp colour and texture contrast between woody vegetation and the surrounding area when a satellite image is displayed using true colour bands or a combination of bands (see Figure 2). Although manual digitising has the potential to produce accurate data, it remains error prone and some of the errors common in this study were related to the subjective interpretation and small island polygons. While care was taken when interpreting a satellite image, the process itself is subjective and the accuracy is based on the interpreter's knowledge and experience. Island polygons are those with external borders and one or more internal islands (Gong & Li 2000) and these are often not desired in the output map. While large islands can be easily masked out when digitising, it remains a challenge for small island polygons because digitising is scale dependent. All these errors may be negligible when working on a large area, but they tend to increase the chance of overestimation and underestimation of the infested area when working on small study areas like railway corridors.

To solve the problems of island polygons and subjectivity in the output map associated with manual digitising, the unsupervised classification technique was implemented in ArcMap using the digitised polygons as a mask. Unsupervised classification uses the Iterative Self-Organising Data Analysis

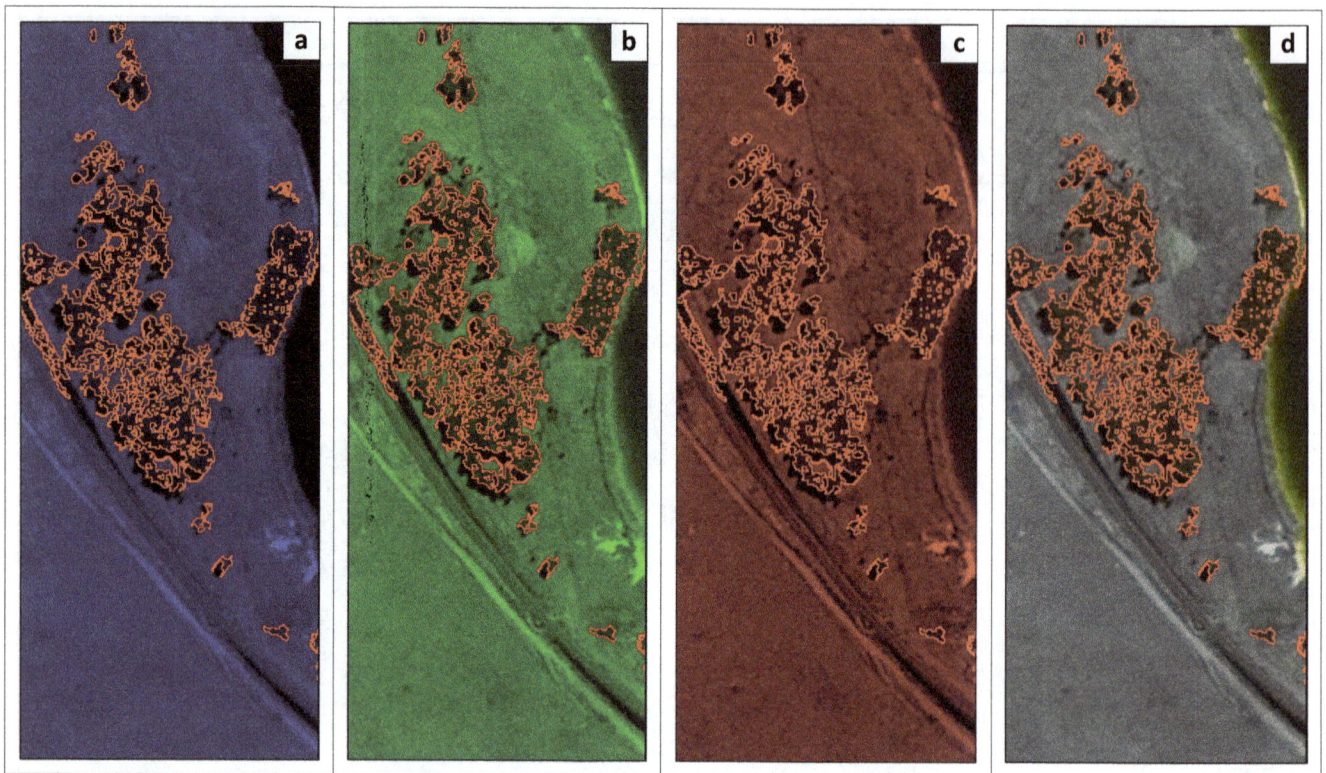

(a), blue band; (b), green band; (c), red band; (d), true colour band.

FIGURE 2: An example of the colour contrast between woody vegetation (red outline) and the surrounding non-woody vegetation using the three selected colour bands and their combination (true colour) of the Satellite Pour l'Observation de la Terre 5 image in the area between Maviristad and Ermelo.

technique clustering algorithm that aggregates unknown pixels in an image, based on their natural groupings (Lillesand, Kiefer & Chipman 2008). It is one of the oldest methods of feature extraction from satellite imagery that has been used effectively for mapping vegetation across the globe (e.g. Anchang, Ananga & Pu 2016; Peerbhay, Mutanga & Ismail 2015). The use of unsupervised classification enables the exclusion of island polygons and eliminates subjectivity in the interpretation and outlining of the vegetation boundary in the output spectral classes.

Species identification

A rapid reconnaissance survey along the service road running parallel to the railway lines in our study area was carried out from November 2015 to February 2016 to identify the species occurring in each polygon of the unsupervised thematic maps. Only the species visible from a moving vehicle were recorded within a distance of up to 100 m on either side of the road travelled. A Garmin nüvi Global Positioning System (GPS) was used to record the co-ordinates for each woody vegetation stand or single tree. The co-ordinates were later loaded into the ArcGIS 10.3.1 (ArcGIS ESRI 2015) software for correlation with the unsupervised classification spectral classes. The spatial join technique in ArcGIS 10.3.1 was used to link GPS data with the spectral classes map. Unlike the quadrat inventory method used often in ecology (e.g. Appiah 2013; Houeto et al. 2013; Molano-Flores et al. 2015) which provides a record of species in a grid format (e.g. 25 km × 27 km grid used in SAPIA and PRECIS) (see Richardson et al. 2005), the use of a GPS allowed

for the recording of species' location which could be linked to the spectral classes. Sometimes field observation revealed more than one alien plant species in a vegetation cluster. Because of the limitation of SPOT 5 imagery in terms of spectral bands for separation of individual species, it was decided to record all the species present in a cluster during field observations and spatial joining.

Interpretation of the satellite imagery was biased towards woody plant species while grasses, succulents, shrubs and herbaceous plants were difficult to detect. However, field observations revealed some alien grass, succulent, shrub and herbaceous species in the study area. An attempt was made to record their location by estimating the extent of the invaded area in the field and by maximising the use of colour and texture contrast during manual digitising to delineate the invaded area. Field data also revealed the occurrence of indigenous woody species, mainly *Diospyros lycioides* Desf. and *Vachelia karroo* (Hayne) Banfi & Galasson, along the Ermelo to Machadodorp railway line. The associated spectral class polygons were subsequently removed from the final alien plant distribution map.

Results
Alien plant distribution

Figure 3 illustrates the spatial distribution of alien plants in the study area. The area with noticeably dense alien vegetation is near Lothair railway station and is characterised by commercial plantations of the *Pinus patula* Schltdl. & Cham. and *Eucalyptus* spp., which are spreading rapidly

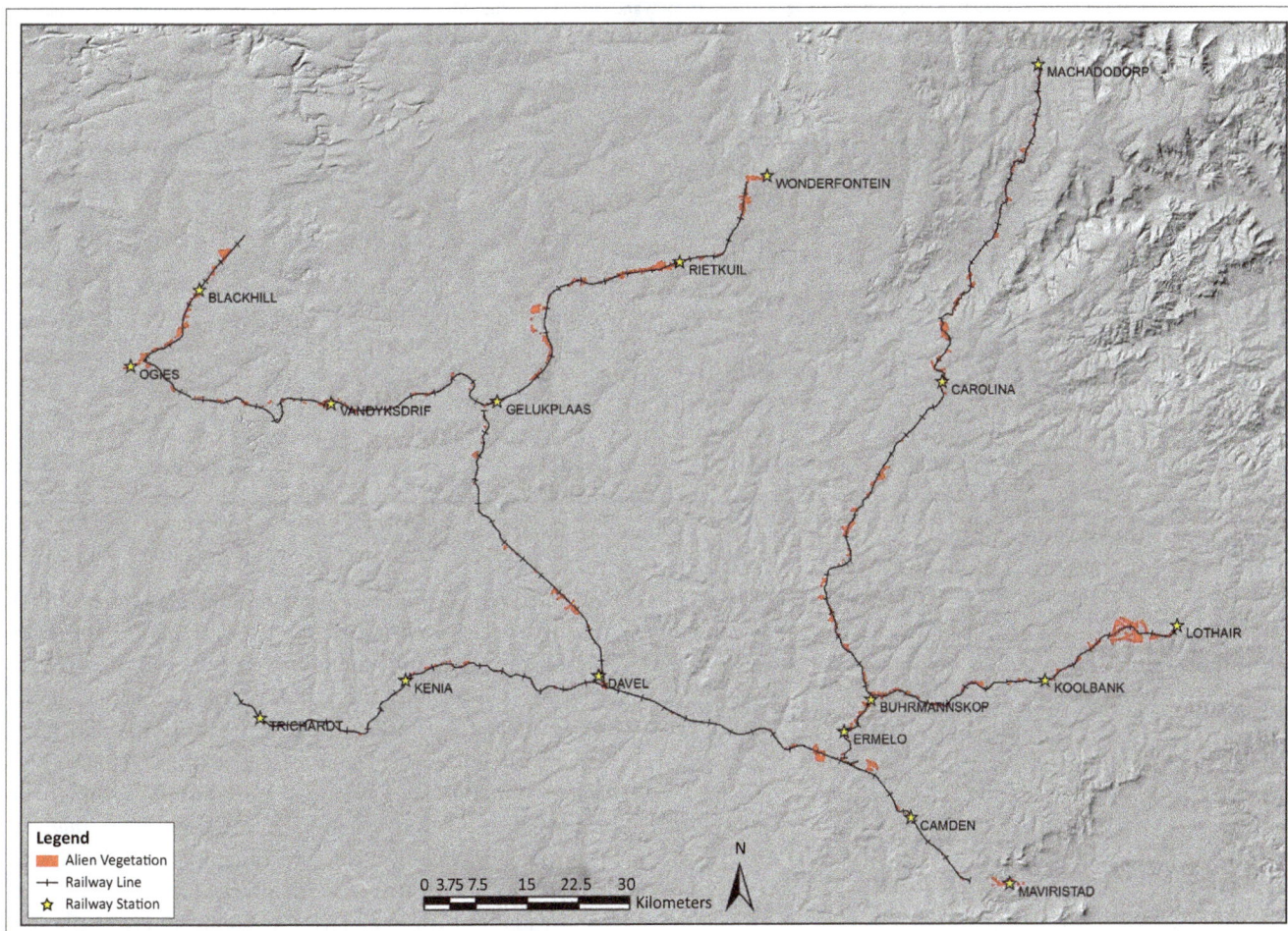

Red polygons indicate alien vegetation cluster or individual tree. Only polygons with the closest boundary from railway line of up to 100 m on either side were considered because they are likely to spread and interfere with rail operations.

FIGURE 3: Spatial distribution of alien vegetation along the railway line in Mpumalanga province.

along the railway corridor. Other sections of the railway line with dense alien vegetation include the Ermelo to Machadodorp route, Gelukplaas to Wonderfontein and Ogies to Blackhill. These are areas where *Eucalyptus* spp. and *Acacia* spp. dominate. Although the main purpose of this study was to map mainly woody plants, field observation revealed the occurrence of two herbaceous (*Argemone mexicana* L. and *Verbena bonariensis* L.), two succulent (*Agave sisalana* Perrine and *Opuntia ficus-indica* (L.) Mill.), two shrub (*Solanum mauritianum* Scop. and *Rubus* spp.) and one grass (*Arundo donax* L.) species, which were subsequently added to the spatial distribution map.

Species identified and area extent

The species observed belong to 12 different families (Table 1). The widespread invaders prevalent in more than four of the railway networks are the Myrtaceae (*Eucalyptus* spp.), Fabaceae (*Acacia* spp. and *Caesalpinia decapetala* (Roth) Alston), Pinaceae (*Pi. patula*), Salicaceae (*Salix babylonica* L. and *Populus alba*) and Cactaceae (*O. ficus-indica*).

It was difficult to quantify precisely the area invaded by each individual species from the satellite images because some polygons have more than one woody species and SPOT 5

spectral bands cannot be used to separate individual species. However, Table 2 shows that of major concern among the widespread invaders are the *Eucalyptus* and *Acacia* species, which affect a combined area of more than 1300 ha and have the highest occurrence or number of individual polygons. *Populus alba* and *Pi. patula* affect 10 and 9 ha, respectively, while the areal extent of *Sa. babylonica* and *C. decapetala* is negligible (3.1 and 1.4 ha, respectively). In the case of the latter two species, they are represented in a large number of polygons, but this is as isolated individuals or in a few stands.

Other species are not prominent and only occur in isolated stands of trees, or they are succulents, shrubs and herbaceous plants. Table 1 provides details of the sections of the railway along which *Ag. sisalana*, *Aru. donax*, *So. mauritianum*, *Melia azedarach* L., *Prunus persica* (L.) Batsch, *Arg. mexicana*, *V. bonariensis* and *Rubus* spp. were observed.

Discussion
Identified species

The most dominant alien plant species in terms of the number of times they were observed are the *Eucalyptus* spp., *Acacia* spp., *C. decapetala*, *Pi. patula*, *Sa. babylonica* and *Po. alba*. *Eucalyptus* spp., *Acacia* spp. and *Pi. patula* are the

TABLE 1: Observed alien plant species organised by family, including their growth form and localities in which they were found along a 479 km railway line and on adjacent farms.

Family	Species	Common name	Growth form	Locality
Agavaceae	Agave sisalana	Sisal	Succulent	TD, OB
Cactaceae	Opuntia ficus-indica	Prickly pear	Succulent	OG, GW, TD, BL, EM
Fabaceae	Caesalpinia decapetala	Mauritius thorn	Tree	OB, GW, DG, TD, EM
	Acacia spp.	Wattle	Tree	OB, OG, GW, DG, TD, ME, BL, EM
Meliaceae	Melia azedarach	Seringa	Tree	OG
Myrtaceae	Eucalyptus spp.	Gum tree	Tree	OB, OG, GW, DG, TD, ME, BL, EM, ED
Papaveraceae	Argemone mexicana	Mexican poppy	Herb	OB
Pinaceae	Pinus patula	Pine	Tree	OB, OG, GW, TD, ED, ME, BL, EM
Poaceae	Arundo donax	Water reeds	Grass	OB, OG, TD, EM
Rosaceae	Prunus persica	Peach	Tree	OG
	Rubus spp.	Wild berries	Shrub	OG, GW
Salicaceae	Populus alba	Poplar tree	Tree	OB, OG, GW, DG, TD, BL, EM
	Salix babylonica	Weeping willow	Tree	OG, TD, ED, ME, BL, EM
Solanaceae	Solanum mauritianum	Bugweed	Shrub	OB, OG
Verbenaceae	Verbena bonariensis	Desel	Herb	GW

EM, Ermelo to Machadodorp; BL, Buhrmanskop to Lothair; TD, Trichardt to Davel; ED, Ermelo to Davel; ME, Maviristad to Ermelo; DG, Davel to Gelukplaas; GW, Gelukplaas to Wonderfontein; OG, Ogies to Gelukplaas; OB, Ogies to Blackhill.

TABLE 2: Species observed and the extent of the area invaded calculated using the ArcMap calculate areas tool after converting the spectral classes derived from unsupervised classification of the satellite imagery to polygons.

Species (scientific name)	Occurrence (no. of individual polygons)	Area (hectares)
Eucalyptus spp.	125 458	813.62
Acacia spp.	46 081	456.73
Eucalyptus spp./Acacia spp.	9666	46.46
Populus alba	2534	10.29
Pinus patula	494	9.00
Argemone mexicana	19	7.78
Salix babylonica	443	3.10
Pinus patula/Eucalyptus spp.	8	2.80
Eucalyptus spp./Acacia spp./Solanum mauritianum	1	2.28
Acacia spp./Solanum mauritianum	15	1.79
Salix babylonica/Populus alba	23	1.66
Caesalpinia decapetala	90	1.43
Arundo donax	19	0.90
Eucalyptus spp./Acacia spp./Caesalpinia decapetala/Salix babylonica	332	0.60
Pinus patula/Eucalyptus spp./Acacia spp.	1	0.57
Pinus patula/Acacia spp.	2	0.42
Eucalyptus spp./Opuntia ficus-indica	1	0.28
Opuntia ficus-indica	8	0.25
Eucalyptus spp./Populus alba	44	0.22
Rubus spp.	5	0.21
Melia azedarach	1	0.07
Verbena bonariensis	1	0.07
Acacia spp./Opuntia ficus-indica	1	0.03
Solanum mauritianum	15	0.02
Agave sisalana	2	0.02
Prunus persica	2	0.01

The grouped species indicate polygons where more than one species occur and where species were not separated because of the lack of spatial and spectral resolutions of Satellite Pour l'Observation de la Terre 5 imagery utilised.

associated with reduction in stream flow (Henderson 1991). All the species considered to be dominant have been classified as major invaders previously by Nel et al. (2004), including *C. decapetala* which falls in category 1 of the *Conservation of Agricultural Resources Act* (CARA) No. 43 of 1983 regulation (Department of Agriculture 2001), which means that the species must be controlled under all circumstances. *Eucalyptus* spp., *Acacia* spp., *Pi. patula*, *Sa. babylonica* and *Po. alba* fall into category 2 of the CARA regulation (Department of Agriculture 2001), which means that because of their commercial value they may be planted in demarcated areas subject to a permit that requires that steps are taken to control their spread to adjacent areas. Other species may not be regarded as dominant in this study, but they raise concerns. Nel et al. (2004) categorised *M. azedarach*, *V. bonariensis*, *O. ficus-indica*, *So. mauritianum* and *Aru. donax* as major invaders because they are well established and are already causing substantial impacts on the natural and semi-natural ecosystems. All these species except *V. bonariensis* appear on the list of species prioritised and ranked by Robertson et al. (2003) based on their potential invasiveness, spatial characteristics, potential impact and potential for control. Additionally, all the species recorded except *Sa. babylonica* and *Pr. persica* have been listed in various categories of the most recent alien and invasive species regulations (Department of Environmental Affairs 2014a) of the *National Environmental Management: Biodiversity Act* (NEM:BA), 2004 (Act No. 10 of 2004) (Department of Environmental Affairs 2014b). This includes category 1b species such as *Arg. mexicana*, *Aru. donax*, *C. decapetala*, *M. azedarach*, *O. ficus-indica*, *Rubus* spp., *So. mauritianum* and *V. bonariensis*, which the NEM:BA regulations require be controlled. *Eucalyptus* spp. are also categorised as 1b if they are in a listed ecosystem. Other species such as *Pi. patula* and *Po. alba* are category 2 species, which requires that a permit is obtained by the landowner and that the spread of these species from that land is prevented. It is therefore proposed that these species should be prioritised when planning local scale management.

most extensively planted exotic plants for wood and timber in the whole of the southern hemisphere including South Africa (Wingfield et al. 2002). However, since their introduction, some have escaped from cultivation and spread into other areas where they compete with indigenous vegetation. The Salicaceae family which includes *Po. alba* and *Sa. babylonica* are rapid-growing deciduous trees that largely occur along watercourses and marshy areas and are

Limitations of the satellite image

The spatial distribution map is limited to large trees because of the limitation of spectral bands and spatial resolution of SPOT 5 satellite imagery to detect smaller trees. Field observations revealed the occurrence of young trees in other parts of the study area. These young trees, mainly *Eucalyptus* species blend with the surrounding grassland area, which makes them difficult to separate by visual image interpretation and unsupervised classification, given the limited spectral bands and spatial resolution. Interpretation was also made difficult because the plants may have been too young or non-existent when the satellite image was taken. Figure 4 illustrates the example of young or resprouting *Eucalyptus* spp. in the study area, which could not be detected from a satellite image. It was observed that young *Eucalyptus* spp. are spreading rapidly to and from the neighbouring agricultural lands. Generally, *Eucalyptus* spp. are recognised as one of the fastest growing trees that may reach up to 6 m in 4 years (Bennett 2011). It is therefore important to control the spread of these young plants before they reach maturity and the costs of control increase substantially.

Some of the land parcels next to the railway line have been planted with *Eucalyptus* spp. and *Pi. patula* for commercial purposes and this appears to be the initiation point for potential spread (see Figure 5). Once they spread into the railway corridor, they spread further into other areas if not controlled. Richardson (1998) claimed that invasion from commercial plantations is a bigger problem than from other forms of introduction because in plantations, species occur in greater numbers over a large area. However, not all alien plant species along the railway corridor in this study can be associated with commercial plantations. Railway development itself is a disturbance of the environment and has the potential for increasing dissemination of propagules, which makes the ecosystems more susceptible to invasion by alien species (Richardson & Van Wilgen 2004). The disturbed area is particularly important during the early stage of invasion because it creates vacant niches that alien plants can colonise (Masters & Sheley 2001). Rutkovska et al. (2013) found that some of the alien taxa in Latvia were associated with topsoil disturbance because of the construction and maintenance of railway lines. It is therefore possible that the development of railway networks in the study area may have promoted the distribution and spread of some of the alien plants.

Comparison with other studies

Our study was not intended to provide a complete list of alien plant species occurring in the area and we did not record all species identified in other databases. This is mainly because the current inventory was only obtained along the railway line and adjacent farms, whereas other studies have included observations in environments such as watercourses, grassland, savannah and human-modified habitats (e.g. Henderson 2007). For easier comparison with other comprehensive databases on alien plants such as SAPIA and PRECIS, alien species were summarised per quarter degree grid as shown in Table 3.

The current study and SAPIA have more similar species recorded per quarter grid than the PRECIS database. The

Source: Photo by Ndifelani Mararakanye

FIGURE 4: An example of young or resprouting *Eucalyptus* species along the railway line and adjacent farms that could not be detected from the satellite image.

Source: Photo by Ndifelani Mararakanye

FIGURE 5: *Eucalyptus* species that appear to be planted but poor control measures have resulted in some plants growing close to the railway infrastructures.

most prominent species in the current study list and SAPIA database include the *Acacia* spp., *Eucalyptus* spp., *Sa. babylonica*, *So. mauritianum*, *Po. alba*, *O. ficus-indica*, *Agave* spp., *Pi. patula*, *Aru. donax* and *Rubus* spp. (Henderson 1999). For years, SAPIA has been the most important data source on the distribution of alien plant species; thus, a good correlation with SAPIA would suggest that the current study could also be used when planning for implementation of control measures in Mpumalanga province. The reason there is more similarity in species observed is probably because the SAPIA database also includes observations made along railway lines in grids 2529CC, 2529DC, 2530CA, 2629AA, 2629AB, 2629AC, 2629AD, 2629DB and 2630CA. There are other species listed by SAPIA as occurring along railway lines in our study site that were not observed in the current study and these include herbaceous (*Campuloclinium macrocephalum* (Less.) DC., *Datura ferox* L., *Cuscuta campestris* Yunck., *Cuscuta suaveolens* Ser., *Solanum elaeagnifolium* Cav. and *Cirsium vulgare* (Savi) Ten.), shrub (*Pyracantha angustifolia* (Franch.) C.K.Schneid., *Pyracantha crenulata* (D.Don) M.Roem.) and grass (*Pennisetum setaceum* (Forssk.) Chiov., *Pennisetum villosum* R.Br. ex Fresen.) species. The reason these species were not observed in the current study could be that they are not widespread or they may have been temporarily or permanently cleared during railroad maintenance work. Some species may also have been dormant at the time the field survey was undertaken.

The inventory of alien plants in this study is notably different to the list from the PRECIS database. This is probably because PRECIS has limited data for alien species, and it has a bias towards herbaceous and shrub species, whereas this study

primarily focused on woody plants. Similar species were only observed in grid 2529DD (*Acacia* spp. and *V. bonariensis*) and 2629DB (*Sa. babylonica*). The differences justify the importance of carrying out a field study at the local level for planning and implementation of control measures. Both SAPIA and PRECIS are useful at the national level when planning and prioritising species and areas for management action (see Nel et al. 2004), whereas the current study will facilitate the implementation of control measures at the local scale.

Conclusion

By combining the maps from remote sensing with field observations, it helped to improve our knowledge of alien plant species' distribution and the extent of the invaded area along the railway line in the Mpumalanga province. This approach can be used in other regions where woody plants dominate. This study found that just like many other disturbed environments which include roadsides, plantations, heavily grazed and cultivated lands, railway lines are invaded by a variety of alien plant species in the Mpumalanga province. The biggest threat is posed by *Eucalyptus* spp., *Acacia* spp., *C. decapetala*, *Pi. patula*, *Sa. babylonica* and *Po. alba* because they cover the most surface area around the railway lines and are represented in a noteworthy number of individual polygons. Other species such as *M. azedarach*, *V. bonariensis*, *O. ficus-indica*, *So. mauritianum* and *Aru. donax* may not necessarily be dominant in this study, but they need to be prioritised for control purposes because they have been identified elsewhere (e.g. Nel et al. 2004; Robertson et al. 2003) as major invaders that are well established and are already

TABLE 3: Comparison of the current species observation and alien plant species records in Southern African Plant Invaders Atlas and National Herbarium's Pretoria Computerised Information System databases.

Grid number	Species observed in this study	Species recorded in the SAPIA database (2016)	Alien plant species recorded in the PRECIS database (2016)
2529CC	*Eucalyptus* spp., *Acacia* spp., **Arundo donax**, *Pinus patula*, **Argemone mexicana**	*Acacia dealbata, A. decurrens, A. mearnsii,* **Cirsium vulgare, Cotoneaster pannosus, Crotalaria agatiflora,** *Eucalyptus* sp., **Pennisetum setaceum, P. villosum, Pontederia cordata,** *Populus alba, P. canescens, Salix babylonica,* **Solanum mauritianum, S. sisymbriifolium,** *Tipuana tipu*	*Solanum sisymbriifolium, Xanthium strumarium*
2529DC	*Eucalyptus* spp., *Acacia* spp., *Populus alba*	*Acacia dealbata, A. decurrens, A. mearnsii,* **Azolla filiculoides,** *Cotoneaster coriaceus,* **C. pannosus,** *Eucalyptus camaldulensis, Eucalyptus* sp., *Jasminum humile,* **Myriophyllum aquaticum,** *Opuntia ficus-indica,* **Populus alba, P. canescens, Prunus persica, Pyracantha angustifolia, P. crenulata,** *Salix babylonica, Schinus molle*	-
2529DD	*Rubus* spp., *Acacia* spp., **Verbena bonariensis, Opuntia ficus-indica, Caesalpinia decapetala,** *Eucalyptus* spp.	*Acacia dealbata, A. decurrens,* **A. longifolia,** *A. mearnsii, Agave americana americana,* **Arundo donax, Cirsium vulgare, Cotoneaster pannosus, Crotalaria agatiflora, Datura ferox,** *Eucalyptus* sp., *Jasminum humile, Populus alba, P. canescens, Prunus persica, Salix babylonica,* **Xanthium strumarium**	*Acacia dealbata,* **Datura ferox,** *Verbena brasiliensis*
2530CA	*Eucalyptus* spp., *Acacia* spp.	*Acacia dealbata, A. decurrens, A. mearnsii, A. melanoxylon,* **Arundo donax, Echium plantagineum,** *Eucalyptus* sp., *Pinus patula, Populus alba, P. canescens, Prunus persica,* **Pyracantha angustifolia, P. crenulata,** *Salix babylonica, S. fragilis,* **Verbena bonariensis, V. brasiliensis, Xanthium strumarium**	*Argemone ochroleuca,* **Verbena bonariensis**
2530CB	*Acacia* spp., *Populus alba, Pinus patula,* **Caesalpinia decapetala,** *Eucalyptus* spp.	*Acacia baileyana, A. dealbata, A. decurrens, A. mearnsii, A. melanoxylon, Achyranthes aspera, Agave americana americana,* **Araujia sericifera, Arundo donax, Cardiospermum grandifloru,** *C. halicacabum, Casuarina cunninghamiana, C. equisetifolia,* **Cereus jamacaru, Citrus limon, Cotoneaster franchetii, C. pannosus, Crotalaria agatiflora,** *Eucalyptus camaldulensis,* **E. grandis,** *Grevillea robusta,* **Jacaranda mimosifolia, Lantana camara, Ligustrum lucidum, Melia azedarach,** *Morus alba,* **Myriophyllum aquaticum,** *Nasturtium officinale,* **Opuntia ficus-indica,** *Pennisetum purpureum, Pinus elliottii, P. taeda, P. patula, Populus alba, P. canescens, Prunus persica, Psidium guajava,* **Pyracantha angustifolia, P. crenulata,** *Ricinus communis, Rubus* sp., *Salix babylonica, Schinus molle,* **Sesbania punicea, Solanum mauritianum, S. sisymbriifolium,** *Tithonia rotundifolia*	*Bidens pilosa, Casuarina equisetifolia,* **Cuscuta campestris, Cuscuta suaveolens, Datura stramonium, Lantana camara,** *Ricinus communis,* **Solanum mauritianum, Solanum sisymbriifolium, Verbena bonariensis**
2530CC	*Salix babylonica, Populus alba, Acacia* spp., **Arundo donax,** *Pinus patula,* **Eucalyptus** spp., **Opuntia ficus-indica**	*Acacia dealbata, A. decurrens, Eucalyptus* sp., *Pinus* sp., *Populus alba, P. canescens, Pyracantha angustifolia, Salix babylonica*	-
2629AA	**Solanum mauritianum,** *Salix babylonica, Acacia* spp., **Eucalyptus spp., Arundo donax,** *Populus alba, Pinus patula,* **Argemone mexicana, Melia azedarach, Caesalpinia decapetala**	*Acacia dealbata, A. decurrens, A. mearnsii, Agave americana americana,* **Arundo donax, Datura ferox,** *Eucalyptus* sp., **Pennisetum setaceum, P. villosum,** *Populus alba, P. canescens, Salix babylonica*	-
2629AB	*Rubus* spp., *Salix babylonica, Acacia* spp., **Caesalpinia decapetala, Eucalyptus spp., Arundo donax, Opuntia ficus-indica,** *Populus alba, Pinus patula, Prunus persica*	*Acacia dealbata, A. mearnsii, Bidens bipinnata, B. pilosa,* **Cirsium vulgare,** *Cosmos bipinnatus,* **Cuscuta campestris, Datura ferox,** *Hibiscus trionum, Hypochaeris radicata, Oenothera rosea,* **Opuntia ficus-indica,** *Oxalis corniculata, Paspalum dilatatum, Pennisetum clandestinum, Persicaria lapathifolia, Plantago lanceolata, P. virginica, Populus alba, P. canescens, Richardia humistrata, Rumex crispus, Salix babylonica, Schkuhria pinnata,* **Solanum mauritianum,** *Sonchus oleraceus, Symphyotrichum subulatum, Tagetes minuta,* **Verbena bonariensis, Xanthium strumarium**	-
2629AC	**Eucalyptus** spp.	*Acacia mearnsii,* **Datura ferox, D. stramonium,** *Eucalyptus* sp., *Pennisetum* sp., *Populus alba, P. canescens, Salix babylonica,* **Solanum elaeagnifolium**	-
2629AD	**Eucalyptus** spp., *Acacia* spp., **Arundo donax, Opuntia ficus-indica,** *Populus alba,* **Caesalpinia decapetala**	*Acacia baileyana, A. dealbata, A. melanoxylon, Agave americana americana,* **Cuscuta campestris, C. suaveolens, Datura ferox,** *Eucalyptus* sp., **Pontederia cordata,** *Salix babylonica*	*Convolvulus arvensis, Datura stramonium, Pontederia cordata*
2629BA	*Acacia* spp., **Eucalyptus spp., Caesalpinia decapetala,** *Populus alba,* **Opuntia ficus-indica, Arundo donax,** *Acacia* spp.	*Acacia dealbata,* **A. longifolia,** *A. mearnsii,* **Eucalyptus** sp., **Opuntia ficus-indica,** *Populus canescens*	-
2629BC	*Agave sisalana, Salix babylonica,* **Opuntia ficus-indica,** *Pinus patula,* **Eucalyptus** spp.	-	-
2629BD	**Eucalyptus spp.,** *Pinus patula, Populus alba,* **Opuntia ficus-indica,** *Acacia* spp.	*Acacia dealbata, A. mearnsii,* **Eucalyptus** sp., **Opuntia ficus-indica,** *Prunus persica,* **Pyracantha angustifolia, P. crenulata**	-
2629CA	**Eucalyptus** spp., *Salix babylonica, Prunus persica*	*Salix babylonica*	*Bidens pilosa*
2629CB	*Agave sisalana,* **Opuntia ficus-indica, Eucalyptus spp.**	-	-
2629DB	*Salix babylonica, Pinus patula,* **Eucalyptus** spp.	*Acacia mearnsii,* **Cestrum laevigatum, Datura ferox,** *Eucalyptus* **grandis,** *Salix babylonica*	*Bidens pilosa,* **Cestrum laevigatum, Datura stramonium,** *Salix babylonica*
2630AA	**Eucalyptus** spp., *Pinus patula, Acacia* spp.	*Acacia dealbata, A. decurrens, A. mearnsii,* **Eucalyptus** sp., *Fraxinus velutina,* **Opuntia ficus-indica,** *Pinus* sp., *Populus alba, P. canescens, Rubus rosifolius, Salix babylonica*	*Leptospermum laevigatum,* **Senna bicapsularis**
2630AC	*Salix babylonica,* **Opuntia ficus-indica,** *Acacia* spp., *Populus alba, Pinus patula,* **Eucalyptus** spp.	*Acacia dealbata, A. decurrens, A. mearnsii,* **Datura stramonium,** *Eucalyptus* sp.	-
2630AD	*Salix babylonica, Acacia* spp., *Pinus patula,* **Eucalyptus** spp.	*Acacia dealbata, A. mearnsii, A. melanoxylon, Achyranthes aspera,* **Arundo donax, Cirsium vulgare,** *Pinus patula, Populus canescens,* **Rubus cuneifolius,** *Salix babylonica,* **Solanum mauritianum**	*Achyranthes aspera*
2630CA	**Eucalyptus** spp., *Acacia* spp., *Salix babylonica*	*Acacia dealbata, A. mearnsii,* **Campuloclinium macrocephalum,** *Eucalyptus* sp., **Opuntia ficus-indica,** *Prunus persica,* **Pyracantha angustifolia, P. crenulata,** *Salix babylonica*	*Bidens pilosa*

In bold are category 1 (1a and 1b) species that must be eradicated or controlled in terms of the *National Environmental Management: Biodiversity Act* alien and invasive species regulations. SAPIA, Southern African Plant Invaders Atlas; PRECIS, National Herbarium's Pretoria Computerised Information System.

causing substantial impacts on the natural and semi-natural ecosystems. They may also pose a threat to biodiversity and railway management in future in the study area if their population is not properly managed.

This study should be regarded as a first step towards understanding alien plant invasion along the railway line in Mpumalanga province. While the current study only focused on the mapping and identification of species, future studies should look into the role of railway lines in dispersal of alien plants. It is well understood that commercial forestry and other means of introduction are the primary sources of alien plants in South Africa, but railway lines may have played a role as driver pathway for spread of alien plants from one locality to the other because not all invaded areas have a history of commercial forestry (Nyoka 2003). The spatial distribution map of alien plant species will be useful for effective planning, management and monitoring of the spatial changes of the infestation.

Acknowledgements

The authors would like to thank the Department of Agriculture, Rural Development, Land and Environmental Affairs for the provision of software, data and imagery used in this study. This article benefited from comments by two anonymous reviewers and those of the editor, Dr Michelle Hamer.

The authors would also like to thank DEA – Working for Water Programme for their willingness to facilitate and fund the clearing of alien plants in the study area.

Competing interests

The authors declare that they have no financial or personal relationship(s) that may have inappropriately influenced them in writing this article.

Authors' contributions

N.M. was the project leader and was responsible for drafting the manuscript as well as geospatial interpretation. M.N.M., M.C.R., N.N.M. and S.R.N. were responsible for both field data collection and revising the scope and content of the draft manuscript.

References

Anchang, J.Y., Ananga, E.O. & Pu, R., 2016, 'An efficient unsupervised index based approach for mapping urban vegetation from IKONOS imagery', *International Journal of Applied Earth Observation and Geoinformation* 50, 211–220. https://doi.org/10.1016/j.jag.2016.04.001

Appiah, M., 2013, 'Tree population inventory, diversity and degradation analysis of a tropical dry deciduous forest in Afram Plains, Ghana', *Forest Ecology and Management* 295, 145–154. https://doi.org/10.1016/j.foreco.2013.01.023

ArcGIS ESRI version 10.3.1, 2015, computer software, Environmental Systems Research Institute, Inc., Redlands, CA.

Barbosa, N.P.U., Fernandes, G.W., Carneiro, M.A.A. & Júnior, L.A.C., 2010, 'Distribution of non-native invasive species and soil properties in proximity to paved roads and unpaved roads in a quartzitic mountainous grassland of Southeastern Brazil (rupestrian fields)', *Biological Invasions* 12(11), 3745–3755. https://doi.org/10.1007/s10530-010-9767-y

Bennett, B.M., 2011, 'Naturalising Australian trees in South Africa: Climate, exotics and experimentation', *Journal of Southern African Studies* 37(2), 265–280. https://doi.org/10.1080/03057070.2011.579434

Christen, D. & Matlack, G., 2006, 'The role of roadsides in plant invasions: A demographic approach', *Conservation Biology* 20(2), 385–391. https://doi.org/10.1111/j.1523-1739.2006.00315.x

Crimmins, T.M., Mauzy, M.S. & Studd, S.E, 2008, 'Assessing exotic plant distribution, abundance, and impact at Montezuma Castle and Tuzigoot National Monuments in Arizona', *Ecological Restoration* 26(1), 44–50. https://doi.org/10.3368/er.26.1.44

Dar, P.A., Reshi, Z.A. & Shah, M.A, 2015, 'Roads act as corridors for the spread of alien plant species in the mountainous regions: A case study of Kashmir Valley, India', *Tropical Ecology* 56(2), 183–190.

Department of Agriculture, 2001, *Conservation of Agricultural Resources Act (CARA) 43 of 1983*, Government Notice R280, Government Gazette 22166, amended 2001, viewed 06 June 2016, from http://www.nda.agric.za/

Department of Environmental Affairs, 2014a, *National Environmental Management: Biodiversity Act (NEMBA) (10/2004): Alien and Invasive Species Regulations*, Government Notice R598, Government Gazette 37885, viewed 10 September 2016, from http://www.gpwonline.co.za

Department of Environmental Affairs, 2014b, *National Environmental Management: Biodiversity Act (NEMBA) (10/2004): Alien and Invasive Species List*, Government Notice R599, Government Gazette 37886, viewed 10 September 2016, from http://www.gpwonline.co.za

Ferrar, A.A. & Lotter, M.C., 2007, *Mpumalanga biodiversity conservation plan handbook*, The Mpumalanga Tourism and Parks Agency, Nelspruit.

Geological Survey, 1986a, *2528 Pretoria, 1:250 000 Geological Series*, Department of Mineral and Energy Affairs, Pretoria.

Geological Survey, 1986b, *2530 Nelspruit, 1:250 000 Geological Series*, Department of Mineral and Energy Affairs, Pretoria.

Geological Survey, 1986c, *2628 East Rand, 1:250 000 Geological Series*, Department of Mineral and Energy Affairs, Pretoria.

Geological Survey, 1986d, *2630 Mbabane, 1:250 000 Geological Series*, Department of Mineral and Energy Affairs, Pretoria.

Gong, J. & Li, D., 2000, 'Object-oriented and integrated spatial data model for managing image, DEM, and vector data', *Photogrammetric Engineering & Remote Sensing* 66(5), 619–623.

Henderson, L., 1991, 'Alien invasive *Salix* spp. (willows) in the grassland biome of South Africa', *South African Forestry Journal* 157(1), 91–95. https://doi.org/10.1080/00382167.1991.9629105

Henderson, L., 1999, 'The Southern African Plant Invaders Atlas (SAPIA) and its contribution to biological weed control', *African Entomology Memoir* 1, 159–163.

Henderson, L., 2007, 'Invasive, naturalized and casual alien plants in southern Africa: A summary based on the Southern African Plant Invaders Atlas (SAPIA)', *Bothalia* 37(2), 215–248. https://doi.org/10.4102/abc.v37i2.322

Houeto, G., Kakai, G.R., Salako, V., Fandohan, B., Assogbadjo, A.E., Sinsin, B. et al., 2013, 'Effect of inventory plot patterns in the floristic analysis of tropical woodland and dense forest', *African Journal of Ecology* 52(3), 257–264. https://doi.org/10.1111/aje.12112

Jeffrey, L.S., 2005, 'Characterization of the coal resources of South Africa', *The Journal of the South African Institute of Mining and Metallurgy* 105(2), 95–102.

Johnson, M.R., 1976, 'Stratigraphy and sedimentology of the Cape and Karoo sequences in the Eastern Cape province', PhD thesis, Dept. of Geology, Rhodes University.

Kotzé, I., Beukes, H., Van den Berg, E. & Newby, T., 2010, *National invasive alien plant survey*. Agricultural Research Council, Institute for Soil, Climate and Water, Pretoria. Report No. GW/A/2010/21, viewed 27 March 2017, from http://sites.google.com/site/wfwplanning/assessment

Land Types of South Africa and Soil Inventory Databases [computer file], 1984, ARC-Institute for Soil, Climate and Water, Pretoria, viewed 18 May 2016, from the Department of Agriculture, Rural Development, Land and Environmental Affairs, GIS_SERVER\Data\Vector.gdb\landtype_polygon\.

Le Maitre, D.C., Versfeld, D.B. & Chapman, R.A., 2000, 'The impact of invading alien plants on surface water resources in South Africa : A preliminary assessment', *Water SA* 26(3), 397–408.

Lillesand, T.M., Kiefer, R.W. & Chipman, J.W., 2008, *Remote sensing and image interpretation*, 6th edn., John Wiley & Sons Ltd, Hoboken, NJ.

Masters, R.A. & Sheley, R., 2001, 'Invited synthesis paper : Principles and practices for managing rangeland invasive plants', *Journal of Range Management* 54(5), 502–517. https://doi.org/10.2307/4003579

Masubelele, M.L., Foxcroft, L.C. & Milton, S.J., 2009, 'Alien plant species list and distribution for Camdeboo National Park, Eastern Cape province, South Africa', *Koedoe* 51(1), Art. #515, 1–10. https://doi.org/10.4102/koedoe.v51i1.515

Milton, S.J. & Dean, W.R.T., 1998, 'Alien plant assemblages near roads in arid and semi-arid South Africa', *Diversity and Distributions* 4(4), 175–187. https://doi.org/10.1046/j.1472-4642.1998.00024.x

Molano-Flores, B., Phillippe, L.R., Marcum, P.B., Carroll-Cunningham, C., Ellis, J.L., Busemeyer, D.T. et al., 2015, 'A floristic inventory and vegetation survey of three dolomite prairies in northeastern Illinois', *Castanea* 80(3), 153–170. https://doi.org/10.2179/14-040

Mucina, L. & Rutherford, M.C. (eds.), 2006, *The vegetation of South Africa, Lesotho and Swaziland*, Strelitzia 19, South African National Biodiversity Institute, Pretoria.

National Herbarium Pretoria Computerised Information System (PRECIS) database, 2016, *Plants of southern Africa: An online checklist*, viewed 06 June 2016, from http://posa.sanbi.org

Nel, J.L., Richardson, D.M., Rouget, M., Mgidi, T.N., Mdzeke, N., Le Maitre, D.C. et al., 2004, 'A proposed classification of invasive alien plant species in South Africa: Towards prioritizing species and areas for management action', *South African Journal of Science* 100(1), 53–64.

Nyoka, B.I., 2003, *Biosecurity in forestry: A case study on the status of invasive forest trees species in Southern Africa*, Forest Biosecurity Working Paper FBS/1E, FAO, Rome, viewed 03 March 2017, from http://www.fao.org/docrep/005/ac846e/ac846e00.htm#Contents

Peerbhay, K.Y., Mutanga, O. & Ismail, R., 2015, 'Random forests unsupervised classification: The detection and mapping of solanum mauritianum infestations in plantation forestry using hyperspectral data', *IEEE Journal of Selected Topics in Applied Earth Observations and Remote Sensing* 8(6), 3107–3122. https://doi.org/10.1109/JSTARS.2015.2396577

Pysek, P., Danihelka, J., Sádlo, J., Chrtek, J., Chytrý, M., Jarošík, V. et al., 2012, 'Catalogue of alien plants of the Czech Republic (2nd edition): Checklist update, taxonomic diversity and invasion patterns', *Preslia* 84(2), 155–255.

Richardson, D.M., 1998, 'Forestry trees as invasive aliens', *Conservation Biology* 12(1), 18–26. https://doi.org/10.1046/j.1523-1739.1998.96392.x

Richardson, D.M., Pysek, P., Rejmanek, M., Barbour, M.G., Dane Panetta, F. & West, C.J., 2000, 'Naturalization and invasion of alien plants : Concepts and definitions', *Diversity and Distributions* 6(2), 93–107. https://doi.org/10.1046/j.1523-1739.1998.96392.x

Richardson, D.M., Rouget, M., Ralston, S.J., Cowling, R.M., Van Rensburg, B.J. & Thuiller, W., 2005, 'Species richness of alien plants in South Africa: Environmental correlates and the relationship with indigenous plant species richness', *Ecoscience* 12(3), 391–402. https://doi.org/10.2980/i1195-6860-12-3-391.1

Richardson, D.M. & Van Wilgen, B.W., 2004, 'Working for Water: Invasive alien plants in South Africa: How well do we understand the ecological impacts?', *South African Journal of Science* 100(1), 45–52.

Robertson, M.P., Villet, M.H., Fairbaks, D.H.K., Henderson, L., Higgins, S.I., Hoffmann, J.H. et al., 2003, 'A proposed prioritization system for the management of invasive alien plants in South Africa', *South African Journal of Science* 99, 37–43.

Rutkovska, S., Pučka, I., Evarts-Bunders, P. & Paidere, J., 2013, 'The role of railway lines in the distribution of alien plant species in the territory of Daugavpils City (Latvia)', *Estonian Journal of Ecology* 62(3), 212–225. https://doi.org/10.3176/eco.2013.3.03

Southern African Plant Invaders Atlas (SAPIA) database, 2016, *Plant Protection Research Institute, Agricultural Research Council*, viewed 06 June 2016, from http://www.agis.agric.za/agisweb/agis.html

Turner, D.P., 2000, 'Soils of KwaZulu-Natal and Mpumalanga: Recognition of natural soil bodies', PhD thesis, Dept. of Plant Production and Soil Science, University of Pretoria.

Van Wilgen, B.W., Richardson, D.M., Le Maitre, D.C., Marais, C. & Magadlela, D., 2001, 'The economic consequences of alien plant invasions: Examples of impacts and approaches to sustainable management in South Africa', *Environment, Development and Sustainability* 3(2), 145–168. https://doi.org/10.1023/A:1011668417953

Wells, M.J., Balsinhas, A.A., Joffe, H., Engelbrecht, V.M., Harding, G. & Stirton, C.H., 1986, *A catalogue of problem plants in southern Africa: Incorporating the national weed list of southern Africa*, Memoirs of the Botanical Survey of South Africa, No. 53, Botanical Research Institute, Pretoria.

Williams, V.L. & Crouch, N.R., 2017, 'Locating sufficient plant distribution data for accurate estimation of geographic range: The relative value of herbaria and other sources', *South African Journal of Botany* 109, 116–127. https://doi.org/10.1016/j.sajb.2016.12.015

Wingfield, M.J., Coutinho, T.A., Roux, J. & Wingfield, B.D., 2002, 'The future of exotic plantation forestry in the tropics and southern Hemisphere: Lessons from pitch canker', *Southern African Forestry Journal* 195, 79–82. https://doi.org/10.1080/20702620.2002.10434607

16

Assessing the effectiveness of invasive alien plant management in a large fynbos protected area

Authors:
Tineke Kraaij[1] ⓘ
Johan A. Baard[2]
Diba R. Rikhotso[2]
Nicholas S. Cole[3] ⓘ
Brian W. van Wilgen[4] ⓘ

Affiliations:
[1]School of Natural Resource Management, Nelson Mandela Metropolitan University, South Africa

[2]Scientific Services, South African National Parks, South Africa

[3]Biodiversity Social Projects, South African National Parks, South Africa

[4]Centre for Invasion Biology, Department of Botany and Zoology, University of Stellenbosch, South Africa

Corresponding author:
Tineke Kraaij,
tineke.kraaij@nmmu.ac.za

Background: Concern has been expressed about the effectiveness of invasive alien plant (IAP) control operations carried out by Working for Water (WfW). South African legislation now also requires reporting on the effectiveness of IAP management interventions.

Objectives: We assessed the effectiveness of IAP management practices in a large fynbos protected area, the Garden Route National Park, South Africa.

Methods: We undertook field surveys of pre-clearing IAP composition and the quality of treatments applied by WfW during 2012–2015 in 103 management units, covering 4280 ha. We furthermore assessed WfW data for evidence of change in IAP cover after successive treatments, and adherence to industry norms.

Results: Despite the development of detailed management plans, implementation was poorly aligned with plans. The quality of many treatments was inadequate, with work done to standard in only 23% of the assessed area. Problems encountered included (1) a complete absence of treatment application despite the payment of contractors (33% of assessed area); (2) treatments not being comprehensive in that select areas (38%), IAP species (11%) or age classes (8%) were untreated; (3) wrong choice of treatment method (9%); and (4) treatments not applied to standard (7%). Accordingly, successive follow-up treatments largely did not reduce the cover of IAPs. Inaccurate (or lack of) infield estimation of IAP cover prior to contract generation resulted in erroneous estimation of effort required and expenditure disparate with WfW norms.

Conclusions: We advocate rigorous, compulsory, infield assessment of IAP cover prior to contract allocation and assessment of the quality of treatments applied prior to contractors' payment. This should improve the efficiency of control operations and enable tracking of both the state of invasions and effectiveness of management.

Introduction

Invasive alien plants (IAPs) are globally considered to be a significant threat to biodiversity conservation and the sustained delivery of ecosystem services (Dukes & Mooney 1999; Vilà et al. 2011; Vitousek et al. 1996; see also Clusella-Trullas & Garcia 2017). Accordingly, considerable resources are expended in attempts to address this problem (Van Wilgen et al. 2012). South Africa has one of the largest government-funded programmes in the world aimed at managing IAPs, that is, the Working for Water Programme (WfW) (Van Wilgen et al. 2012). This programme was initiated in 1995 with the dual objectives of (1) clearing IAPs to increase water delivery and improve ecological integrity and (2) job creation to alleviate poverty (Van Wilgen, Le Maitre & Cowling 1998).

Strategic assessments of WfW have repeatedly expressed concern about the efficiency of the programme at various levels of operation (Common Ground 2003; Van Wilgen et al. 2012; Van Wilgen & Wannenburgh 2016). Recommendations put forward by these assessments included the prioritisation of IAP species and areas for management (i.e. better planning) (Roura-Pascual et al. 2009), improved coordination, efficiency and professionalism of interventions, and the development and implementation of a monitoring programme (Van Wilgen et al. 2012). National legislation under the Alien and Invasive Species Regulations of the National Environmental Management Biodiversity Act (Act 10 of 2004) requires regular (every 3 years) reporting on the status and impact of invasions, and the effectiveness of management and policy interventions (Wilson et al. 2017).

Globally, inadequate attention has been paid to assessing the effectiveness of control interventions (Kettenring & Adams 2011). In South Africa, studies that have evaluated particular aspects of IAP control operations include assessments of (1) the efficiency of WfW in the Cape Floristic Region by determining what would have happened had the programme not intervened (using counterfactuals) (McConnachie et al. 2016), (2) cost-benefit of IAP control in terms of water gains (Hosking & Du Preez 1999, 2002), (3) the effects of clearing treatments on IAP seedbanks (Holmes et al. 1987) and recovery of indigenous vegetation (Holmes & Marais 2000; Parker-Allie et al. 2004), (4) the use of adaptive management in IAP management in national parks (Loftus 2013) and (5) the cost-efficiency of WfW at biome scale (Van Wilgen et al. 2012) and project (local) scale (McConnachie et al. 2012). The latter study compared records of IAP cover before and after multiple control treatments during a defined study period to determine whether treatments effected a reduction in IAP cover. To our knowledge, no study has undertaken targeted assessments of treatment efficacy through field observations.

We report here on a case study in a large protected area of the Cape Floristic Region, the Garden Route National Park (GRNP), where we assessed the efficiency of WfW's IAP management practices in the field. Parts of the GRNP have a long history of WfW operations, while comprehensive, strategic planning, prioritisation and improved monitoring have only recently been initiated. In particular, we considered the following aspects:

(1) the alignment of implementation with management plans
(2) the effectiveness of alien plant clearing practices in the field
(3) the relationships between IAP species, age classes and cover, and treatment effort.

These investigations allowed us to identify challenges experienced by WfW projects during different stages of implementation (planning, costing and execution) and to produce recommendations towards improving the effectiveness of IAP management practices, which may be widely applicable.

Methods

Study area

The study area is the GRNP (33.80°S 22.50°E – 34.15°S 24.20°E), situated along the southern Cape coast of South Africa between the Indian Ocean in the south and the watershed of the Outeniqua and Tsitsikamma Mountains in the north. The park extends over 152 500 ha of which ca. 78 000 ha comprise fire-prone fynbos shrublands and ca. 41 500 ha comprise Afrotemperate forest. A more detailed biophysical description of the park is given by Kraaij, Cowling and Van Wilgen (2011), and an account of the alien flora is given by Baard and Kraaij (2014). The GRNP was only recently (2009) proclaimed, and the proclamation was preceded by approximately 20 years of neglect in terms of fire and IAP management in most of the mountain catchment areas that now form a part of the GRNP (Kraaij, Cowling & Van Wilgen 2011). More than 244 species of alien plants occur in the park (Baard & Kraaij 2014), the most common invasive genera being *Pinus* and *Hakea*, both

estimated to occur over > 90% of the park's fynbos vegetation at various densities, and *Acacia* over almost 30% (Van Wilgen et al. 2016). IAPs are accordingly considered the leading ecological threat to the GRNP (SANParks 2010), with considerable expenditure (approximately ZAR 20 million, ~US $1.5 million in 2015) allocated annually to IAP clearing operations undertaken by WfW (SANParks 2010).

Procedures followed in IAP management at park/project level

WfW has been involved in IAP control in the area of the GRNP (prior to proclamation) since the programme's inception in 1995 albeit initially at a small scale (Hosking & Du Preez 1999). Up until 2013, the selection of areas to treat during any particular year largely did not follow a strategic plan or prioritisation process, and the tendency was to mostly do follow-up treatments in areas that had been previously treated. This trend partially stemmed from a prominent, and financially rational, drive in the WfW programme to maintain areas that have been worked previously (Loftus 2013), but also from a general lack of a strategy to guide its operations and the selection of projects (Common Ground 2003; Van Wilgen & Wannenburgh 2016). At project operational level, it is furthermore convenient to keep working in accessible areas and under familiar conditions.

Since 2014/2015, the scientific services department of South African National Parks (in consultation with park and WfW staff) developed a strategic medium-term plan for clearing IAPs from the GRNP. This plan was based on the principles of sound prioritisation of area and IAP species (Forsyth et al. 2012; Nel et al. 2004), accurate costing of clearing requirements as per WfW norms (Neethling & Shuttleworth 2013), alignment of treatment approaches and practices with ecological and biological attributes of systems and species (Table 1), and monitoring of changes in IAP distribution and cover over time (Working for Water Programme 2003), *inter alia* to monitor the success of control operations. There was general acceptance of this plan by park and WfW staff alike, and mutual agreement to translate the strategic plan into annual plans of operation (APOs) (which are formulated by WfW project management staff) and to implement these plans.

We briefly outline the procedures involved in implementation of WfW projects by implementing agents, including SANParks, but more detail is provided by Loftus (2013). Annual project level funding is based primarily on historical allocations, but has grown steadily in the GRNP since 2010. Annual funding requirements per project are outlined in an APO, and once funding has been secured, the APO is approved and the operational targets (in terms of expenditure, hectares to be cleared and effort required expressed as person days) are captured into the WfW information management system. Prior to awarding contracts to service providers, WfW project managers are required to do infield inspections in each management unit, collecting data on IAP species present, their cover, age classes and appropriate treatment methods.

TABLE 1: Principles associated with a strategic medium-term plan to control invasive alien plants in extensive and often remote fynbos areas of the Garden Route National Park.

Principle	Rationale
Fynbos at post-fire ages of 1–2 years should be given first priority.	- At 1–2 years after fire, treatment occurs before reproductive maturity of most IAPs and, thus, largely prevents seed set. - 2 years allows for some seedling mortality because of self-thinning (Geldenhuys 2004). - Young vegetation is readily accessible and treatment methods are cheaper, thus reducing the cost.
Fynbos at post-fire ages of 3–10 years are given second priority.	- At 3–10 years post-fire, vegetation is still reasonably accessible, reducing the cost. - Some alien species may have reached reproductive maturity, but seed banks will be relatively smaller than in older vegetation.
Follow-up treatments should take place at 4-year intervals where pines are the dominant invaders.	- Pines are the dominant invaders in large tracts of mountain fynbos (Van Wilgen et al. 2016) and 4-year intervals should ensure that treatment occurs before regrowth (seedlings) reaches reproductive maturity [juvenile periods in *Pinus pinaster* are 6 years and in *Pinus radiata* 5 years, Richardson, Cowling and Le Maitre (1990)]. - 4 years after the previous treatment, the regrowth is still small enough that simple treatment methods and equipment can be used.
Follow-up treatments should take place at 2-year intervals in areas where acacias or other re-sprouters with large persistent soil-stored seed banks are the dominant invaders (in addition to ensuring that biological control agents are present, if available).	- 2 years allows for some seedling mortality because of self-thinning (Geldenhuys 2004). - After 2 years, seedlings are still small enough and total biomass low enough that the required treatment methods are simple and relatively cheap; biomass does not yet cause a high fuel load and fire risk, neither does it have to be removed from riparian zones. - Although many invasive acacias reach reproductive maturity within a year, they usually do not produce large numbers of seeds at an early age, that is < 2 years (Milton & Hall 1981). - Follow-up intervals of < 2 years (albeit potentially more effective to treat acacias) would result in less resources being available for clearing of extensive catchment areas with low-cover pine invasions (Van Wilgen et al. 2016).
Fell mature pines with chainsaws instead of ringbarking, as an initial clearing treatment (at densities where felled biomass does not create excessive fuel loads).	- Ringbarking facilitates wind dispersal of pine seeds from slow-dying, standing trees, whereas seeds do not disperse from felled trees. - Felling results in 100% mortality (in the non-sprouting species, *P. pinaster* and *P. radiata*, common to the study area), as opposed to ringbarking 20% – 90% mortality (Pers. Obs.). - Felling enables rapid verification of the extent and the quality of work done after treatment application, whereas it is difficult to assess whether ringbarking has been done properly, especially in inaccessible areas. - Felling (unlike dead standing trees) enhances landscape aesthetics (Barendse et al. 2016).
Prioritising areas where IAPs occur at low levels of cover.	- Provides the best return for investment (Van Wilgen et al. 2016).

IAP, invasive alien plant.

Source: South African National Parks, unpublished information

Theoretically, cover can exceed 100% if different IAP species or age classes form multiple strata at high cover classes, although it seldom happens in the GRNP. These data are then cross-referenced with the APO prior to contracts being awarded. Contracts for work are generated through the WfW information system, and the effort requirement for each contract is automatically calculated using the WfW norms (Neethling & Shuttleworth 2013). Project managers then estimate the cost of each contract based on the allocated effort (person days). WfW-registered contractors tender for the contracts through a bidding system, and contracts are awarded based on these tenders. Protocol requires that a minimum of three joint site inspections per contract are undertaken by the WfW project manager and the contractor prior to the start of the contract, during execution, and before contracts are signed off for payment. Once a contract has been signed off for payment, the completed contract document (which includes time-sheets reflecting actual days worked on the contract, start and end dates, as well as actual expenses) is captured into the WfW information system.

Data collection and analysis

We restricted our analyses to the areas covered with fynbos vegetation within the GRNP, and to work carried out by WfW between 2004 and 2015 with an emphasis on the 2014/2015 financial year (1 April 2014 to 31 March 2015). Our unit of assessment was the management unit as spatially delineated by WfW and reported on as cases in their information system [see Working for Water Programme (2003) and Loftus (2013) for details of information recorded in this information system]. 'Cover' throughout this paper refers to the percentage projected canopy cover of IAPs, which is recorded as 'density' by WfW in their information system, and theoretically determined by WfW from stem counts of IAPs and then converted to a percentage cover value (Working for Water Programme 2003).

To assess whether implementation was aligned with the strategic medium-term plan, we compared, in terms of geographic overlap and total extent, the areas that were planned to be treated during 2014/2015 with those included in the 2014/2015 APO. We furthermore compared the strategic medium-term plan and APO with records of areas cleared during 2014/2015 according to the WfW information system.

To evaluate the effectiveness of clearing practices, we inspected 103 management units in the field (covering 4280 ha worked by WfW during 2012–2015) where we (1) estimated pre-treatment cover and size classes of IAP species and (2) assessed the quality of IAP management treatments. Field surveys were mostly done within 6 months after the execution of control treatments, which then provided information on both these aspects (as ringbarked and felled trees and shrubs remain *in situ* enabling estimation of pre-treatment cover). We compared our field estimates of pre-treatment IAP cover with those estimated by WfW prior to contract allocation of that particular treatment in the same 103 management units using a Wilcoxon matched pairs test. To assess the quality of IAP treatment practices, we recorded for each management unit the degree to which the treatments were satisfactorily carried out, and if not, how the treatments deviated from acceptable standards. We subsequently classified these deviations by type (Table 2) and calculated their rates of incidence, both in terms of the number of management units affected and by the areas affected.

We also assessed whether IAP cover recorded in WfW's information system decreased with successive follow-up treatments. For these analyses, we considered the complete

TABLE 2: Types of deviations from acceptable standards of treatment application, and their rates of incidence in terms of the number of management units affected, and by area. A total of 103 management units, and 4280 ha, were assessed. More than one deviation type could pertain to a management unit.

Type of deviation	Incidence (% of number of units examined)	Incidence (% of total area examined)
Inaccurate estimation of alien plant cover, causing inaccurate allocation of person days and cost	25 on par; 5 underestimated; 70 overestimated	40 on par; 7 underestimated; 53 overestimated
Incorrect identification of dominant species (often listing re-sprouters, requiring greater treatment effort, instead of non-sprouters)	22	36

No evidence of work done during last treatment
    – Aliens present were not cleared, yet contractor paid
    – No aliens present, yet contract generated and contractor paid
	44	33

Some individuals of target species or part of management unit not treated	34	38
Some age groups of target species not treated (e.g. adults treated, saplings untreated)	5	8
Some species not treated (e.g. *Pinus* treated, *Hakea* untreated)	11	11
Wrong choice of treatment method (e.g. ringbarking of trees < 10 cm diameter; ringbarked AND felled trees; ringbarking of dead trees)	11	9

Treatment not applied to standard
    – Ringbarked strip too narrow
    – Ringbarked on only one (the most visible) side of the tree
    – Re-sprouting plants felled but no herbicide applied, or wrong choice/ concentration of herbicide
	7	7

Work done well	15	23

treatment history (spanning the period 2004–2015) of management units ($n = 764$) that received their last treatment during 2012–2015. Most management units received multiple treatments (mean = 5, maximum = 12); we discarded data for treatments beyond the seventh follow-up treatment, as there were too few cases for meaningful analysis. We assessed the efficiency of treatments by comparing change in cover between successive treatments on the same site. Cover was recorded prior to each treatment, and we expressed the treatment effect as the ratio of cover before a specific treatment to the cover before the prior treatment; this measure we refer to as 'proportional change', where ratios of < 1 represent reductions in cover, and ratios of > 1 represent increases in cover. We also calculated 'absolute change' as the difference between the cover before a specific treatment and the cover before the prior treatment, where negative values represent reductions in cover and positive values represent increases in cover. We calculated these measures for each treatment (up until the seventh follow-up treatment) within each management unit (total of 2738 treatments on 764 management units), and thereafter, calculated mean and median change in cover achieved per treatment cycle across all management units.

The large and persistent soil-stored seed banks of the dominant *Acacia* species in the study area continue to fuel vigorous recruitment even after several successive clearing treatments (Holmes et al. 1987; Milton & Hall 1981; Richardson & Kluge 2008). Most of the other common woody IAP taxa (*Pinus* and *Hakea*) do not have persistent soil-stored seed banks, and recruitment in these taxa should consequently be much less after successive treatments, provided seeds are not released from cones onto open ground such as post-fire (Macdonald, Clark & Taylor 1989). Given this differential response to treatment, we expected a much slower decrease in the cover of *Acacia* compared with that of *Pinus* and *Hakea* after successive treatments. We tested whether the two groups (differentiated based on indication of dominant species per management unit in the WfW information system) differed in terms of change in cover following successive follow-up treatments (using Mann–Whitney U-test).

To assess whether allocation of effort to treatments was aligned with the dominant IAP species, age classes and cover, we compared the relationship between these variables with the WfW norms (Neethling & Shuttleworth 2013). For this comparison, we limited our analysis to, and distinguished between, *Acacia* and *Pinus* or *Hakea*, as differential norms pertain to these groups (Neethling & Shuttleworth 2013). *Acacia* is more costly to treat than *Pinus* or *Hakea* as the former requires herbicide application to prevent re-sprouting. We used 'modelled person days' (which is the number of person days allocated at the time of contract generation to the treatment of a management unit) in the WfW information system as a measure of effort allocated (effort is a better measure than cost as it is not subject to inflation) and expressed that per unit area (i.e. person days per hectare). Our dataset comprised 2092 treatments on 738 management units. The relationship between effort (person days per hectare) and IAP cover was established by means of regression for *Acacia* and *Pinus* or *Hakea* (differentiated based on indications in the WfW information system of the dominant species treated), and the residuals compared by a t-test to determine whether the effort-cover relationship differed between the two groups. All statistical analyses were performed in Statistica (version 13, 1984–2015, Dell Inc.).

Results

Alignment of implementation with planning

The work carried out during the 2014/2015 financial year deviated to a large degree from the agreed priorities in the medium-term strategic plan. Only 47% of the area on the strategic medium-term plan for that year was carried forward to the APO. Furthermore, there was a failure to complete all of the work scheduled in the APO. Only 19% of the area on the strategic plan for 2014/2015 and 37% of the area on the APO were actually carried out (see Appendix 1), but not necessarily treated to standard (see subsequent results). In addition, work was carried out in non-priority areas that were not included in either the medium-term strategic plan or the APO, equivalent in size to 12% of the annual plan.

Quality of work and discrepancies in cover estimates

We found evidence for widespread ineffective treatment of IAPs in the field. Field surveys of recently treated areas (103 management units, total area 4280 ha) showed that in 85% of the assessed units (77% of the assessed area), work was not done to standard (Table 2). Various types of problems contributed to this finding, including an apparent complete absence of work despite the payment of contractors, partial work done, not all the IAP species or age classes present being treated, wrong choice of treatment method and treatments not being applied to standard.

We found evidence that cover was regularly overestimated prior to the awarding of contracts. The IAP cover estimates recorded by WfW (mean 54.8% ± SD 56.5%; median 32.5%) prior to contract allocation were more than double ($Z = 7.06$, $p < 0.001$, df = 100) those of our estimates from field surveys (13.8% ± 16.9%; 8.0%). WfW estimates also exceeded 100% in 15% of cases (Figure 1). The frequency distribution of IAP cover recorded by WfW (in 764 management units treated between 2012 and 2015) shows that more than half (54%) of cover estimates were > 25%, and 17% exceeded 100% (Figure 2).

We also found, contrary to expectation, that repeated treatments of management units more frequently led to increases, rather than decreases, in cover. In addition, and also contrary to expectation, we found no significant difference ($Z = -1.11$, $p = 0.27$, df = 1988) between *Acacia* (where increases in cover may potentially be explained by germination from large and persistent soil seed banks) and other IAP genera (where there are no persistent seed banks to fuel continuous germination and regrowth). For this reason, we present pooled data for all IAP taxa in Figure 3. Due to

large numbers of extreme outliers (positively skewed data; note the respective sizes of 75 percentiles vs. 25 percentiles in Figure 3, and outliers in Figure 3b), means and medians presented varied results. We present medians which modulate the influence of outliers, although the substantial influence of outliers as revealed in means, should also be noted. Across all treatments (irrespective of follow-up treatment cycle), the median absolute change in cover was 0, while the mean showed an increase in cover of 7%. When considering proportional change in IAP cover resulting from all treatments, the median was 0.94 (a slight reduction), while the mean was an eightfold increase in cover. The only treatments resulting in reductions in the median cover were treatments 0 (initial), 1 and 4 (Figure 3), while none of the treatments led to reductions in mean cover. Contrary to expectation, both measures of change in cover furthermore showed a trend of rising increases in IAP cover as more follow-up treatment cycles were applied.

Deviation of actual expended effort from norms

The effort required to control IAPs increased with IAP cover ($F_{1,2090} = 2781$, $p < 0.001$, $R^2 = 0.57$), but a large proportion of the variation in the data was not explained by the regression model (Figure 4). In addition, and contrary to expectation, there was no difference in this relationship between cover and effort for *Acacia* and *Pinus* or *Hakea* ($t = 1.24$, $p = 0.21$, df = 1593). A large difference should be expected, as the costs of clearing *Acacia* are much higher than the costs of clearing *Pinus* or *Hakea* at comparable levels of cover. There was large variation in the data with cover ranging up to 318% and effort up to 93 person days per hectare, also revealing substantial deviation from (both above and below) the norms. Given that these data do not include treatments undertaken by specialised high-altitude teams (which are very costly), high effort allocations may only be accounted for in a limited number of cases where remoteness and difficult terrain could have demanded increased effort of up to 150% of the norms (Van Wilgen et al. 2016). Yet, three- to five fold inflations of effort compared with the norms were common.

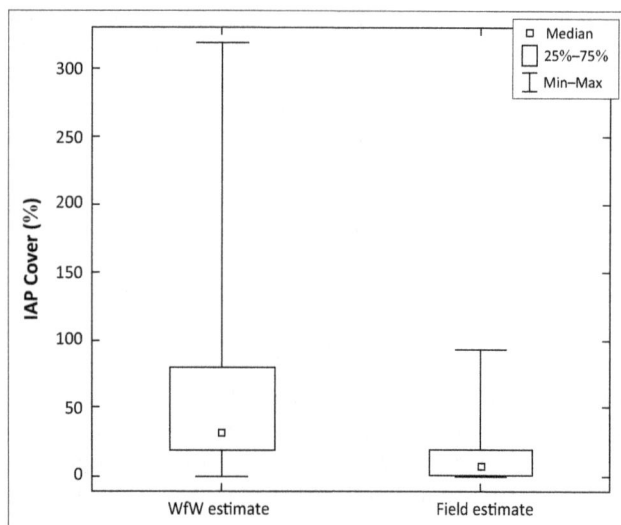

FIGURE 1: Cover (%) of IAPs as estimated by Working for Water prior to contract allocation and during our field surveys in the same 103 management units (total area 4280 ha).

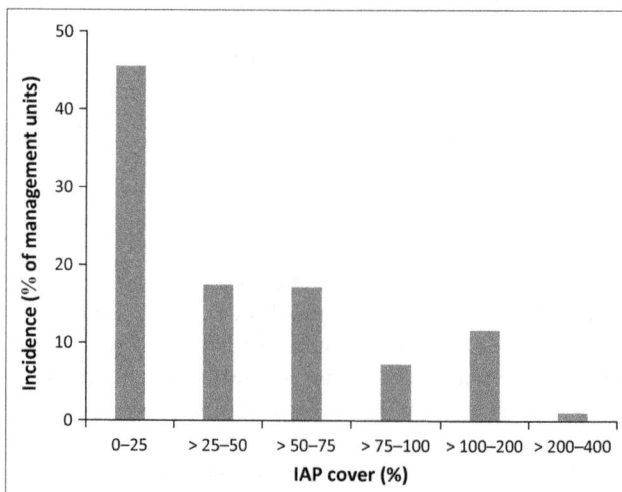

FIGURE 2: Frequency distribution of cover (%) of IAPs estimated by WfW in 764 management units last treated between 2012 and 2015 (total area 21 760 ha).

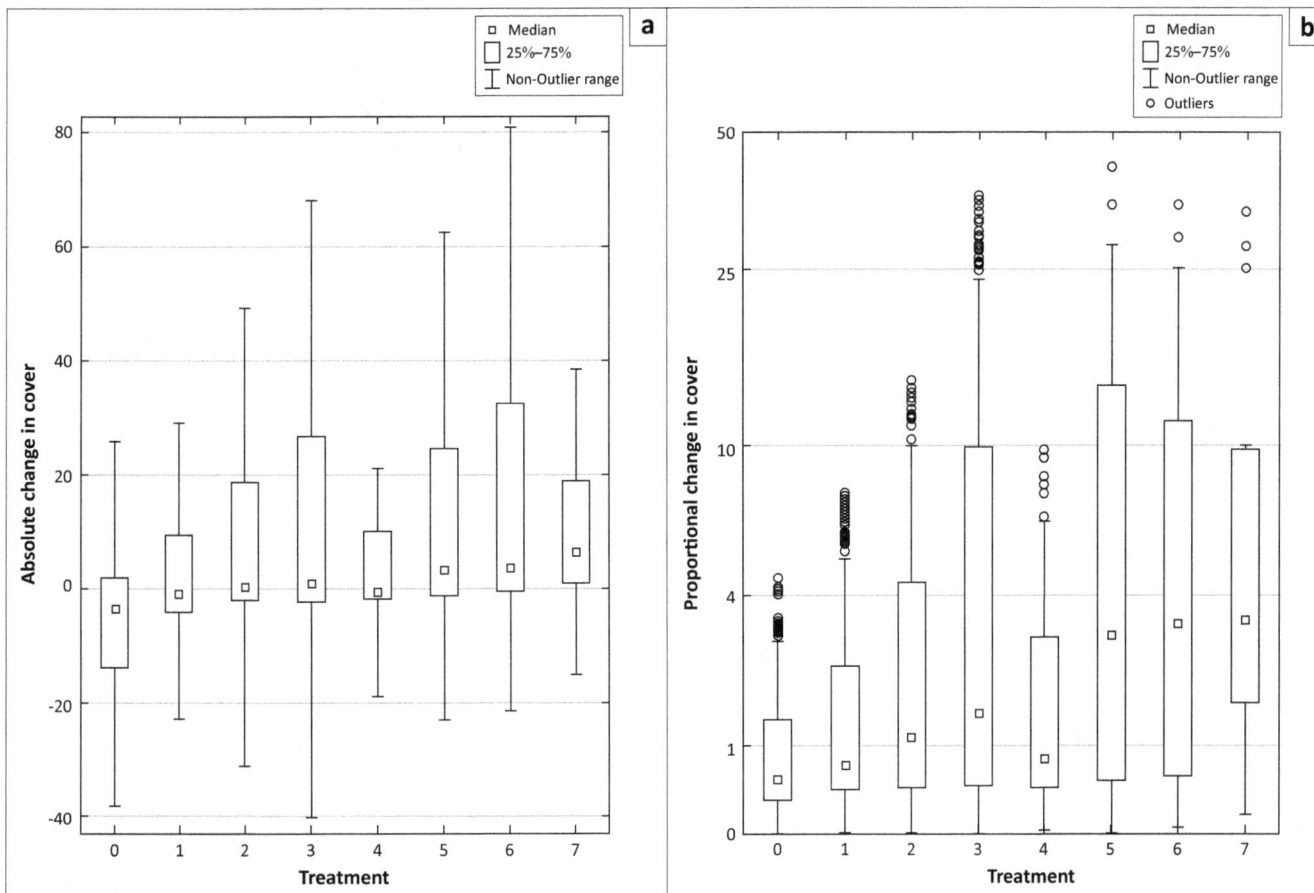

FIGURE 3: (a) Absolute and (b) proportional change in cover of IAPs effected by successive follow-up treatments, as recorded by Working for Water in 764 management units (21 760 ha). Absolute change was calculated as the difference between cover before a specific treatment and cover before the prior treatment; proportional change was calculated as the ratio of cover before a specific treatment to the cover before the prior treatment. Treatment = 0 represents the effect of the initial treatment. Outliers are not shown in (a) to enhance figure clarity, while the y-axis in (b) is spaced on a log-scale to accommodate outliers.

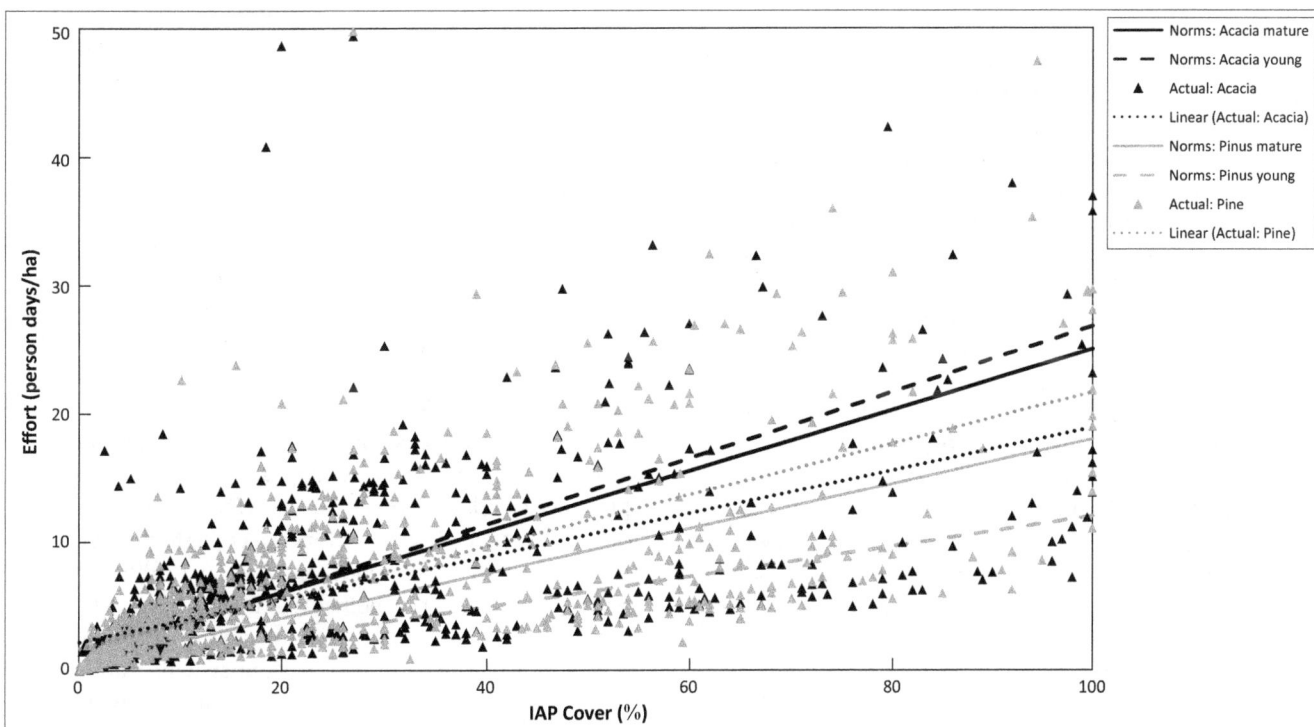

IAP, invasive alien plant.

FIGURE 4: The relationship between actual allocated treatment effort (expressed as person days per hectare) and IAP cover estimates, compared to Working for Water norms. Distinction is made between *Acacia vs. Pinus* or *Hakea* as dominant invaders. Outliers were removed to enhance figure clarity.

Discussion

A lack of planning had previously been identified as a factor reducing the effectiveness of control (Van Wilgen et al. 2012). To address this, the GRNP developed guidelines for the prioritisation and planning of control operations. To date, there has been a failure to implement these plans. To our knowledge, this study presents a first targeted assessment, largely based on field observations, of the effectiveness of IAP management interventions at WfW project level. We critically evaluated the work flow and procedures employed during the implementation of WfW projects, which allowed us to identify the challenges experienced by such projects during different stages of implementation (planning, costing and execution) and to make recommendations towards improving the efficiency of IAP management practices.

Effectiveness of IAP management interventions

The GRNP proved exemplary in having heeded the calls (Van Wilgen et al. 2012) for prioritisation of species and areas for treatment, and better planning and coordination. A strategic medium-term plan, based on sound ecological principles, was developed and agreed to by park- and WfW management authorities. However, there appeared to be major challenges with implementation and coordination, with < 20% and < 40% achievement of annual area targets according to the medium-term and annual plans, respectively, as well as considerable effort expended in areas not aligned with the plans. Apart from substantial underachievement of area targets, this furthermore resulted in disjointed, and thus less efficient, allocation of treatment effort in space and time (Roura-Pascual et al. 2009).

Our assessments of the quality of infield treatment applications revealed equally disappointing results with no evidence of work done over a third of the assessed area, and deviations from acceptable standards of treatment application occurring over an additional 44% of the area. Treatments have, thus, been applied to standard in less than 15% of the assessed units. Our findings are similar to those of McConnachie et al. (2012) in the Krom and Kouga catchments of the Eastern Cape, where 'many sites [24% of sites in Kouga and 4% in Krom] that were recorded as treated in the information system were in fact never treated'. The poor quality of treatment applications observed during our field surveys can also partially explain the lack of a consistent decline in IAP cover after successive follow-up treatments evident in the WfW information system (Figure 3). Apart from a complete absence of work (despite payment), a diverse array of problem types were apparent in our study (Table 2). Jointly, these different types of problems suggest that ignorance, inappropriate equipment, inadequate skills and training, as well as deliberate negligence and even fraudulent behaviour could have contributed to the poor standard of treatment application. Such diverse issues are expected to require considerable and varied interventions to try to correct.

Overall, less than 10% of the strategic medium-term plan that was designed to 'effectively' reduce IAP infestations in the fynbos of the GRNP has been achieved, when considering that approximately a third of the area targets as per the planning products have been 'implemented', of which less than a quarter has been treated to standard. Although better planning and prioritisation are often recommended to improve the efficiency of IAP management practices, our case study suggests that sound planning by itself does not ensure efficiency. Accordingly, Van Wilgen et al. (2011) proposed 'that the available management activities and practices be appropriately combined for each management category and strategically implemented collaboratively by affected parties at appropriate scales'. The general failure with reducing IAP infestations in our sample, which spanned a considerable period, significant and sustained investment (i.e. eight or more successive treatments), a diversity of projects, personnel, IAP species and environmental conditions, is disconcerting and suggests that these results are not unique (Fill et al. 2017; McConnachie et al. 2012).

Cover estimates and application of norms

IAP cover mostly appeared to be overestimated by WfW. A substantial proportion of the IAP cover records in WfW's information system greatly exceeded 100% (Figure 2), which is unlikely given that the IAP species that are commonly treated in the study area rarely form overlapping canopies or multiple strata. Compared to our field estimates, WfW substantially overestimated IAP cover over more than half the area that was assessed (Table 2; Figure 1). A large portion of WfW's records, furthermore, fell in high IAP cover classes (Figure 2), while another study (Van Wilgen et al. 2016) estimated that IAP cover in the fynbos of the GRNP was less than 25% over more than 85% of the area. Overestimation of IAP cover by WfW, thus, appears to be a perennial source of error with knock-on effects on various aspects of IAP management operations (discussed below).

Inaccurate IAP cover estimation is likely exacerbated by the lack of a clearly defined method for determining cover (or density), inconsistent application of any such method and observer subjectivity that may account for up to sevenfold variances in cover estimates for the same area (Loftus 2013; Neethling & Shuttleworth 2013). IAP cover estimates are mostly based on coarse infield visual assessments and 'gut feel', or worse, are often desktop-derived and are based on previous records of cover as per the WfW information system, or the deliberate allocation of incorrect (higher) cover values in order to disburse funds to meet expenditure targets, without concern as to whether the funds are used effectively (Loftus 2013; Neethling & Shuttleworth 2013). Accurate information on IAP cover is the basis upon which most aspects of IAP management rest. Without it, norms and workload-/cost-estimations become meaningless, while trends in IAP distributions and densities and the efficiency of management efforts cannot be measured (cf. Loftus 2013).

The observed relationship between workload/effort allocations and IAP cover estimates did not reflect a close correlation or stringent adherence to norms, with three- to fivefold deviations from the norms being common (Figure 4). Loftus (2013) likewise found large variation in the cost of IAP clearing per condensed hectare among five South African National parks, with the GRNP being the most expensive and most variable. The lack of adherence to norms was surprising in light of WfW operational staff often alleging that inappropriate norms are causing poor standard of work delivery. However, if norms are not rigorously applied (as is suggested by our results), and effort mostly gets allocated as requested (cf. Loftus 2013) regardless of IAP cover, then inappropriate norms cannot account for the poor standard of work observed. Moreover, if norms are not adhered to, and IAP cover records are unreliable, then the data on effort allocations (as captured in WfW's information system) cannot be used to evaluate effectiveness. In contrast, it is evident in the commercial forestry industry that the norms and standards relevant to weed treatments can be applied successfully and can contribute to effective operations (Rolando & Little 2009; Wagner et al. 2006).

The lack of a discrepancy observed in the effort – IAP cover relationship between two main IAP species groups (*Acacia vs. Pinus* or *Hakea*) (Figure 4) with disparate treatment requirements and distinct ecological responses, provides additional evidence that effort allocations do not match the nature of infestations and, thus, that norms are not adhered to. It, furthermore, suggests that cover, cost and effort estimations and allocations are often inaccurate and do not get adequately audited.

Recommendations to improve effectiveness

Interventions essential to improving the effectiveness of IAP management practices at project level largely relate to (1) improvement in IAP cover estimations, (2) additional quality control in terms of infield operations and (3) auditing of data captured in the WfW information system. In particular, we recommend the following:

- Increased alignment of project level annual plans and implementation with strategic planning.
- Implementation of effective protocol for IAP cover estimations that are relevant to specific biomes/regions and IAP species to be managed (cf. Loftus 2013; Neethling & Shuttleworth 2013).
- Compulsory infield assessments of IAP cover prior to contract generation and increased investment (in terms of travel, time, staff skills/training allowances) to this end. Outsourcing of this function should be considered as that may reduce subjectivity and scope for fraud, improve professionalism, specialisation and standardisation of this function, and yield data that are more comparable in space and time.
- Auditing of field data submitted for generation of contracts through the WfW information system, and in particular in terms of deviations from historical data in IAP cover.

- Ensuring that the WfW information system correctly applies the norms when calculating effort allocations in relation to IAP infestation attributes.
- Compulsory infield inspections of the quality of treatments applied during contract implementation, that is, mid-term and prior to contract payment.
- Allocation of funding by WfW to implementing agents specifically for monitoring and research, including compliance monitoring, ecological monitoring, and applied research applicable to the challenges faced by conservation agencies in the management of IAPs.

Many challenges experienced in the management of IAPs as revealed by this study relate to functions performed at WfW project management level. These include the development of annual plans that are aligned with the medium-term strategic plan, coordination of different processes involved in implementation, infield identification of IAP species and age classes, IAP cover estimations, application of norms in the generation and costing of contracts, choice of best treatment methods and infield inspections of the standard of treatment applications. In reality, the project management function in WfW is a daunting task, requiring a considerable and varied skillset and experience, including ecological, social and financial, with project managers typically managing annual budgets of ZAR 3–8 million. Due to poor remuneration, most applicants for project manager positions do not meet the tertiary education requirements, resulting in managers appointed being inadequately skilled. Furthermore, the mechanisms necessary to mentor and support these positions are not present within the implementing agents (cf. Coetzer & Louw 2012). A lack of career opportunities and succession planning also leads to a lack of sustainable management capacity, which in turn compromises the efficiency of project implementation.

We strongly advocate greater involvement of implementing agents (particularly in the case of conservation agencies) in all the major processes involved in IAP management, including planning, monitoring of IAP distributions, quality control of infield operations and training of all staff involved. The current approach that uses poverty-relief funds for alien plant control projects is politically attractive, and it has been the main reason that the control projects have received high levels of funding. However, as currently configured, the model imposes exacting requirements and demands that employment be maximised. These demands come at the cost of effectively achieving ecological goals that in the longer term would arguably support greater economic development (Van Wilgen & Wannenburgh 2016). In addition, the practice of issuing short-term contracts for clearing and follow-up (instituted as a developmental opportunity to disadvantaged contractors) requires cumbersome procedures to approve and implement and results in delays to work schedules and late payments to intended beneficiaries, substantially diluting the intended social benefits (Coetzer & Louw 2012; Hough & Prozesky 2012). It would be better to employ fewer, better-trained personnel on a more permanent basis. The current model also does not allow for capacity to be built within the conservation authorities who are ultimately mandated to manage protected areas. Hence, a scenario in which this funding is phased out, or channelled elsewhere, would leave the conservation agencies without embedded capacity and experience to manage invasions. We would, therefore, recommend that the funding be made available directly to conservation agencies to reduce the problems outlined herein.

Acknowledgements

We thank Working for Water for providing access to data and the anonymous reviewers for their suggestions which enabled improvement of the manuscript.

Funding was provided by South African National Parks, the Nelson Mandela Metropolitan University, the DST-NRF Centre of Excellence for Invasion Biology and the National Research Foundation (grant no. 87 550 for B.W.v.W.).

Competing interests

The authors declare that they have no financial or personal relationship(s) that may have inappropriately influenced them in writing this article.

Authors' contributions

T.K. was the project leader. T.K., J.A.B. and D.R.R. undertook field data collection. T.K., J.A.B. and N.S.C. were responsible for data processing and analyses. All authors, including B.W.v.W., made conceptual contributions.

References

Baard, J.A. & Kraaij, T., 2014, 'Alien flora of the Garden Route National Park, South Africa', *South African Journal of Botany* 94, 51–63. https://doi.org/10.1016/j.sajb.2014.05.010

Barendse, J., Roux, D., Erfmann, W., Baard, J.A., Kraaij, T. & Niewoudt, C., 2016, 'Viewshed and sense of place as conservation features: A case study and research agenda for South Africa's national parks', *Koedoe* 58(1), a1357. https://doi.org/10.4102/koedoe.v58i1.1357

Clusella-Trullas, S. & Garcia, R.A., 2017, 'Impacts of invasive plants on animal diversity in South Africa: A synthesis', *Bothalia* 47(2), a2166. https://doi.org/10.4102/abc.v47i2.2166

Coetzer, A. & Louw, J., 2012, 'An evaluation of the Contractor Development Model of Working for Water', *Water SA* 38, 793–802. https://doi.org/10.4314/wsa.v38i5.19

Common Ground, 2003, *Working for Water external evaluation synthesis report*, Common Ground, Cape Town, South Africa.

Dukes, J.S. & Mooney, H.A., 1999, 'Does global change increase the success of biological invaders?', *Trends in Ecology & Evolution* 14, 135–139. https://doi.org/10.1016/S0169-5347(98)01554-7

Fill, J.M., Forsyth, G.G., Kritzinger-Klopper, S., Le Maitre, D.C. & Van Wilgen, B.W., 2017, 'An assessment of the effectiveness of a long-term ecosystem restoration project in a fynbos shrubland catchment in South Africa', *Journal of Environmental Management* 185, 1–10. https://doi.org/10.1016/j.jenvman.2016.10.053

Forsyth, G.G., Le Maitre, D.C., O'Farrell, P.J. & Van Wilgen, B.W., 2012, 'The prioritisation of invasive alien plant control projects using a multi-criteria decision model informed by stakeholder input and spatial data', *Journal of Environmental Management* 103, 51–7. https://doi.org/10.1016/j.jenvman.2012.01.034

Geldenhuys, C.J., 2004, 'Concepts and process to control invader plants in and around natural evergreen forest in South Africa', *Weed Technology* 18, 1386–1391. https://doi.org/10.1614/0890-037X(2004)018[1386:CAPTCI]2.0.CO;2

Holmes, P. & Marais, C., 2000, 'Impacts of alien plant clearance on vegetation in the mountain catchments of the Western Cape', *Southern African Forestry Journal* 189, 113–117. https://doi.org/10.1080/10295925.2000.9631286

Holmes, P.M., Macdonald, I.A.W. & Juritz, J., 1987, 'Effects of clearing treatment on seed banks of the alien invasive shrubs *Acacia saligna* and *Acacia cyclops* in the southern and south-western Cape, South Africa', *Journal of Applied Ecology* 24, 1045–1051. https://doi.org/10.2307/2404000

Hosking, S. & Du Preez, M., 1999, 'A cost-benefit analysis of removing alien trees in the Tsitsikamma mountain catchment', *South African Journal of Science* 95, 442–448.

Hosking, S. & Du Preez, M., 2002, 'Valuing water gains in the Eastern Cape's Working for Water Programme', *Water SA* 28, 23–28. https://doi.org/10.4314/wsa.v28i1.4863

Hough, J.A. & Prozesky, H., 2012, 'Beneficiaries' aspirations to permanent employment within the South African Working for Water Programme', *Social Dynamics* 38, 331–349. https://doi.org/10.1080/02533952.2012.719395

Kettenring, K.M. & Adams, C.R., 2011, 'Lessons learned from invasive plant control experiments: A systematic review and meta-analysis', *Journal of Applied Ecology* 48, 970–979. https://doi.org/10.1111/j.1365-2664.2011.01979.x

Kraaij, T., Cowling, R.M. & Van Wilgen, B.W., 2011, 'Past approaches and future challenges to the management of fire and invasive alien plants in the new Garden Route National Park', *South African Journal of Science* 107, 1–11.

Loftus, W.J., 2013, 'Strategic adaptive management and the efficiency of invasive alien plant management in South African National Parks', MTech, Nelson Mandela Metropolitan University.

Macdonald, I., Clark, D. & Taylor, H., 1989, 'The history and effects of alien plant control in the Cape of Good Hope Nature Reserve 1941–1987', *South African Journal of Botany* 55, 56–75. https://doi.org/10.1016/S0254-6299(16)31233-9

Mcconnachie, M.M., Cowling, R.M., Van Wilgen, B.W. & Mcconnachie, D.A., 2012, 'Evaluating the cost-effectiveness of invasive alien plant clearing: A case study from South Africa', *Biological Conservation* 155, 128–135. https://doi.org/10.1016/j.biocon.2012.06.006

Mcconnache, M.M., Van Wilgen, B.W., Ferraro, P.J., Forsyth, A.T., Richardson, D.M., Gaertner, M. et al., 2016, 'Using counterfactuals to evaluate the cost-effectiveness of controlling biological invasions', *Ecological Applications* 26(2), 475–483. https://doi.org/10.1890/15-0351

Milton, S.J. & Hall, A.V., 1981, 'Reproductive biology of Australian Acacias in the south-western Cape Province, South Africa', *Transactions of the Royal Society of South Africa* 44, 465–487. https://doi.org/10.1080/00359198109520589

Neethling, H. & Shuttleworth, B., 2013, *Revision of the Working for Water workload norms*, Forestry Solutions, White River, South Africa.

Nel, J.L., Richardson, D.M., Rouget, M., Mgidi, T.N., Mdzeke, N., Le Maitre, D.C. et al., 2004, 'A proposed classification of invasive alien plant species in South Africa: Towards prioritizing species and areas for management action', *South African Journal of Science* 100, 53–64.

Parker-Allie, F., Richardson, D., Holmes, P. & Musil, C., 2004, 'The effects of past management practices for invasive alien plant control on subsequent recovery of fynbos on the Cape Peninsula, South Africa', *South African Journal of Botany* 70, 804–815. https://doi.org/10.1016/S0254-6299(15)30183-6

Richardson, D.M., Cowling, R.M. & Le Maitre, D.C., 1990, 'Assessing the risk of invasive success in *Pinus* and *Banksia* in South African mountain fynbos', *Journal of Vegetation Science* 1, 629–642. https://doi.org/10.2307/3235569

Richardson, D.M. & Kluge, R.L., 2008, 'Seed banks of invasive Australian Acacia species in South Africa: Role in invasiveness and options for management', *Perspectives in Plant Ecology, Evolution and Systematics* 10, 161–177. https://doi.org/10.1016/j.ppees.2008.03.001

Rolando, C. & Little, K., 2009, 'Regional vegetation management standards for commercial pine plantations in South Africa', *Southern Forests* 71, 187–199. https://doi.org/10.2989/SF.2009.71.3.3.915

Roura-Pascual, N., Richardson, D.M., Krug, R.M., Brown, A., Chapman, R.A., Forsyth, G.G. et al., 2009, 'Ecology and management of alien plant invasions in South African fynbos: Accommodating key complexities in objective decision making', *Biological Conservation* 142, 1595–1604. https://doi.org/10.1016/j.biocon.2009.02.029

Sanparks, 2010, *The management plan of the Garden Route National Park*, South African National Parks, Knysna, South Africa.

Van Wilgen, B.W., Dyer, C., Hoffmann, J.H., Ivey, P., Le Maitre, D.C., Moore, J.L. et al., 2011, 'National-scale strategic approaches for managing introduced plants: Insights from Australian acacias in South Africa', *Diversity and Distributions* 17, 1060–1075. https://doi.org/10.1111/j.1472-4642.2011.00785.x

Van Wilgen, B.W., Fill, J.M., Baard, J.A., Cheney, C., Forsyth, A.T. & Kraaij, T., 2016, 'Historical costs and projected future scenarios for the management of invasive alien plants in protected areas in the Cape Floristic Region', *Biological Conservation* 200, 168–177. https://doi.org/10.1016/j.biocon.2016.06.008

Van Wilgen, B.W., Forsyth, G.G., Le Maitre, D.C., Wannenburgh, A., Kotzé, J.D.F., Van Den Berg, E. et al., 2012, 'An assessment of the effectiveness of a large, national-scale invasive alien plant control strategy in South Africa', *Biological Conservation* 148, 28–38. https://doi.org/10.1016/j.biocon.2011.12.035

Van Wilgen, B.W., Le Maitre, D.C. & Cowling, R.M., 1998, 'Ecosystem services, efficiency, sustainability and equity: South Africa's Working for Water programme', *Trends in Ecology & Evolution* 13, 378. https://doi.org/10.1016/j.cosust.2015.08.012

Van Wilgen, B.W. & Wannenburgh, A., 2016, 'Co-facilitating invasive species control, water conservation and poverty relief: Achievements and challenges in South Africa's Working for Water programme', *Current Opinion in Environmental Sustainability* 19, 7–17. https://doi.org/10.1111/j.1461-0248.2011.01628.x

Vilà, M., Espinar, J.L., Hejda, M., Hulme, P.E., Jarošík, V., Maron, J.L. et al., 2011, 'Ecological impacts of invasive alien plants: A meta-analysis of their effects on species, communities and ecosystems', *Ecology Letters* 14, 702–708.

Vitousek, P.M., Antonio, C.M., Loope, L.L. & Westbrooks, R., 1996, 'Biological invasions as global environmental change', *American Scientist* 84, 218–228.

Wagner, R.G., Little, K.M., Richardson, B. & MCNABB, K., 2006, 'The role of vegetation management for enhancing productivity of the world's forests', *Forestry* 79, 57–79. https://doi.org/10.1093/forestry/cpi057

Wilson, J.R.U., Gaertner, M., Richardson, D.M. & Van Wilgen, B.W., 2017, 'Contributions to the National Status Report on Biological Invasions in South Africa', *Bothalia* 47(2).

Working for Water Programme, 2003, *Standards for mapping and management of alien vegetation and operational data*, Department of Water Affairs and Forestry, Pretoria, South Africa.

Appendix 1

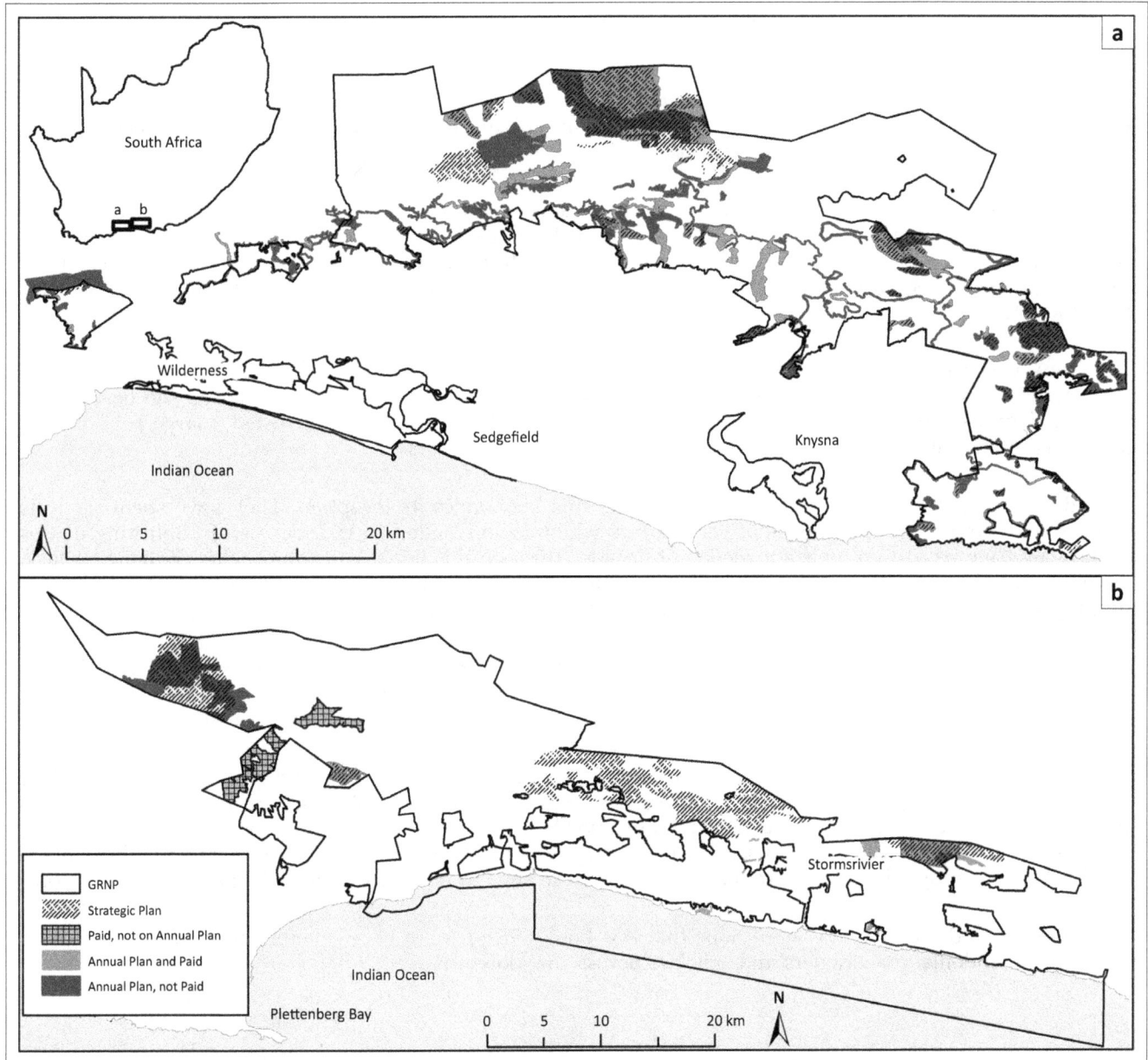

FIGURE1-A1: Geographic correlation between strategic medium-term plan ('Strategic Plan'), annual plan of operation ('Annual Plan'), and work recorded in Working for Water information system as contracted and paid ('Paid') during the 2014/2015 financial year in the Garden Route National Park ('GRNP').

PERMISSIONS

All chapters in this book were first published in BOTHALIA, by African Online Scientific Information Systems (Pty) Ltd; hereby published with permission under the Creative Commons Attribution License or equivalent. Every chapter published in this book has been scrutinized by our experts. Their significance has been extensively debated. The topics covered herein carry significant findings which will fuel the growth of the discipline. They may even be implemented as practical applications or may be referred to as a beginning point for another development.

The contributors of this book come from diverse backgrounds, making this book a truly international effort. This book will bring forth new frontiers with its revolutionizing research information and detailed analysis of the nascent developments around the world.

We would like to thank all the contributing authors for lending their expertise to make the book truly unique. They have played a crucial role in the development of this book. Without their invaluable contributions this book wouldn't have been possible. They have made vital efforts to compile up to date information on the varied aspects of this subject to make this book a valuable addition to the collection of many professionals and students.

This book was conceptualized with the vision of imparting up-to-date information and advanced data in this field. To ensure the same, a matchless editorial board was set up. Every individual on the board went through rigorous rounds of assessment to prove their worth. After which they invested a large part of their time researching and compiling the most relevant data for our readers.

The editorial board has been involved in producing this book since its inception. They have spent rigorous hours researching and exploring the diverse topics which have resulted in the successful publishing of this book. They have passed on their knowledge of decades through this book. To expedite this challenging task, the publisher supported the team at every step. A small team of assistant editors was also appointed to further simplify the editing procedure and attain best results for the readers.

Apart from the editorial board, the designing team has also invested a significant amount of their time in understanding the subject and creating the most relevant covers. They scrutinized every image to scout for the most suitable representation of the subject and create an appropriate cover for the book.

The publishing team has been an ardent support to the editorial, designing and production team. Their endless efforts to recruit the best for this project, has resulted in the accomplishment of this book. They are a veteran in the field of academics and their pool of knowledge is as vast as their experience in printing. Their expertise and guidance has proved useful at every step. Their uncompromising quality standards have made this book an exceptional effort. Their encouragement from time to time has been an inspiration for everyone.

The publisher and the editorial board hope that this book will prove to be a valuable piece of knowledge for researchers, students, practitioners and scholars across the globe.

LIST OF CONTRIBUTORS

Duilio Iamonico
Laboratory of Phytogeography and Applied Geobotany, Department DPTA, Section Environment and Landscape, Sapienza University of Rome, Italy

Ridha El Mokni
Faculty of Sciences of Bizerta, University of Carthage, Tunisia
Laboratory of Plant Biology, Faculty of Pharmacy of Monastir, University of Monastir, Tunisia

Michelle Greve, Rabia Mathakutha and Christien Steyn
Department of Plant and Soil Sciences, University of Pretoria, South Africa

Steven L. Chown
School of Biological Sciences, Monash University, Australia

Carel Jongkind
Botanic Garden Meise, Belgium

Haylee Kaplan, Phetole Manyama and Philip Ivey
Invasive Species Programme, South African National Biodiversity Institute, Kirstenbosch Research Centre, South Africa

John R.U. Wilson and Ana Novoa
Invasive Species Programme, South African National Biodiversity Institute, Kirstenbosch Research Centre, South Africa
Centre for Invasion Biology, Department of Botany and Zoology, Stellenbosch University, South Africa

Hildegard Klein and Lesley Henderson
Agricultural Research Council – Plant Protection Research Institute, South Africa

Helmuth G. Zimmermann
Helmuth Zimmermann and Associates, South Africa

Martin P. Hill
Department of Zoology and Entomology, Rhodes University, South Africa

Julie Coetzee
Department of Botany, Rhodes University, South Africa

Louise Stafford
Environmental Resource Management Department (ERMD), City of Cape Town, Westlake Conservation Office, South Africa

Ulrike M. Irlich, Mirijam Gaertner and Luke Potgieter
Environmental Resource Management Department (ERMD), City of Cape Town, Westlake Conservation Office, South Africa
Centre for Invasion Biology, Department of Botany and Zoology, Stellenbosch University, South Africa

Katelyn T. Faulkner
Invasive Species Programme, South African National Biodiversity Institute, Kirstenbosch Research Centre, South Africa
Centre for Invasion Biology, Department of Zoology and Entomology, University of Pretoria, South Africa

Brett P. Hurley
Forestry and Agricultural Biotechnology Institute (FABI), University of Pretoria, South Africa
Department of Zoology and Entomology, University of Pretoria, South Africa

Mark P. Robertson
Centre for Invasion Biology, Department of Zoology and Entomology, University of Pretoria, South Africa

Mathieu Rouget
Centre for Invasion Biology, School of Agricultural, Earth and Environmental Sciences, University of KwaZulu-Natal, South Africa

John R.U. Wilson
Invasive Species Programme, South African National Biodiversity Institute, Kirstenbosch Research Centre, South Africa
Centre for Invasion Biology, Department of Botany and Zoology, Stellenbosch University, South Africa

Johan A. Baard and Nicolas S. Cole
South African National Parks (SANParks), South Africa

Llewellyn C. Foxcroft and Nicola J. van Wilgen
South African National Parks (SANParks), South Africa
Centre for Invasion Biology, Department of Botany and Zoology, Stellenbosch University, South Africa

Tsungai Zengeya
South African National Biodiversity Institute, Kirstenbosch Research Centre, South Africa

Darragh J. Woodford
Centre for Invasion Biology, Animal, Plant and Environmental Sciences, University of the Witwatersrand, South Africa
Centre for Invasion Biology, South African Institute for Aquatic Biodiversity (SAIAB), South Africa

Olaf Weyl
Centre for Invasion Biology, South African Institute for Aquatic Biodiversity (SAIAB), South Africa

Ana Novoa, Ross Shackleton, David Richardson and Brian van Wilgen
Centre for Invasion Biology, Department of Botany and Zoology, Stellenbosch University, South Africa

Susana Clusella-Trullas and Raquel A. Garcia
Centre for Invasion Biology, Department of Botany and Zoology, Stellenbosch University, South Africa

Vernon Visser
Seec – Statistics in Ecology, the Environment and Conservation, Department of Statistical Sciences, University of Cape Town, South Africa
African Climate and Development Initiative, University of Cape Town, South Africa
Centre for Invasion Biology, Department of Botany and Zoology, Stellenbosch University, South Africa
Invasive Species Programme, South African National Biodiversity Institute, Kirstenbosch Research Centre, South Africa

John R.U. Wilson and Sabrina Kumschick
Centre for Invasion Biology, Department of Botany and Zoology, Stellenbosch University, South Africa
Invasive Species Programme, South African National Biodiversity Institute, Kirstenbosch Research Centre, South Africa

Ingrid Nänni and Philip Ivey
Invasive Species Programme, South African National Biodiversity Institute, Kirstenbosch Research Centre, South Africa

Kim Canavan
Department of Zoology and Entomology, Rhodes University, South Africa

Susan Canavan and David M. Richardson
Centre for Invasion Biology, Department of Botany and Zoology, Stellenbosch University, South Africa

Lyn Fish
National Herbarium, South African National Biodiversity Institute (SANBI), South Africa

David Le Maitre
Centre for Invasion Biology, Department of Botany and Zoology, Stellenbosch University, South Africa
Natural Resources and the Environment, Stellenbosch, South Africa

Caroline Mashau
National Herbarium, South African National Biodiversity Institute (SANBI), South Africa

Tim G. O'Connor
South African Environmental Observation Network (SAEON), Pretoria, South Africa

Costas Zachariades
Plant Protection Research Institute, Agricultural Research Council, South Africa
School of Life Sciences, University of KwaZulu-Natal, South Africa

Iain D. Paterson
Department of Zoology and Entomology, Rhodes University, South Africa

Lorraine W. Strathie
Plant Protection Research Institute, Agricultural Research Council, South Africa

Brian W. van Wilgen
Centre for Invasion Biology, Department of Botany and Zoology, Stellenbosch University, South Africa

Reuben P. Keller
Institute of Environmental Sustainability, Loyola University, United States

Alan R. Wood
Weeds Division, ARC-Plant Protection Research Institute, South Africa

Ndifelani Mararakanye
Directorate: Information Services, Department of Agriculture, Rural Development, Land and Environmental Affairs, South Africa

Modau N. Magoro and Matome C. Rabothata
Directorate: Veld, Pasture Management and Nutrition, Department of Agriculture, Rural Development, Land and Environmental Affairs, South Africa

Nomakhazi N. Matshaya and Sthembele R. Ncobeni
Rail Network Division, Transnet Freight Rail, South Africa

Tineke Kraaij
School of Natural Resource Management, Nelson Mandela Metropolitan University, South Africa

Johan A. Baard and Diba R. Rikhotso
Scientific Services, South African National Parks, South Africa

Nicholas S. Cole
Biodiversity Social Projects, South African National Parks, South Africa

Index